Classroom Manual for

Automotive Engine
Performance

Second Edition

NOTICE TO THE READER

Cover illustration courtesy of David Kimble

DELMAR STAFF

Publisher: Alar Elken
Acquisitions Editor: Vernon R. Anthony
Developmental Editor: Catherine A. Wein
Project Editor: Cori Filson
Production Coordinator: Karen Smith
Art/Design Coordinator: Michele Canfield
Editorial Assistant: Betsy Hough

COPYRIGHT © 1998
By Delmar Publishers
an International Thomson Publishing Company
The ITP logo is a trademark under license

Printed in the United States of America

For information, contact:

Delmar Publishers
3 Columbia Circle, Box 15015
Albany, New York 12212-5015

International Thomson Editores
Campos Eliseos 385, Piso 7
Col Polanco
11560 Mexico DF Mexico

International Thomson Publishing Europe
Berkshire House
168-173 High Holborn
London, WC1V7AA
England

International Thomson Publishing GmbH
Königswinterer Strasse 418
53227 Bonn
Germany

Thomas Nelson Australia
102 Dodds Street
South Melbourne, 3205
Victoria, Australia

International Thomson Publishing Asia
221 Henderson Road
#05-10 Henderson Building
Singapore 0315

Nelson Canada
1120 Birchmount Road
Scarborough, Ontario
Canada M1K 5G4

International Thomson Publishing–Japan
Hirakawacho Kyowa Building, 3F
2-2-1 Hirakawacho
Chiyoda-ku, Tokyo 102
Japan

3 4 5 6 7 8 9 10 XXX 03 02 01 00 99 98

Library of Congress Cataloging-in-Publication Data

Knowles, Don.
    Automotive engine performance / Don Knowles.
       p.   cm. -- (Today's technician)
    Includes bibliographical references and index.
    Contents: [1]. Classroom manual -- [2]. Shop manual.
    ISBN 0-8273-8519-6 (alk. paper)
    1. Automobiles -- Motors -- Modification. 2. Automobiles -
-Performance. I. Erjavec, Jack II. Title. III. Series.
TL210.K56 1998
629.25'04`0288 -- dc21
                                      97-29055
                                       CIP

Classroom Manual for
Automotive Engine Performance
Second Edition

Don Knowles

Knowles Automotive Training
Moose Jaw, Saskatchewan
CANADA

Jack Erjavec
Series Advisor
Columbus State Community College
Columbus, Ohio

Delmar Publishers

I(T)P® International Thomson Publishing

Albany • Bonn • Boston • Cincinnati • Detroit • London • Madrid
Melbourne • Mexico City • New York • Pacific Grove • Paris • San Francisco
Singapore • Tokyo • Toronto • Washington

CONTENTS

PREFACE

Thanks to the support the *Today's Technician* series has received from those who teach automotive technology, Delmar Publishers is able to live up to its promise to provide new editions every three years. We have listened to our critics and our fans and present this new revised edition. By revising our series every three years, we can and will respond to changes in the industry, changes in the certification process, and to the ever-changing needs of those who teach automotive technology.

The *Today's Technician* series, by Delmar Publishers, features textbooks that cover all mechanical and electrical systems of automobiles and light trucks. Principal titles correspond with the eight major areas of ASE (National Institute for Automotive Service Excellence) certification. Additional titles include remedial skills and theories common to all of the certification areas and advanced or specialized subject areas that reflect the latest technological trends.

Each title is divided into two manuals: a Classroom Manual and a Shop Manual. Dividing the material into two manuals provides the reader with the information needed to begin a successful career as an automotive technician without interrupting the learning process by mixing cognitive and performance-based learning objectives.

Each Classroom Manual contains the principles of operation for each system and subsystem. It also discusses the design variations used by different manufacturers. The Classroom Manual is organized to build upon basic facts and theories. The primary objective of this manual is to allow the reader to gain an understanding of how each system and subsystem operates. This understanding is necessary to diagnose the complex automobile systems.

The understanding acquired by using the Classroom Manual is required for competence in the skill areas covered in the Shop Manual. All of the high priority skills, as identified by ASE, are explained in the Shop Manual. The Shop Manual also includes step-by-step instructions for diagnostic and repair procedures. Photo Sequences are used to illustrate many of the common service procedures. Other common procedures are listed and are accompanied with fine-line drawings and photographs that allow the reader to visualize and conceptualize the finest details of the procedure. The Shop Manual also contains the reasons for performing the procedures, as well as when that particular service is appropriate.

The two manuals are designed to be used together and are arranged in corresponding chapters. Not only are the chapters in the manuals linked together, the contents of the chapters are also linked. Both manuals contain clear and thoughtfully selected illustrations. Many of the illustrations are original drawings or photos prepared for inclusion in this series. This means that the art is a vital part of each manual.

The page layout remains the same. The main body of the text includes all of the "need-to-know" information and illustrations. In the side margins are many of the special features of the series. These are provided to the side of the text so they do not break the flow of the text, making it easier to read and follow.

Highlights of this Edition-Shop Manual

The text was updated throughout, to include the latest developments. Some of these new topics include the use of lab scopes to test various components, the testing of central port fuel injection systems and direct ignition systems, and I/M testing procedures. We also added a new chapter that covers OBD II system diagnostic features and procedures. In addition to this new chapter, there is also a chapter that covers the testing of systems not directly related to the engine that could be the source of driveability complaints.

Located at the end of each chapter are two new features: Job Sheets and ASE Challenge Questions. The Job Sheets provide a format for students to perform some of the tasks covered in the chapter. In addition to walking students through a procedure, step-by-step, these Job Sheets challenge students by asking why or how something should be done, thereby making the students think about what they are doing.

Speaking of challenging questions, each chapter ends with a group of questions that reflect the content of an ASE exam. These questions are not merely end-of-chapter questions, they represent the contents of an actual ASE test. These questions, of course, are in addition to the ASE style end-of-chapter questions that were in the first edition.

Highlights of this Edition-Classroom Manual

The text was updated throughout, to include the latest developments. Some of these new topics include lab scopes, Electronic Ignition systems, central port fuel injection systems, and I/M program initiatives. We also added a new chapter that covers OBD II systems, including the rationale, features, and system-specific modifications of these mandated engine control systems. There is also a new chapter that covers those systems that affect the driveability of a vehicle but are not directly related to the engine, such as transmissions and air conditioning.

Jack Erjavec, Series Advisor

Classroom Manual

To stress the importance of safe work habits, the Classroom Manual dedicates one full chapter to safety. Included in this chapter are common safety practices, safety equipment, and safe handling of hazardous materials and wastes. This includes information on MSDS sheets and OSHA regulations. Other features of this manual include:

Cognitive Objectives

These objectives define the contents of the chapter and define what the student should have learned upon completion of the chapter. *Each topic is divided into small units to promote easier understanding and learning.*

References to the Shop Manual

Reference to the appropriate page in the Shop Manual is given whenever necessary. Although the chapters of the two manuals are synchronized, material covered in other chapters of the Shop Manual may be fundamental to the topic discussed in the Classroom Manual.

Cautions and Warnings

Throughout the text, cautions are given to alert the reader to potentially hazardous materials or unsafe conditions. Warnings are also given to advise the student of things that can go wrong if instructions are not followed or if a nonacceptable part or tool is used.

Marginal Notes

New terms are pulled out and defined. Common trade jargon also appears in the margin and gives some of the common terms used for components. This allows the reader to speak and understand the language of the trade, especially when conversing with an experienced technician.

A Bit of History

This feature gives the student a sense of the evolution of the automobile. This feature not only contains nice-to-know information, but also should spark some interest in the subject matter.

(Sample page 1 — Chapter 12)

CHAPTER 12

Distributor Ignition Systems

Upon completion and review of this chapter, you should be able to:

- Describe the operation of distributor-based ignition systems.
- Explain the purpose of the electronic control unit.
- Describe the various types of spark timing systems, including electronic switching systems and their related engine position sensors.
- Describe the operation of the various switching devices used in distributors.
- Describe the major differences between a Dura-Spark II and a Dura-Spark I ignition system.
- Explain the term variable dwell as it relates to a high energy ignition (HEI) system.
- Describe the differences in operation of a thick film integrated system and a high energy ignition system.
- Explain the basic operation of a computer-controlled ignition system.
- Explain how the fuel injection system may rely on components of the ignition system.

Introduction

One of the requirements for an efficient running engine is the correct amount of heat delivered into the cylinders at the right time. This requirement is the responsibility of the ignition system. The ignition system supplies properly timed high-voltage surges to the spark plugs. These voltage surges cause combustion inside the cylinder. For each cylinder in an engine, the ignition system has three main jobs. First, it must generate an electrical spark that has enough heat to ignite the air-fuel mixture in the combustion chamber. Secondly, it must maintain that spark long enough to allow for the combustion of all the air and fuel in the cylinders. Lastly, it must deliver the spark to each cylinder so that combustion can begin at the right time during the compression stroke of each cylinder.

The job of the ignition system is not easy. When it fails to provide the correct amount of heat at the correct time, exhaust emissions increase and engine performance and fuel economy decrease. Through the years, the ignition system has become more precise, reliable, and durable.

From the fully mechanical breaker point system, ignition technology progressed to basic electronic triggering and switching devices. The electronic switching components are normally inside a separate housing known as an electronic control module or control unit (Figure 12-1). The original electronic ignitions still relied on mechanical and vacuum advance mechanisms in the distributor.

In order to have an efficient running engine there must be the correct amount of air mixed with the correct amount of fuel, in a sealed container, shocked by the correct amount of heat at the right time.

Figure 13-25 Crankshaft position sensor. *(Courtesy of Cadillac Motor Division—GMC)*

Shop Manual
Chapter 13, page 574

is in the crankshaft sensor. These other systems have a notched reluctor ring positioned near the center of the crankshaft (Figure 13-26). This ring is permanently cast on the crankshaft. A magnetic sensor containing a permanent magnet and a winding is mounted in an opening in the engine block. The tip of this sensor is 0.050 in. (1.27 mm) from the reluctor ring outer surface. This gap between the magnetic sensor tip and the reluctor ring is not adjustable. The magnetic sensor is retained in the engine block with a bolt and clamp. The coil assembly and coil module are similar to those used on other slow-start and fast-start EI systems.

On the 2.0L four-cylinder engine and the 2.8L and 3.1L V6 engines, the coil and module assembly is positioned on one side of the engine block and the magnetic sensor is located in the opposite side of the block. The magnetic sensor is mounted on the back of the coil module on the 2.5L four-cylinder engine. Therefore, the magnetic sensor is not visible until the coil and module assembly is removed from the engine block.

The reluctor ring has seven notches on four-cylinder or V6 engines. Six of these notches are spaced 60 degrees apart, and the seventh notch is positioned 10 degrees from the sixth notch. A signal from the seventh notch is referred to as a SYNC signal, which is used by the coil module for coil sequencing. On four-cylinder engines, the coil module is programmed to recognize the SYNC notch and count notch 1. When notch 2 passes the sensor tip, the coil module opens the 2-3 coil primary circuit and fires spark plugs 2 and 3. After this event, the coil module counts notches 3 and

---

(Sample page 2)

A BIT OF HISTORY

From the 1920s to the 1960s, intake manifold vacuum was supplied to two components: the distributor vacuum advance and the windshield wipers. From the 1970s to the present time, a wide variety of vacuum and electric/vacuum emission and computer system components have been added to the average automobile. Intake manifold vacuum is now responsible for such items as brake boosting, cruise control, air conditioning, computer input sensor signals, computer output control devices, and emission components.

Shop Manual
Chapter 7, page 290

CAUTION: Exhaust system components are extremely hot if the engine has been running. Wear protective gloves to avoid burns when servicing the exhaust system.

The exhaust system is responsible for collecting the exhaust gas from each cylinder and discharging this gas at the rear of the vehicle. While performing this function, the exhaust system must silence the exhaust flow to an acceptable level outside and inside the vehicle. Catalytic converters in the exhaust system reduce emission levels. The main components in a typical exhaust system are these:

1. Exhaust manifolds
2. Exhaust pipe and seal
3. Catalytic converter
4. Muffler
5. Resonator
6. Tailpipe
7. Heat shields
8. Hangers, brackets, and clamps

All the parts of the system are designed to confirm to the available space of the vehicle's undercarriage and yet be a safe distance above the road.

WARNING: When inspecting or working on the exhaust system, remember that its components get very hot when the engine is running and contact with them could cause a severe burn. Also, always wear safety glasses or goggles when working under a vehicle.

Exhaust Manifolds

Many exhaust manifolds are made from cast iron or nodular iron. Some exhaust manifolds are made from stainless steel or heavy-gauge steel. The exhaust manifold contains an exhaust port for each exhaust port in the cylinder head, and a flat machined surface on this manifold fits against a matching surface on the exhaust port area in the cylinder head. Some exhaust manifolds have a gasket between the manifold and the cylinder head (Figure 7-20). In other applications, the machined surface fits directly against the matching surface on the cylinder head. The exhaust passages from each port in the manifold join into a common single passage before they reach the manifold flange. An exhaust pipe is connected to the exhaust manifold flange. On a V-type engine an exhaust manifold is bolted to each cylinder head.

Exhaust system components are designed for a specific engine. The pipe diameter, component length, **catalytic converter** size, muffler size, and exhaust manifold design are engineered to provide proper exhaust flow, silencing, and emission levels on a particular engine. Exhaust headers are used in place of exhaust manifolds on some engines (Figure 7-21). Each time a power stroke occurs and an exhaust valve opens, a positive pressure occurs in the exhaust manifold. A

A positive pressure may be defined as a pressure higher than atmospheric pressure.

212

x

Summaries

Each chapter concludes with summary statements that contain the important topics of the chapter. These are designed to help the reader review the contents.

Terms to Know

A list of new terms appears next to the Summary. Definitions for these terms can be found in the Glossary at the end of the manual.

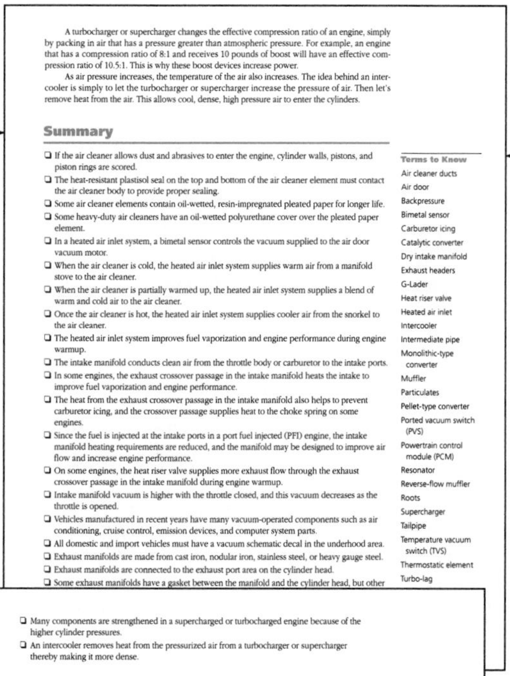

A turbocharger or supercharger changes the effective compression ratio of an engine, simply by packing in air that has a pressure greater than atmospheric pressure. For example, an engine that has a compression ratio of 8:1 and receives 10 pounds of boost will have an effective compression ratio of 10.5:1. This is why these boost devices increase power.

As air pressure increases, the temperature of the air also increases. The idea behind an intercooler is simply to let the turbocharger or supercharger increase the pressure of air. Then let's remove heat from the air. This allows cool, dense, high pressure air to enter the cylinders.

Summary

- If the air cleaner allows dust and abrasives to enter the engine, cylinder walls, pistons, and piston rings are scored.
- The heat-resistant plastisol seal on the top and bottom of the air cleaner element must contact the air cleaner body to provide proper sealing.
- Some air cleaner elements contain oil-wetted, resin-impregnated pleated paper for longer life.
- Some heavy-duty air cleaners have an oil-wetted polyurethane cover over the pleated paper element.
- In a heated air inlet system, a bimetal sensor controls the vacuum supplied to the air door vacuum motor.
- When the air cleaner is cold, the heated air inlet system supplies warm air from a manifold stove to the air cleaner.
- When the air cleaner is partially warmed up, the heated air inlet system supplies a blend of warm and cold air to the air cleaner.
- Once the air cleaner is hot, the heated air inlet system supplies cooler air from the snorkel to the air cleaner.
- The heated air inlet system improves fuel vaporization and engine performance during engine warmup.
- The intake manifold conducts clean air from the throttle body or carburetor to the intake ports.
- In some engines, the exhaust crossover passage in the intake manifold heats the intake to improve fuel vaporization and engine performance.
- The heat from the exhaust crossover passage in the intake manifold also helps to prevent carburetor icing, and the crossover passage supplies heat to the choke spring on some engines.
- Since the fuel is injected at the intake ports in a port fuel injected (PFI) engine, the intake manifold heating requirements are reduced, and the manifold may be designed to improve air flow and increase engine performance.
- On some engines, the heat riser valve supplies more exhaust flow through the exhaust crossover passage in the intake manifold during engine warmup.
- Intake manifold vacuum is higher with the throttle closed, and this vacuum decreases as the throttle is opened.
- Vehicles manufactured in recent years have many vacuum-operated components such as air conditioning, cruise control, emission devices, and computer system parts.
- All domestic and import vehicles must have a vacuum schematic decal in the underhood area.
- Exhaust manifolds are made from cast iron, nodular iron, stainless steel, or heavy gauge steel.
- Exhaust manifolds are connected to the exhaust port area on the cylinder head.
- Some exhaust manifolds have a gasket between the manifold and the cylinder head, but other

Terms to Know

Air cleaner ducts
Air door
Backpressure
Bimetal sensor
Carburetor icing
Catalytic converter
Dry intake manifold
Exhaust headers
G-Lader
Heat riser valve
Heated air inlet
Intercooler
Intermediate pipe
Monolithic-type converter
Muffler
Particulates
Pellet-type converter
Ported vacuum switch (PVS)
Powertrain control module (PCM)
Resonator
Reverse-flow muffler
Roots
Supercharger
Tailpipe
Temperature vacuum switch (TVS)
Thermostatic element
Turbo-lag

Review Questions

Short answer essay, fill-in-the-blank, and multiple-choice type questions follow each chapter. These questions are designed to accurately assess the student's competence in the stated objectives at the beginning of the chapter.

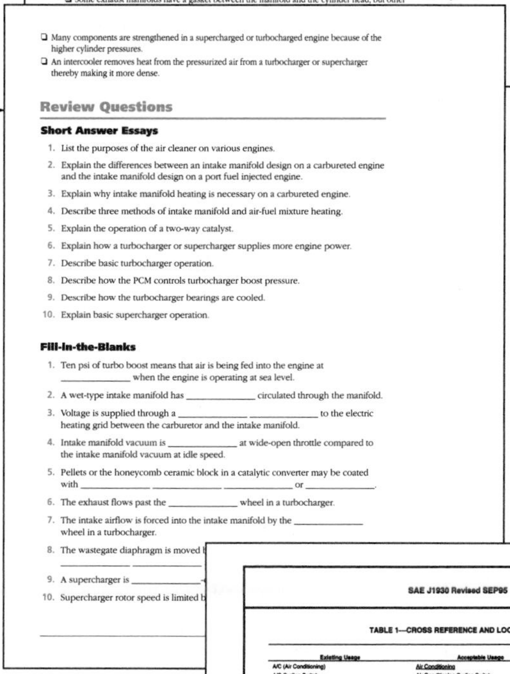

- Many components are strengthened in a supercharged or turbocharged engine because of the higher cylinder pressures.
- An intercooler removes heat from the pressurized air from a turbocharger or supercharger thereby making it more dense.

Review Questions

Short Answer Essays

1. List the purposes of the air cleaner on various engines.
2. Explain the differences between an intake manifold design on a carbureted engine and the intake manifold design on a port fuel injected engine.
3. Explain why intake manifold heating is necessary on a carbureted engine.
4. Describe three methods of intake manifold and air-fuel mixture heating.
5. Explain the operation of a two-way catalyst.
6. Explain how a turbocharger or supercharger supplies more engine power.
7. Describe basic turbocharger operation.
8. Describe how the PCM controls turbocharger boost pressure.
9. Describe how the turbocharger bearings are cooled.
10. Explain basic supercharger operation.

Fill-in-the-Blanks

1. Ten psi of turbo boost means that air is being fed into the engine at ______________ when the engine is operating at sea level.
2. A wet-type intake manifold has ______________ circulated through the manifold.
3. Voltage is supplied through a ______________ ______________ to the electric heating grid between the carburetor and the intake manifold.
4. Intake manifold vacuum is ______________ at wide-open throttle compared to the intake manifold vacuum at idle speed.
5. Pellets or the honeycomb ceramic block in a catalytic converter may be coated with ______________ ______________ ______________ or ______________.
6. The exhaust flows past the ______________ wheel in a turbocharger.
7. The intake airflow is forced into the intake manifold by the ______________ wheel in a turbocharger.
8. The wastegate diaphragm is moved
9. A supercharger is ______________
10. Supercharger rotor speed is limited

The J1930 List of Terminology

Located in the appendix, this list serves as a reference to the acceptable industry terms as defined by SAE.

SAE J1930 Revised SEP95

TABLE 1—CROSS REFERENCE AND LOOK UP

Existing Usage	Acceptable Usage	Acceptable Acronized Usage
A/C (Air Conditioning)	Air Conditioning	A/C
A/C Cycling Switch	Air Conditioning Cycling Switch	A/C Cycling Switch
A/T (Automatic Transaxle)	Automatic Transaxle[1]	A/T[1]
A/T (Automatic Transmission)	Automatic Transmission[1]	A/T[1]
AAT (Ambient Air Temperature)	Ambient Air Temperature	AAT
AC (Air Conditioning)	Air Conditioning	A/C
ACC (Air Conditioning Clutch)	Air Conditioning Clutch	A/C Clutch
Accelerator	Accelerator Pedal	AP
Accelerator Pedal Position	Accelerator Pedal Position[1]	APP[1]
ACCS (Air Conditioning Cyclic Switch)	Air Conditioning Cycling Switch	A/C Cycling Switch
ACH (Air Cleaner Housing)	Air Cleaner Housing[1]	ACL Housing1
ACL (Air Cleaner)	Air Cleaner[1]	ACL[1]
ACL (Air Cleaner) Element	Air Cleaner Element[1]	ACL Element[1]
ACL (Air Cleaner) Housing	Air Cleaner Housing[1]	ACL Housing[1]
ACL (Air Cleaner) Housing Cover	Air Cleaner Housing Cover[1]	ACL Housing Cover[1]
ACS (Air Conditioning System)	Air Conditioning System	A/C System
ACT (Air Charge Temperature)	Intake Air Temperature[1]	IAT[1]
Adaptive Fuel Strategy	Fuel Trim[1]	FT[1]
AFC (Air Flow Control)	Mass Air Flow	MAF
AFC (Air Flow Control(	Volume Air Flow	VAF
AFS (Air Flow Sensor)	Mass Air Flow Sensor	MAF Sensor
AFS (Air Flow Sensor)	Volume Air Flow Sensor	VAF Sensor
After Cooler	Charge Air Cooler[1]	CAC[1]
AI (Air Injection)	Secondary Air Injection[1]	AIR[1]
AIP (Air Injection Pump)	Secondary Air Injection Pump[1]	AIR Pump[1]
AIR (Air Injection Reactor)	Pulsed Secondary Air Injection[1]	PAIR[1]
AIR (Air Injection Reactor)	Secondary Air Injection[1]	AIR[1]
AIRB (Secondary Air Injection Bypass)	Secondary Air Injection Bypass[1]	AIR Bypass[1]
AIRD (Secondary Air Injection Diverter)	Secondary Air Injection Diverter[1]	AIR Diverter[1]
Air Cleaner	Air Cleaner[1]	ACL[1]
Air Cleaner Element	Air Cleaner Element[1]	ACL Element[1]
Air Cleaner Housing	Air Cleaner Housing[1]	ACL Housing[1]
Air Cleaner Housing Cover	Air Cleaner Housing Cover[1]	ACL Housing Cover[1]
Air Conditioning	Air Conditioning	A/C
Air Conditioning Sensor	Air Conditioning Sensor	A/C Sensor
Air Control Valve	Secondary Air Injection Control Valve[1]	AIR Control Valve[1]
Air Flow Meter	Mass Air Flow Sensor[1]	MAF Sensor[1]
Air Flow Meter	Volume Air Flow Sensor[1]	VAF Sensor[1]
Air Intake System	Intake Air System[1]	IA System[1]
Air Flow Sensor	Mass Air Flow Sensor[1]	MAF Sensor[1]
Air Management 1	Secondary Air Injection Bypass[1]	AIR Bypass[1]
Air Management 2	Secondary Air Injection Diverter[1]	AIR Diverter[1]
Air Temperature Sensor	Intake Air Temperature Sensor[1]	IAT Sensor[1]
Air Valve	Idle Air Control Valve[1]	IAC Valve[1]
AIV (Air Injection Valve)	Pulsed Secondary Air Injection[1]	PAIR[1]
ALCL (Assembly Line Communication Link)	Data Link Connector[1]	DLC[1]
Alcohol Concentration Sensor	Flexible Fuel Sensor[1]	FF Sensor[1]
ALDL (Assembly Line Diagnostic Link)	Data Link Connector[1]	DLC[1]

Reprinted with permission from SAE J1930 © 1995 Society of Automotive Engineers, Inc.

504

Shop Manual

To stress the importance of safe work habits, the Shop Manual also dedicates one full chapter to safety. Other important features of this manual include:

Tools Lists

Each chapter begins with a list of the Basic Tools needed to perform the tasks included in the chapter. Whenever a Special Tool is required to complete a task, it is listed in the margin next to the procedure.

Marginal Notes

Page numbers for cross-referencing appear in the margin. Some of the common terms used for components, and other bits of information, also appear in the margin. This provides an understanding of the language of the trade and helps when conversing with an experienced technician.

Cautions and Warnings

Throughout the text, cautions are given to alert the reader to potentially hazardous materials or unsafe conditions. Warnings are also given to advise the student of things that can go wrong if instructions are not followed or if a nonacceptable part or tool is used.

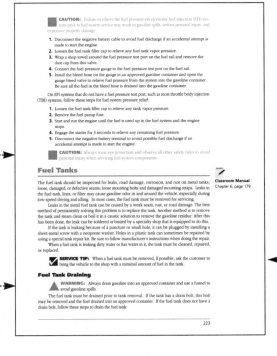

Performance Objectives

These objectives define the contents of the chapter and define what the student should have learned upon completion of the chapter. These objectives also correspond with the list of required tasks for ASE certification. *Each ASE task is addressed.*

Although this textbook is not designed to simply prepare someone for the certification exams, it is organized around the ASE task list. These tasks are defined generically when the procedure is commonly followed and specifically when the procedure is unique for specific vehicle models. Imported and domestic model automobiles and light trucks are included in the procedures.

Service Tips

Whenever a short-cut or special procedure is appropriate, it is described in the text. These tips are generally those things commonly done by experienced technicians.

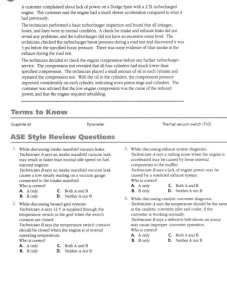

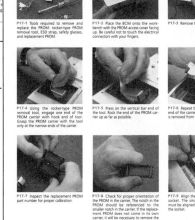

References to the Classroom Manual

Reference to the appropriate page in the Classroom Manual is given whenever necessary. Although the chapters of the two manuals are synchronized, material covered in other chapters of the Classroom Manual may be fundamental to the topic discussed in the Shop Manual.

Photo Sequences

Many procedures are illustrated in detailed Photo Sequences. These detailed photographs show the students what to expect when they perform particular procedures. They also can provide a student a familiarity with a system or type of equipment, which the school may not have.

Case Studies

Case Studies concentrate on the ability to properly diagnose the systems. Each chapter ends with a case study in which a vehicle has a problem, and the logic used by a technician to solve the problem is explained.

ASE Style Review Questions

Each chapter contains ASE style review questions that reflect the performance objectives listed at the beginning of the chapter. These questions can be used to review the chapter as well as to prepare for the ASE certification exam.

◄--- Customer Care

This feature highlights those little things a technician can do or say to enhance customer relations.

◄--- Terms to Know

Terms in this list can be found in the Glossary at the end of the manual.

ASE Challenge Questions

At the end of each technical chapter of the Shop Manual are challenging ASE style questions. The questions are written with the same criteria and rigor as actual ASE certification questions. These questions are designed to make the reader think and use critical thinking skills, along with the knowledge gained by studying the chapter.

Diagnostic Chart

Chapters include detailed diagnostic charts linked with the appropriate ASE task. These charts list common problems and most probable causes. They also list a page reference in the Classroom Manual for better understanding of the system's operation and a page reference in the Shop Manual for details on the procedure necessary for correcting the problem.

Job Sheets

Located at the end of each chapter, the Job Sheets provide a format for students to perform procedures covered in the chapter. A reference to the ASE Task addressed by the procedure is referenced on the Job Sheet.

ASE Practice Examination

A 50 question ASE practice exam, located in the appendix, is included to test students on the content of the complete Shop Manual.

---

7. While discussing an engine compression test:
Technician A says the engine should be cranked until four compression strokes occur during the compression test on each cylinder.
Technician B says the gasoline fuel injection system should be left in operation during a compression test.
Who is correct?
A. A only   C. Both A and B
B. B only   D. Neither A nor B

8. While discussing the cylinder leakage test:
Technician A says during a cylinder leakage test, if the air escapes from the tailpipe, an intake valve is leaking.
Technician B says during a cylinder leakage test the piston must be at TDC on the exhaust stroke.
Who is correct?
A. A only   C. Both A and B
B. B only   D. Neither A nor B

9. While discussing engine noise diagnosis:
Technician A says loose main bearings cause a light rapping noise during acceleration.
Technician B says piston slap causes a hollow rapping noise that is most noticeable on acceleration with a cold engine.
Who is correct?
A. A only   C. Both A and B
B. B only   D. Neither A nor B

10. While discussing engine oil pressure:
Technician A says low oil pressure may be caused by a leaking oil pump pick-up tube.
Technician B says low oil pressure may be caused by loose camshaft bearings.
Who is correct?
A. A only   C. Both A and B
B. B only   D. Neither A nor B

ASE Challenge Questions

1. A V-6 engine has oil leaking from the vicinity of the lower front engine covers. The most probable cause of the leak is:
A. Worn front main seal
B. Timing belt cover
C. Valve cover(s)
D. Oil pressure sending unit

2. The customer states the coolant level must be topped off every few days. There are no visible leaks.
Technician A says this may be caused by a bad radiator cap.
Technician B says a stuck open thermostat will cause this problem.
Who is correct?
A. A only   C. Both A and B
B. B only   D. Neither A nor B

3. Main bearing wear is being discussed.
Technician A says a heavy thumping knock at irregular intervals during acceleration may be caused by worn main thrust bearings.
Technician B says this noise indicates loose/worn main bearings.
Who is correct?
A. A only   C. Both A and B
B. B only   D. Neither A nor B

4. During an engine vacuum test, the gauge reading fluctuates between 11 and 16 in. Hg. This indicates service should be performed on:
A. Carburetor idle mixture screws
B. The valve and valve springs
C. Fuel injectors
D. Either A or C

5. An emission test indicates normal O_2 reading with low CO reading. This may be caused by:
A. Clogged PCV system
B. Restricted fuel filter
C. Leaking fuel injectors
D. None of the above

153

Job Sheet 25

(25)

Name _____ Date _____

Check the Operation of a TP Sensor

Upon completion of this job sheet, you will be able to test the operation of a throttle position sensor with a variety of test instruments.

ASE Correlation

This job sheet is related to the ASE Engine Performance Test's content area: computerized engine controls diagnosis and repair; task: inspect, test, adjust, and replace computerized engine control system sensors, powertrain control module, actuators, and circuits.

Tools and Materials

DMM   Lab scope

Procedure

1. Describe the vehicle being worked on:
   Year _____ Make_____   VIN _____
   Model _____

2. Connect the lab scope across the TP sensor.

3. With the ignition on, move the throttle from closed to fully open and then allow it to close slowly.

4. Observe the trace on the scope while moving the throttle. Describe what the trace looked like.

   _____
   _____

5. Based on the waveform of the TP sensor, what can you tell about the sensor?

   _____

6. With a voltmeter, measure the reference voltage to the TP sensor. The reading should be _____ volts. The reading is _____ volts.

7. What is the output voltage from the sensor when the throttle is closed? _____ volts

8. What is the output voltage of the sensor when the throttle is opened? _____ volts

9. Move the throttle from closed to fully open and then allow it to close slowly. Describe the

441

Table 8-1 ASE TASK

Diagnose rough idle, stalling, hesitation on acceleration, detonation, and loss of power problems and determine needed repairs.

Problem Area	Symptoms	Possible Causes	Classroom Manual	Shop Manual
ENGINE PERFORMANCE	Rough idle operation, stalling	PCV valve sticking open	251	331
	Hesitation on acceleration	Inoperative spark control system, reduced spark advance	254	334
	Hesitation on acceleration during warmup	1. Inoperative mixture heater system	266	348
		2. Heat riser valve always open	266	348
	Detonation	1. Defective spark control system, excessive spark advance	254	334
		2. EGR valve remaining closed all the time	256	339
		3. EGR valve's passage plugged by carbon		
		4. EVR solenoid faulty, vacuum hose to EVR		
		5. Defective wires EVR so to PCM		
	Loss of engine power, detonation	Heat riser valve remains closed at all times		
	Rough idle, surging at low speed	1. EGR valve remaining o at idle, low speed		
		2. EVR solenoid vent plug holding EGR valve ope		
		3. Defective input sensor improper EGR valve o		
		4. Canister purging at idl and low speed		
	Stalling on deceleration or after a cold start	1. EGR valve remaining open at idle, low speed		
		2. EVR solenoid vent plug holding EGR valve ope		
FUEL CONSUMPTION	Excessive fuel consumption	1. AIR pump airflow cont directed upstream		
		2. AIR pump airflow not to exhaust ports during		

APPENDIX A

ASE PRACTICE EXAMINATION

1. *Technician A* says a hydrometer reading of 1.200 at 80°F means the battery must be recharged before performing a capacity test.
Technician B says a capacity test can be correctly performed with the battery cables connected.
Who is correct?
A. A only   C. Both A and B
B. B only   D. Neither A nor B

2. Technician A says engine detonation may be caused by a stuck closed heat riser valve.
Technician B says a bad heat riser valve thermal vacuum switch may cause the valve to remain open.
Who is correct?
A. A only   C. Both A and B
B. B only   D. Neither A nor B

3. The PCV system is being discussed.
Technician A says oil in the crankcase breather filter will confirm that the PCV valve is clogged.
Technician B says the PCV valve must be disconnected to be checked with the engine operating.
Who is correct?
A. A only   C. Both A and B
B. B only   D. Neither A nor B

4. A distributor equipped engine has battery voltage at the coil's positive terminal at all times when the ignition switch is on.
Technician A says this may be caused by an open ballast resistor.
Technician B says the ballast resistor is normally not used during starting.
Who is correct?
A. A only   C. Both A and B
B. B only   D. Neither A nor B

5. Electronic fuel injection systems are being discussed.
Technician A says a hard to start engine may have an open electrical fuel pump relay bypass circuit.
Technician B says a hesitation when accelerating from idle may be caused by dirty injectors.
Who is correct?
A. A only   C. Both A and B
B. B only   D. Neither A nor B

6. A vehicle backfires during almost every deceleration.
Technician A says a secondary air injection diverter valve may be causing the backfire.
Technician B says this condition may be caused by a low performing secondary air injection pump.
Who is correct?
A. A only   C. Both A and B
B. B only   D. Neither A nor B

7. A vehicle has a consistent slow- or no-crank condition. This may be caused by:
A. A slipping AC generator drive belt
B. A defective AC generator
C. A low regulator voltage
D. Any of the above

8. An engine runs rough at idle during high ambient air temperatures.
Technician A says this may be caused by a stuck closed EVAP vacuum regulator.
Technician B says continuous vacuum to the fuel tank pressure control valve may cause this condition.
Who is correct?
A. A only   C. Both A and B
B. B only   D. Neither A nor B

9. *Technician A* says static electricity generated by clothing in contact with vehicle upholstery must be discharged before beginning work on electronic devices.
Technician B says the best control for static electricity is the wearing of a waist or wrist grounding strap.
Who is correct?
A. A only   C. Both A and B
B. B only   D. Neither A nor B

10. *Technician A* says a camshaft installed one tooth retarded can be compensated by adjusting the ignition's base timing.
Technician B says to use the starter to crank the engine with the timing belt removed to determine if the engine is causing interference or free-wheeling.
Who is correct?
A. A only   C. Both A and B
B. B only   D. Neither A nor B

11. Cold enrichment systems are being discussed.
Technician A says excessive resistance in the ECT sensor may cause the cold start system of a fuel injection system to malfunction.
Technician B says the cold start enrichment system on a carburetor can be checked by removing the air filter and snapping the throttle open with the engine off.
Who is correct?
A. A only   C. Both A and B
B. B only   D. Neither A nor B

655

Instructor's Guide

The Instructor's Guide is provided free of charge as part of the *Today's Technician Series* of automotive technology textbooks. It contains Lecture Outlines, Answers to Review Questions, Pretest and Test Bank including ASE style questions.

Classroom Manager

The complete ancillary package is designed to aid the instructor with classroom preparation and provide tools to measure student performance. For an affordable price, this comprehensive package contains:

Instructor's Guide

200 Transparency Masters

Answers to Review Questions

Lecture Outlines and Lecture Notes

Printed and Computerized Test Bank

Laboratory Worksheets and Practicals

Reviewers

I would like to extend a special thanks to those who saw things I overlooked and for their contributions:

Frank Barrows
Monterey Peninsula College
Monterey, CA

Walter M. Bertotti
Porter and Chester Institute
Watertown, CT

Ronnie Bush
Tennessee Technology Center
Jackson, TN

Albert L. Dent
South Puget Sound Community College
Olympia, WA

Rick Escalambre
Skyline College
San Bruno, CA

Dennis L. Gaddis
IVY Tech State College
Muncie, IN

Anthony Hoffman
Arizona Western College
Yuma, AZ

James J. Morton
Automotive Training Center
Exton, PA

Robert VerMuelen
Jackson Community College
Jackson, MI

John E. Wood
Ranken Technical College
St. Louis, MO

Patrick E. Yancey
San Jacinto College–Central
Pasadena, TX

Contributing Companies

I would also like to thank these companies who provided technical information and art for this edition:

AAMCO Tools
Actron Manufacturing Co.
Allen Test Products Division
American Honda Motor Company, Inc.

Contributing Companies (continued)

Aspire Inc.
Automotive Diagnostics, A division of SPX Corporation
Blackhawk Automotive, Inc.
Chilton Professional Automotive Manuals
Chrysler Corporation
Counterman Magazine
CRC Industries, Inc.
Detroit Diesel Allison
DuPont Automotive Products
EDGE Diagnostics Systems
Echlin Manufacturing Co.
Environmental Systems Products, Inc.
EPA
Fluke Corporation
Ford Motor Company
General Fire Extinguisher Corporation
Goodson Shop Supplies
Gutman Advertising Agency
Intel Corporation
Johnson Matthey, Catalytic Systems Division
Kent-Moore Division, SPX Corporation
Kleer–Flo Company
Lincoln, St. Louis
L.S. Starrett Company
Mac Tools, Inc.
Mazda
Motor Publications
National Institute of Automotive Service Excellence
Nissan Motors
OTC Division, SPX Corporation
Robert Bosch Corporation
The Sherwin Williams Company
Siebe North
Snap-on Tools Corporation
Society of Automotive Engineers, Inc. (SAE)
Southern Illinois University, Automotive Technology Department
SPX Corporation, Aftermarket Tool and Equipment Group
Sun Electric Corporation
Thexton Manufacturing Company
TIF Instruments, Inc.
The Valvoline Company
Western Emergency Equipment

Portions of materials contained herein have been reprinted with permission of General Motors Corporation, Service Technology Group.

Safety Practices

Upon completion and review of this chapter, you should be able to:

❏ Recognize shop hazards and take the necessary steps to avoid personal injury or property damage.

❏ Explain the purposes of the Occupational Safety and Health Act.

❏ Identify the necessary steps for personal safety in the automotive shop.

❏ Describe the reasons for prohibiting drug and alcohol use in the shop.

❏ Explain the steps required to provide electrical safety in the shop.

❏ Define the steps required to provide safe handling and storage of gasoline.

❏ Describe the necessary housekeeping safety steps.

❏ Explain the essential general shop safety practices.

❏ Define the steps required to provide fire safety in the shop.

❏ Describe typical fire extinguisher operating procedure.

❏ Explain four different types of fires, and the type of fire extinguisher required for each type of fire.

❏ Describe three other pieces of shop safety equipment other than fire extinguishers.

❏ Follow proper safety precautions while handling hazardous waste materials.

❏ Dispose of hazardous waste materials in accordance with state and federal regulations.

Introduction

Safety is extremely important in the automotive shop! The knowledge and practice of safety precautions prevents serious personal injury and expensive property damage. Automotive students and technicians must be familiar with shop hazards and all types of safety, including personal, gasoline handling, housekeeping, general shop, fire, and hazardous material. The first step in providing a safe shop is learning about all types of safety precautions. However, the second, and most important step in this process is applying our knowledge of safety precautions while working in the shop. In other words, we must actually develop safe working habits in the shop from our understanding of various safety precautions. When shop employees have a careless attitude toward safety, accidents are more likely to occur. All shop personnel must develop a serious attitude toward safety in the shop. The result of this serious attitude is that shop personnel will learn and adopt all shop safety rules.

Shop personnel must be familiar with their rights regarding hazardous waste disposal. These rights are explained in the right-to-know laws. Secondly, shop personnel must be familiar with hazardous materials in the automotive shop and the proper disposal methods for these materials according to state and federal regulations.

Occupational Safety and Health Act

The Occupational Safety and Health Act (OSHA) was passed by the United States government in 1970. The purposes of this legislation are these:

1. To assist and encourage the citizens of the United States in their efforts to assure safe and healthful working conditions by providing research, information, education, and training in the field of occupational safety and health.

2. To assure safe and healthful working conditions for working men and women by authorizing enforcement of the standards developed under the Act.

The **Occupational Safety and Health Act (OSHA)** regulates working conditions in the United States.

1

Since approximately 25% of workers are exposed to health and safety hazards on the job, the OSHA is necessary to monitor, control, and educate workers regarding health and safety in the workplace. Employers and employees should be familiar with workplace hazardous materials information systems (WHMIS).

Shop Hazards

Shop hazards must be recognized and avoided to prevent personal injury.

Service technicians and students encounter many hazards in an automotive shop. When these hazards are known, basic shop safety rules and procedures must be followed to avoid personal injury. Some of the hazards in an automotive shop are these:

1. Flammable liquids such as gasoline and paint must be handled and stored properly.
2. Flammable materials such as oily rags must be stored properly to avoid a fire hazard.
3. Batteries contain a corrosive sulfuric acid solution and produce explosive hydrogen gas while charging.
4. Loose sewer and drain covers may cause foot or toe injuries.
5. Caustic liquids such as those in hot cleaning tanks are harmful to skin and eyes.
6. High-pressure air in the shop compressed air system can be very dangerous if it penetrates the skin and enters the bloodstream.
7. Frayed cords on electric equipment and lights may result in severe electrical shock.
8. Hazardous waste material such as batteries and the caustic cleaning solution from a hot or cold cleaning tank must be handled properly to avoid harmful effects.
9. Carbon monoxide from vehicle exhaust is poisonous.
10. Loose clothing, jewelry, or long hair may become entangled in rotating parts on equipment or vehicles, resulting in serious injury.
11. Dust and vapors generated during some repair jobs are harmful. Asbestos dust generated during brake lining service and clutch service is a contributor to lung cancer.
12. High noise levels from shop equipment such as an air chisel may be harmful to the ears.
13. Oil, grease, water, or parts cleaning solutions on shop floors may cause someone to slip and fall, resulting in serious injury.

Safety in the Automotive Shop

Personal injury, vehicle damage, and property damage must be avoided by following safety rules regarding personal protection, substance abuse, electrical safety, gasoline safety, housekeeping safety, fire safety, and general shop safety.

Each person in an automotive shop must follow certain basic shop safety rules to remove the danger from shop hazards. When all personnel in the automotive shop follow these basic shop safety rules, personal injury, vehicle damage, and property damage may be prevented.

Personal Protection

1. Always wear safety glasses or a face shield in the shop (Figure 1-1).
2. Wear ear plugs or covers if high noise levels are encountered.
3. Always wear boots or shoes that provide adequate foot protection. Safety work boots or shoes with steel toe caps are best for working in the automotive shop. Most safety shoes also have slip-resistant soles. Footwear must protect against heavy falling objects, flying sparks, and corrosive liquids. Soles on footwear must protect against punctures by sharp objects. Sneakers and street shoes are not recommended in the shop.
4. Do not wear watches, jewelry, or rings when working on a vehicle. Severe burns occur when jewelry makes contact between an electric terminal and ground. Jewelry may catch on an object, resulting in painful injury.
5. Do not wear loose clothing, and keep long hair tied behind your head. Loose clothing or long hair is easily entangled in rotating parts.
6. Wear a respirator to protect your lungs when working in dusty conditions.

Figure 1-1 Shop safety equipment, including safety goggles, respirator, welding shield, proper work clothes, ear protection, welding gloves, work gloves, and safety shoes. *(Courtesy of Oldsmobile Division—GMC)*

Smoking, Alcohol, and Drugs in the Shop

Do not smoke when working in the shop. If the shop has designated smoking areas, smoke only in these areas. Do not smoke in customers' cars. Nonsmokers may not appreciate cigarette odor in their cars. A spark from a cigarette or lighter may ignite flammable materials in the workplace. The use of drugs or alcohol must be avoided while working in the shop. Even a small amount of drugs or alcohol affects reaction time. In an emergency situation, slow reaction time may cause personal injury. If a heavy object falls off the workbench, and your reaction time is slowed by drugs or alcohol, you may not get your foot out of the way in time, resulting in foot injury. When a fire starts in the workplace, and you are a few seconds slower getting a fire extinguisher into operation because of alcohol or drug use, it could make the difference between extinguishing a fire and having expensive fire damage.

The improper or excessive use of alcoholic beverages and/or drugs may be referred to as substance abuse.

Electrical Safety

1. Frayed cords on electrical equipment must be replaced or repaired immediately.
2. All electric cords from lights and electric equipment must have a ground connection. The ground connector is the round terminal in a three-prong electrical plug. Do not use a two-prong adaptor to plug in a three-prong electrical cord. Three-prong electrical outlets should be mandatory in all shops.
3. Do not leave electrical equipment running and unattended.

Gasoline Safety

Gasoline is a very explosive liquid! One exploding gallon of gasoline has a force equal to 14 sticks of dynamite. It is the expanding vapors from gasoline that are extremely dangerous. These vapors are present even in cold temperatures. Vapors formed in gasoline tanks on cars are controlled, but vapors from a gasoline storage container may escape from the can, resulting in a hazardous situation. Therefore, gasoline storage containers must be placed in a well-ventilated space.

Approved gasoline storage cans have a flash-arresting screen at the outlet (Figure 1-2). These screens prevent external ignition sources from igniting the gasoline within the can while the gasoline is being poured. Follow these safety precautions regarding gasoline containers:

1. Always use approved gasoline containers that are painted red for proper identification.
2. Do not fill gasoline containers completely full. Always leave the level of gasoline at least one inch from the top of the container. This action allows expansion of the gasoline at higher temperatures. If gasoline containers are completely full, the gasoline will

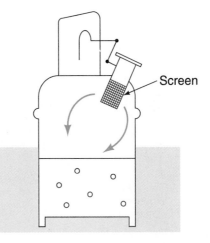

Screen

Figure 1-2 Approved gasoline container.

expand when the temperature increases. This expansion forces gasoline from the can and creates a dangerous spill.

3. If gasoline containers must be stored, place them in a well-ventilated area such as a storage shed. Do not store gasoline containers in your home or in the trunk of a vehicle.

4. When a gasoline container must be transported, be sure it is secured against upsets.

5. Do not store a partially filled gasoline container for long periods of time, because it may give off vapors and produce a potential danger.

6. Never leave gasoline containers open except while filling or pouring gasoline from the container.

7. Do not prime an engine with gasoline while cranking the engine.

8. Never use gasoline as a cleaning agent.

CAUTION: Never siphon gasoline or diesel fuel with your mouth. These liquids are poisonous and can make you sick or fatally ill.

Housekeeping Safety

1. Keep shop floors clean! Always clean shop floors immediately after a spill.

2. Store paint and other flammable liquids in a closed steel cabinet (Figure 1-3).

Figure 1-3 Paints and combustible material containers must be kept in an approved safety cabinet. *(Courtesy of the Sherwin Williams Co.)*

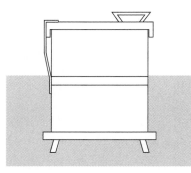

Figure 1-4 Oily rags must be stored in approved airtight containers.

3. Oily rags must be stored in approved, covered airtight containers (Figure 1-4). A slow generation of heat occurs from oxidation of oil on these rags. Heat may continue to be generated until the ignition temperature is reached. The oil and the rags then begin to burn, causing a fire. This action is called spontaneous combustion. However, if the oily rags are in an airtight approved container, the fire cannot get enough oxygen to cause burning.
4. Keep the shop neat and clean. Always pick up tools and parts, and do not leave creepers lying on the floor.
5. Keep the workbenches clean. Do not leave heavy objects such as used parts on the bench after you are finished with them.

General Shop Safety

1. All sewer covers must fit properly and be kept securely in place.
2. Always wear a face shield, protective gloves, and protective clothing when necessary. Gloves should be worn when working with solvents and caustic solutions, handling hot metal, or grinding metal. Various types of protective gloves are available. Shop coats and coveralls are the most common types of protective clothing.
3. Never direct high-pressure air from an air gun against human flesh. If this action is allowed, air may penetrate the skin and enter the bloodstream, causing serious health problems or death. Always keep air hoses in good condition. If an end blows off an air hose, the hose may whip around and result in personal injury. Use only Occupational Safety and Health Act (OSHA) approved air gun nozzles.
4. Handle all hazardous waste materials according to state and federal regulations. (These regulations are explained later in this chapter.)
5. Always place a shop exhaust hose on the tailpipe of a vehicle if the engine is running in the shop, and be sure the shop exhaust fan is turned on.
6. Keep hands, long hair, jewelry, and tools away from rotating parts such as fan blades and belts on running engines. Remember that an electric-drive fan may start turning at any time.
7. When servicing brakes or clutches from manual transmissions, always clean asbestos dust from these components with an approved asbestos dust parts washer.
8. Always use the correct tool for the job. For example, never strike a hardened steel component, such as a piston pin, with a steel hammer. This type of component may shatter, and fragments may penetrate eyes or skin.
9. Follow the car manufacturer's recommended service procedures.
10. Be sure that the shop has adequate ventilation.
11. Make sure the work area has adequate lighting.
12. Use trouble lights with steel or plastic cages around the bulb. If an unprotected bulb breaks, it may ignite flammable materials in the area.

13. When servicing a vehicle, always apply the parking brake and place the transmission in park with an automatic transmission, or neutral with a manual transmission, if the engine is running. When the engine is stopped, place the transmission in park with an automatic transmission, or reverse with a manual transmission.
14. Avoid working on a vehicle parked on an incline.
15. *Never work under a vehicle unless the vehicle chassis is supported securely on jack stands.*
16. When one end of a vehicle is raised, place wheel chocks on both sides of the wheels remaining on the floor.
17. Be sure that you know the location of shop first-aid kits, eyewash fountains, and fire extinguishers.
18. Collect oil, fuel, brake fluid, and other liquids in the proper safety containers.
19. Use only approved cleaning fluids and equipment. Do not use gasoline to clean parts.
20. Obey all state and federal safety, fire, and hazardous material regulations.
21. Always operate equipment according to the equipment manufacturer's recommended procedure.
22. Do not operate equipment unless you are familiar with the correct operating procedure.
23. Do not leave running equipment unattended.
24. Be sure the safety shields are in place on rotating equipment.
25. All shop equipment must have regularly scheduled maintenance and adjustment.
26. Some shops have safety lines around equipment. Always work within these lines when operating equipment.
27. Be sure that shop heating equipment is well ventilated.
28. Do not run in the shop or engage in horseplay.
29. Post emergency phone numbers near the phone. These numbers should include a doctor, ambulance, fire department, hospital, and police.
30. Do not place hydraulic jack handles where someone can trip over them.
31. Keep aisles clear of debris.

Fire Safety

1. Familiarize yourself with the location and operation of all shop fire extinguishers.
2. If a fire extinguisher is used, report it to management so the extinguisher can be recharged.
3. Do not use any type of open flame heater to heat the work area.
4. Do not turn on the ignition switch or crank the engine with a gasoline line disconnected.
5. Store all combustible materials such as gasoline, paint, and oily rags in approved safety containers.
6. Clean up gasoline, oil, or grease spills immediately.
7. Always wear clean shop clothes. Do not wear oil-soaked clothes.
8. Do not allow sparks and flames near batteries.
9. Welding tanks must be securely fastened in an upright position.
10. Do not block doors, stairways, or exits.
11. Do not smoke when working on vehicles.
12. Do not smoke or create sparks near flammable materials or liquids.
13. Store combustible shop supplies such as paint in a closed steel cabinet.
14. Gasoline must be kept in approved safety containers.
15. If a gasoline tank is removed from a vehicle, do not drag the tank on the shop floor.
16. Know the approved fire escape route from your classroom or shop to the outside of the building.
17. If a fire occurs, do not open doors or windows. This action creates extra draft, which makes the fire worse.

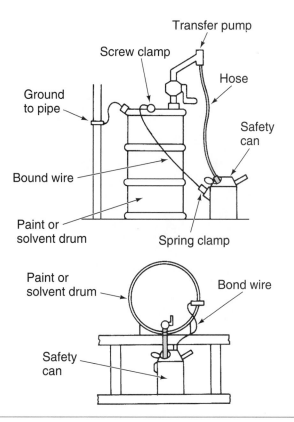

Figure 1-5 Safe procedures for flammable liquid transfer. *(Courtesy of DuPont Automotive Products)*

18. Do not put water on a gasoline fire, because the water will make the fire worse.
19. Call the fire department as soon as a fire begins, and then attempt to extinguish the fire.
20. If possible, stand 6 to 10 feet from the fire and aim the fire extinguisher nozzle at the base of the fire with a sweeping action.
21. If a fire produces a lot of smoke in the room, remain close to the floor to obtain oxygen and avoid breathing smoke.
22. If the fire is too hot or the smoke makes breathing difficult, get out of the building.
23. Do not re-enter a burning building.
24. Keep solvent containers covered except when pouring from one container to another. When flammable liquids are transferred from bulk storage, the bulk container should be grounded to a permanent shop fixture such as a metal pipe. During this transfer process, the bulk container should be grounded to the portable container (Figure 1-5). These ground wires prevent the buildup of a static electric charge, which could result in a spark and disastrous explosion. Always discard or clean empty solvent containers, because fumes in these containers are a fire hazard.
25. Familiarize yourself with different types of fires and fire extinguishers, and know the type of extinguisher to use on each fire.

Shop Manual
Chapter 1, page 12

Shop Safety Equipment

Fire Extinguishers

Fire extinguishers are one of the most important pieces of safety equipment. All shop personnel must know the location of the fire extinguishers in the shop. If you have to waste time looking for an extinguisher after a fire starts, the fire could get out of control before you get the extinguisher

Shop safety equipment must be easily accessible and in good working condition.

Figure 1-6 Types and sizes of fire extinguishers. *(Courtesy of General Fire Extinguisher Corporation)*

into operation. Fire extinguishers should be located where they are easily accessible at all times. Everyone working in the shop must know how to operate the fire extinguishers. There are several different types of fire extinguishers, but their operation usually involves these steps:

1. Get as close as possible to the fire without jeopardizing your safety.
2. Grasp the extinguisher firmly and aim the extinguisher at the fire.
3. Pull a pin from the extinguisher handle.
4. Squeeze the handle to dispense the contents of the extinguisher.
5. Direct the fire extinguisher nozzle at the base of the fire, and dispense the contents of the extinguisher with a sweeping action back and forth across the fire. Most extinguishers discharge their contents in 8 to 25 seconds.
6. Always be sure the fire is extinguished.
7. Always keep an escape route open behind you so a quick exit is possible if the fire gets out of control.

A decal on each fire extinguisher identifies the type of chemical in the extinguisher and provides operating information (Figure 1-6). Shop personnel should be familiar with the following types of fires and fire extinguishers:

1. Class A fires are those involving ordinary combustible materials such as paper, wood, clothing, and textiles. **Multipurpose dry chemical fire extinguishers** are used on these fires.
2. Class B fires involve the burning of flammable liquids such as gasoline, oil, paint, solvents, and greases. These fires may be extinguished with multipurpose dry chemical extinguishers. **Fire extinguishers** containing **halogen**, or **halon**, may be used to extinguish class B fires. The chemicals in this type of extinguisher attach to the hydrogen, hydroxide, and oxygen molecules to stop the combustion process almost instantly. However, the resultant gases from the use of halogen-type extinguishers are very toxic and harmful to the operator of the extinguisher.
3. Class C fires involve the burning of electrical equipment such as wires, motors, and switches. These fires may be extinguished with multipurpose dry chemical extinguishers.

Table 1-1 FIRE EXTINGUISHER SELECTION

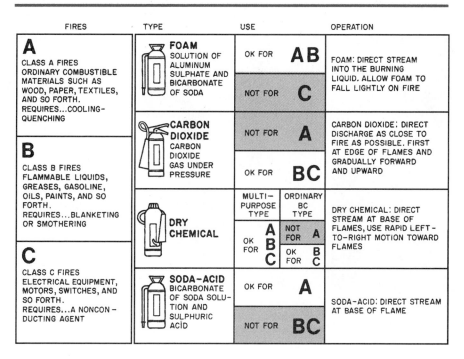

FIRES	TYPE	USE		OPERATION
A CLASS A FIRES ORDINARY COMBUSTIBLE MATERIALS SUCH AS WOOD, PAPER, TEXTILES, AND SO FORTH. REQUIRES...COOLING-QUENCHING	**FOAM** SOLUTION OF ALUMINUM SULPHATE AND BICARBONATE OF SODA	OK FOR **AB**		FOAM: DIRECT STREAM INTO THE BURNING LIQUID. ALLOW FOAM TO FALL LIGHTLY ON FIRE
		NOT FOR **C**		
	CARBON DIOXIDE CARBON DIOXIDE GAS UNDER PRESSURE	NOT FOR **A**		CARBON DIOXIDE: DIRECT DISCHARGE AS CLOSE TO FIRE AS POSSIBLE. FIRST AT EDGE OF FLAMES AND GRADUALLY FORWARD AND UPWARD
B CLASS B FIRES FLAMMABLE LIQUIDS, GREASES, GASOLINE, OILS, PAINTS, AND SO FORTH. REQUIRES...BLANKETING OR SMOTHERING		OK FOR **BC**		
	DRY CHEMICAL	MULTI-PURPOSE TYPE OK FOR **ABC**	ORDINARY BC TYPE NOT FOR **A** OK FOR **BC**	DRY CHEMICAL: DIRECT STREAM AT BASE OF FLAMES, USE RAPID LEFT-TO-RIGHT MOTION TOWARD FLAMES
C CLASS C FIRES ELECTRICAL EQUIPMENT, MOTORS, SWITCHES, AND SO FORTH. REQUIRES...A NONCONDUCTING AGENT	**SODA-ACID** BICARBONATE OF SODA SOLUTION AND SULPHURIC ACID	OK FOR **A**		SODA-ACID: DIRECT STREAM AT BASE OF FLAME
		NOT FOR **BC**		

4. Class D fires involve the combustion of metal chips, turnings, and shavings. Dry chemical extinguishers are the only type of extinguisher recommended for these fires.

Additional information regarding types of extinguishers for various types of fires is provided in Table 1-1.

Eyewash Fountains

Eye injuries may occur in various ways in an automotive shop. Some of the common eye accidents are these:

1. Thermal burns from excessive heat
2. Irradiation burns from excessive light such as from an arc welder
3. Chemical burns from strong liquids such as battery electrolyte
4. Foreign material in the eye
5. Penetration of the eye by a sharp object
6. A blow from a blunt object

Wearing safety glasses and observing shop safety rules will prevent most eye accidents. If a chemical gets in your eyes, it must be washed out immediately to prevent a chemical burn. An eyewash fountain is the most effective way to wash the eyes, and every shop should be equipped with some eyewash facility (Figure 1-7). Be sure you know the location of the eyewash fountain in the shop.

Safety Glasses and Face Shields

The mandatory use of eye protection with safety glasses or a face shield is one of the most important safety rules in a shop. Many shop insurance policies require the use of eye protection in the shop. Some automotive technicians have been blinded in one or both eyes because they did not bother to wear safety glasses. All safety glasses must be equipped with safety glass, and they should provide some type of side protection (Figure 1-8). When selecting a pair of safety glasses, they should feel comfortable on your face. If they are uncomfortable, you may tend to take them off, leaving the eyes unprotected. A face shield should be worn when handling hazardous chemicals or when using an electric grinder or buffer (Figure 1-9).

Figure 1-7 Eyewash fountain. *(Courtesy of Western Emergency Equipment)*

Figure 1-8 Safety glasses with side protection must be worn in the automotive shop. *(Courtesy of Siebe North)*

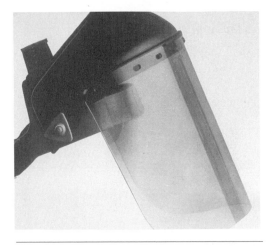

Figure 1-9 Face shield. *(Courtesy of Siebe North)*

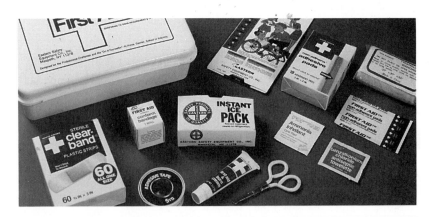

Figure 1-10 First-aid kit.

First-aid Kits

First-aid kits should be clearly identified and conveniently located (Figure 1-10). These kits contain such items as bandages and ointment required for minor cuts. All shop personnel must be familiar with the location of first-aid kits. At least one of the shop personnel should have basic first-aid training, and this person should be in charge of administering first aid and keeping first-aid kits filled.

Hazardous Waste Disposal

CAUTION: When handling hazardous waste material, always wear proper protective clothing and equipment detailed in the right-to-know laws. This includes respirator equipment (Figure 1-11). All recommended procedures must be followed accurately.

Figure 1-11 Wear recommended safety clothing and equipment when handling hazardous materials. *(Courtesy of DuPont Automotive Products)*

Hazardous waste materials in automotive shops are chemicals or components that the shop no longer needs and that pose a danger to the environment and people if they are disposed of in ordinary garbage cans or sewers. However, it should be noted that no material is considered hazardous waste until the shop has finished using it and is ready to dispose of it. The **Environmental Protection Agency (EPA)** publishes a list of hazardous materials that is included in the Code of Federal Regulations. Waste is considered hazardous if it is included on the EPA list of hazardous materials, or if it has one or more of these characteristics:

1. *Reactive*. Any material that reacts violently with water or other chemicals is considered hazardous. When exposed to low-pH acid solutions, if a material releases cyanide gas, hydrogen sulphide gas, or similar gases, it is hazardous.
2. *Corrosive.* If a material burns the skin or dissolves metals and other materials, it is considered hazardous.
3. *Toxic.* **Materials** are hazardous if they leach one or more of eight heavy metals in concentrations greater than 100 times primary drinking water standard.
4. *Ignitable.* A liquid is hazardous if it has a flash point below 140°F (60°C), and a solid is hazardous if it ignites spontaneously.

⚠ **WARNING:** Hazardous waste disposal laws include serious penalties for anyone responsible for breaking these laws.

The automotive service industry is considered a generator of hazardous wastes. However, the vehicles it services are the real generators. New oil is not a hazardous waste, used oil is. Once you drain oil from an engine, you have generated the waste and now become responsible for the proper disposal of this hazardous waste. There are many other wastes that need to be handled properly after you have removed them, such as batteries, brake fluid, and transmission fluid.

Engine coolant should not be allowed to go down sewage drains. This is also true for all liquids drained from a car. Coolant should be captured and recycled or disposed of properly.

Filters for fluids (transmission, fuel, and oil filters) also need to be handled in designated ways. Used filters need to be drained and then crushed or disposed of in a special shipping barrel. Most regulations demand that oil filters be drained for at least 24 hours before they are disposed of or crushed.

Figure 1-12 Hazardous waste hauler. *(Courtesy of DuPont Automotive Products)*

Federal and state laws control the disposal of hazardous waste materials. Every shop employee must be familiar with these laws. Hazardous waste disposal laws include the Resource Conservation and Recovery Act (RCRA). This law basically states that hazardous material users are responsible for hazardous materials from the time they become a waste until the proper waste disposal is completed. Many automotive shops hire an independent hazardous waste hauler to dispose of hazardous waste material (Figure 1-12). The shop owner or manager should have a written contract with the hazardous waste hauler. Rather than have hazardous waste material hauled to an approved hazardous waste disposal site, a shop may choose to recycle the material in the shop. Therefore, the user must store hazardous waste material properly and safely, and be responsible for the transportation of this material until it arrives at an approved hazardous waste disposal site and is processed according to the law.

The RCRA controls these types of automotive waste:

1. Paint and body repair products waste
2. Solvents for parts and equipment cleaning
3. Batteries and battery acid
4. Mild acids used for metal cleaning and preparation
5. Waste oil, engine coolants, or antifreeze
6. Air conditioning refrigerants
7. Engine oil filters

Never, under any circumstances, use these methods to dispose of hazardous waste material:

1. Pour hazardous wastes on weeds to kill them.
2. Pour hazardous wastes on gravel streets to prevent dust.
3. Throw hazardous wastes in a dumpster.
4. Dispose of hazardous wastes anywhere but an approved disposal site.
5. Pour hazardous wastes down sewers, toilets, sinks, or floor drains.
6. Bury hazardous wastes in the ground.

The **right-to-know laws** state that employees have a right to know when the materials they use at work are hazardous. The right-to-know laws started with the **Hazard Communication Standard** published by the Occupational Safety and Health Administration (OSHA) in 1983. This document was originally intended for chemical companies and manufacturers that required employees to handle hazardous materials in their work situation. At the present time, most states have established their own right-to-know laws. Meanwhile, the federal courts have decided to apply these laws to all companies, including automotive service shops. Under the right-to-know laws, the employer has three responsibilities regarding the handling of hazardous materials by its employees.

First, all employees must be trained about the types of hazardous materials they will encounter in the workplace. The employees must be informed about their rights under legislation regarding the handling of hazardous materials. All hazardous materials must be properly labeled, and information about each hazardous material must be posted on material safety data sheets (MSDS) available from the manufacturer (Figure 1-13). In Canada, MSDS sheets are called **workplace hazardous materials information systems (WHMIS)**.

Material safety data sheets (MSDS) provide extensive information about hazardous materials.

The employer has a responsibility to place MSDS sheets where they are easily accessible by all employees. The MSDS sheets provide extensive information about the hazardous material such as:

1. Chemical name
2. Physical characteristics
3. Protective equipment required for handling
4. Explosion and fire hazards
5. Other incompatible materials
6. Health hazards such as signs and symptoms of exposure, medical conditions aggravated by exposure, and emergency and first-aid procedures
7. Safe handling precautions
8. Spill and leak procedures

CRC MATERIAL SAFETY DATA SHEET SILOO

CRC Industries, Inc. • 885 Louis Drive • Warminster, PA 18974 • (215) 674-4300

PRODUCT NAME CLEAN-R-CARB (AEROSOL) #-MSDS05079
PRODUCT- 5079,5079T,5081,5081T
(Page 1 of 2)

1. INGREDIENTS	CAS #	ACGIH TLV	OSHA PEL	OTHER LIMITS	%
Acetone	67-64-1	750 ppm	750 ppm		2-5
Xylene	1330-20-7	100 ppm	100 ppm		68-75
2-Butoxy Ethanol	111-76-2	25 ppm	25 ppm	(skin)	3-5
Methanol	67-56-1	200 ppm	200 ppm		3-5
Detergent	-	NA	NA		0-1
Propane	74-98-6	NA	1000 ppm		10-20
Isobutane	75-28-5	NA	NA	1000ppm	10-20

2. PHYSICAL DATA : (without propellent)
Specific Gravity : 0.865
Vapor Pressure : ND
% Volatile : > 99
Boiling Point : 176 F initial
Evaporation Rate : Moderately fast
Freezing Point : ND
Vapor Density : ND
Appearance and Odor:
A clear colorless liquid, aromatic odor
pH: NA

Solubility : Partially soluble in water.

3. FIRE AND EXPLOSION DATA
Flashpoint : -40 F Method : TCC
Flammable Limits : propellent LEL:1.8 UEL:9.5
Extinguishing Media : CO2, dry chemical, foam
Unusual Hazards : Aerosol cans may explode when heated above 120 F.

4. REACTIVITY AND STABILITY
Stability : Stable
Hazardous decomposition products
: CO2, carbon monoxide (thermal)

Materials to avoid : Strong oxidizing agents and sources of ignition.

5. PROTECTION INFORMATION
Ventilation : Use mechanical means to insure vapor conc. is below TLV.

Respiratory : Use self-contained breathing apparatus above TLV.

Gloves : Solvent resistant Eye & Face : Safety glasses
Other Protective Equipment: Not normally required for aerosol product usage.

Figure 1-13 Material safety data sheets (MSDS) inform employees about hazardous materials. *(Courtesy of CRC Industries, Inc.)*

Secondly, the employer has a responsibility to make sure that all hazardous materials are properly labeled. The label information must include health, fire, and reactivity hazards posed by the material, and the protective equipment necessary to handle the material. The manufacturer must supply all warning and precautionary information about hazardous materials, and this information must be read and understood by the employee before handling the material.

Thirdly, employers are responsible for maintaining permanent files regarding hazardous materials. These files must include information on hazardous materials in the shop, proof of employee training programs, and information about accidents such as spills or leaks of hazardous materials. The employer's files must also include proof that employees' requests for hazardous material information such as MSDS sheets have been met. A general right-to-know compliance procedure manual must be maintained by the employer.

Shop Manual
Chapter 1, page 15

Summary

Terms to Know

Corrosive

Environmental
Protection Agency
(EPA)

Halogen and halon
fire extinguishers

Hazard
Communication
Standard

Ignitable

Material Safety Data
Sheets (MSDS)

Multipurpose dry
chemical fire
extinguisher

Occupational Safety
and Health Act
(OSHA)

Reactive

Resource
Conservation and
Recovery Act (RCRA)

Right-to-know laws

Toxic

Workplace
Hazardous Materials
Information Systems
(WHMIS)

❏ The United States Occupational Safety and Health Act of 1970 assured safe and healthful working conditions, and authorized enforcement of safety standards.

❏ Many hazardous materials and conditions can exist in an automotive shop, including flammable liquids and materials, corrosive acid solutions, loose sewer covers, caustic liquids, high-pressure air, frayed electric cords, hazardous waste materials, carbon monoxide, improper clothing, harmful vapors, high noise levels, and spills on shop floors.

❏ Workplace hazardous material information systems (WHMIS) provide information regarding the labeling and handling of hazardous materials.

❏ The danger regarding the labeling and handling of hazardous conditions and materials may be avoided by applying the necessary safety precautions. These precautions include all areas of safety such as personal safety, gasoline safety, housekeeping safety, general shop safety, fire safety, and hazardous waste handling safety.

❏ The automotive shop must supply the necessary shop safety equipment, and all shop personnel must be familiar with the location and operation of this equipment. Shop safety equipment includes gasoline safety cans, steel storage cabinets, combustible material containers, fire extinguishers, eyewash fountains, safety glasses and face shields, first-aid kits, and hazardous waste disposal containers.

Review Questions

Short Answer Essays

1. Explain the purposes of the Occupational Safety and Health Act.

2. Define 12 shop hazards, and explain why each hazard is dangerous.

3. Describe five steps that are necessary for personal protection in the automotive shop.

4. Explain why smoking is dangerous in the shop.

5. Describe the danger in drug or alcohol use in the shop.

6. Explain three safety precautions related to electrical safety in the shop.

7. Define six essential safety precautions regarding gasoline handling.

8. Describe five steps required to provide housekeeping safety in the shop.

9. Describe how used oil filters need to be disposed of.

10. Describe typical fire extinguisher operation.

Fill-in-the-Blanks

1. The poisonous gas in vehicle exhaust is _____ _____.

2. Safety boots with _____ toe caps are best for working in the shop.

3. One gallon of gasoline has a force equal to _____ sticks of dynamite.

4. Breathing asbestos dust may cause _____ _____.

5. Class C fires involve the burning of _____ equipment.

6. Irradiation eye burns may be caused by excessive light from a(n) _____ .

7. Hazardous wastes in an automotive shop include:

   A. _____   B. _____   C. _____

   D. _____   E. _____   F. _____

8. The right-to-know laws state that employees have a right to know when the _____ they handle at work are _____.

9. Material safety data sheets (MSDS) supply specific information regarding _____ _____.

10. Hazardous materials must never be dumped in _____, _____, _____, or _____.

ASE Style Review Questions

1. While discussing shop hazards:
   Technician A says high-pressure air from an air gun may penetrate the skin.
   Technician B says air in the bloodstream may be fatal.
   Who is correct?
   A. A only   **C.** Both A and B
   B. B only   **D.** Neither A nor B

2. While discussing personal protection in the shop:
   Technician A says jewelry, such as rings or watches may cause serious burns if they make contact between an electric terminal and ground on the vehicle.
   Technician B says sneakers are suitable footwear in the automotive shop.
   Who is correct?
   A. A only   **C.** Both A and B
   B. B only   **D.** Neither A nor B

3. While discussing fire fighting:
   Technician A says halogen-type fire extinguishers produce no harmful gases when they are used to extinguish a fire.
   Technician B says that multipurpose dry chemical fire extinguishers may only be used on type D fires.
   Who is correct?
   A. A only
   B. B only
   C. Both A and B
   D. Neither A nor B

4. While discussing fire fighting:
   Technician A says water should be sprayed on a gasoline fire.
   Technician B says if a fire occurs inside a building, the doors and windows should be opened.
   Who is correct?
   A. A only
   B. B only
   C. Both A and B
   D. Neither A nor B

5. While discussing hazardous waste disposal:
   Technician A says the right-to-know laws require employers to train employees regarding hazardous waste materials.
   Technician B says the right-to-know laws do not require employers to keep permanent records regarding hazardous waste materials.
   Who is correct?
   A. A only
   B. B only
   C. Both A and B
   D. Neither A nor B

6. While discussing material safety data sheets (MSDS):
   Technician A says these sheets explain employers' and employees' responsibilities regarding hazardous material handling and disposal.
   Technician B says these sheets contain specific information about hazardous materials.
   Who is correct?
   A. A only
   B. B only
   C. Both A and B
   D. Neither A nor B

7. While discussing hazardous materials:
   Technician A says a solid that ignites spontaneously is considered a hazardous material.
   Technician B says a liquid with a flash point below 140°F (60°C) is considered a hazardous material.
   Who is correct?
   A. A only
   B. B only
   C. Both A and B
   D. Neither A nor B

8. While discussing hazardous waste disposal:
   Technician A says certain types of hazardous waste may be poured down a floor drain.
   Technician B says hazardous waste users are responsible for hazardous waste materials from the time they become waste until the proper waste disposal is completed.
   Who is correct?
   A. A only
   B. B only
   C. Both A and B
   D. Neither A nor B

9. While discussing hazardous waste disposal:
   Technician A says hazardous waste materials may be hauled to an approved hazardous waste disposal site or recycled in the shop.
   Technician B says the disposal of all hazardous waste materials must be done in accordance with federal, state, and local laws.
   Who is correct?
   A. A only
   B. B only
   C. Both A and B
   D. Neither A nor B

10. While discussing hazardous waste disposal:
    Technician A says air conditioning refrigerants are considered hazardous waste materials.
    Technician B says information about spill and leak procedures for hazardous waste materials is contained in material safety and data sheets (MSDS).
    Who is correct?
    A. A only
    B. B only
    C. Both A and B
    D. Neither A nor B

Basic Theories

Upon completion and review of this chapter, you should be able to:

- ❏ Define atoms and elements.
- ❏ Define compounds and molecules.
- ❏ Describe the parts of an atom.
- ❏ Explain Newton's Laws of Motion.
- ❏ Define work and force.
- ❏ List the most common types of energy and energy conversions.
- ❏ Define inertia and momentum.

- ❏ Explain friction.
- ❏ Define mass, weight, and volume.
- ❏ Define power.
- ❏ Explain the compressibility of gases and the noncompressibility of liquids.
- ❏ Describe atmospheric pressure and vacuum.
- ❏ Explain venturi operation.
- ❏ Explain what happens during combustion.

Introduction

An understanding of the basics is absolutely essential before a study of complex systems and components is attempted. Basic engine theories such as force, work, torque, and power must be understood prior to a study of engine components, systems, and diagnosis in this book. To truly understand how an engine works, you must have a basic understanding of these principles of chemistry and physics. Other important concepts that you must understand are those related to electricity and electronics. These topics are covered in much detail in the next chapter of this book. If you have studied engine theory and physics previously, the information in this chapter may be used as a review. A thorough study of this chapter will provide all the necessary background information before you study the information in this book.

The Basics

Atoms and Elements

An **atom** may be defined as the smallest particle of an element in which all the chemical characteristics of the element are present. An **element** is defined as a liquid, solid, or gas that contains only one type of atom. For example, copper contains only copper atoms.

Compounds and Molecules

A **compound** is a liquid, solid, or gas that contains two or more types of atoms. Water is a compound that contains hydrogen and oxygen. A **molecule** is the smallest particle of a compound in which all the chemical characteristics of the compound are present.

Electrons, Protons, and Neutrons

Protons are small positively charged particles located at the center, or nucleus, of each atom.

Electrons are small, very light particles with a negative electrical charge. These electrons move in orbits around the nucleus of an atom.

Neutrons do not have an electrical charge. These particles add weight to an atom. Neutrons are positioned in the nucleus of an atom.

Electron Movement

The outer ring on an atom is called a **valence ring**. The number of electrons on the valence ring determines the electrical characteristics of the element.

Newton's Laws of Motion

First Law

A body in motion remains in motion, and a body at rest remains at rest, unless some outside force acts on it. When a car is parked on a level street, it remains stationary unless it is driven or pushed. If the gas pedal is depressed with the engine running and the transmission in drive, the engine delivers power to the drive wheels and this force moves the car.

Second Law

A body's acceleration is directly proportional to the force applied to it, and the body moves in a straight line away from the force. For example, if the engine power supplied to the drive wheels increases, the vehicle accelerates faster.

Third Law

For every action there is an equal and opposite reaction. A practical application of this law occurs when the wheel on a vehicle strikes a bump in the road surface. This action drives the wheel and suspension upward with a certain force, and a specific amount of energy is stored in the spring. After this action occurs, the spring forces the wheel and suspension downward with a force equal to the initial upward force.

Work and Force

Work is defined as the result of applying a force.

When a force moves a certain mass a specific distance, work is produced. When work is accomplished, the mass may be lifted or slid on a surface (Figure 2-1). Since force is measured in pounds and distance is measured in feet, the measurement for work is foot-pounds (ft.-lb.). In the metric system work, it is measured in Newton meters (Nm). If a force moves a 3,000-pound vehicle for 50 feet, 150,000 foot-pounds of work are produced. Mechanical force acts on an object to start, stop, or change the direction of the object. It is possible to apply force to an object and not move the object. Under this condition, no work is done. Work is only accomplished when an object is started, stopped, or redirected by a force.

Energy

Energy may be defined as the ability to do work.

When energy is released to do work, it is called kinetic energy. This type of energy may also be referred to as energy in motion. Stored energy may be called potential energy. Energy is available in one of these six forms:

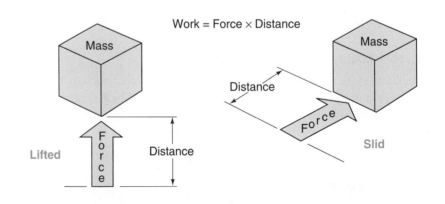

Figure 2-1 Work is accomplished when a mass is lifted or slid on a surface.

1. Chemical energy is contained in the molecules of different atoms. In the automobile, chemical energy is contained in the molecules of gasoline, and also in the molecules of electrolyte in a battery.
2. Electrical energy is required to move electrons through an electric circuit. In the automobile, the battery is capable of producing electrical energy to start the vehicle, and the alternator produces electrical energy to power the electrical accessories and recharge the battery.
3. Mechanical energy is defined as the ability to move objects. In the automobile, the battery supplies electrical energy to the starting motor, and this motor converts the electrical energy to mechanical energy to crank the engine.
4. Thermal energy may be defined as energy produced by heat. When gasoline burns, thermal energy is released.
5. Radiant energy is defined as light energy. In the automobile, radiant energy is produced by the lights.
6. Nuclear energy is defined as the energy within atoms when they are split apart or combined. Nuclear power plants generate electricity with this principle. This type of energy is not used in the automobile.

Energy Conversion

Energy conversion occurs when one form of energy is changed to another form. Since energy is not always in the desired form, it must be converted to a form we can use. Some of the most common automotive energy conversions are these:

Chemical to Thermal Energy Conversion

Chemical energy in gasoline or diesel fuel is converted to thermal energy when the fuel burns in the engine cylinders.

Thermal to Mechanical Energy Conversion

Mechanical energy is required to rotate the drive wheels and move the vehicle. The piston and crankshaft in the engine and the drive train are designed to convert the thermal energy produced by the burning fuel into mechanical energy (Figure 2-2).

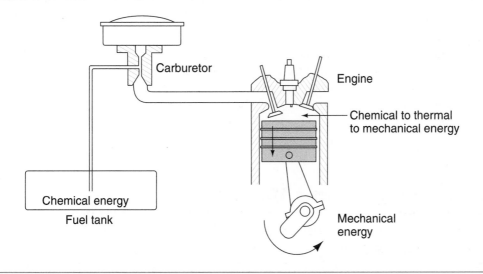

Figure 2-2 Thermal energy in the fuel is converted to mechanical energy in the engine cylinders. The piston, crankshaft, and drivetrain deliver this mechanical energy to the drive wheels.

Electrical to Mechanical Energy Conversion

The windshield wiper motor converts electrical energy from the battery or alternator to mechanical energy to drive the windshield wipers.

Mechanical to Electrical Energy Conversion

The alternator is driven by mechanical energy from the engine. The alternator converts this energy to electrical energy, which powers the electrical accessories on the vehicle.

Mechanical to Thermal Energy Conversion

The brakes of a vehicle change the mechanical energy of the moving vehicle to heat energy, in the form of friction, created by the action of the brake linings rubbing against the rotating brake drums or rotors.

Inertia

The inertia of an object at rest is called static inertia, whereas dynamic inertia refers to the inertia of an object in motion. Inertia exists in liquids, solids, and gases. When you push and move a parked vehicle, you overcome the static inertia of the vehicle. If you catch a ball in motion, you overcome the dynamic inertia of the ball.

Momentum

Momentum is the product of an object's weight times its speed. Momentum is a type of mechanical energy. An object loses momentum if another force overcomes the dynamic inertia of the moving object.

Friction

Friction may occur in solids, liquids, and gases. When a car is driven down the road, friction occurs between the air and the car's surface. This friction opposes the momentum, or mechanical energy, of the moving vehicle. Since friction creates heat, some of the mechanical energy from the vehicle's momentum is changed to heat energy in the air and body components. The mechanical energy from the engine must overcome the vehicle inertia and the friction of the air striking the vehicle. Body design has a very dramatic effect on the amount of friction developed by the air striking the vehicle. The total resistance to motion caused by friction between a moving vehicle and the air is referred to as coefficient of drag (Cd).

The study of Cd is very complicated, and also very important. At 45 miles per hour (mph), or 72 kilometers per hour (kph), half of the engine's mechanical energy is used to overcome air friction, or resistance. Therefore, reducing a vehicle's Cd can be a very effective method of improving fuel economy.

Mass, Weight, and Volume

A lawn mower is much easier to push than a 2,500-pound vehicle, because the lawn mower has very little inertia compared to the vehicle. A space ship might weigh 100 tons here on earth where it is affected by the earth's gravitational pull. In outer space beyond the earth's gravity and atmosphere, the space ship is almost weightless. Here on Earth, mass and weight are measured in pounds and ounces in the English system. In the metric system, mass and weight are measured in kilograms.

Friction is the resistance to motion when one object is moved over another object. Friction generates heat.

Inertia is defined as the tendency of an object at rest to remain at rest, or the tendency of an object in motion to stay in motion.

When a force overcomes static inertia and moves an object, the object gains momentum.

Coefficient of drag (Cd) may also be called aerodynamic drag.

Mass is the measurement of an object's inertia.

Weight is the measurement of the earth's gravitational pull on the object.

Volume is a measurement of size, and it is related to mass and weight. For example, a pound of gold and a pound of feathers both have the same weight, but the pound of feathers occupies a much larger volume. In the English system, volume is measured in cubic inches, cubic feet, cubic yards, or gallons. The measurement for volume in the metric system is cubic centimeters or liters.

Volume is the length, width, and height of a space occupied by an object.

Torque

When you pull a wrench to tighten a bolt, you supply torque to the bolt. This torque, or twisting force, is calculated by multiplying the force and the radius. For example, if you supply 10 pounds of force on the end of a 2-foot wrench to tighten a bolt, the torque is 10 times 2 = 20 foot-pounds (ft.-lb.) (Figure 2-3). If the bolt turns during torque application, work is done. When a bolt does not rotate during torque application, no work is accomplished.

Torque is a force that does work with a twisting, or turning, force.

Power

James Watt, a Scotsman, is credited with being the first person to calculate power. He measured the amount of work that a horse could do in a specific amount of time. Watt calculated that a horse could move 330 pounds for 100 feet in 1 minute. If you multiply 330 pounds by 100 feet, the answer is 33,000 foot-pounds of work. Watt determined that one horse could do 33,000 foot-pounds of work in 1 minute. Thus, one horsepower is equal to 33,000 foot-pounds per minute, or 550 foot-pounds per second (Figure 2-4). Two horsepower could do this same amount of work in one-half minute, or four horsepower would be capable of completing this work in one-quarter minute. If you push a 3,000 pound, 1,360 kilograms (kg), car for 11 feet (3.3 meters) in one-quarter minute, you produce four horsepower. From this brief discussion about horsepower, we can understand that as power increases, speed also increases, or the time to do work decreases.

Power is a measurement for the rate, or speed, at which work is done.

Principles Involving Liquids and Gases

Molecular Energy

Remember that kinetic energy refers to energy in motion. Since electrons are constantly in motion around the nucleus in atoms or molecules, kinetic energy is present in all matter. Kinetic energy in atoms and molecules increases as the temperature increases. A decrease in temperature reduces this kinetic energy. Molecules in solids move slowly compared to those in liquids or gases. Gas molecules move quickly compared to liquid molecules. Since gas molecules are in constant motion, they spread out to fill all the space available. At higher temperatures, gas molecules spread out more, whereas lower temperatures cause gas molecules to move closer together.

Molecular energy may be defined as the kinetic energy available in atoms and molecules because of the constant electron movement within these molecules.

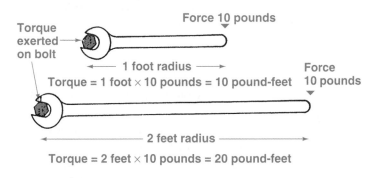

Figure 2-3 Force applied to a wrench produces torque. If the bolt turns, work is accomplished.

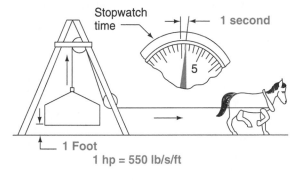

Figure 2-4 One horsepower is produced when 550 pounds are moved 1 foot in 1 second.

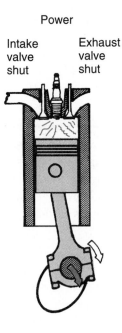

Power

Intake valve shut

Exhaust valve shut

Figure 2-5 Hot expanding gases push the piston downward and rotate the crankshaft. *(Courtesy of Sun Electric Corporation)*

Temperature

Temperature affects all liquids, solids, and gases. The volume of any matter increases as the temperature increases. Conversely, the volume decreases in relation to a reduction in temperature. When the gases in an engine cylinder are burned, the sudden temperature increase causes rapid gas expansion, which pushes the piston downward and causes engine rotation (Figure 2-5).

Pressure and Compressibility

Pressure may be defined as a force exerted on a given surface area.

Pressure and temperature are directly related. If you increase one, you also increase the other.

Pressure and volume are inversely proportional. If volume is decreased, pressure is increased.

Liquids are not compressible.

Since liquids and gases are both substances that flow, they may be classified as fluids. If a nail punctures an automotive tire, the air escapes until the pressure in the tire is equal to atmospheric pressure outside the tire. When the tire is repaired and inflated, air pressure is forced into the tire. If the tire is inflated to 32 pounds per square inch (psi), or 220 kilopascals (kPa), this pressure is applied to every square inch on the inner tire surface. Pressure is always supplied equally to the entire surface of a container. Since air is a gas, the molecules have plenty of space between them. When the tire is inflated, the pressure in the tire increases, and the air molecules are squeezed closer together, or compressed. Under this condition, the air molecules cannot move as freely, but extra molecules of air can still be forced into the tire. Therefore, gases such as air are said to be **compressible**.

The air in the tire may be compared to a few balls on a billiard table without pockets. If a few more balls are placed on the table, the balls are closer together, but they can still move freely (Figure 2-6).

If the vehicle is driven at high speed, friction between the road surface and the tires heats the tires and the air in the tires. When air temperature increases, the pressure in the tire also increases. Conversely, a temperature decrease reduces pressure.

If 100 cubic feet (2.8 cubic meters) of air is forced into a large truck tire, and the same amount of air is forced into a much smaller car tire, the pressure in the car tire is much greater.

Molecules in a liquid may be compared to a billiard table without pockets that is completely filled with balls. These balls can roll around, but no additional balls can be placed on the table because the balls cannot be compressed. Similarly, liquid molecules cannot be compressed (Figure 2-7).

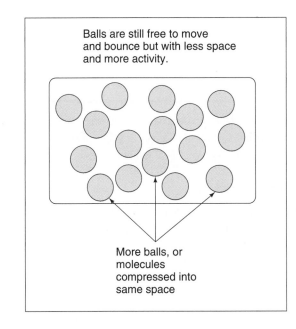

Balls are still free to move
and bounce but with less space
and more activity.

More balls, or
molecules
compressed into
same space

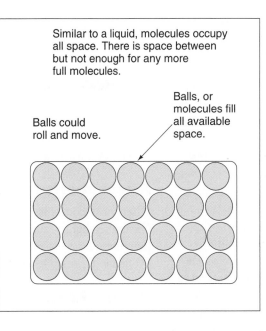

Similar to a liquid, molecules occupy
all space. There is space between
but not enough for any more
full molecules.

Balls could
roll and move.

Balls, or
molecules fill
all available
space.

Figure 2-6 Gases can be compressed much like more balls can be placed on a billiard table containing a few balls. *(Courtesy of Chrysler Corporation)*

Figure 2-7 Liquids are noncompressible, just as more balls cannot be added to a billiard table with no pockets that is completely filled with balls. *(Courtesy of Chrysler Corporation)*

Liquid Flow

If a tube is filled with billiard balls and the outlet is open, more balls may be added to the inlet. When each ball is moved into the inlet, a ball is forced from the outlet. If the outlet is closed, no more balls can be forced into the inlet (Figure 2-8).

The billiard balls in the tube may be compared to molecules of power steering fluid in the line between the power steering pump and steering gear. Since **noncompressible** fluid fills the line and gear chamber, the force developed by the pump pressure is transferred through the line to the gear chamber (Figure 2-9).

This pressure is applied equally to every square inch in the gear chamber. This pressure applied to the rack piston in the gear chamber helps to move the rack piston. Since the rack is

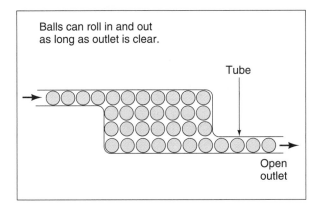

Balls can roll in and out
as long as outlet is clear.

Tube

Open
outlet

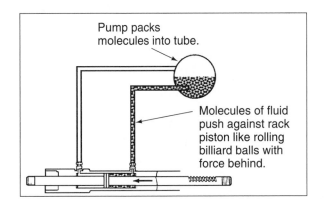

Pump packs
molecules into tube.

Molecules of fluid
push against rack
piston like rolling
billiard balls with
force behind.

Figure 2-8 Billiard ball movement in a tube filled with balls compared to liquid flow. *(Courtesy of Chrysler Corporation)*

Figure 2-9 Power steering pump pressure supplied to the steering gear chamber. *(Courtesy of Chrysler Corporation)*

connected through steering linkages and arms to the front wheels, the force on the rack piston helps the driver to move the front wheels to the left or right during a turn (Figure 2-10).

Atmospheric Pressure

Since air is gaseous matter with mass and weight, it exerts pressure on the earth's surface. A one-square-inch column of air extending from the earth's surface to the outer edge of the atmosphere weighs 14.7 psi at sea level. Therefore, atmospheric pressure is 14.7 psi at sea level (Figure 2-11).

Atmospheric Pressure and Temperature

When air becomes hotter, it expands. This hotter air is lighter than an equal volume of cooler air. This hotter, lighter, air exerts less pressure on the earth's surface compared to cooler air. If the temperature decreases, air contracts and becomes heavier. Therefore, an equal volume of cooler air exerts more pressure on the earth's surface compared to hotter air.

Atmospheric Pressure and Height

As you climb above sea level, atmospheric pressure decreases. At 5,000 feet (1,524 meters) above sea level, a one-square-inch column of air from the earth's surface to the outer edge of the atmosphere is 5,000 feet (1,524 meters) less than the same column of air at sea level. Therefore, the weight of this column of air is less at an elevation of 5,000 feet (1,524 meters) than at sea level. As altitude continues to increase, atmospheric pressure continues to decrease. At an altitude of several hundred miles above sea level, the earth's atmosphere ends. There is zero atmospheric pressure at that point.

Vacuum

When air has a pressure higher than atmospheric pressure, a **positive pressure** exists. When air has a lower pressure than atmospheric, it has a **negative pressure**. **Vacuum** is the term commonly used to describe negative pressures. Vacuum is best described as any pressure less than atmospheric. A total or complete vacuum is the total or complete absence of air pressure.

Atmospheric pressure may be defined as the total weight of the earth's atmosphere. However, if the pressure is measured with a pressure gauge, the gauge would read zero psi. A pressure gauge measures only the difference in pressure between what surrounds the gauge and what is applied to the gauge. To avoid confusion, pressures measured by the gauge are listed as "psig." The "g" means gauge. True or absolute (such as atmospheric pressures) pressures are given as psia.

An equal volume of hot air weighs less and exerts less pressure on the earth's surface compared to cold air.

Atmospheric pressure decreases as altitude increases.

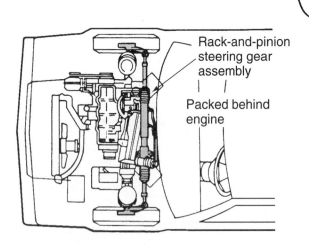

Figure 2-10 Steering gear, linkages, and arms connected to the front wheels. *(Courtesy of Chrysler Corporation)*

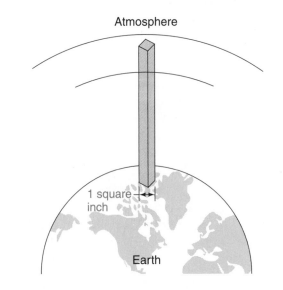

Figure 2-11 A column of air one-inch square extending from the earth's surface at sea level to the outer edge of the atmosphere weighs 14.7 lb.

The understanding
engine works. It also exp
supposed to. The reaction
components. Whenever tl
higher pressure will move

We all are quite fam
we are increasing the pre
the lower pressure outsid

Vacuum could be n
for this measurement. L
atmospheric pressure is a
end of the "U" tube, the
movement occurs, the m
the mercury moves dow
sure is supplied, and upv
plied to the "U" tube. Th
is 0 psi of pressure.

Vacuum and atmos
pheric pressure is availab
intake valve open, a vacu
the high pressure outside

Pumps use high an
rotates, it creates a high
difference causes coolan

Venturi Prir

Pressures of a liquid or g
relieving some of the pr

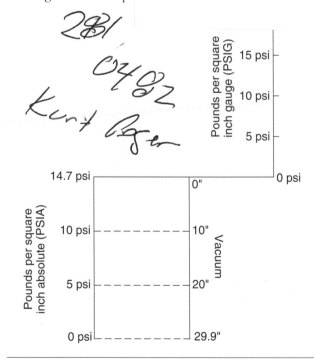

Figure 2-12 A comparison of pressure and vacuum readings.

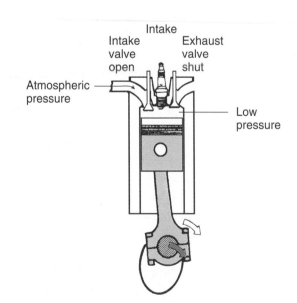

Figure 2-13 Air moves from the outside atmosphere, which is a high pressure, to the lower pressure in the cylinder during the engine's intake stroke. *(Courtesy of Sun Electric Co.)*

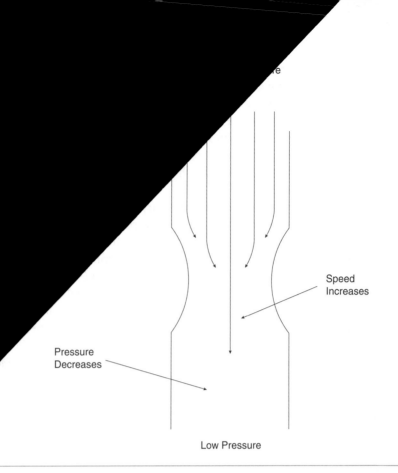

Figure 2-14 A venturi increases the speed of the incoming air and creates a vacuum below the venturi.

tion in whatever the liquid or gas is flowing through. At the restriction, the pressure increases. After the restriction, the pressure is lower. A venturi is a restriction designed to increase the speed of the flow and decrease its pressure.

Let's assume we have a tube with air flowing through it. The air is moving from the high pressure side and the low pressure side. In the tube we have a restriction or venturi. As the moving air enters the venturi, the center column of air has no direct contact with the restriction. The air surrounding the center column is directed, by the venturi, toward the center column. This pushes the column through the venturi faster than it would on its own.

On the other side of the restriction, the size of the tube returns to normal. Although the speed of the air increased through the venturi, the volume of air entering into this larger area is less than the amount of air on the other side of the venturi. As a result, a lower pressure or vacuum is present at the outlet of the venturi. The venturi principle (Figure 2-14) is the basic theory of operation for a carburetor, as well as other automotive components.

Combustion

An engine converts the energy found in fuel into heat energy, which is changed to mechanical energy that can be used to move a vehicle. The conversion of energy results from the mechanical design of the engine and through combustion. **Combustion** is a burning process. During this burning process, a chemical reaction takes place between the air and the fuel that entered the engine's cylinders. During this chemical reaction heat energy is released. With the release of the heat, the pressure of the air in the cylinders increases. This high pressure pushes on the engine's piston causing it to move. The movement of the piston is converted to move the vehicle.

The energy of the moving piston is dependent upon the amount of pressure there is on the piston. The amount of pressure on the piston depends on the amount of heat generated by the

```
                    INITIALIZATION HEAT
AIR + FUEL          ───────────────►  WATER + CARBON DIOXIDE + AIR + HEAT ENERGY
```

Figure 2-15 The ideal combustion process.

combustion process. Complete combustion of the air and fuel will provide for the maximum amount of heat from that amount of air and fuel.

In order to have complete combustion, four things must be present: 1) the correct amount of air must be mixed with, 2) the correct amount of fuel in, 3) a sealed container, and this mixture must be 4) shocked by the correct amount of heat at the correct time (Figure 2-15). Although there are other things that can affect combustion, these are the absolute most important factors. As you will see, the engine has systems that attempt to meet these requirements and provide for complete combustion or maximum efficiency. When an engine doesn't run well, it is because one or more of these requirements have not been met.

All components of the engine are designed in an attempt to achieve total combustion. Although no engine can achieve complete combustion during all operating conditions, total combustion is what is strived for. Incomplete or inefficient combustion results from not having a sealed cylinder or the right amounts of air, fuel, and/or heat in the cylinder. The more incorrect any of these requirements are, the less complete the combustion will be.

The result of total combustion is the generation of great heat and the conversion of all of the cylinder's air and fuel into water and oxygen. When total combustion does not take place, full conversion of the fuel and air also does not take place. This results in the release of pollutants from the engine.

SUMMARY

❏ An element contains only one type of atom.

❏ An atom is the smallest particle of an element.

❏ Compounds contain two or more types of atoms.

❏ A molecule is the smallest particle in a compound.

❏ Protons are positively charged particles at the center, or nucleus, of each atom.

❏ Neutrons have no electric charge, and they are located in the nucleus of most atoms.

❏ Electrons are negatively charged particles found in various orbits around the nucleus of an atom.

❏ Work is the result of applying a force.

❏ Force is measured in pounds and distance.

❏ Energy is the ability to do work, and there are six basic types of energy.

❏ Inertia is the tendency of an object at rest to remain at rest, or the tendency of an object in motion to remain in motion.

❏ An object gains momentum when force overcomes static energy and moves the object.

❏ Friction is the resistance to motion when one object is moved over another object.

❏ Mass is a measurement of an object's inertia.

❏ Weight is a measurement of the earth's gravitational pull on an object.

❏ Volume is the length, width, and height of a space occupied by an object.

❏ Power is a measurement for the speed at which work is done.

Terms to know

Atmospheric pressure

Atom

Combustion

Compound

Compressible

Electron

Element

Energy

Friction

Inertia

Mass

Molecule

Momentum

Negative pressure

Neutron

Noncompressible

Positive pressure

Power

❏ Torque is a twisting force that does work.

❏ Atmospheric pressure is the total weight of the earth's atmosphere.

❏ Vacuum is often used to describe a condition where air pressure is lower than atmospheric pressure.

❏ In nature, a high pressure always moves toward a lower pressure.

❏ Complete combustion takes place when there is the correct amount of air mixed with the correct amount of fuel in a sealed container, which is shocked by the correct amount of heat at the correct time.

Review Questions

Short Answer Essays

1. Define an element and a compound.

2. Name the four things that are necessary for complete combustion.

3. Define a molecule.

4. Describe three particles found in an atom, including the electrical charge and location of each particle.

5. Explain why a venturi decreases pressure.

6. What is atmospheric pressure?

7. What happens to air pressure as the temperature of the air changes?

8. Describe Newton's three laws of motion.

9. Describe six different forms of energy.

10. Describe four different types of energy conversion.

Fill-in-the-Blanks

1. At sea level, atmospheric pressure is _____ psi.

2. In nature, a _____ pressure always moves toward a _____ pressure.

3. The nucleus of an atom is comprised of a(n) _____ and a(n) _____.

4. Work is calculated by multiplying _____ times _____ .

5. Energy may be defined as the ability to do_____ .

6. When one object is moved over another object the resistance to motion is called _____ .

7. Weight is the measurement of the earth's _____ _____ on an object.

8. Torque is a force that does work with a _____ action.

9. Power is a measurement for the rate at which _____ is done.

10. Negative pressure may be called _____ .

ASE Style Review Questions

1. While discussing the parts of an atom:
   Technician A says the term valence ring applies to the outer ring on an atom.
   Technician B says the second ring on a copper atom is the valence ring.
   Who is correct?
   A. A only
   B. B only
   C. Both A and B
   D. Neither A nor B

2. While discussing combustion:
   Technician A says combustion is the process of burning.
   Technician B says a chemical reaction takes place during combustion.
   Who is correct?
   A. A only
   B. B only
   C. Both A and B
   D. Neither A nor B

3. While discussing work:
   Technician A says James Watt is credited as being the person to discover work.
   Technician B says power is a measurement of the rate or speed at which work is done.
   Who is correct?
   A. A only
   B. B only
   C. Both A and B
   D. Neither A nor B

4. While discussing different types of energy:
   Technician A says when energy is released to do work it is called potential energy.
   Technician B says stored energy is referred to as potential energy.
   Who is correct?
   A. A only
   B. B only
   C. Both A and B
   D. Neither A nor B

5. While discussing friction in matter:
   Technician A says that friction generates heat.
   Technician B says that friction occurs in liquids, solids, and gases.
   Who is correct?
   A. A only
   B. B only
   C. Both A and B
   D. Neither A nor B

6. While discussing mass and weight:
   Technician A says that mass is the measurement of an object's inertia.
   Technician B says that mass is the measurement of an object's momentum.
   Who is correct?
   A. A only
   B. B only
   C. Both A and B
   D. Neither A nor B

7. While discussing torque:
   Technician A says torque is calculated by multiplying force times distance.
   Technician B says work is accomplished if a bolt turns when torque is applied.
   Who is correct?
   A. A only
   B. B only
   C. Both A and B
   D. Neither A nor B

8. While discussing horsepower:
   Technician A says one horsepower is equal to 400 foot-pounds per second.
   Technician B says when horsepower increases, the time required to do work remains the same.
   Who is correct?
   A. A only
   B. B only
   C. Both A and B
   D. Neither A nor B

9. While discussing the compressibility of liquids and gases:
   Technician A says that liquids and gases are both compressible.
   Technician B says that gases are compressible, but liquids are noncompressible.
   Who is correct?
   A. A only
   B. B only
   C. Both A and B
   D. Neither A nor B

10. While discussing atmospheric pressure:
    Technician A says as air becomes colder it contracts and becomes heavier.
    Technician B says atmospheric pressure is higher at sea level compared to a 10,000 foot (3,048 meters) elevation.
    Who is correct?
    A. A only
    B. B only
    C. Both A and B
    D. Neither A nor B

Electricity and Electronics

Upon completion and review of this chapter, you should be able to:

❏ Explain the basic principles of electricity.

❏ Define the terms voltage, current, and resistance.

❏ Name the various electrical components and their uses in electrical circuits.

❏ Use Ohm's law formula to calculate volts, amperes, or ohms in a circuit.

❏ Explain the differences between series, parallel, and series-parallel circuits.

❏ Define an electromagnet.

❏ Explain electromagnetic induction.

❏ Describe the operation of a diode on forward and reverse bias.

❏ Explain the operation of a transistor.

❏ Describe briefly how an integrated chip is manufactured.

❏ Explain the purposes of the battery in the vehicle electrical system.

❏ Describe the design of a cell group in an automotive battery.

❏ Describe the chemical changes that occur while the battery is charging and discharging.

❏ Explain four different battery ratings.

❏ Describe the operation of the neutral safety switch, starter relay, and theft deterrent computer.

❏ Explain the current flow through the solenoid windings and starting motor while the starter is engaging.

❏ Describe the current flow through the solenoid windings and starting motor while the engine is cranking.

❏ Explain the starter drive operation when the engine is cranking and while the starting motor is disengaging.

❏ Explain how a voltage regulator limits the AC generator voltage.

❏ Describe how the voltage is induced in the AC generator stator windings, and explain the current flow from the stator windings through the charging circuit when the engine is running.

Introduction

Many complicated electronic systems such as electronic fuel injection and distributorless ignition systems are described in this text. A study of electric and electronic fundamentals is necessary before you begin to learn about these systems. If you have studied the fundamentals of electricity before, the information in this chapter may be used as a review. To understand the operation of many of the components that affect engine performance, you must have a good understanding of electricity and electronics.

There is often confusion concerning the terms electrical and electronic. In this book, electricity and electrical systems will refer to wiring and electrical parts. Electronics will mean computers and other "black box" type items used to control engine and vehicle systems.

A basic understanding of electrical principles is important to proper diagnosis of any system that is monitored, controlled, or operated by electricity. Although the subject is normally covered in a separate course, a quick overview of electricity and its principles is presented here. This chapter also covers the battery, starting motor, and AC generator which are very important components of an automobile. The battery supplies the voltage and current to operate the starting motor to crank the engine. If the electrical accessories are turned on when the engine is not running, the battery supplies the power to operate these accessories. When the engine is running and the charging system cannot meet the power requirements of the electrical system, some power is supplied by the battery.

Basic Electricity

All things are made up of atoms. An atom is the smallest particle of something. Atoms are very small and cannot be seen with your eye. This may be the reason many technicians struggle to understand electricity. The basics of electricity focus on atoms. Understanding the structure of the atom is the first step in understanding how electricity works. The following principles describe atoms, which are the building blocks of all materials.

- ❏ In the center of every atom is a nucleus.
- ❏ The nucleus contains positively charged particles called protons and particles called neutrons that have no charge.
- ❏ Negatively charged particles called electrons orbit around every nucleus.
- ❏ Every type of atom has a different number of protons and electrons, but each atom has an equal number of protons and electrons. Therefore, the total electrical charge of an atom is zero, or neutral.

In all atoms, the electrons are arranged in different orbits, called shells. Each shell contains a specific number of electrons. The outer shell of electrons is called a valence ring. The number of electrons on the valence ring determines the electrical characteristics of the element. For example, a copper atom, which is a good conductor of electricity, has 29 electrons and 29 protons (Figure 3-1). The 29 electrons are arranged in shells. The outer shell has only 1 electron. This outer shell needs 32 electrons to be completely full. This means that the 1 electron in the outer shell is loosely tied to the atom and can be easily removed.

The looseness or tightness of the electrons in orbit around the neutron of an atom explains the behavior of electricity. Electricity is caused by the flow of electrons from one atom to another. The release of energy as one electron leaves the orbit of one atom and jumps into the orbit of another is **electricity**. The key behind creating electricity is to give a reason for the electrons to move.

A BIT OF HISTORY

Electricity was discovered by the Greeks over 2,500 years ago. They noticed when amber was rubbed with other materials, it was charged with an unknown force that had the power to attract objects, such as dry leaves and feathers. The Greeks called amber "elektron." The word electric was derived from the word and meant "to be like amber."

There is a natural attraction of electrons to protons. Electrons have a negative charge and are attracted to something with a positive charge. When an electron leaves the orbit of an atom, the

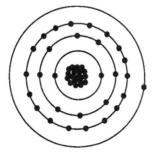

Figure 3-1 Basic structure of a copper atom.

atom then has a positive charge. An electron moves from one atom to another because the atom next to it appears to be more positive than the one it is orbiting around. An electrical power source provides for a more positive charge and in order to allow for a continuous flow of electricity, it supplies free electrons. In order to have a continuous flow of electricity, three things must be present: an excess of electrons in one place, a lack of electrons in another place, and a path between the two places.

Two power or energy sources are used in an automobile's electrical system; these are based on a chemical reaction and on magnetism. A car's battery is a source of chemical energy. A chemical reaction in the battery provides for an excess of electrons and a lack of electrons in another place. Batteries have two terminals, a positive and a negative. Basically the negative terminal is the outlet for the electrons and the positive terminal is the inlet for the electrons to get to the protons. The chemical reaction in a battery causes a lack of electrons at the positive (+) terminal and an excess at the negative (–) terminal. This creates an electrical imbalance, causing the electrons to flow through the path provided by a wire. A simple example of this process is shown in the battery and light arrangement in Figure 3-2.

The chemical process in the battery continues to provide electrons until the chemicals become weak. At that time, either the battery has run out of electrons or all of the protons are matched with an electron. When this happens, there is no longer a reason for the electrons to want to move to the positive side of the battery. It no longer looks more positive. Fortunately, the vehicle's charging system restores the battery's supply of electrons. This allows the chemical reaction in the battery to continue indefinitely.

Electricity and magnetism are interrelated. One can be used to produce the other. Moving a wire (a conductor) through an already existing magnetic field (such as a permanent magnet) can produce electricity. This process of producing electricity through magnetism is called induction. In a generator, a coil of wire is moved through a magnetic field. In an AC generator, a magnetic field is moved through a coil of wire. In both cases, electricity is produced. The amount of electricity that is produced depends on a number of factors, including the strength of the magnetic field, the number of wires that pass through the field, and the speed at which the wire moves through the magnetic field.

Measuring Electricity

Shop Manual
Chapter 3, page 72

Electrical **current** is a term used to describe the movement or flow of electricity. The greater the number of electrons flowing past a given point in a given amount of time, the more current the circuit has. This current, like the flow of water or any other substance, can be measured. **Voltage** is electrical pressure. Voltage is the force developed by the attraction of the electrons to the protons.

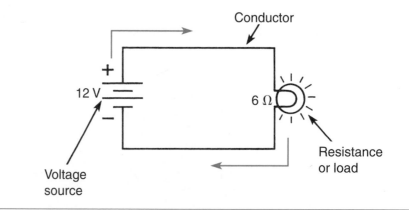

Figure 3-2 A simple light circuit consists of a voltage source, conductors, and a resistance or load.

The more positive one side of the circuit is, the more voltage is present in the circuit. Voltage does not flow, rather it is the pressure that causes current flow. When any substance flows, it meets resistance. The resistance to electrical flow can be measured.

Electrical Flow (Current)

The unit for measuring electrical current is the **ampere**, usually called an amp. The instrument used to measure electrical current flow in a circuit is called an **ammeter**.

In the flow of electricity, millions of electrons are moving past any given point at the speed of light. The electrical charge of any one electron is extremely small. It takes millions of electrons to make a charge that can be measured.

There are two types of electrical flow, or current: **direct current** (DC) and **alternating current** (AC). In direct current, the electrons flow in one direction only. The example of the battery and light shown earlier is based upon direct current. In alternating current, the electrons change direction at a fixed rate. Most automobile circuits operate on DC current while the current in homes and buildings is AC.

> 1 ampere means that 6.25 billion electrons are flowing past a given point in 1 second.
>
> The ampere is named after André Ampere who worked with magnetism and current flow in the 1700s.

Resistance

In every atom, the electrons resist being moved out of their shell. The amount of resistance depends on the type of atom. As explained earlier, in some atoms (such as those in copper) there is very little resistance to electron flow because the outer electron is loosely held. In other substances there is more resistance to flow, because the outer electrons are tightly held.

The resistance to current flow produces heat. This heat can be measured to determine the amount of resistance. A unit of measured resistance is called an **ohm**. Resistance can be measured by an instrument called an **ohmmeter**.

Pressure

In electrical flow, some force is needed to move the electrons between atoms. This force is the pressure that exists between a positive and negative point within an electrical circuit. This force, also called electromotive force (EMF), is measured in units called **volts**. One volt is the amount of pressure (force) required to move 1 ampere of current through a resistance of 1 ohm. Voltage is measured by an instrument called a **voltmeter**.

Voltage Drop

As current passes through a resistance, heat is generated. With this generation of heat comes a loss in voltage. The voltage that is converted to heat by a resistance is called the voltage drop. In all circuits, all of the voltage provided by the source of power is dropped across the circuit. When a circuit has only one resistance, source voltage is dropped across that resistance. If a circuit has more than one resistance, the voltage drop across each resistance depends upon the resistance value of each resistor.

For example, if a 12-volt circuit has two 3-ohm resistors, each resistor will drop 6 volts. If we add a 6-ohm resistor to the circuit, each 3-ohm resistor will drop 3 volts and the 6-ohm resistor will drop 6 volts.

Circuits

When electrons are able to flow along a path (wire) between two points, an electrical circuit is formed. An electrical circuit is considered complete when there is a path that connects the positive and negative terminals of the electrical power source. Somewhere in the circuit there must be a load or resistance to control the amount of current in the circuit. Most automobile electrical circuits use the chassis as the path to the negative side of the battery. Electrical components have a lead that connects them to the chassis. These are called the chassis ground connections. In a complete circuit, the flow of electricity can be controlled and applied to do useful work, such as light a headlamp or turn over a starter motor. Components that use electrical power put a load on the circuit and consume electrical energy.

The amount of current that flows in a circuit is determined by the resistance in that circuit. As resistance goes up, the current goes down. The energy used by a load is measured in volts. Amperage stays constant in a circuit but the voltage is dropped as it powers a load. Measuring voltage drop determines the amount of energy consumed by the load.

Ohm's Law

George S. Ohm was a German scientist in the 1800s who discovered that all electrical quantities are proportional to each other and therefore have a mathematical relationship.

To understand the relationship between current, voltage, and resistance in a circuit you need to know the basic law of electricity, **Ohm's Law**. This law states that it takes one volt of electrical pressure to push one ampere of electrical current through one ohm of resistance. As such, the law provides a mathematical formula for determining the amount of current, voltage, or resistance in a circuit when two of these are known. The basic formula is: Voltage = Current multiplied by Resistance.

Although the basic premise of this formula is calculating unknown values in an electrical circuit (Figure 3-3), it also helps to define the behaviors of electrical circuits. A knowledge of these behaviors is important to an automotive technician.

If voltage does not change, but there is a change in the resistance of the circuit, the current will change. If resistance increases, current decreases. If resistance decreases, current will increase. If voltage changes, so must the current or resistance. If the resistance stays the same and current decreases, so will voltage. Likewise, if current increases, so will the voltage.

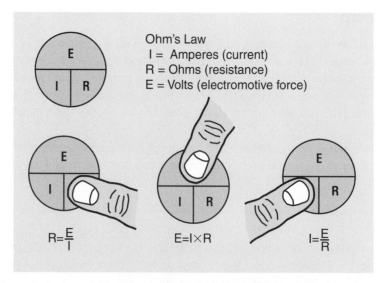

Figure 3-3 Ohm's Law. *(Reprinted with the permission of Ford Motor Company)*

Circuits

A complete electrical circuit exists when electrons flow along a path between two points. In a complete circuit, resistance must be low enough to allow the available voltage to push electrons between the two points. Most automotive circuits contain four basic parts.

1. **Power sources**, such as a battery or AC generator that provides the energy needed to create electron flow.

2. **Conductors**, such as copper wires and the vehicle's frame that provide a path for current flow.

3. **Loads**, which are devices that use electricity to perform work, such as light bulbs, electric motors, or resistors.

4. **Controllers**, such as switches or relays that direct the flow of electrons.

A complete circuit must have a complete path from the power source to the load and back to the source. With the many circuits on an automobile, this would require hundreds of wires connected to both sides of the battery. To avoid this, automobiles are equipped with power distribution centers or fuse blocks that distribute battery voltage to various circuits. The positive side of the battery is connected to the fuse block and power is distributed from there.

As a common return circuit, auto manufacturers use a wiring style that involves using the vehicle's metal frame as part of the return circuit. The load is often grounded directly to the metal frame. The metal frame then acts as the return wire in the circuit. Current passes from the battery, through the load, and into the frame. The frame is connected to the negative terminal of the battery through the battery's ground wire. This completes the circuit (Figure 3-4).

An electrical component such as an AC generator is often mounted directly to the engine block, transmission case, or frame. This direct mounting effectively grounds the component without the use of a separate ground wire. In other cases, however, a separate ground wire must be run from the component to the frame or another metal part to ensure a sound return path. The increased use of plastics and other nonmetallic materials in body panels and engine parts has made electrical grounding more difficult. To assure good grounding back to the battery, some manufacturers now use a network of common grounding terminals and wires.

Types of Circuits

In a **series circuit**, the resistances are connected one after the other in the circuit, and the same current flows through all resistances (Figure 3-5). These facts may be stated about a series circuit:

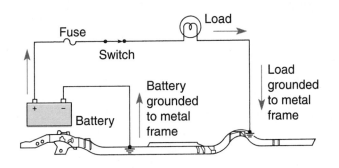

Figure 3-4 In most vehicles, the metal frame, engine block, or transmission case is used as a source of ground to complete the circuit back to the battery.

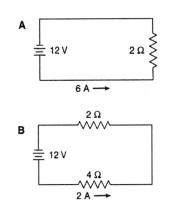

Figure 3-5 In a series circuit, the same amount of current flows through the entire circuit.

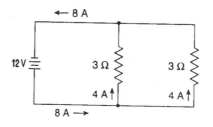

Figure 3-6 A simple parallel circuit.

1. The total resistance is the sum of all resistors in the circuit.
2. The same current flows through each resistor.
3. Part of the source voltage is dropped across each resistor.
4. The sum of the voltage drops across each resistor equals the source voltage.

In a **parallel circuit**, each resistance is connected across the circuit, and each resistance is a separate path for current flow (Figure 3-6). These facts may be stated about a parallel circuit:

1. The current flow through each resistor depends on the ohms in that resistor.
2. The total current is the sum of the current flow through each branch.
3. Full source voltage is supplied to each branch and full source voltage is dropped across each resistor.
4. The total resistance is less than the value of any resistor in the circuit.

In a **series-parallel circuit**, a resistor is connected in series with some parallel resistors in the circuit (Figure 3-7). Series-parallel circuits are used in many automotive applications. For example, in the instrument panel lamp circuit, the variable resistor in the headlamp switch is connected in series with the parallel instrument panel lamps.

Circuit Components

Automotive electrical circuits contain a number of different types of electrical devices. The more common components are outlined in the following sections.

Resistors are used to limit current flow (and thereby voltage) in circuits where full current flow and voltage are not needed, or where too much voltage may cause damage. Resistors are devices specially constructed to introduce a measured amount of electrical resistance into a circuit (Figure 3-8). In addition, some other components use resistance to produce heat and even light. An electric window defroster is a specialized type of resistor that produces heat. Electric lights are resistors that get so hot they produce light.

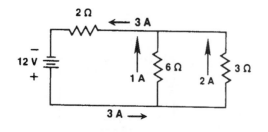

Figure 3-7 In a series-parallel circuit, the sum of the currents through the parallel legs must equal the current through the series part of the circuit.

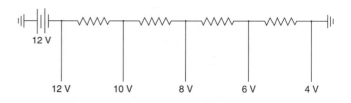

Figure 3-8 The use of a stepped resistor assembly is a common application of fixed resistors.

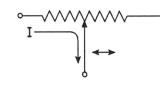

Figure 3-9 A rheostat.

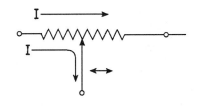

Figure 3-10 A potentiometer.

Resistors in common use in automotive circuits are of three types: fixed value, stepped or tapped, and variable.

Fixed value resistors are designed to have only one rating, which should not change. These resistors are used to control voltage such as in an automotive ignition system.

Tapped or stepped resistors are designed to have two or more fixed values, available by connecting wires to the several taps of the resistor. Heat motor resistor packs, which provide for different fan speeds, are an example of this type of resistor.

Variable resistors are designed to have a range of resistances available through two or more taps and a control. Two examples of this type of resistor are rheostats and potentiometers. **Rheostats** have two connections (Figure 3-9), one to the fixed end of a resistor and one to a sliding contact with the resistor. Turning the control moves the sliding contact away from or toward the fixed end tap, increasing or decreasing the resistance. **Potentiometers** have three connections (Figure 3-10), one at each end of the resistance and one connected to a sliding contact with the resistor. Turning the control moves the sliding contact away from one end of the resistance, but toward the other end.

Another type of variable resistor is the thermistor. This resistor is designed to change in values as its temperature changes. Although most resistors are carefully constructed to maintain their rating within a few ohms through a range of temperatures, the thermistor is designed to change its rating. Thermistors are used to provide compensating voltage in components or to determine temperature. As a temperature sender, the thermistor is connected to a voltmeter calibrated in degrees. As the temperature rises or falls, the resistance also changes. This changes the reading on the meter.

Circuit Protective Devices

When overloads or shorts in a circuit cause too much current to flow, the wiring in the circuit heats up, the insulation melts, and a fire can result unless the circuit has some kind of protective device. Fuses, fuse links, maxi-fuses, and circuit breakers are designed to provide protection from high current (Figure 3-11). These protection devices open the circuit when high current is present. As a result, the circuit no longer works but the wiring and the components are saved from damage.

Switches

Electrical circuits are usually controlled by a switch of some type. Switches do two things. They turn the circuit on or off and they direct the flow of current in a circuit. Switches can be under the control of the driver or can be self-operating through a condition of the circuit, the vehicle, or the environment.

Contacts in a switch can be of several types, each named for the job they do or the sequence in which they work. A hinged-pawl switch is the simplest type of switch. It either makes or breaks the current in a single conductor or circuit. It is a single-pole, single-throw (SPST) switch. The throw refers to the number of output circuits, and the pole refers to the number of input circuits made by the switch (Figure 3-12).

Shop Manual
Chapter 3, page 92

There are two types of thermistors: Negative Temperature Co-efficient and Positive Temperature Co-efficient. These names describe how the resistance of the thermistor changes with a change in temperature.

Shop Manual
Chapter 3, page 91

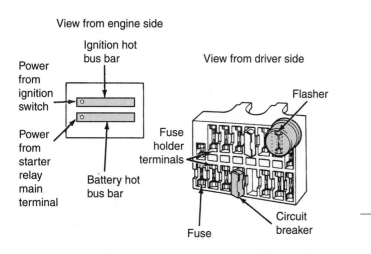

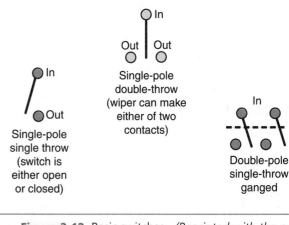

Figure 3-12 Basic switches. *(Reprinted with the permission of Ford Motor Company)*

Figure 3-11 A typical fuse block. *(Reprinted with the permission of Ford Motor Company)*

Another type of SPST switch is the momentary contact switch. The spring-loaded contact on this switch keeps it from making the circuit except when pressure is being applied to the button. A horn is a switch of this type. Because the spring holds the contacts open, the switch has a further designation: normally open. In the case where the contacts are held closed except when the button is pressed, the switch is designated normally closed.

Single-pole, double-throw switches have one wire in and two wires out. This type of switch allows the driver to select between two circuits, such as high-beam or low-beam headlights.

Switches can be designed with a great number of poles and throws. The transmission neutral start switch may have two poles and six throws and is referred to as a multiple-pole, multiple-throw (MPMT) switch. It contains two movable wipers that move in unison across two sets of terminals. The dotted line shows that the wipers are mechanically linked, or ganged. The switch closes a circuit to the starter in either P (park) or N (neutral) and to the back-up lights in R (reverse).

Most switches are combinations of hinged-pawl and push-pull switches, with different numbers of poles and throws. Some special switches are required, however, to satisfy the circuits of modern automobiles. A mercury switch is sometimes used to detect motion in a component, such as the one used in the engine compartment to turn on the compartment light.

A temperature-sensitive switch usually contains a bimetallic element heated either electrically or by some component where the switch is used as a sensor. When engine coolant is below or at normal operating temperature, the engine coolant temperature sensor is in its normally open condition (Figure 3-13). If the coolant exceeds the temperature limit, the bimetallic element bends the

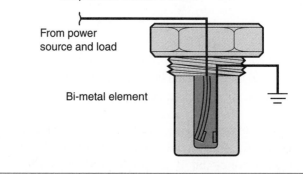

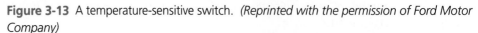

Figure 3-13 A temperature-sensitive switch. *(Reprinted with the permission of Ford Motor Company)*

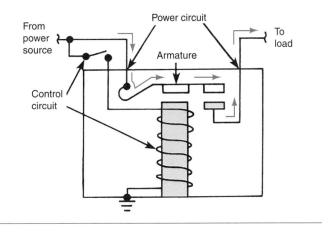

Figure 3-14 A typical relay.

two contacts together and the switch is closed to the indicator or the instrument panel. Other applications for heat-sensitive switches are time delay switches and flashers.

Relays

A **relay** (Figure 3-14) is an electric switch that allows a small amount of current to control a much larger one. It consists of a control circuit. When the control circuit switch is open, no current flows to the coil, so the windings are de-energized. When the switch is closed, the coil is energized, turning the soft iron core into an electromagnet and drawing the armature down. This closes the power circuit contacts, connecting power to the load circuit. When the control switch is opened, the current stops slowing in the coil, the electromagnet disappears, and the armature is released, which breaks the power circuit contacts.

Solenoids

Solenoids are also electromagnets with movable cores used to translate electrical current flow into mechanical movement. The movement of the core causes something else to move, such as lever. They can also close electrical contacts, acting as a relay at the same time.

Electromagnetism Basics

Electricity and magnetism are related. One can be used to create the other. Current flowing through a wire creates a magnetic field around the wire. Moving a wire through a magnetic field creates current flow in the wire.

Many automotive components, such as AC generators, ignition coils, starter solenoids, and magnetic pulse generators operate using principles of electromagnetism.

Although almost everyone has seen magnets at work, a simple review of magnetic principles is in order to ensure a clear understanding of electromagnetism.

A substance is said to be a magnet if it has the property of magnetism—the ability to attract such substances as iron, steel, nickel, or cobalt. These are called magnetic materials.

A magnet has two points of maximum attraction, one at each end of the magnet. These points are called poles, with one being designated the North pole and the other the South pole. When two magnets are brought together, opposite poles attract, while similar poles repel each other.

A magnetic field, called a **field of flux**, exists around every magnet (Figure 3-15). The field consists of imaginary lines along which a magnetic force acts. These lines emerge from the North pole and enter the South pole, returning to the North pole through the magnet itself. All lines of force leave the magnet at right angles to the magnet. None of the lines cross each other. All lines are complete.

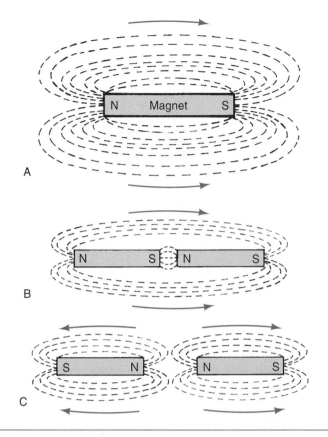

Figure 3-15 Magnetic principles: (A) field of flux around a magnet, (B) unlike poles attract each other, and (C) like poles repel.

Magnets can occur naturally in the form of a mineral called magnetite. Artificial magnets can also be made by inserting a bar of magnetic material inside a coil of insulated wire and passing a heavy direct current through the coil. This principle is very important in understanding certain automotive electrical components. Another way of creating a magnet is by stroking the magnetic material with a bar magnet. Both methods force the randomly arranged molecules of the magnetic material to align themselves along North and South poles.

Artificial magnets can be either temporary or permanent. Temporary magnets are usually made of soft iron. They are easy to magnetize but quickly lose their magnetism when the magnetizing force is removed. Permanent magnets are difficult to magnetize, but once magnetized they retain this property for very long periods.

Four metals (iron, steel, nickel, and cobalt) have magnetic qualities. When one of these metals is not magnetized, the molecules in the metal are randomly arranged, and the magnetic strength of each molecule does not add together. If a metal is magnetized, the molecules are aligned so their magnetic strength adds together.

A BIT OF HISTORY

The force of a magnet was first discovered over 2,000 years ago by the Greeks. They found that a type of rock, magnetite, was attracted to iron. During the Dark Ages, the strange powers of the magnetite were believed to be caused by evil spirits.

When current flows through a conductor, an invisible field of force surrounds the wire. This magnetic field is concentric to the conductor, and an increase in current flow results in a stronger magnetic field. The direction of the magnetic field around the conductor is determined by the direction of current flow through the conductor. If the current flow is reversed, the magnetic field is also reversed. If an iron core is placed in the center of the coil, the magnetic lines of force are strengthened because iron is a better conductor for lines of force compared with air.

The magnetic strength of an **electromagnet** is determined mainly by the number of turns and the current flow through the winding. To calculate the strength of an electromagnet, multiply the number of turns and the current flow through the winding. Thus, we have the formula: number of turns times amperes = ampere turns of magnetic strength.

Induced Voltage

Now that we have explained how current can be used to generate a magnetic field, it is time to examine the opposite effect of how magnetic fields can produce electricity. Consider a straight piece of conducting wire with the terminals of a voltmeter attached to both ends. If the wire is moved across a magnetic field, the voltmeter registers a small voltage reading (Figure 3-16). A voltage has been induced in the wire.

It is important to remember that the conducting wire must cut across the flux lines to induce a voltage. Moving the wire parallel to the lines of flux does not induce voltage. The wire need not be the moving component in this setup. Holding the conducting wire still and moving the magnetic field at right angles to it also induces voltage in the wire.

The wire or conductor becomes a source of electricity and has a polarity or distinct positive and negative end. However, this polarity can be switched depending on the relative direction of movement between the wire and magnetic field. This is why an AC generator produces alternating current.

When voltage is induced in a conductor and the conductor is connected to a complete circuit, current flows through the circuit. This process of inducing a voltage in a conductor with a moving magnetic field is referred to as electromagnetic **induction**. These requirements are necessary to induce a voltage in a conductor:

1. A conductor or conductors

2. A magnetic field

3. Relative motion

Generators and alternators produce AC volts but the AC voltage is converted to DC before it is released into the rest of the electrical system.

During electromagnetic induction, the amount of voltage induced in the conductor is determined by these factors:

1. The strength of the magnetic field

2. The number of conductors

3. The speed of motion

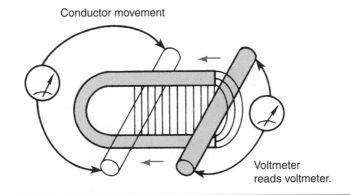

Figure 3-16 Moving a conductor through a magnetic field induces a voltage in the conductor.

Basics of Electronics

Computerized engine controls and other features of today's cars would not be possible if it were not for electronics. For purposes of clarity, let us define electronics as the technology of controlling electricity. Electronics has become a special technology beyond electricity. Transistors, diodes, semiconductors, integrated circuits, and solid-state devices are all considered to be part of electronics rather than just electrical devices. But keep in mind that all the basic laws of electricity apply to electronic controls.

Semiconductors

A **semiconductor** is a material or device that can function as either a conductor or an insulator, depending on how its structure is arranged. Semiconductor materials have less resistance than an insulator but more resistance than a conductor. Some common semiconductor materials include silicon (Si) and germanium (Ge).

In semiconductor applications, materials have a crystal structure. This means that their atoms do not lose and gain electrons as the atoms in conductors do. Instead, the atoms in these semiconductor materials share outer electrons with each other. In this type of atomic structure, the electrons are tightly held and the element is stable.

Because the electrons are not free, crystals cannot conduct current. These materials are called electrically inert materials. In order to function as semiconductors, a small amount of trace element must be added. The addition of these traces, called impurities, allows the material to function as a semiconductor. The type of impurity added determines what type of semiconductor will be produced.

The **diode** is the simplest semiconductor device. A diode allows current to flow in one direction, but not in the opposite direction. Therefore, it can function as a switch, acting as either conductor or insulator, depending on the direction of current flow.

Silicon has four valence electrons. When two pieces of silicon are joined together, the outer rings on the atoms join together, and this places eight electrons on the valence rings. This joining process between two pieces of silicon is called covalent bonding. A silicon crystal is the result of this action. When one side of a silicon crystal is mixed, or doped, with phosphorus, a negative-type material, with an excess of electrons is formed. Doping the opposite side of the silicon crystal with boron creates a positive-type material with a lack of electrons.

If the positive polarity from a voltage source is connected to the positive side of a diode, and the negative polarity is connected to the negative side of the diode, the current flows through the diode. This type of connection to a diode is called forward bias.

When the positive polarity from a voltage source is connected to the negative side of a diode, and the negative polarity is connected to the positive side of the diode, the diode blocks current flow. Reverse bias is the term applied to this type of diode connection. A diode may be defined as a one-way electronic control device.

A diode connected with reverse bias is capable of blocking a certain amount of voltage. The maximum voltage that a diode blocks is called peak inverse voltage (**PIV**). If the PIV is exceeded, the diode may break down in the reverse direction. This action ruins the diode.

A variation of the diode is the **Zener diode**. This device functions like a standard diode until a certain voltage is reached. A Zener diode is doped more heavily in the manufacturing process. This type of diode breaks down at a specific voltage in the reverse direction. For example, an 8 V Zener diode breaks down and conducts current when an 8 V reverse-bias voltage is supplied to the diode. If a reverse-bias voltage of 7.99 V is supplied to the diode, it does not conduct current. A Zener diode is not damaged by a normal amount of reverse current flow.

A **transistor** is an electronic device produced by joining three sections of semiconductor materials. Like the diode, it is very useful as a switching device, functioning as either a conductor or an insulator.

A conductor may be defined as an element with one, two, or three valence electrons. When an element has one, two, or three valence electrons, these electrons can be moved easily to other atoms in the conductor. Therefore, this type of element is classified as a good conductor.

A semiconductor may be defined as an element with four valence electrons. Semiconductors with four valence electrons have unusual characteristics when they are combined with other elements, and these semiconductors are used to manufacture diodes and transistors. Silicon is one of the most common semiconductors used in diodes and transistors.

Shop Manual
Chapter 3, page 92

Insulators are elements with five or more valence electrons. When the valence ring on each atom in an element contains five or more electrons, the electrons do not move easily from one atom to another. Therefore, this type of element is called an insulator.

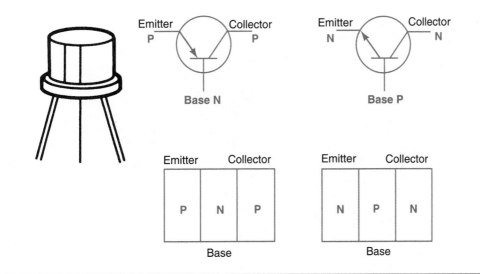

Figure 3-17 NPN and PNP transistors. *(Courtesy of Chrysler Corporation)*

A transistor is manufactured by joining three positive and/or negative materials. These positive and negative materials may be arranged as NPN or PNP. The terminals on a transistor are referred to as the emitter, base, and collector. In an NPN transistor, the emitter is negative, the base is positive, and the collector is negative (Figure 3-17).

The base material is sandwiched between the emitter and the collector. The base material is much thinner than the emitter and collector materials. If positive circuit polarity is connected to the positive emitter on the PNP transistor, and the negative base is connected to negative circuit polarity, the emitter-base is forward biased. This type of connection results in current flow through the emitter-base circuit. If the emitter-base is forward biased and the collector terminal is also connected to the negative circuit polarity, a higher current flows from the emitter through the base material and out the collector terminal.

The relationship between the lower emitter-base current and the higher emitter-collector current is called the gain in a transistor. If a set of contacts in the base circuit is opened, current flow in the emitter-base circuit is stopped. Under this condition, the current flow is also stopped in the emitter-collector circuit. A very low current in the emitter-base circuit may be used to switch off a much higher current in the emitter-collector circuit.

The emitter-base circuit of a PNP transistor is reverse biased when positive circuit polarity is connected to the negative base material, and negative circuit polarity is connected to the positive emitter material. If the emitter-base circuit is reverse biased, the emitter-base current flow is stopped. Under this condition, the emitter-collector current is also stopped.

One transistor or diode is limited in its ability to do complex tasks. However, when many semiconductors are combined into a circuit, they can perform complex functions.

An **integrated circuit** is simply a large number of diodes, transistors, and other electronic components such as resistors and capacitors, all mounted on a single piece of semiconductor material. This type of circuit has a tremendous size advantage. It is extremely small. Circuitry that used to take up entire rooms can now fit into a pocket. The principles of semiconductor operation remain the same in integrated circuits—only the size has changed.

In the IC manufacturing process, an etching and photographic procedure is used to form thousands of components on a silicon slice (Figure 3-18). A single chip may contain more than 1 million transistors and other components. As more transistors are added to a computer chip, the computer is capable of performing more instructions per second. Currently, we have computers that handle more than 1 million instructions per second. By the end of this decade, computers will be capable of processing 10 million instructions per second (Figure 3-19). These facts indicate that computer response time and the control of output functions are increasing dramatically.

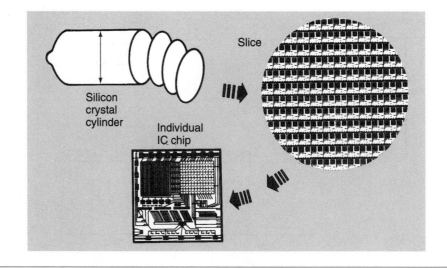

Figure 3-18 An integrated circuit etched on a silicon slice. *(Courtesy of Chrysler Corporation)*

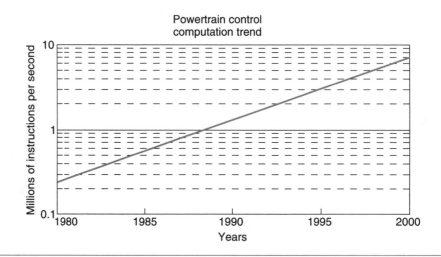

Figure 3-19 Microprocessor instruction processing capabilities: past, present, and future. *(Courtesy of Intel Corporation)*

While the cost of electronics per vehicle has increased gradually, the price of chips has decreased consistently. Therefore, the car manufacturers can actually purchase much smarter and smaller chips for their electronic dollar. This is one reason why the number of electronic systems on cars has increased and will continue to increase.

Because of the small size of ICs, electronics are no longer confined to simple tasks such as rectifying AC generator current. Enough transistors, diodes, and other solid-state components can be installed in a car to make logic decisions and issue commands to other areas of the engine. This is the foundation of computerized control systems.

The computer has taken over many of the tasks in cars and trucks that were formerly performed by vacuum, electromechanical, or mechanical devices. When properly programmed, they can carry out explicit instructions with blinding speed and almost flawless consistency.

A typical electronic control system is made up of sensors, actuators, and related wiring that is tied to a computer.

The computer is called the PCM, central processor, or microprocessor.

Sensors

All sensors perform the same basic function. They detect a mechanical condition (movement or position), chemical state, or temperature condition and change it into an electrical signal that can

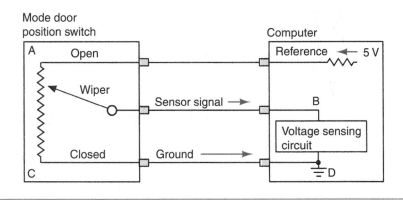

Figure 3-20 A potentiometer alters the reference voltage in response to the movement of the wiper and sends a signal back to the voltage sensing circuit of the computer.

be used by the computer to make decisions. The computer makes decisions based on information it receives from sensors. Each sensor used in a particular system has a specific job to do. Together these sensors provide enough information to help the computer form a complete picture of vehicle operation. Even though there are a variety of different sensor designs, they all fall under one of two operating categories: reference voltage sensors or voltage generating sensors.

Reference voltage (Vref) sensors provide input to the computer by modifying or controlling a constant, predetermined voltage signal (Figure 3-20). This signal, which can have a reference value from 5 to 9 volts, is generated and sent out to each sensor by a reference voltage regulator located inside the processor. The term processor is used to describe the actual metal box that houses the computer and its related components. Because the computer knows that a certain voltage value has been sent out, it can indirectly interpret things like motion, temperature, and component position, based on what comes back. For example, consider the operation of the throttle position sensor (TPS). During acceleration (from idle to wide-open throttle), the computer monitors throttle plate movement based on the changing reference voltage signal returned by the TPS. (The TPS is a type of variable resistor known as a rotary potentiometer that changes circuit resistance based on throttle shaft rotation.) As TPS resistance varies, the computer is programmed to respond in a specific manner (for example, increase fuel delivery or alter spark timing) to each corresponding voltage change.

Most sensors presently in use are variable resistors or potentiometers. They modify a voltage to or from the computer, indicating a constantly changing status that can be calculated, compensated for, and modified. That is, most sensors simply control a voltage signal from the computer. When varying internal resistance of the sensor allows more or less voltage to ground, the computer senses a voltage change on a monitored signal line. The monitored signal line may be the output signal from the computer to the sensor (one- and two-wire sensors), or the computer may use a separate return line from the sensor to monitor voltage changes (three-wire sensors).

While most sensors are variable resistance/reference voltage, there is another category of sensors—the **voltage generating devices** (Figure 3-21). These sensors include components like the Hall effect switch, oxygen sensor (zirconium dioxide), and knock sensor (piezoelectric), which are capable of producing their own input voltage signal. This varying voltage signal, when received by the computer, enables the computer to monitor and adjust for changes in the computerized engine control system.

In addition to variable resistors, two other commonly used reference voltage sensors are switches and thermistors. Switches provide the necessary voltage information to the computer so that vehicles can maintain the proper performance and driveability. Thermistors are special types of resistors that convert temperature into a voltage. Regardless of the type of sensors used in electronic control systems, the computer is incapable of functioning properly without input signal voltage from sensors.

Shop Manual
Chapter 3, page 92

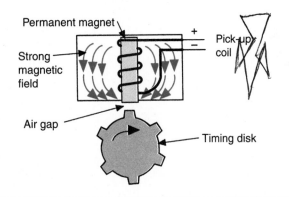

Figure 3-21 A strong magnetic field is produced in the pick-up coil as the teeth align with the core. The field weakens as the teeth pass the core.

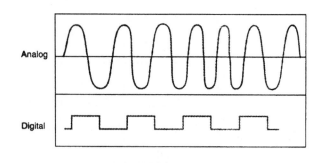

Figure 3-22 Analog signals are constantly variable, whereas digital signals are either on or off, or high or low.

Communication Signals

Most input sensors are designed to produce a voltage signal that varies within a given range (from high to low, including all points in between). A signal of this type is called an **analog** signal. Unfortunately, the computer does not understand analog signals. It can only read a **digital** binary signal, which is a signal that has only two values—on or off (Figure 3-22).

To overcome this communication problem, all analog voltage signals are converted to a digital format by a device known as an analog-to-digital converter (A/D converter). The A/D converter is located in a section of the processor called the input signals. However, some sensors like the Hall effect switch produce a digital or square wave signal that can go directly to the computer as input (Figure 3-23).

A computer's memory holds the programs and other data, such as vehicle calibrations, which the microprocessor refers to in performing calculations. To the computer, the program is a set of instructions or procedures that it must follow. Included in the program is information that tells the microprocessor when to retrieve input (based on temperature, time, etc.), how to process the input, and what to do with it once it has been processed.

Actuators

After the computer has assimilated the information and the tools used by it to process this information it sends output signals to control devices called actuators. These actuators which are solenoids, switches, relays, or motors physically act or carry out a decision the computer has made.

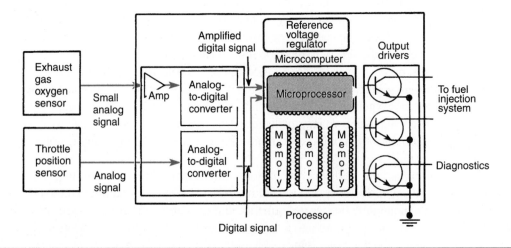

Figure 3-23 The A/D converter prepares input signals for the computer.

Actually, an **actuator** is an electromechanical device that converts an electrical current into mechanical action. This mechanical action can then be used to open and close valves, control vacuum to other components, or open and close switches. When the microcomputer receives an input signal indicating a change in one or more of the operating conditions, the microcomputer determines the best strategy for handling the conditions. The microcomputer then controls a set of actuators to achieve a desired effect or strategy goal. In order for the computer to control an actuator, it must rely on a component called an **output driver**.

Output drivers are also located in the processor (along with the input conditioners, microprocessor, and memory) and operate by the digital commands issued by the microcomputer. Basically, the output driver is nothing more than an electronic on/off switch that the computer uses to control the ground circuit of a specific actuator.

Battery, Starting, and Charging Systems

The battery supplies voltage and current to operate the starting motor while cranking the engine. This voltage and current must be supplied at all operating temperatures. If the electrical accessories are turned on when the engine is not running, the battery supplies voltage and current to operate these accessories. When the engine is running and the charging system cannot meet the current flow requirements of the electrical system, some current is supplied from the battery to the electrical system.

The starting motor must crank the engine with enough cranking speed to provide easy starting at all operating temperatures. The AC generator must provide voltage and current to recharge the battery and power the electrical accessories while the engine is running. If any of these three components does not function properly, a no-start condition will likely occur.

Battery Construction

Case

Shop Manual
Chapter 3, page 93

The battery case holds and protects the battery components and electrolyte (Figure 3-24). Separating walls in the case form a separate reservoir for each cell. Most battery cases are made from translucent plastic, and the electrolyte level is visible through the case, which allows the electrolyte level to be checked without removing the cell vent caps. Some cases have sediment spaces in the bottom of the case, which provides spaces for material shed from the plates.

Cover

The battery cover is permanently sealed to the top of the case. The cover has openings for the terminal posts and vent holes for the venting of gases. Some batteries have the terminals extending through the side of the case, and other batteries have terminals in the sides and top. Some covers have removable vent caps, which may be removed to add water or test the electrolyte. These vent caps may be individual, strip-type, or box-type.

Plates

Battery plates contain a grid or coarse screen. In some batteries, the plate grids were made from lead mixed with antimony, which stiffened the lead. Many batteries manufactured today have plate grids made from lead and calcium. The active plate materials are pasted on the grids. The material on the positive plates is lead peroxide, and the negative plate material is sponge lead. A tab on the top of each plate grid allows the plate to be connected to the cell connector.

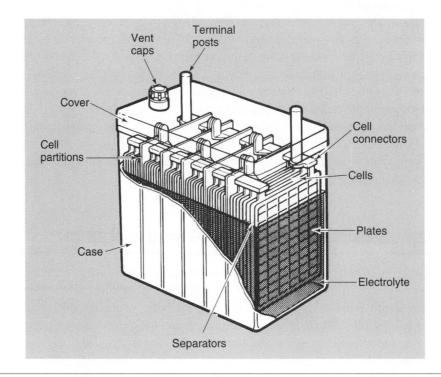

Figure 3-24 Battery construction.

Cell Group

A **cell group** contains a number of alternately spaced negative and positive plates. There is a negative plate positioned on both sides of each cell group, and thus each group has an odd number of plates. For example, a cell group may have five positive plates and six negative plates.

Thin porous separators are positioned between the negative and positive plates. These separators must be porous to allow electrolyte to contact the plates. Separators may be made from fiberglass sheets or plastic envelopes that fit over the plates. The separators prevent the plates from touching each other and shorting together electrically.

Cell connectors are lead burned to the positive and negative plate tabs in each cell group. These cell connectors also extend through the partitions between the cells, and the negative plates are connected to the positive plates in the next cell. The negative plates in one end cell and the positive plates in the opposite end cell are connected to the terminal posts that extend through the top or sides of the battery.

Terminal Posts

On top-terminal batteries, large round terminal posts extend through the top of the battery. The positive post is larger than the negative post. These terminal posts are sealed into the battery cover to prevent electrolyte leakage around the posts. POS or NEG are usually stamped on the battery cover beside the proper terminal post. The battery cable ends are clamped to the terminal posts. On side-terminal batteries, the battery terminals are threaded, and the cables are bolted to the terminals.

Electrolyte

CAUTION: The sulfuric acid in battery electrolyte is a very strong, corrosive acid. It is very damaging to human skin and eyes. Always wear eye and face protection, and protective gloves and clothing when handling batteries or electrolyte. If electrolyte contacts your skin or eyes, flush with clean water immediately and obtain medical attention.

The **electrolyte** is a mixture of approximately 64% water (H_2O) and 36% sulfuric acid (H_2SO_4). The electrolyte reacts with the plate materials to produce voltage. The electrolyte must contain enough sulfuric acid so it does not freeze in extremely cold temperatures. If the battery contained a higher percentage of sulfuric acid, this acid would attack and deteriorate the plate grids too quickly.

Battery Operation

If one negative and one positive plate are placed in an electrolyte solution, a chemical reaction takes place that causes approximately 2.1 V difference between the plates. Since a 12 V battery contains six cells, the total fully charged battery voltage is approximately 12.6 V. When an electrical resistance such as a light bulb is connected to these two plates, the higher voltage on one plate forces current through the light to the other plate with the lower voltage. This action is called battery discharging (Figure 3-25). A battery cell with two plates will not supply a lot of current flow. When more plates are added, the current flow capability increases.

When a voltage source such as the AC generator in the charging system is connected to the battery plates, the AC generator forces current flow in one battery plate and out of the other battery plate (Figure 3-26). This action is called battery charging. The AC generator voltage must be higher than the battery voltage to allow the AC generator to charge the battery.

A charged battery has lead peroxide (PbO_2) on the positive plates and sponge lead (Pb) on the negative plates. The electrolyte contains water (H_2O) and sulfuric acid (H_2SO_4) (Figure 3-27).

When an electrical load is connected across the battery terminals, the voltage difference between the plate materials forces current through the electrical load. As the battery discharges these chemical changes occur:

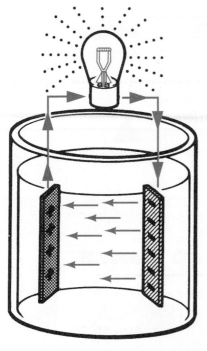

Discharging

Figure 3-25 Current flows from one battery plate to the other when a conductor with some electrical resistance is connected between the plates.

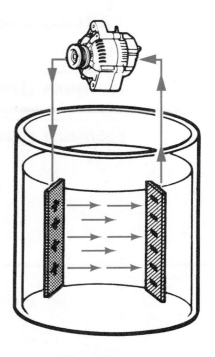

Charging

Figure 3-26 When the engine is running, current flows from the alternator through the battery plates.

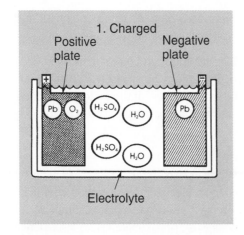

Figure 3-27 Plate materials and electrolyte content in a charged battery.

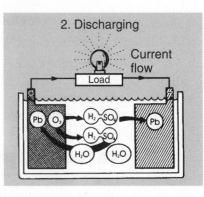

Figure 3-28 Chemical action in a battery while discharging.

1. The H_2SO_4 breaks up in the electrolyte, and the SO_4 goes to both plates where it joins with the Pb to form lead sulfate ($PbSO_4$) on both plates.

2. The O_2 on the positive plates joins with the hydrogen (H) in the electrolyte to form H_2O (Figure 3-28). The percentage of water in the electrolyte increases as the battery discharges.

In a discharged battery, both sets of plates are coated with lead sulfate ($PbSO_4$), and the electrolyte contains a high percentage of H_2O (Figure 3-29).

When a charging source such as an AC generator is charging the battery, these chemical changes occur in the battery:

1. The SO_4 comes off both plates and joins with the H in the electrolyte to form H_2SO_4 (Figure 3-30).

2. The H_2O breaks up and the O goes to the positive plates where it joins with the Pb to form PbO_2.

While a battery is charging, H gas escapes at the negative plates, and O gas escapes at the positive plates. These two gases combine to form H_2O. The SO_4 is always in the electrolyte or on the plates; thus, it does not escape from the battery. Since H_2O is the only chemical that escapes from the battery, it is the only chemical that should be added to the battery.

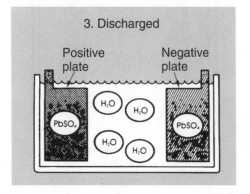

Figure 3-29 Plate materials and electrolyte content in a discharged battery.

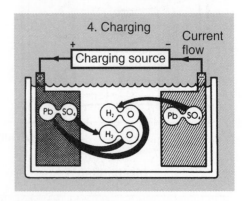

Figure 3-30 Chemical action in a battery while charging.

Battery Classifications

Low-Maintenance Batteries

Low-maintenance batteries are designed to reduce internal heat and water loss. The addition of water may be required at 15,000 miles (24,000 km).

Maintenance-Free Batteries

Maintenance-free batteries are designed to reduce internal heat and water loss so the addition of water is not required during the life of the battery. The battery cover is vented to allow gases to escape, but the vent caps cannot be removed to add water or test the electrolyte.

Battery Ratings

Shop Manual
Chapter 3, page 96

Many different battery rating methods have been used by industry. What follows are the most common.

Cold Cranking Amperes

The Cold Cranking Amperes (CCA) rating indicates the amperes a battery will deliver at 0°F (-18°C) for 30 seconds while maintaining a voltage of at least 1.2 V per cell or 7.2 V in the complete battery. Many automotive batteries have a CCA rating of 350 to 600.

Reserve Capacity Rating

The reserve capacity rating indicates, in minutes, the length of time a fully charged battery at 80°F (27°C) will deliver 25 amperes while the voltage remains above 1.75 V per cell or 10.5 V in the complete battery. Reserve capacity ratings are usually between 55 and 115 minutes.

Ampere-Hour Rating

The ampere-hour rating indicates the amount of current a battery will deliver for 20 hours with the voltage remaining above 1.75 V per cell or 10.5 V for the complete battery. If a battery delivers 5 amperes for 20 hours with the voltage above the specified value, the battery is rated at 5 X 20 = 100 ampere-hours.

Power (Watt) Rating

The watt rating is determined by multiplying the available battery voltage and the current flow delivered at 0°F (-18°C). Watt, or power battery ratings are usually between 2,000 and 4,000 watts.

Starting the Motor

Shop Manual
Chapter 3, page 100

Any motor converts electrical energy into a rotating mechanical energy. When two like magnetic poles face each other, they tend to push away. In a motor, one of the magnetic poles is stationary in the starter's case. The other pole is fixed to a shaft that is fitted in the center of the case. The shaft is allowed to move. When the magnetic poles of the windings in the case are the same as the magnetic poles of the shaft, they push apart. This pushing causes the shaft to rotate. By connecting the shaft of the starter motor to the flywheel of the engine, the engine cranks as the shaft or armature spins.

All motors are made up of the same basic parts: a case, armature, brushes, and field windings (Figure 3-31).

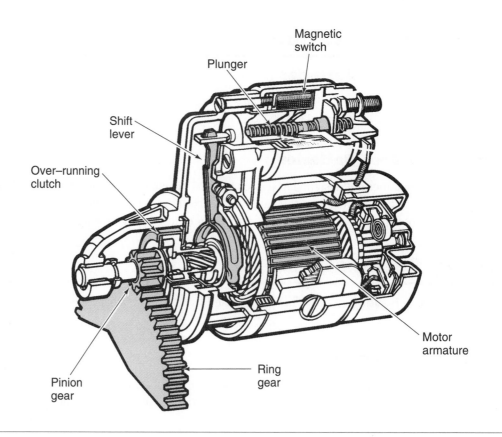

Figure 3-31 Starting motor components.

Armature

A steel shaft is the basis of an armature and it runs through the center of the armature assembly. A laminated iron core is pressed onto this shaft. Heavy insulated windings are mounted in the armature core slots.

A commutator is mounted on an insulating sleeve near the front of the armature. The commutator contains a series of copper bars that are insulated from each other. The armature windings are soldered to the commutator bars. In most starting motors, four brushes are mounted in holders, and springs keep the brushes in contact with the commutator bars. In many starting motors, two of the brushes are connected to ground on the starter case, and the other two brushes are connected to the field coils.

The starter drive is mounted on splines on the back of the armature shaft. A pinion gear on the drive is pulled into mesh with the flywheel ring gear by a shift lever. Most starter drives are the overrunning clutch-type. Bushings in the starter end housings support the armature shaft.

Field Windings

Some starting motors contain four insulated field windings mounted on steel pole shoes that are bolted to the starter case. The field coils in many starting motors are connected through an insulated terminal in the starter case to the solenoid terminal. In some starting motors, the field windings are replaced with strong permanent magnets.

Solenoid

In many starting motors, the solenoid is mounted on the starter housing. A plunger is mounted in a bore at the center of the solenoid and two windings surround the plunger bore. The rear of the plunger is connected to the pivoted shift lever, and the lower end of this lever is mounted in the drive collar. A heavy copper disc is mounted in front of the plunger. The large starting motor terminal and the battery cable terminal are mounted in front of the heavy copper disc. A plunger return spring holds the plunger toward the rear of the starting motor, and a smaller disc return spring holds the disc away from the terminals. A small terminal on the front of the solenoid is connected to the solenoid windings.

Some starting circuits are fitted with a relay or magnetic switch that completes the battery positive circuit to the starter when the ignition key is moved to the start position. These systems may or may not have a solenoid mounted to the starter.

As the contacts in a relay or solenoid open and close, some arcing may occur across the contacts. This causes voltage spikes to be present in the electrical system. Some manufacturers install clamping diodes at the solenoid and/or relay to prevent these voltage spikes.

Some mistakenly call this relay a solenoid. It is not! A solenoid is something that performs a mechanical act due to the control of its magnetic field.

Neutral Safety Switch Operation

In many starting motor circuits, the starter switch is part of the ignition switch. When the ignition switch is turned to the start position, voltage is supplied to the neutral safety switch. If the transmission is in neutral or park, the neutral safety switch is closed, and current flows through this switch to the solenoid terminal. When the transmission is in any gear except neutral or park, the neutral safety switch is open, and the starting circuit is inoperative.

The theft deterrent computer prevents starting motor operation if someone breaks into the car—for example, if a door is forced open or a window is broken to enter the vehicle. Most theft deterrent computers also flash the headlights and taillights and blow the horn if someone breaks into the car.

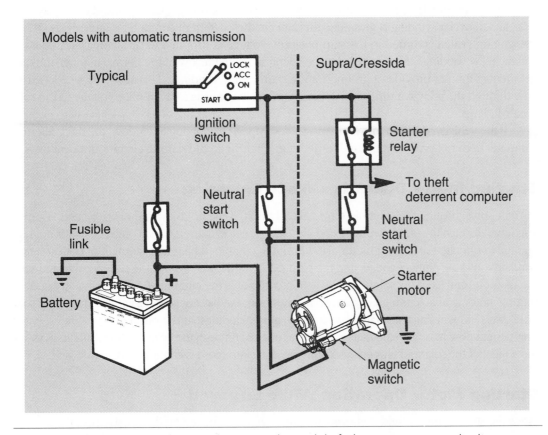

Figure 3-32 Neutral safety switch, starter relay, and theft deterrent computer circuit.

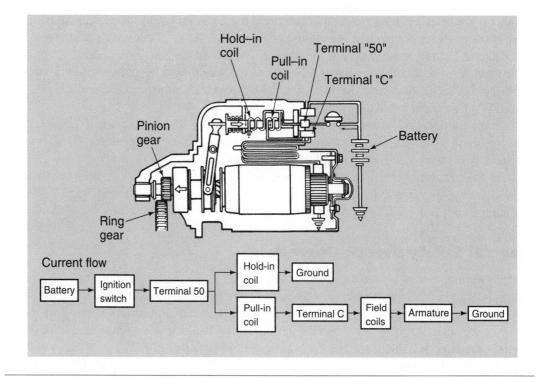

Figure 3-33 Starting motor and solenoid current flow while engaging.

On some vehicles, the neutral safety switch is connected in series with the starter relay contacts, and this relay winding is grounded through the theft deterrent computer (Figure 3-32). When the ignition switch is turned to the start position, voltage is supplied to the starter relay contacts and the relay winding. If the vehicle has been entered with the key in the normal manner, the theft deterrent computer provides a ground for the starter relay winding. This action provides current flow through the relay winding, which closes the relay contacts and supplies voltage to the solenoid windings.

If the vehicle is broken into by forcing a door open or breaking a window, the theft deterrent computer does not provide a ground for the starter relay and the starting circuit is inoperative.

Starting Motor Operation While Engaging

Current flows through the solenoid terminal and through the hold-in winding to ground. Current also flows through the pull-in winding and the starting motor to ground. This current flow is not high enough to operate the starting motor. However, the combined magnetic strength of both solenoid windings pulls the plunger ahead, and the plunger drives the disc against the battery cable and starting motor terminals. This solenoid disc action completes the circuit between the battery and the starting motor, and a very high current now flows from the battery positive cable through the solenoid disc, field windings, insulated brushes, armature windings, and ground brushes (Figure 3-33). This current flow returns to the battery negative terminal through the ground return circuit. Forward movement of the solenoid plunger also pulls the drive into mesh with the flywheel ring gear.

Starting Motor Operation While Engaged

Since the starting motor field coils and armature windings have very low resistance, the average current flow through this circuit is 100 to 200 amperes. This high current flow creates very strong

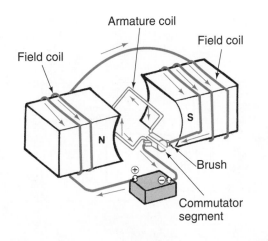

Figure 3-34 Current flow through the field coils and armature windings.

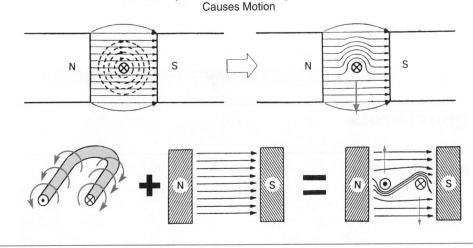

Motor Principles: Interaction of Magnetic Fields
Causes Motion

Figure 3-35 Interaction of the magnetic fields between the field coils and armature windings causes the armature to rotate.

magnetic fields between the field coils and around the armature windings (Figure 3-34). The inter-action of the magnetic field between the field coils and the magnetic field around the armature windings creates a strong armature turning force to crank the engine (Figure 3-35). The starting motor is designed to deliver very high horsepower for short time periods, such as 30 seconds.

WARNING: Cranking the engine continually for more than 30 seconds may dam-age the starting motor.

Once the solenoid disc contacts the terminals, equal voltage is supplied to both ends of the solenoid pull-in winding, and the current flow in this winding is stopped. Current continues to flow through the hold-in winding, and the magnetic field around this winding holds the plunger ahead (Figure 3-36).

Starting Motor Operation While Disengaging

When the driver releases the ignition switch from the start to the on position, the circuit is opened between the ignition switch and the solenoid windings. When current stops flowing through the

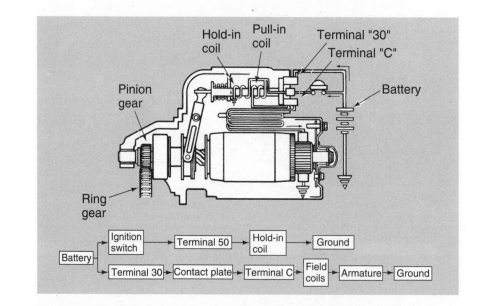

Figure 3-36 Current flow while the starting motor is engaged.

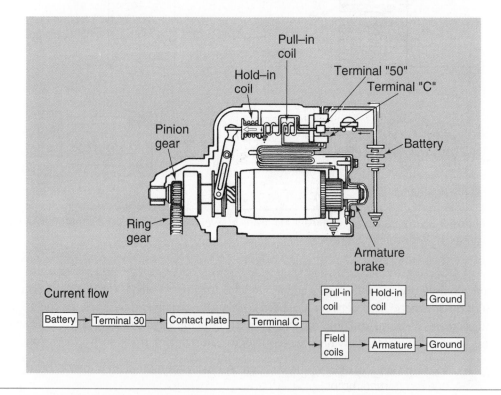

Figure 3-37 Starting motor operation while disengaging.

hold-in winding, the plunger spring moves the plunger rearward. This action pushes the drive out of mesh with the flywheel ring gear and allows the disc return spring to move the disc away from the solenoid terminals (Figure 3-37).

After the ignition switch is released to the on position, and before the disc moves away from the terminals, current flows momentarily through the disc, pull-in winding, and hold-in winding to ground. However, current flow in the pull-in winding is reversed, and the magnetic fields of the

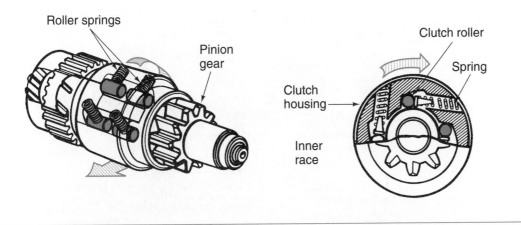

Figure 3-38 Starter drive operation while cranking the engine.

two windings cancel each other. Once the disc moves away from the solenoid terminals, the circuit is open from the battery to the starting motor and the motor stops turning.

Starter Drive Operation

The starter drive clutch housing is splined to the armature shaft and must rotate with the shaft. Four spring-loaded steel rollers are mounted in tapered grooves in the drive housing. A pinion gear is mounted in the end of the drive housing, and the steel rollers contact a machined steel surface on this gear.

The solenoid shift lever moves the drive into mesh with the flywheel ring gear before the armature begins turning. When the solenoid disc contacts the terminals and the armature starts turning, the steel rollers are wedged into the narrow part of the tapered grooves where they are jammed between the drive housing and the pinion gear machined surface (Figure 3-38). Under this condition, the pinion gear must rotate with the drive housing to crank the engine.

Once the engine starts, the ring gear drives the pinion gear faster than the armature. This action moves the steel rollers against the spring tension, and the rollers are now positioned in the wider part of the tapered grooves (Figure 3-39). Under this condition, the starter drive overruns so the flywheel ring gear does not drive the armature at high speed.

This overrunning action of the drive is very important to protecting the armature from excessive speed. The average flywheel ring gear-to-pinion gear ratio is 15:1. If the drive did not overrun and the engine was running at 1,000 rpm, the armature speed would be 15,000 rpm. At this rpm,

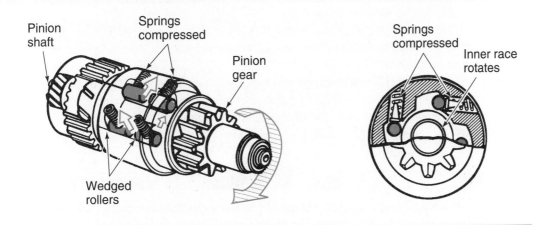

Figure 3-39 Starter drive operation while overrunning.

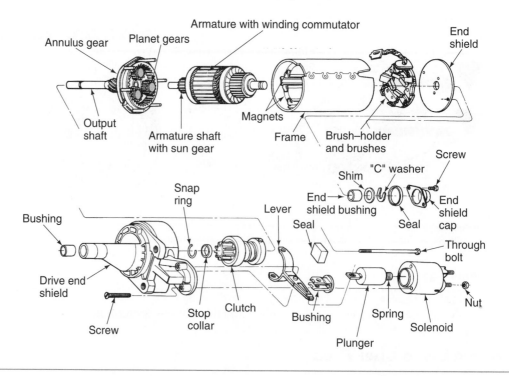

Figure 3-40 Starting motor with planetary reduction gears. *(Courtesy of Chrysler Corporation)*

centrifugal force would throw the armature windings from their slots in the core and destroy the starting motor.

Gear Reduction Starting Motors

Many starting motors have a gear reduction between the armature and the drive. The reduction gears may be conventional or planetary (Figure 3-40). Planetary gearsets use a combination of external and internal gears. The reduction gears allow a smaller, more compact starting motor to provide the same cranking power as a larger starting motor without reduction gears.

AC Generators

Prior to the passing of J1930 terms with OBD-II, AC generators were commonly called alternators.

A BIT OF HISTORY

In the 1960s, the average AC generator output was about 40 to 50 amperes. Since that time, the electrical accessory load has gradually increased as more electrical and electronic systems have been installed on most vehicles. In the 1990s, the average AC generator is rated at 70 to 140 amperes, and the demand for electrical power continues to increase with each model year. If this demand continues to increase, automotive manufacturers may have to design electrical systems with a higher voltage, such as 24 V. When the voltage is doubled, the same amount of electrical power (watts) is available with one-half the current flow.

AC generators keep a vehicle's battery charged by sending electrical energy, which was converted from mechanical energy, to the battery. The AC generator's inner shaft is rotated by the engine via a drive belt. This inner shaft is called the rotor and is a rotating magnetic field. The rotor and its

Shop Manual
Chapter 3, page 103

magnetic field spins inside the stator winding which are fastened to the generator case. As the magnetic field moves past the stator windings, AC voltage is induced. A rectifier circuit, comprised of several diodes, inside the generator converts the AC voltage to DC voltage before it is sent to the battery.

Rotor

A steel shaft is mounted in the center of the rotor assembly, and an insulated field winding is positioned on this shaft. Steel pole pieces with interlaced fingers are mounted over the field coil. Two insulated slip rings are positioned on the end of the rotor shaft, and the ends of the field coil are connected to the slip rings. The two slip rings are insulated from each other and from the rotor shaft. A small spring-loaded brush contacts each slip ring.

The drive end of the rotor shaft is supported on a ball bearing in the drive end housing, and the slip ring end of the shaft is mounted on a needle bearing in the slip ring end frame. A pulley is bolted or pressed on the drive end of the rotor shaft, and a cooling fan is positioned behind the pulley (Figure 3-41). The cooling fan circulates air through the AC generator to cool the internal components.

Stator and End Housings

The stator windings are mounted in a laminated iron frame, and this frame surrounds the rotor assembly. Many stators contain three windings, and the ends of these windings are connected to the diodes. The stator frame is positioned between the two end housings and through bolts extend from the drive end housing into the slip ring end housing to hold the assembly together. Bolt openings in the end housings allow the AC generator to be mounted on the engine.

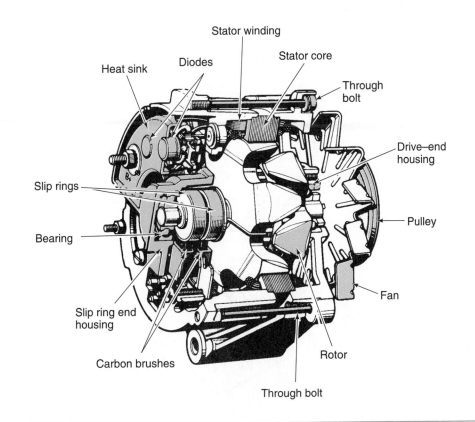

Figure 3-41 Components of a typical AC generator.

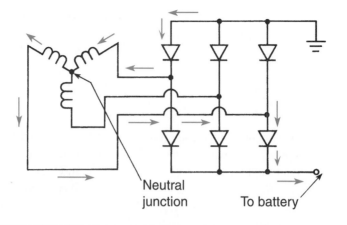

Neutral junction   To battery

Figure 3-42 A Wye-wound stator wired to six diodes for rectification.

Diodes

Shop Manual
Chapter 3, page 107

Many AC generators contain three negative diodes and three positive diodes (Figure 3-42). The positive diodes are mounted in an insulated heat sink to dissipate heat from the diodes. These diodes are connected to the insulated battery terminal in the end housing. Three negative diodes are connected to the end housing.

Voltage Regulators

Shop Manual
Chapter 3, page 105

Some AC generators have an electromechanical regulator, but most vehicles are equipped with integrated circuit (IC) regulators (Figure 3-43). Most IC regulators are mounted inside the AC generator or are bolted to the AC generator slip ring end housing. Electromechanical regulators are usually mounted on the fender shield or cowl.

The voltage regulator is always connected in the AC generator field circuit, regardless of the type of voltage regulator. The purpose of the voltage regulator is to limit the AC generator voltage to protect the electrical accessories and battery on the vehicle. If the voltage regulator allows

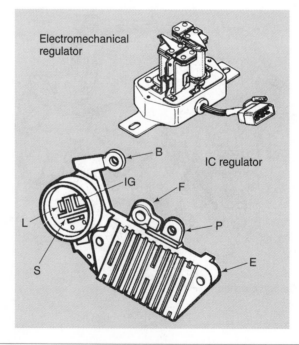

Figure 3-43 Electromechanical and integrated circuit (IC) voltage regulators.

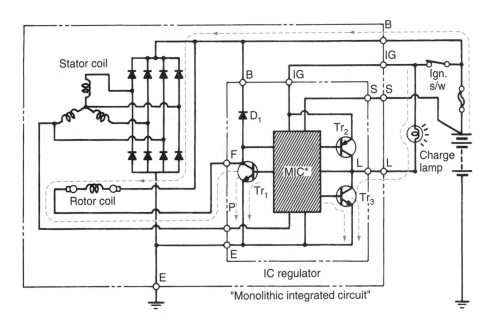

Figure 3-44 Current flow through the AC generator with the ignition switch on and the engine stopped.

excessive AC generator voltage, this voltage forces high current flow through the electrical accessories and battery. This high current flow may burn out the electrical accessories and cause excessive battery gassing, which may damage the battery.

The voltage regulator controls field current through the field winding, and this action regulates the magnetic strength around the rotor. The induced voltage in the stator windings is limited by the magnetic strength of the rotor and the speed at which the rotor spins.

AC Generator Operation with the Ignition Switch On, Engine Stopped

When the ignition switch is turned on, current flows through the ignition switch and charge indicator light to the AC generator L terminal. From this location, the current flows through a transistor in the IC regulator to ground. Once this current flow occurs, current also flows from the battery positive terminal through the AC generator battery terminal and the field coil, slip rings, and brushes. This current flows through another transistor in the IC regulator to ground (Figure 3-44). The current flow through the field winding creates a magnetic field around the rotor, which induces voltage in the stator windings when the rotor starts turning.

AC Generator Operation with the Engine Running

When the engine starts, the rotor magnetic field induces an alternating current (AC) voltage in the stator windings. This AC voltage is changed to a direct current (DC) voltage by the diodes. Current flows from the diodes through the AC generator battery terminal to the battery and electrical accessories. This current flows through the ground return on the vehicle to the AC generator end housing and negative diodes.

Once the engine starts, equal voltage is supplied to both sides of the charge indicator light, and this light remains off. If the AC generator becomes defective and stops supplying current to the battery and accessories, the charge indicator light will be on with the engine running.

Some current also flows from the AC generator battery terminal through the brushes, slip rings, field winding, and IC regulator (Figure 3-45). The regulator controls this current to regulate the magnetic strength around the rotor and limit the voltage in the stator windings.

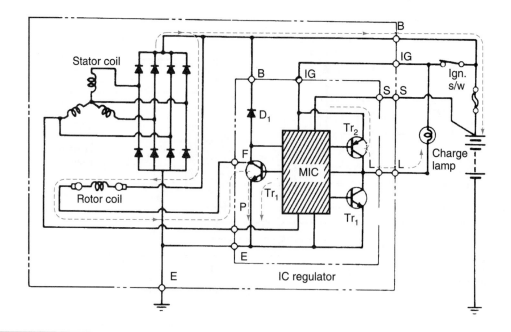

Figure 3-45 Current flow through the AC generator with the engine running

Summary

❏ The release of energy as one electron leaves the orbit of an atom and jumps into the orbit of another is electricity.

❏ Two power or energy sources are used in an automobile's electrical system, these are based on a chemical reaction and on magnetic principles.

❏ A car's battery is a source of chemical energy. A chemical reaction in the battery provides for an excess of electrons and a lack of electrons in another place.

❏ The flow of electricity is called current and is measured in amperes. There are two types of electrical flow: direct current (DC) and alternating current (AC).

❏ Resistance to current flow produces heat. The amount of resistance is measured in ohms.

❏ Voltage is electrical pressure and it is measured in volts.

❏ For electrical flow to occur there must be an excess of electrons in one place, a lack of electrons in another, and a path between the two places.

❏ Ohm's law states that it takes one volt of electrical pressure to push one ampere of electrical current through one ohm of resistance.

❏ If voltage does not change, but there is a change in the resistance of the circuit, the current will change. If resistance increases, current decreases. If resistance decreases, current will increase. If voltage changes, so must the current or resistance. If the resistance stays the same and current decreases, so will voltage. Likewise, if current increases, so will the voltage.

❏ The mathematical relationship between current, resistance, and voltage is expressed in Ohm's law, $E = IR$, where voltage is measured in volts, current in amperes, and resistance in ohms.

❏ Resistors in common use in automotive circuits are of three types: fixed value, stepped or tapped, and variable.

❏ Fixed value resistors are designed to have only one rating, which should not change. These resistors are used to control voltage such as in an automotive ignition system.

❏ Rheostats have two connections, one to the fixed end of a resistor and one to a sliding contact with the resistor.

❏ Potentiometers have three connections, one at each end of the resistance and one connected to a sliding contact with the resistor. Turning the control moves the sliding contact away from one end of the resistance, but toward the other end.

❏ Electrical schematics are diagrams with electrical symbols that show the parts and how electrical current flows through the vehicle's electrical circuits. They are used in troubleshooting.

❏ The strength of an electromagnet depends on the number of current carrying conductors, and what is in the core of the coil. Inducing a voltage requires a magnetic field producing lines of force, conductors that can be moved, and movement between the conductors and the magnetic field so that the lines of force are cut.

❏ Fuses, fuse links, maxi-fuses, and circuit breakers protect circuits against overloads. Switches control on/off and direct current flow in a circuit. A relay is an electric switch. A solenoid is an electromagnet that translates current flow into mechanical movement. Resistors limit current flow.

❏ A semiconductor is a material or device that can function as either a conductor or an insulator, depending on how its structure is arranged.

❏ An integrated circuit is simply a large number of diodes, transistors, and other electronic components such as resistors and capacitors, all mounted on a single piece of semiconductor material.

❏ The diode allows current to flow in one direction but not in the opposite direction.

❏ Transistors are used as switching devices.

❏ Computers are electronic decision-making centers. Input devices called sensors feed information to the computer. The computer processes this information and sends signals to controlling devices.

❏ A typical electronic control system is made up of sensors, actuators, microcomputer, and related wiring.

❏ Most input sensors are variable resistance/reference types, switches, and thermistors.

❏ All sensors detect a mechanical condition, chemical state, or temperature condition and change it into an electrical signal that can be used by the computer to make decisions.

❏ All sensors are either reference voltage sensors or voltage generating sensors.

❏ Most input sensors are designed to produce a voltage signal that varies within a given range called an analog signal.

❏ A computer does not understand analog signals. It can only read a digital binary signal, which is a signal that has only two values—on or off.

❏ After the computer has assimilated the information and the tools used by it to process this information it sends output signals to control devices called actuators which are solenoids, switches, relays, or motors that physically act or carry out a decision the computer has made.

❏ In a series circuit, the same current flows through each resistance.

❏ Each resistance in a parallel circuit is a separate path for current flow.

❏ The battery supplies voltage and current to operate the starting motor while cranking the engine, to operate the electrical accessories if the engine is not running, and to overcome any charging inefficiencies of the AC generator.

❏ The electrolyte contains 64% water and 36% sulfuric acid.

❏ When a positive and a negative plate are placed in electrolyte, a chemical reaction takes place, and the pressure difference between the plates is about 2.13 V.

❏ When a battery is discharging, the sulfate comes from the electrolyte and joins with the lead on the plates to form lead sulfate. The oxygen comes off the positive plates and joins with the hydrogen in the electrolyte to form water.

❏ The battery voltage depends on the chemical condition of the plates.

❏ When a battery is charging, the sulfate comes off both plates and joins with the hydrogen in the electrolyte to form sulfuric acid. The oxygen breaks away from the hydrogen in the electrolyte and joins with the lead on the positive plates to form lead peroxide.

❏ Batteries may be rated in cold cranking amperes, reserve capacity, ampere-hours, or watts.

❏ A neutral safety switch is connected in series in the circuit between the ignition switch and the solenoid windings. This switch is closed if the gear selector is in neutral or park, and it is open in other selector positions.

❏ When the solenoid plunger forces the disc against the terminals, a very high current flows from the battery through the starting motor.

❏ The rotating magnetic field on the rotor induces an AC voltage in the AC generator stator windings, and this AC voltage is changed to a DC voltage by the diodes.

❏ Many AC generators have an integrated circuit (IC) regulator, while other AC generators have an electromechanical regulator.

❏ Either type of voltage regulator controls field current and magnetic strength around the rotor to limit the voltage in the stator windings.

❏ If the AC generator is allowed to produce excessive voltage, this voltage forces high current through the battery and electrical accessories on the vehicle.

Review Questions

Short Answer Essays

1. Name the two energy sources used in automobile electrical systems.
2. For electrical flow to occur, what must be present?
3. Describe the chemical changes that occur while a battery is discharging.
4. What is the difference between voltage and current?
5. Explain the design of a starting motor armature.
6. State Ohm's law.
7. Describe the differences between a rheostat and a potentiometer.
8. What is the difference between a fixed resistor and a variable resistor?
9. What types of sensors are typically used in an automotive computer system?
10. Explain how the voltage is induced in the AC generator stator windings, and describe the type of voltage induced.

Fill-in-the-Blanks

1. The two power or energy sources used in an automobile's electrical system are based on a _____ _____ and on _____.

2. Current is measured in _____, electrical voltage is measured in _____, and electrical resistance is measured in _____.

3. Many automotive battery cases are made from _____ _____.

4. _____, _____, _____ _____ and _____ _____ are used to protect circuits against current overloads.

5. _____ control on/off and direct current flow in a circuit.

6. Many starting motors have two _____ brushes and two _____ brushes.

7. The ends of the rotor field winding are connected to the _____ _____.

8. A _____ is a material or device that can function as either a conductor or an insulator, depending on how its structure is arranged.

9. A computerized circuit depends on two types of signals: _____ and _____.

10. The AC generator voltage regulator is connected in the _____ circuit.

ASE Style Review Questions

1. *Technician A* says that magnetism is a source of electrical energy in an automobile.
   Technician B says that chemical reaction is a source of electrical energy in an automobile.
   Who is correct?
   A. A only   **C.** Both A and B
   B. B only   **D.** Neither A nor B

2. While discussing the behavior of electricity:
   Technician A says that if voltage does not change but there is a change in the resistance of the circuit, the current will change.
   Technician B says that if resistance increases, current decreases.
   Who is correct?
   A. A only   **C.** Both A and B
   B. B only   **D.** Neither A nor B

3. *Technician A* says that rheostats have three connections, one at each end of the resistance and one connected to a sliding contact with the resistor.
   Technician B says that potentiometers have two connections, one at the power end of the resistance and one connected to a sliding contact with the resistor.
   Who is correct?
   A. A only   **C.** Both A and B
   B. B only   **D.** Neither A nor B

4. *Technician A* says that electrical resistance is the pressure that causes current to flow in a circuit.
   Technician B says if there is zero resistance in a circuit, a maximum amount of current will flow in the circuit.
   Who is correct?
   A. A only   **C.** Both A and B
   B. B only   **D.** Neither A nor B

5. *Technician A* says that a diode allows current to flow in one direction but not in the opposite direction.
   Technician B says transistors are used as switching devices.
   Who is correct?
   A. A only   **C.** Both A and B
   B. B only   **D.** Neither A nor B

6. *Technician A* says that the flow of electricity is called current.
   Technician B says the flow of electricity is measured in amperes.
   Who is correct?
   A. A only   **C.** Both A and B
   B. B only   **D.** Neither A nor B

7. *Technician A* says that if the resistance in a circuit changes, so must the voltage or current.
Technician B says if voltage changes, so must the current or resistance.
Who is correct?
A. A only
B. B only
C. Both A and B
D. Neither A nor B

8. While discussing battery design:
Technician A says the positive plates in one cell are connected to the negative plates in the adjoining cell.
Technician B says each cell group contains the same number of positive and negative plates.
Who is correct?
A. A only
B. B only
C. Both A and B
D. Neither A nor B

9. *Technician A* says there are two types of electrical flow: direct current (DC) and alternating current (AC).
Technician B says that DC is used to operate most automotive electrical circuits.
Who is correct?
A. A only
B. B only
C. Both A and B
D. Neither A nor B

10. While discussing AC generator operation:
Technician A says the voltage regulator controls field current and the magnetic strength around the rotor.
Technician B says the voltage regulator controls the current flow in the stator windings.
Who is correct?
A. A only
B. B only
C. Both A and B
D. Neither A nor B

Engine Design and Operation

Upon completion and review of this chapter, you should be able to:

❏ Define the methods used for engine classification.

❏ Describe the four strokes in the four-stroke cycle.

❏ Explain compression ratio.

❏ Explain the purpose of the camshaft, pushrods, and rocker arms.

❏ Describe the difference between an overhead cam engine and an overhead valve engine.

❏ Describe four different types of engine block design.

❏ Briefly describe the different engine systems.

❏ Define cylinder bore and stroke.

❏ Explain how to calculate engine displacement.

❏ Describe three different methods of measuring engine efficiency.

❏ Name and describe the components of a typical lubricating system.

❏ Describe the purpose of a crankcase ventilation system.

❏ Explain oil service and viscosity ratings.

❏ List and describe the major components of the cooling system.

❏ Describe the operation of the cooling system.

❏ Describe the function of the water pump, radiator, radiator cap, and thermostat in the cooling system.

Introduction

Modern engines are highly engineered power plants. These engines are designed to meet the performance and fuel efficiency demands of the public. The days of the heavy, cast-iron V8 engine with its poor gas mileage are quickly drawing to a close. Today, these engines have been replaced by compact, lightweight, and fuel efficient engines (Figure 4-1). Modern engines are made of lightweight engine castings and stampings, non-iron materials (for example, aluminum, magnesium, fiber-reinforced plastics), and fewer and smaller fasteners to hold things together. These fasteners

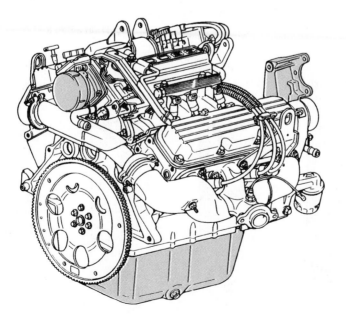

Figure 4-1 A typical modern engine. *(Courtesy of Oldsmobile Division—GMC)*

are made possible through computerized joint designs that optimize loading patterns. Each of these newer engine designs has its own distinct personality, based on construction materials, casting configurations, and design.

These modern engine building techniques have changed how technicians make their money. Before these changes can be explained, it is important to explain the "basics" of engine design and operation.

Engine Classifications

Today's automotive engines can be classified in several ways, depending on the following design features:

- ❑ *Operational cycles.* Most technicians will generally come in contact with only four- stroke cycle engines (Figure 4-2). However, a few older cars have used and some cars in the future will use a two-stroke engine.
- ❑ *Number of cylinders.* Current engine designs include 3-, 4-, 5-, 6-, 8-, 10-, and 12- cylinder engines.
- ❑ *Cylinder arrangement.* An engine can be flat (opposed), in-line, or V-type. Other more complicated designs have also been used.
- ❑ *Displacement.* The volume of the engine's cylinders added together.
- ❑ *Valve train type.* Engine valve trains can be either the **overhead camshaft** (**OHC**) type or the camshaft in-block **overhead valve** (**OHV**) type. Some engines separate camshafts for the intake and exhaust valves. These are based on the OHC design and are called **dual over-head camshaft** (**DOHC**) engines. V-type DOHC engines have four camshafts—two on each side.
- ❑ *Ignition type.* There are two types of ignition systems: spark and compression. Gasoline engines use a spark ignition system. In a spark ignition system, the air-fuel mixture is ignited by an electrical spark. Diesel engines, or compression ignition engines, have no spark plugs. An automotive diesel engine relies on the heat generated as air is compressed to ignite the air-fuel mixture for the power stroke.
- ❑ *Cooling systems.* There are both air-cooled and liquid-cooled engines in use. Nearly all of today's engines have liquid-cooling systems.
- ❑ *Fuel type.* Several types of fuel are currently used in automobile engines include gasoline, natural gas, diesel, and propane. The most commonly used is gasoline. Recently electric vehicles have been introduced for sale to the public. These vehicles do not rely on a fuel for power.

An engine with two exhaust valves and two intake valves in each cylinder is commonly referred to as a four-valve engine. These are typically called by their total number of valves, e.g., a four-valve eight-cylinder engine may be called a 32-valve engine.

The only major manufacturer that uses air-cooled engines is Porsche, which (after more than 50 years) is equipping new models with water-cooled engines.

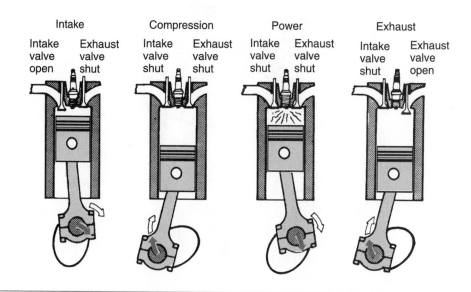

Figure 4-2 The four-stroke cycle. *(Courtesy of Sun Electric Corporation)*

Four-Stroke Gasoline Engines

In a passenger car or truck, the engine provides the rotating power to drive the wheels through the transmission and driving axles. All automobile engines, both gasoline and diesel, are classified as internal combustion because the combustion or burning takes place inside the engine. These systems require an air fuel mixture that arrives in the combustion chamber at the correct time and an engine constructed to withstand the temperatures and pressures created by the burning of thousands of fuel droplets.

The **combustion chamber** is the space between the top of the piston and cylinder head. It is an enclosed area in which the gasoline and air mixture is burned. The piston is a hollow metal tube with one end closed that moves up and down in the cylinder. This reciprocating motion is caused by an increase of pressure, due to combustion, in the cylinder.

The reciprocating motion must be converted to a rotary motion before it can drive the wheels of a vehicle. This change of motion is accomplished by connecting the piston to a crankshaft with a connecting rod (Figure 4-3). The upper end of the connecting rod moves with the piston as it moves up and down in the cylinder. The lower end of the connecting rod is attached to the crankshaft and moves in a circle. The end of the crankshaft is connected to the flywheel, which transfers the engine's power through the drivetrain to the wheels.

In order to have complete combustion in an engine, the right amount of fuel must be mixed with the right amount of air. This mixture must be compressed in a sealed container then shocked by the right amount of heat at the right time. When these conditions exist, all of the fuel that enters a cylinder is burned and converted to power. This power is used to move the vehicle. The ignition system is responsible for providing the heat and the air and the fuel system is responsible for providing the air-fuel mixture. The construction of the engine and its different parts are responsible for providing a container for the combustion process. This container is the cylinder of an engine. Automotive engines have more than one cylinder. Each cylinder should receive the same amounts of air, fuel, and heat if the engine is to run efficiently.

Although the combustion must occur in a sealed cylinder, the cylinder must also have some means of allowing heat, fuel, and air into it. There must also be a means to allow the burnt air-fuel mixture out so that a fresh mixture can enter and the engine continue to run. To accommodate this requirements, engines are fitted with valves (Figure 4-4).

There are at least two valves at the top of each cylinder. The air-fuel mixture enters the combustion chamber through an intake valve and leaves (after having been burned) through an exhaust

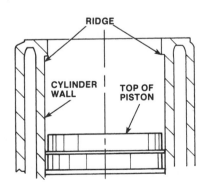

Figure 4-3 The reciprocating motion of the pistons is converted to a rotating motion by the crankshaft.

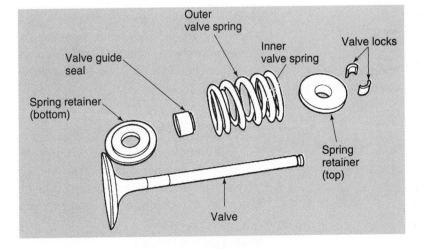

Figure 4-4 An engine valve assembly. *(Courtesy of Chrysler Corporation)*

valve. The valves are accurately machined plugs that fit into machined openings. A valve is said to be seated or closed when it rests in its opening. When the valve is pushed off its seat, it opens.

A rotating camshaft, driven by the crankshaft, opens and closes the intake and exhaust valves. **Cams** are raised sections of a shaft that have high spots called **lobes**. As the camshaft rotates, the lobes rotate and push the valve open by pushing it away from its seat. Once the cam lobe rotates out of the way, the valve, forced by a spring, closes. The camshaft can be located either in the cylinder block or in the cylinder head.

When the action of the valves and the spark plug is properly timed to the movement of the piston, the combustion cycle takes place in four strokes of the piston: the intake stroke, the compression stroke, the power stroke, and the exhaust stroke. A **stroke** is the full travel of the piston either up or down in the cylinder bore.

The up-and-down movement of the piston on all four strokes is converted to a rotary motion by the crankshaft. It takes two full revolutions of the crankshaft to complete the four-stroke cycle.

The piston moves by the pressure produced during combustion only about half a stroke or one-quarter of crankshaft revolution. This explains why a flywheel is needed. The flywheel stores some of the power produced by the engine. This power is used to keep the pistons in motion during the rest of the four-stroke cycle. This power is also used to compress the air-fuel mixture just before combustion.

Intake Stroke

Shop Manual
Chapter 4, page 138

The first stroke of the cycle is the intake stroke. As the piston moves away from top dead center (TDC), the intake valve opens. The downward movement of the piston increases the volume of the cylinder above it. This reduces the pressure in the cylinder. This reduced pressure, commonly referred to as engine vacuum, causes the atmospheric pressure to push a mixture of air and fuel through the open intake valve. (Some engines are equipped with a super- or turbocharger that pushes more air past the valve.) As the piston reaches the bottom of its stroke, the reduction in pressure stops. This causes the intake of air-fuel mixture to slow down. It does not stop because of the weight and movement of the air-fuel mixture. It continues to enter the cylinder until the intake valve closes. The intake valve closes after the piston has reached bottom dead center (BDC). This delayed closing of the valve increases the volumetric efficiency of the cylinder by packing as much air and fuel into it as possible.

Compression Stroke

Shop Manual
Chapter 4, page 142

The compression stroke begins as the piston starts to move from BDC. The intake valve closes, trapping the air-fuel mixture in the cylinder. The upward movement of the piston compresses the air-fuel mixture, thus heating it up. At TDC, the piston and cylinder walls form a combustion chamber in which the fuel will be burned. The volume of the cylinder with the piston at BDC compared to the volume of the cylinder with the piston at TDC determines the compression ratio of the engine.

Power Stroke

The power stroke begins as the compressed fuel mixture is ignited. An electrical spark across the electrodes of a spark plug ignites the air-fuel mixture. The burning fuel rapidly expands, creating a very high pressure against the top of the piston. This drives the piston down toward BDC. The downward movement of the piston is transmitted through the connecting rod to the crankshaft.

Exhaust Stroke

The exhaust valve opens just before the piston reaches BDC on the power stroke. Pressure within the cylinder causes the exhaust gas to rush past the open valve and into the exhaust system. Movement of the piston from BDC pushes most of the remaining exhaust gas from the cylinder. As the piston nears TDC, the exhaust valve begins to close as the intake valve starts to open. The exhaust stroke completes the four-stroke cycle. The opening of the intake valve begins the cycle again. This cycle occurs in each cylinder and is repeated over and over, as long as the engine is running.

The period of time that the exhaust and intake valves are both open during the exhaust stroke is called valve overlap.

Two-Stroke Gasoline Engines

In the past, several imported vehicles have used two-stroke engines. As the name implies, this engine requires only two strokes of the piston to complete all four operations: intake, compression, power, and exhaust. As shown in Figure 4-5, this is accomplished as follows.

1. Movement of the piston from BDC to TDC completes both intake and compression.
2. When the piston nears TDC, the compressed air-fuel mixture is ignited, causing an expansion of the gases. Note that the reed valve is closed and the piston is blocking the intake port.
3. Expanding gases in the cylinder force the piston down, rotating the crankshaft.
4. With the piston at BDC, the intake and exhaust ports are both open, allowing exhaust gases to leave the cylinder and air-fuel mixture to enter.

Although the two-stroke-cycle engine is simple in design and lightweight because it lacks a valve train, it has not been widely used in automobiles. They tend to be less fuel efficient and have dirtier exhaust than four-stroke engines. Oil is often in the exhaust stream because these engines require constant oil delivery to the cylinders to keep the piston lubricated. Some of these engines require that a certain amount of oil be mixed with the fuel.

In recent years, however, thanks to a revolutionary pneumatic fuel injection system, there has been increased interest in the two-stroke engine. The injection system, which works something like a spray paint gun, uses compressed air to flow highly atomized fuel directly into the top of the combustion chamber. The system becomes the long sought-after answer to the fuel economy and emissions problems of the conventional two-stroke engine. This fuel injection system is the basis for orbital two-stroke direct injection piston engine, which may be used in cars in the future.

Characteristics of Four-Stroke Engine Design

Depending on the vehicle, either an in-line, V-type, slant, or opposed cylinder design can be used. The most popular designs are in-line and V-type engines.

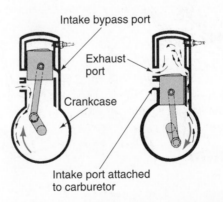

Intake bypass port

Exhaust port

Crankcase

Intake port attached to carburetor

Figure 4-5 The two-stroke cycle. *(Courtesy of Sun Electric Corporation)*

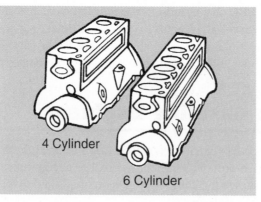

4 Cylinder

6 Cylinder

Figure 4-6 In-line engine designs.

In-Line Engines

In the in-line engine design (Figure 4-6), the cylinders are all placed in a single row. There is one crankshaft and one cylinder head for all of the cylinders. The block is cast so that all cylinders are located in an upright position.

In-line engine designs have certain advantages and disadvantages. They are easy to manufacture and service. However, because the cylinders are positioned vertically, the front of the vehicle must be higher. This affects the aerodynamic design of the car. Aerodynamic design refers to the ease at which the car can move through the air. When equipped with an in-line engine, the front of a vehicle cannot be made as low as it can with other engine designs.

V-Type Engines

The V-type engine design has two rows of cylinders (Figure 4-7) located 60 to 90 degrees away from each other. A V-type engine uses one crankshaft that is connected to the pistons on both sides of the V. This type of engine has two cylinder heads, one over each row of cylinders.

One advantage of using a V-configuration is that the engine is not as high or long as an in-line configuration. The front of a vehicle can now be made lower. This design improves the outside aerodynamics of the vehicle. If eight cylinders are needed for power, a V- configuration makes the engine much shorter, lighter, and more compact. Many years ago, some vehicles had an in-line eight-cylinder engine. The engine was very long and its long crankshaft also caused increased torsional vibrations in the engine.

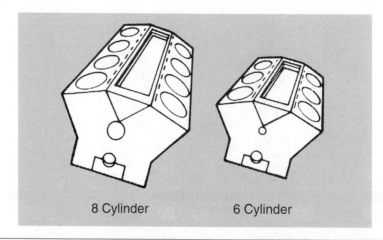

8 Cylinder

6 Cylinder

Figure 4-7 V-type engine designs.

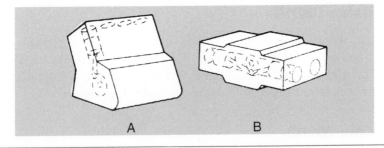

Figure 4-8 (A) Slant engine and (B) opposed cylinder engine designs.

Slant Cylinder Engines

Another way of arranging the cylinders is in a slant configuration (Figure 4-8A). This is much like an in-line engine, except that the entire block has been placed at a slant. The slant engine was designed to reduce the distance from the top to the bottom of the engine. Vehicles using the slant engine can be designed more aerodynamically.

Opposed Cylinder Engines

In this design, two rows of cylinders are located opposite the crankshaft (Figure 4-8B). Opposed cylinder engines are used in applications where there is very little vertical room for the engine. For this reason, opposed cylinder designs are commonly used on vehicles that have the engine in the rear. The angle between the two cylinder heads is typically 180 degrees. One crankshaft is used with two cylinder heads.

Valve and Camshaft Placement Configurations

Two basic valve and camshaft placement configurations of the four-stroke gasoline engines are used in automobiles (Figure 4-9).

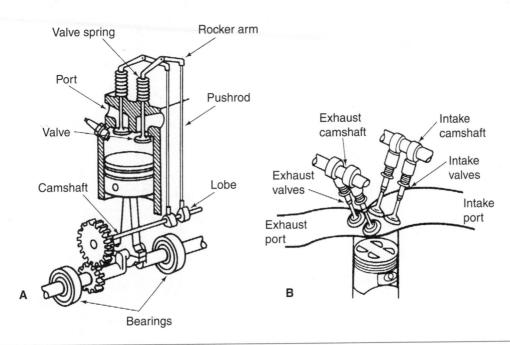

Figure 4-9 Basic valve and camshaft placement configurations: (A) OHV and (B) OHC.

Overhead Valve (OHV)

As the name implies, the intake and exhaust valves on an overhead valve engine are mounted in the cylinder head and are operated by a camshaft located in the cylinder block. This arrangement requires the use of valve lifters, pushrods, and rocker arms to transfer camshaft rotation to valve movement. The intake and exhaust manifolds are attached to the cylinder head.

Overhead Cam (OHC)

An overhead cam engine also has the intake and exhaust valves located in the cylinder head. But as the name implies, the cam is located in the cylinder head. In an overhead cam engine, the valves are operated directly by the camshaft or through cam followers or tappets.

Valve and Camshaft Operation

Shop Manual
Chapter 4, page 148

In OHV engines with the camshaft in the block (Figure 4-10), the valves are operated by valve lifters and pushrods that are actuated by the camshaft. On overhead cam engines, the cam lobes operate the valves directly and there is no need for pushrods or lifters.

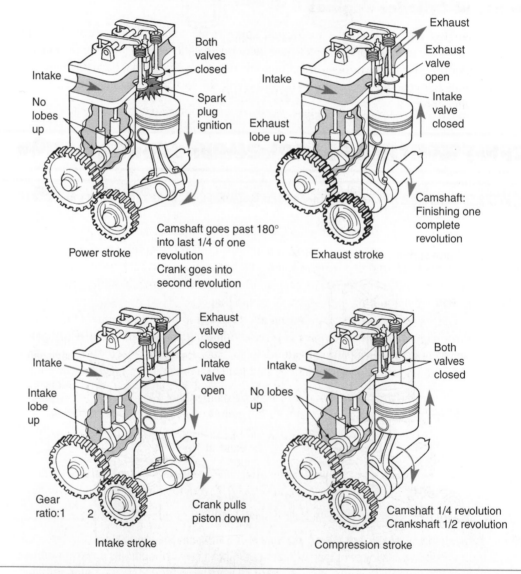

Figure 4-10 Valve operation in an OHV engine.

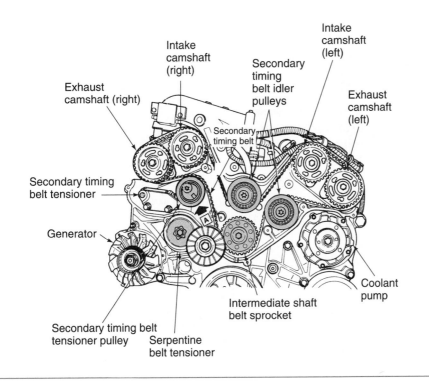

Figure 4-11 Timing belt, sprockets, and pulleys. *(Courtesy of Oldsmobile Division—GMC)*

Cam lobes are oval shaped. The placement of the lobe on the shaft determines when the valve will open. Design of the lobe determines how far the valve will open and how long it will remain open in relation to piston movement.

The camshaft is driven by the crankshaft through gears, or sprockets, and a cogged belt (Figure 4-11), or timing chain. The camshaft turns at half the crankshaft speed and rotates one complete turn during each complete four-stroke cycle.

Engine Location

The engine is typically placed in one of three locations. In the vast majority of vehicles, it is located at the front of the vehicle, forward of the passenger compartment. Front-mounted engines can be positioned either longitudinally or transversely with respect to the vehicle.

The second engine location is a mid-mount position between the passenger compartment and rear suspension. Mid-mount engines are normally transversely mounted. The third, and least common, engine location is in the rear of the vehicle. The engines are typically opposed type engines. Each of these engine locations offers advantages and disadvantages.

Gasoline Engine Systems

The operation of an engine relies on several other systems. The efficiency of these systems affects the overall operation of the engine.

Air-Fuel System

This system makes sure the engine gets the right amount of both air and fuel needed for efficient operation. For many years air and fuel were mixed in a carburetor, which supplied the resulting

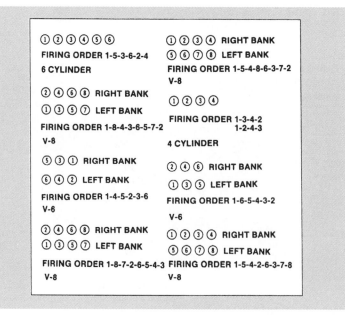

Figure 4-12 Common firing orders.

mixture to the cylinder. Today, most late-model automobiles have a fuel injection system that replaces the carburetor but performs the same basic function.

Ignition System

Shop Manual
Chapter 4, page 121

This system delivers a spark to ignite the compressed air-fuel mixture in the cylinder at the end of the compression stroke. The **firing order** of the cylinders is determined by the engine's manufacturer and can be found in the vehicle's service manual. Typical firing orders are illustrated in Figure 4-12.

Lubrication System

The firing order of an engine is the order in which the pistons reach TDC on the compression stroke, and the spark plug firing must be synchronized with this firing order.

This system supplies oil to the various moving parts in the engine. The oil lubricates all parts that slide in or on other parts, such as the piston, bearings, crankshaft, and valve stems. The oil enables the parts to move easily so that little power is lost and wear is kept to a minimum. The lubrication system also helps transfer heat from one part to another for cooling.

Cooling System

This system is also extremely important. Coolant circulates in jackets around the cylinder and in the cylinder head. This removes part of the heat produced by combustion and prevents the engine from being damaged by overheating.

Exhaust System

This system removes the burned gases from the combustion chamber and limits the noise produced by the engine.

Emission Control System

Several control devices, which are designed to reduce the amount of pollutants released by the engine, have been added to the engine. Engine design changes, such as reshaped combustion chambers and altered valve timing, have also been part of the manufacturers' attempt to reduce emission levels.

Engine Measurement and Performance

Some of the engine measurements and performance characteristics that a technician should be familiar with follow.

Bore and Stroke

The **bore** of a cylinder is simply its diameter measured in inches (in.) or millimeters (mm). The stroke is the length of the piston travel between TDC and BDC. Between them, bore and stroke determine the displacement of the cylinders (Figure 4-13). When the bore of the engine is larger than its stroke, it is said to be oversquare. When the stroke is larger than the bore, the engine is said to be undersquare. Generally, an oversquare engine will provide for high engine speeds, such as for automobile use. An undersquare or long-stroke engine will deliver good low speed power, such as an engine for a truck or tractor.

The **crank throw** is the distance from the crankshaft's main bearing centerline to the crankshaft throw centerline. The stroke of any engine is twice the crank throw.

Displacement

Displacement is the volume of a cylinder between the TDC and BDC positions of the piston. It is usually measured in cubic inches, cubic centimeters, or lifters (Figure 4-14). The total displacement of an engine (including all cylinders) is a rough indicator of its power output. Displacement can be increased by opening the bore to a larger diameter or by increasing the length of the stroke. Total displacement is the sum of displacements for all cylinders in an engine. Cubic inch displacement (CID) may be calculated as follows.

When people refer to an engine's size, they are normally talking about the engine's displacement.

$$CID = pi \times R^2 \times L \times N$$

where pi = 3.1416
R = bore radius or bore diameter/2
L = length of stroke
N = number of cylinders

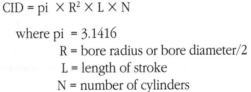

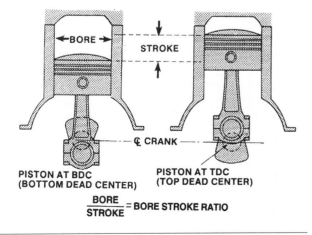

Figure 4-13 The bore and stroke of an engine.

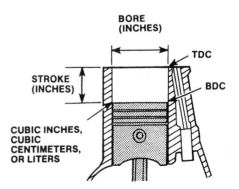

Figure 4-14 Displacement is the volume of a cylinder between TDC and BDC.

Example: Calculate the cubic inch displacement (CID) of a six-cylinder engine with a 3.7-in. bore and 3.4-in. stroke.

$$CID = 3.1416 \times 1.85^2 \times 3.4 \times 6$$
$$CID = 219.66$$

Most of today's engines are described by their metric displacement. Cubic centimeters and liters are determined by using metric measurements in the displacement formula.

Example: Calculate the metric displacement of a four-cylinder engine with a 78.9-mm stroke and a 100-mm bore.

$$Displacement = 3.1416 \times 100^2 \times 78.9 \times 4$$
$$Displacement = 2479 \text{ cubic centimeters (cc)} = 2.5 \text{ liters (L)}$$

Larger, heavier vehicles are provided with large displacement engines. Large displacement engines produce more torque than smaller displacement engines. They also consume more fuel. Smaller, lighter vehicles can be adequately powered by lower displacement engines that use less fuel.

Compression Ratio

The **compression ratio** of an engine expresses how much the air and fuel mixture will be compressed during the compression stroke. The compression ratio is defined as the ratio of the volume in the cylinder above the piston when the piston is at BDC to the volume in the cylinder above the piston when the piston is at TDC (Figure 4-15). The formula for calculating the compression ratio is as follows.

$$\frac{\text{volume above the piston at BDC}}{\text{volume above the piston at TDC}}$$

or

$$\frac{\text{total cylinder volume}}{\text{total combustion chamber volume}}$$

In many engines, the top of the piston is even or level with the top of the cylinder block at TDC. The combustion chamber is in the cavity in the cylinder head above the piston. This is modified slightly by the shape of the top of the piston. The volume of the combustion chamber must be added to each volume in the formula in order to get an accurate calculation of compression ratio.

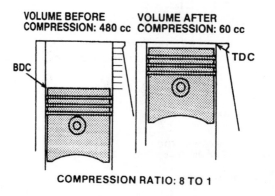

Figure 4-15 Compression ratio measures the amount the air and fuel will be compressed.

Example: Calculate the compression ratio if the total piston displacement is 45 cubic inches and the combustion chamber volume is 5.5 cubic inches.

$$\frac{45 + 5.5}{5.5} = \text{compression ratio}$$

$$9.1 \text{ to } 1 = \text{compression ratio}$$

(Be sure to add the combustion chamber volume to the piston displacement to get the total cylinder volume.)

The higher the compression ratio, the more power an engine theoretically can produce. Also, as the compression ratio increases, the heat produced by the compression stroke also increases. Gasoline with a low octane rating burns fast and may explode rather than burn when introduced to a high compression ratio. This can cause preignition. The higher a gasoline's octane rating, the less likely it is to explode.

As the compression ratio increases, the octane rating of the gasoline also should be increased to prevent abnormal combustion.

Engine Efficiency

Engine **efficiency** is a measure of the relationship between the amount of energy put into the engine and the amount of available energy from the engine. Engine efficiency is expressed in a percentage. The formula for determining efficiency is:

$$\text{efficiency} = \frac{\text{output energy}}{\text{input energy}} \times 100$$

There are other aspects of the engine that are expressed in efficiencies. These include mechanical efficiency, volumetric efficiency, and thermal efficiency. They are expressed as a ratio of input (actual) to output (maximum or theoretical). Efficiencies are expressed as percentages. They are always less than 100%. The difference between the efficiency and 100% is the percentage lost during the process. For example, if there were 100 units of energy put into the engine and 28 units came back, the efficiency would be equal to 28%. This would mean that 72% of the energy received was wasted or lost.

Thermal Efficiency

An engine converts the chemical energy of the fuel into heat energy, which is then converted into mechanical energy by the pistons and connecting rods of the engine. Thermal energy is the relationship between the engine power output and the heat energy available in the fuel. A percentage is used to express this relationship.

As the cooling system and the lubrication system cool the engine components, these systems carry away much of the heat energy in the fuel. However, this cooling action is necessary to prevent the destruction of engine parts. Part of the heat energy in the fuel is dissipated through the exhaust system. Some of the heat energy in the fuel is consumed to overcome the internal friction in the engine and powertrain. From the total energy in the fuel, these average losses may occur:

- ❏ 35% loss to the cooling and lubrication system
- ❏ 35% loss to the exhaust gas
- ❏ 5% loss to engine friction
- ❏ 10% loss to powertrain friction

After these losses, only 15% of the thermal energy in the fuel is left to drive the vehicle. In recent years, engine components have been designed to reduce engine friction, and improved lubricants reduce engine and powertrain friction. However, the average engine thermal efficiency still remains below 30%. Some engine manufacturers are designing ceramic engine components that are capable of operating at higher temperatures than metal and aluminum components. The use of ceramic engine components increases engine thermal efficiency.

Mechanical Efficiency

Mechanical efficiency is the relationship between the engine power delivered and the power that would be delivered if the engine operated without any power loss. The mechanical efficiency of an engine is expressed as a percentage.

Volumetric Efficiency

Volumetric efficiency is the relationship between the amount of air actually taken into the cylinder on the intake stroke compared to the amount of air required to fill the cylinder at atmospheric pressure. Volumetric pressure is also expressed as a percentage. The volumetric efficiency of an engine is affected by intake manifold design, valve timing, and exhaust system design.

Torque and Horsepower

Brake horsepower may be defined as the power developed at the engine's crankshaft.

Torque is a turning or twisting force. The engine's crankshaft rotates with a torque that is transmitted through the drivetrain to turn the drive wheels of the vehicle. **Horsepower** (HP) is the rate at which torque is produced.

Engines produce power by turning a crankshaft in a circular motion. To convert terms of force applied in a straight line to a force applied in a circular motion, the formula is:

$$torque = force \times radius$$

Friction horsepower is the power required to overcome the internal friction of the engine.

Example: A 10-pound force applied to a wrench 1 foot long will produce 10 pound-feet (lb.- ft.) of torque. Imagine that the 1-foot-long wrench is connected to a shaft. If 1 pound of force is applied to the end of the wrench, 1 pound-foot of torque is produced. Ten pounds of force applied to a wrench 2 feet long will produce 20 pound-feet of torque (Figure 4-16).

The technically correct torque measurement is stated in pound-feet (lb.-ft.). However, it is rather common to state torque in terms of foot-pounds (ft.-lb.). In the metric or SI system, torque is stated in Newton-meters (Nm) or kilogram-meters (kg-m).

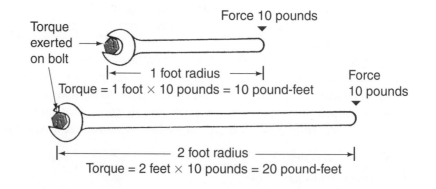

Figure 4-16 Force applied to a wrench produces torque.

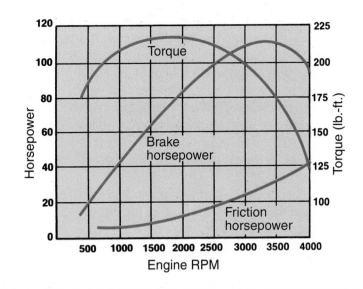

Figure 4-17 The relationship between horsepower and torque.

If the torque output of an engine at a given speed (rpm) is known, horsepower (HP) can be calculated by the following formula.

$$HP = (torque \times rpm) \div 5{,}252$$

An engine produces different amounts of torque based on the rotational speed of the crankshaft and other factors. A mathematical representation, or graph, of the relationship between the horsepower and torque of one engine is shown in Figure 4-17.

This graph shows that torque begins to decrease when the engine's speed reaches about 1,700 rpm. Brake horsepower increases steadily until about 3,500 rpm. Then it drops. The third line on the graph indicates the horsepower needed to overcome the friction or resistance created by the internal parts of the engine rubbing against each other.

Other Engine Designs

The gasoline-powered, internal combustion piston engine has been the primary automotive power plant for many years and probably will remain so for years to come. Present-day social requirements and new technological developments, however, have necessitated searches for ways to modify or replace this time-proven workhorse. This portion of the chapter takes a brief look at the most likely contenders, and how they work. The orbital two-stroke cycle engine was discussed earlier and is not explained in the following. This certainly does not mean that this engine design is not a viable one.

Diesel Engines

Diesel engines represent tested, proven technology with a long history of success. Invented by Dr. Rudolph Diesel, a German engineer, and first marketed in 1897, the diesel engine is now the dominate power plant in heavy-duty trucks, construction equipment, farm equipment, buses, and marine applications.

During the late 1970s and early 1980s, many predicted small diesel engines would replace gasoline engines in passenger vehicles. However, stabilized gas prices and other factors dampened the enthusiasm for diesels in these markets. The use of diesel engines in cars and light trucks is now limited to a few manufacturers.

Diesel engines (Figure 4-18) and gasoline-powered engines share several similarities. They have a number of components in common, such as the crankshaft, pistons, valves, camshaft, and water and oil pumps. They both are available in four-stroke combustion cycle models. However, the diesel engine and four-stroke compression-ignition engine are easily recognized by the absence of an ignition system. Instead of relying on a spark for ignition, a diesel engine uses the heat produced by compressing air in the combustion chamber to ignite the fuel. The systems used in diesel-powered vehicles are essentially the same as those used in gasoline vehicles.

Fuel injection is used on all diesel engines. Injectors spray pressurized fuel into the cylinders just as the piston is completing its compression stroke. The heat of the compressed air ignites the fuel and begins the power stroke.

Glow plugs are used only to warm the combustion chamber when the engine is cold. Cold starting is impossible without these plugs because even the high-compression ratios cannot heat cold air enough to cause combustion.

Diesel combustion chambers are different from gasoline combustion chambers because diesel fuel burns differently. Three types of combustion chambers are used in diesel engines: open combustion chamber, precombustion chamber, and turbulence combustion chamber. The open combustion chamber has the combustion chamber located directly inside the piston. Diesel fuel is injected directly into the center of the chamber. The shape of the chamber and the quench area produces turbulence. The precombustion chamber is a smaller, second chamber connection to the main combustion chamber. On the power stroke, fuel is injected into the small chamber. Combustion is started there and then spreads to the main chamber. This design allows lower fuel injection pressures and simpler injection systems on diesel engines.

The turbulence combustion chamber is designated to create an increase in air velocity or turbulence in the combustion chamber. The fuel is injected into the turbulent air and burns more completely. The prechambers on a diesel engine head must be properly indexed with the head and correctly installed. They must be perfectly flush (not above or below) with the cylinder head. Failure to follow this will cause the head gasket to fail.

Rotary Engine

The **rotary** engine, or **Wankel engine**, is somewhat similar to the standard piston engine in that it is a four-cycle, spark ignition, internal combustion engine. Its mechanical design, however, is quite

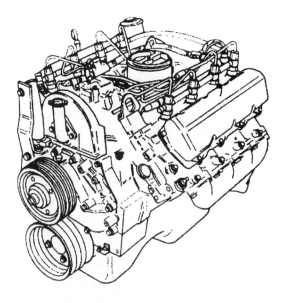

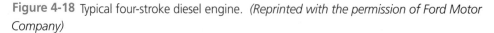

Figure 4-18 Typical four-stroke diesel engine. *(Reprinted with the permission of Ford Motor Company)*

different. For one thing, the rotary engine uses a rotating motion rather than a reciprocating motion. In addition, it uses ports rather than valves for controlling the intake of the air-fuel mixture and the exhaust of the combusted charge.

The heart of a rotary engine is a roughly triangular rotor that "walks" around a smaller, rigidly mounted gear. The rotor is connected to the crankshaft through additional gears in such a manner that for every rotation of the rotor the crankshaft revolves three times. The tips of the triangular rotor move within the housing and are in constant contact with the housing walls. As the rotor moves, the volume between each side of the rotor and the housing walls continually changes.

Referring to Figure 4-19, when the rotor is in position A, the intake port is uncovered and the air-fuel mixture is entering the upper chamber. As the rotor moves to position B, the intake port closes and the upper chamber reaches its maximum volume. When full compression has reached position C, the two spark plugs fire, one after the other, to start the power stroke. At position D, rotor side A uncovers the exhaust port and exhaust begins. This cycle continues until position A is reached where the chamber volume is at minimum and the intake cycle starts once again.

The fact that the rotating combustion chamber engine is small and light for the amount of power it produces makes it attractive for use in automobiles. Using this small, lightweight engine can provide the same performance as a larger engine. But the rotary engine, at present, cannot compete with the piston gasoline on durability, exhaust emission control, and economy.

Stratified Charge Engine

The stratified charge engine (Figure 4-20) combines the features of both the gasoline and diesel engines. It differs from the conventional gasoline engine in that the air-fuel mixture is deliberately stratified to produce a small rich mixture at the spark plug while providing a leaner, more efficient and cleaner burning main mixture. In addition, the air-fuel mixture is swirled for more complete combustion.

On the intake stroke a large amount of very lean mixture is drawn through the main intake valve to the main combustion chamber. At the same time, a small amount of rich mixture is drawn through the auxiliary intake valve into the precombustion chamber. At the end of the compression stroke, the spark plug fires the rich mixture in the precombustion chamber. As the rich mixture ignites, it in turn ignites the lean mixture in the main chamber. The lean mixture minimizes the formation of carbon monoxide during the power stroke. In addition, the peak temperature stays low enough to minimize the formation of oxides of nitrogen, and the mean temperature is maintained

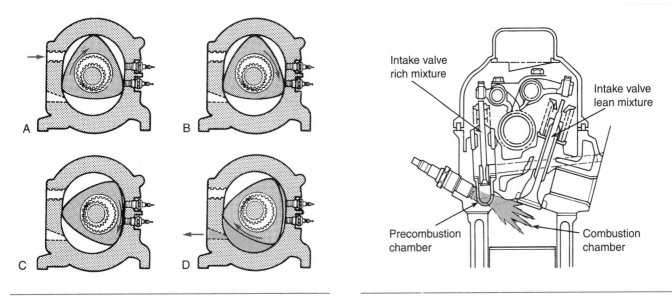

Figure 4-19 Rotary operational cycles.

Figure 4-20 Typical stratified charge engine.

high enough and long enough to reduce hydrocarbon emissions. During the exhaust stroke the hot gases exit through the exhaust valve.

A great deal of automobile engineering research, especially by Japanese and European manufacturers, is being done on these engines. In fact, the Honda CVCC engine uses a stratified charge design. This engine uses a third valve to release the initial charge. The stratified charge combustion chamber has three important advantages. It produces good part-load fuel economy. It can run efficiently on low-octane fuel and has low exhaust emissions.

Miller-Cycle Engines

This engine design is a modification of the four-stroke-cycle engine. During the intake stroke a supercharger feeds highly compressed air to an intercooler. This cooled, but compressed, air is fed to the cylinders. During the compression stroke, the intake valve remains open for a longer than normal time. This prevents compression from occurring until the piston has moved one-fifth of its upward travel. Then the valve closes and compression occurs. The shortened compression stroke keeps the compression ratios and cylinder temperatures low. The power stroke begins as soon as the piston is ready to move down its bore and continues until it reaches BDC. This longer power stroke provides more torque and increased efficiency. The exhaust stroke is much the same as that in a four-stroke-cycle engine.

Electric Motors

In the early days of the automobile, electric cars outnumbered gasoline cars. An electric motor is quiet, has little or no emissions, and has few moving parts. It starts well in the cold, is simple to maintain, and does not burn petroleum products to operate. Its disadvantages are limited speed, power, and range as well as the need for heavy, costly batteries. Experimental vehicles employing solar-charged batteries are being considered as sources of automotive power. Electric vehicles are now available. These use electric motors to move the vehicle and use household current to recharge the batteries.

Engine Lubrication

Shop Manual
Chapter 4, page 129

An engine's lubricating system does several important things. It holds an adequate supply of oil to cool, clean, lubricate, and seal the engine. It also removes contaminants from the oil and delivers oil to all necessary areas of the engine.

Engine Oil

Engine oil is often called motor oil.

Engine oil is a clean or refined form of **crude oil**. Crude oil, when taken out of the ground, is dirty and does not work well as a lubricant for engines. Crude oil must be refined to meet industry standards. Engine oil is just one of the many products that come from crude oil. Engine oil is specially formulated so that it has the following properties.

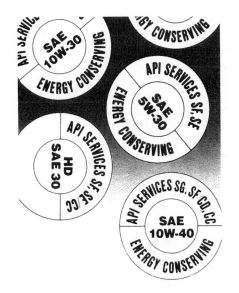

Figure 4-21 Identification label on oil containers. *(Courtesy of The Valvoline Company)*

❏ Prompt circulation through the engine's lubrication system

❏ The ability to lubricate without foaming

❏ The ability to reduce friction and wear

❏ The ability to prevent the formation of rust and corrosion

❏ The ability to cool the engine parts it flows on

❏ The ability to keep internal engine parts clean

To provide these properties, engine oil contains many additives. Because of these additives, choosing the correct oil for each engine application can be a difficult task. However, the **American Petroleum Institute** (**API**) has developed service ratings for motor oil that greatly simplifies oil selection.

The API classifies engine oil as S-class for passenger cars and light trucks with spark ignition and as C-class for heavy-duty diesel (combustion ignition) applications (Figure 4-21). Additionally, various grades of oil within each class are further classified alphabetically according to their ability to meet the engine manufacturers' warranty specifications (Table 4-1).

The API circle and markings on oil containers is commonly called the API doughnut.

Friction Modifiers

Oil refiners add a number of different additives to their oil that improve oil performance in the engine. Friction modifiers in the engine oil reduce friction between moving parts that contact each other. The oil film between two moving surfaces is actually sheared by the moving components. During this oil shearing process, there is a certain amount of friction in the oil. This friction within the oil is referred to as viscous friction, and this type of friction is reduced by friction modifiers in the oil.

Viscous friction refers to the resistance of an oil film to shearing apart.

Antifoaming Agents

Oil foaming may be caused by the rotating action of the crankshaft and connecting rods. Since oil foam contains oil and air, excessive oil foaming causes air in the lubrication system, which partially destroys the oil film on moving components in contact with each other. This action allows these moving components to contact each other, causing excessive friction and wear. Antifoaming agents added to the engine oil reduce oil foaming in the engine. Oil foaming may be caused by overfilling the crankcase; therefore, it is important to maintain the oil at the proper level on the dipstick.

Oils marked "Energy Conserving" can increase fuel economy by more than 1.5%. Those oils marked "Energy Conserving-II" can increase fuel economy by more than 2.7%.

Table 4-1 OIL CLASSIFICATIONS *(Courtesy of The Valvoline Company)*

SG **For 1989 Gasoline Engine Warranty Maintenance Service:**

Service typical of gasoline engines in passenger cars and light truck diesel engines. Oils designed for this service provide greater resistance to sludge, varnish buildup, oxidation, and wear. These oils provide more protection for your engine than oils that are satisfactory for API Engine Service Classifications SF and SE. SG should be used and is recommended when either of these classifications is specified.

SF **For 1980 Gasoline Engine Warranty Maintenance Service:**

Service typical of gasoline engines in passenger cars and some trucks beginning with 1980. Oils provide increased oxidation stability and improved antiwear performance for autos that meet API Engine Service Classification SE. It also provides protection against engine deposits, rust, and corrosion. Oils meeting classification SF may be used when SE, SD, or SC are recommended.

SE **For 1972 Gasoline Engine Warranty Maintenance Service:**

Service typical of gasoline engines in passenger cars and some trucks beginning with 1972, and certain 1971 models. Oils designed for this service provide more protection against oil oxidation, high temperature engine deposits, rust, and corrosion in gasoline engines than oils that are satisfactory for API Engine Service Classifications SD or SC, and may be used when either of these classifications is recommended.

CD **For Severe-Duty Diesel Engine Service:**

Service typical of supercharged diesel engines in high-speed, high-output heavy-duty conditions requiring highly effective control of wear and deposits. Oils designed for this service provide protection from bearing corrosion and from high-temperature deposits in supercharged diesel engines when using fuels of a wide quality range.

CC **For Moderate-Duty Diesel and Gasoline Engine Service:**

Service typical of lightly supercharged diesel engines operated in moderate to severe duty, and certain heavy-duty gasoline engines. Oils designed for this service provide protection from high-temperature deposits and supercharged diesels, and also from rust, corrosion, and low-temperature deposits in gasoline engines. These oils are used in many trucks, industrial and construction equipment, and farm tractors.

Corrosion and Rust Inhibitors

Since the engine experiences severe temperature changes, condensation may form water in the engine. Condensation on engine components may cause rust. When rust inhibitors are added to the oil, these inhibitors help to disperse moisture from the metal surfaces. Certain acids may be formed in small quantities during the combustion process, and these acids get past the rings into the engine oil. Acids cause corrosion on engine components as they circulate through the lubrication system with the oil. Acids in the oil are neutralized by corrosion inhibitors added to the oil.

Extreme Pressure Resistance

When an engine is operating under heavy load, the pistons and connecting rods produce an extremely high downward force on the crankshaft journals. Under this condition, the film of oil in the connecting rod and main bearings may be subjected to pressures up to 1,000 psi (6,900 kPa). This high pressure tends to squeeze the oil out of the crankshaft bearings. If the oil is squeezed out of the bearings and the bearing insert contacts the crankshaft, bearing damage results. Extreme pressure resistance additives help to prevent the oil from being squeezed out of the bearings.

Detergents and Dispersants

Carbon is a normal by product of the combustion process. This carbon collects on the combustion chamber surfaces. However, some of this carbon collects on the top ring lands and compression rings. Detergents added to the oil help to clean these carbon deposits from the piston and rings, and this action keeps the rings working freely in the ring grooves.

A dispersant added to the engine oil helps to prevent carbon particles from sticking together to form larger particles. Oil passages may be plugged by large carbon particles. The smaller carbon particles are removed by the oil filter, while some larger particles may remain in the oil pan. These carbon particles in the oil pan are removed when the oil is changed.

Oxidation Inhibitors

The rotation of the crankshaft causes oil agitation in the crankcase. Air combines with high-temperature oil when the oil is agitated. During this combination of air and high-temperature oil, the oxygen in the air combines with the oil and forms a sticky substance that is similar to tar and contains corrosive compounds. This tar-like material may plug oil passages or destroy the oil film on some of the engine components. The corrosive compounds may attack and erode bearing materials. Oxidation of engine oil is reduced by oxidation inhibitors added to the oil.

Viscosity

In addition to oil additives, oil **viscosity** is equally important in selecting an engine oil. The ability of an oil to flow is its viscosity. Viscosity is affected by temperature. For example, hot oil flows faster than cold oil. The rate of oil flow is important to the life of an engine. Because an engine operates under a wide range of temperatures, selecting the correct viscosity becomes even more important.

Shop Manual
Chapter 4, page 129

To standardize oil viscosity ratings, the **Society of Automotive Engineers** (**SAE**) has established an oil viscosity classification system that is accepted throughout the industry (Figure 4-22). This system is a numeric rating in which the higher viscosity, or heavier weight oils, receive the higher numbers. For example, an oil classified as an SAE 50 weight oil is heavier and flows slower than SAE 10 weight oil. Heavyweight oils are best suited for use in high-temperature regions. Low-weight oils work best in low temperature operations.

To meet the needs of the average motorist who might not want to change oils seasonally, oil manufacturers have developed **multiviscosity** oils. These oils carry a combined classification such

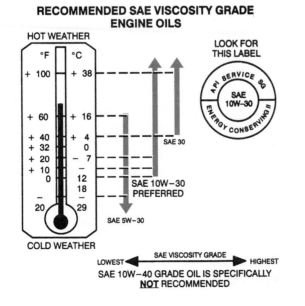

Figure 4-22 SAE oil viscosity ratings in relationship to ambient temperatures. *(Courtesy of Cadillac Motor Division—GMC)*

A - API service
B - SAE viscosity
C - Energy conserving
D - API service
E - Part number

Figure 4-23 Typical oil container. *(Courtesy of The Valvoline Company)*

as 10W-30. This classification means that the oil has a weight of 10 at ambient air temperature. Once the engine builds up heat, the oil warms up to the viscosity of the 30 weight oil. This type of oil allows easy starting in cold weather and adequate protection at higher temperatures.

The SAE classification and the API rating are usually indicated on the oil bottle or on the top of the can (Figure 4-23). Selecting oils that specifically meet or exceed the manufacturer's recommendations and changing the oil on a regular basis will allow the owner to get the maximum service life from an engine.

> ⚠ **WARNING:** Using the incorrect grade or type of oil can cause a variety of problems. Using an oil with the incorrect service rating can result in poor protection for engine bearings and other moving engine parts. This is especially true of oil for today's engines. Because of computerized controls, these engines start quickly. Oil for these engines must be able to circulate through the engine very quickly. This is why manufacturers recommend lower viscosity oils for these engines.

Engine oils can be classified as **energy-conserving** (fuel-saving) **oils**. These are designed to reduce friction, which in turn reduces fuel consumption. Friction modifiers and other additives are used to achieve this. Energy-conserving oil is identified as such on the top of the oil can. Some oils are specially formulated for turbocharged engines. This oil is identified on the container with a marking such as "Turbo Formula."

Synthetic and Recycled Oils

Synthetic oils are recommended by a few engine manufacturers. These oils are manufactured in a laboratory rather than being refined. Synthetic oils have improved stability over a wide temperature range, and they have improved viscous friction qualities. When synthetic oil is used in an engine, there may be a slight improvement in fuel economy and cold weather cranking speed. The cost of synthetic oil is considerably higher than the cost of refined oil.

The introduction of synthetic motor oils dates back to World War II. It is often described as the oil of the future. Synthetic oils are manufactured in a laboratory rather than pumped out of the ground. They offer a variety of advantages over natural oils including better fuel economy, stability over a wide range of temperatures and operating conditions, and longevity. They are more costly than natural oil.

Engine oil that is based on petroleum and synthetic products is available. These oils combine some of the better attributes of synthetics with the lower costing petroleum-based oils. Typically these oils are 80% petroleum and 20% synthetic.

Recycled oil is used oil that has been processed to remove impurities and restore oil to original standards. Many oil recycling plants have been built in North America, and the recycled oil from these plants should be equivalent to new refined oil. Most states have strict laws regarding the dumping of used oil.

Lubricating Systems

The main components of a typical lubricating system are described here.

Oil Pump

The oil pump is the heart of the lubricating system. Just as the heart in a human body circulates blood through veins, an engine's oil pump circulates oil through passages in the engine.

The oil pump is usually located in the oil pan. Its purpose is to supply oil to the various moving parts in the engine. To make sure the parts are lubricated, an adequate amount of oil must be delivered to the parts. To get the oil to move through the engine and into the various parts, the oil must be pressurized. Oil pumps are designed to move a certain volume of oil and pressurize it to a certain amount. The amount of oil flow through the engine depends on the volume of oil available, the pressure of the oil, and the clearance or space the oil must flow through (Figure 4-24).

Oil Pump Pickup

The **oil pump pickup** is a line from the oil pump to the oil stored in the oil pan (Figure 4-25). It usually contains a filter screen, which is submerged in the oil at all times. The screen serves to keep large particles from reaching the oil pump.

Oil Pan or Sump

The oil pan attaches to the crankcase or block. It serves as the reservoir for the engine's oil. It is designed to hold the amount of oil that is needed to lubricate the engine when it is running, plus a reserve. The oil pan helps to cool the oil through its contact with the outside air.

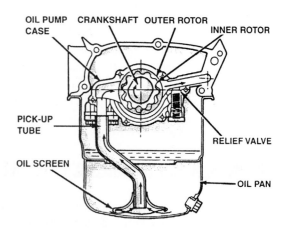

Figure 4-25 Oil pump and pickup assembly. *(Courtesy of Chrysler Corporation)*

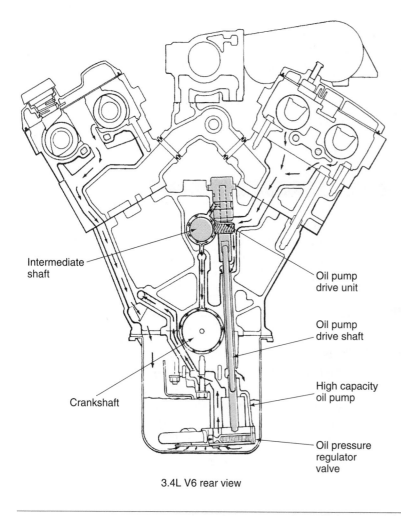

Intermediate shaft

Oil pump drive unit

Oil pump drive shaft

Crankshaft

High capacity oil pump

Oil pressure regulator valve

3.4L V6 rear view

Figure 4-24 Oil pump and drive assembly. *(Courtesy of Oldsmobile Division—GMC)*

Shop Manual
Chapter 4, page 129

Pressure Relief Valve

Since the oil pump is a **positive displacement pump**, an oil **pressure relief valve** (Figure 4-26) is included in the system to prevent excessively high system pressures from occurring as engine speed is increased. Once oil pressure exceeds a preset limit, the spring-loaded pressure relief valve opens and allows the excess oil to bypass the rest of the system and return directly to the sump.

Oil Filter

Shop Manual
Chapter 4, page 128

Under pressure from the oil pump, oil flows through a filter (Figure 4-27) to remove any impurities that might have become suspended in the oil. This prevents impurities from circulating through the engine, which can cause premature wear. Filtering also increases the usable life of the oil. Oil filters are equipped with a bypass valve to allow oil to bypass the filter, if the filter is clogged.

Engine Oil Passages or Galleries

From the filter, the oil flows into the engine oil galleries (Figure 4-28). These galleries consist of interconnecting passages that have been drilled completely through the engine block during man-ufacturing. The outside ends of the passages are blocked off so the oil can be routed through these galleries to various parts of the engine. The crankshaft also contains oil passages (oilways) to route the oil from the main bearings to the connecting rod bearing surfaces.

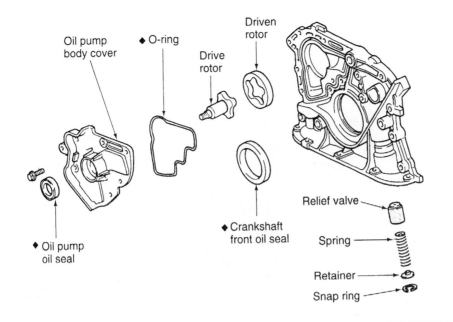

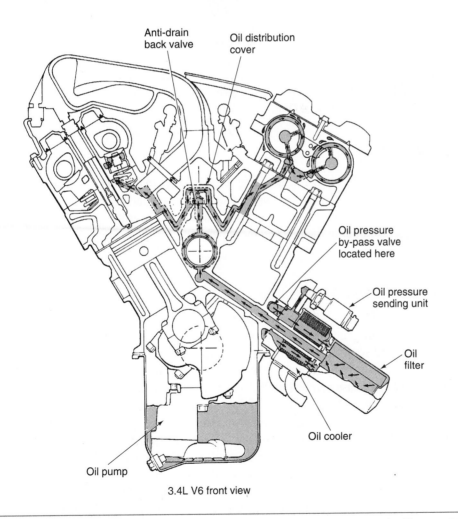

Figure 4-26 Oil pressure relief valve.

3.4L V6 front view

Figure 4-27 Oil passes from the filter through the antidrain back valve to the lifters and the camshaft journals. *(Courtesy of Oldsmobile Division—GMC)*

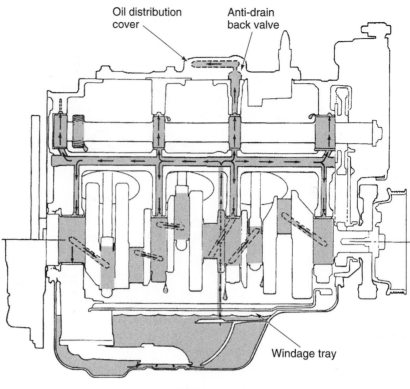

Oil distribution cover

Anti-drain back valve

Windage tray

3.4L V6 side view

Figure 4-28 Oil passes from the main oil gallery to the main bearings and connecting rod journals. *(Courtesy of Oldsmobile Division—GMC)*

Engine Bearings

Since oil is delivered to the engine bearings by an oil gallery, an oil hole is machined in the bearing for alignment with the oil gallery in the engine block. Many engine bearings are manufactured with an oil groove to help distribute the oil over the surface of the bearing. Once the oil has been used by the bearing, it flows out of the oil clearance space and is replenished with a fresh supply of oil under pressure from the oil pump.

Crankcase Ventilation

Shop Manual
Chapter 4, page 128

The true purpose of the PCV system is often overlooked as it is typically considered an emission control device. The PCV is an emission control device but, most importantly, it relieves the engine's crankcase of pressure and vapors.

Crankcase ventilation is necessary because pressure can build in the crankcase due to combustion pressure. This pressure passes by the piston rings. Piston rings do not provide a total seal of the combustion area and some combustion gases are able to reach the oil pan. These gases contaminate the oil and apply unwanted pressure on gaskets and seals.

The positive crankcase ventilation (PCV) system removes gases from the crankcase and delivers them back into the intake manifold and the cylinders. In the PCV system, a clean air hose is connected from the air cleaner to the rocker arm cover. A small filter is mounted on the end of this clean air hose inside the air cleaner. The PCV valve hose is connected from the opposite rocker arm cover to one end of the PCV valve, and the other end of this valve is connected to the intake manifold. Inside the PCV, a spring-loaded, tapered valve controls the flow of crankcase vapors in relation to engine vacuum and load. With the engine idling, the PCV valve is nearly closed. The valve moves toward the open position as the throttle is opened and engine load increases. Intake manifold vacuum moves clean air from the air cleaner into the crankcase and the crankcase gases are moved through the PCV valve into the intake manifold (Figure 4-29).

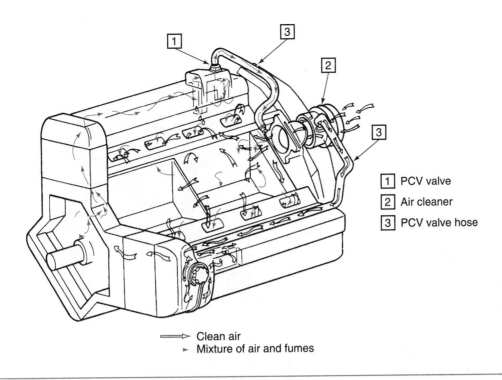

1	PCV valve
2	Air cleaner
3	PCV valve hose

⟹ Clean air
▸ Mixture of air and fumes

Figure 4-29 PCV system operation. *(Courtesy of Cadillac Motor Division—GMC)*

Oil Pressure Indicator

The driver can monitor oil pressure by looking at a gauge, which indicates the engine oil pressure at all times, or it can be a warning light that will come on whenever the engine is running with insufficient oil pressure. The warning light is the most common oil pressure indicator.

⚠ **WARNING:** Never operate an engine when the oil pressure indicator shows low oil pressure. If the oil supply to the engine components is reduced, severe engine damage occurs.

Oil Seals and Gaskets

These are used throughout the engine to prevent both external and internal oil leaks. The most common materials used for sealing are synthetic rubber, soft plastics, fiber, and cork. In critical areas these materials might be bonded to a metal.

Dipstick

The dipstick is used to measure the level of oil in the oil pan. The end of the stick is marked to indicate when the engine oil level is correct. It also has a mark to indicate the need to add oil to the system (Figure 4-30).

Oil Coolers

Some engines, such as diesel engines or turbocharged engines, use an external oil cooler. These typically look like a small radiator mounted near the front of the engine. Heat is removed from the oil as air flows through the cooler. Normal maximum engine oil temperature is considered to be 250°F. Hot oil mixed with oxygen breaks down (oxidizes) and forms carbon and varnish. The higher the temperature, the faster these deposits build. An oil cooler helps keep the oil at its normal operating temperature.

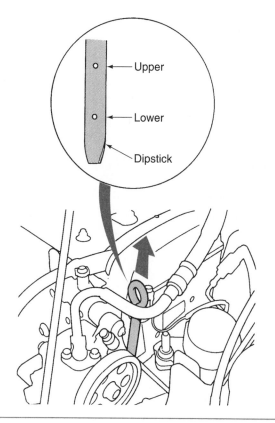

Figure 4-30 Typical dipstick markings. *(Courtesy of American Honda Motor Company, Inc.)*

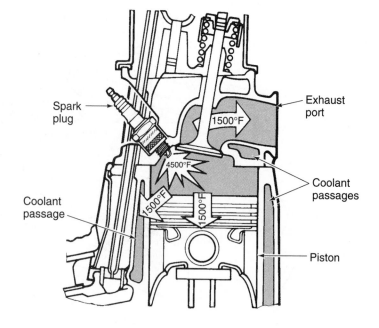

Figure 4-31 Heat is transferred from the cylinders to the coolant circulating through the coolant jackets. *(Courtesy of Chrysler Corporation)*

Cooling Systems

During the combustion process, temperatures can reach 4,500°F (2,468°C). While the engine is idling or operating at moderate speeds, the average combustion chamber temperature is 2,000°F (1,080°C). These temperatures in the cylinder are high enough to melt aluminum pistons, distort the cylinder walls, warp the cylinder head, and destroy the lubricating oil. Coolant circulation through the jackets in the cylinder head and block must absorb heat and lower component temperatures to a safe operating level (Figure 4-31).

Heat may be transferred by conduction, convection, or radiation in an engine cooling system. However, most of the heat is transferred by conduction from the hot engine components to the lower temperature coolant. Coolant temperature also affects combustion and engine efficiency. Some computer system input sensors also depend on correct coolant temperature to produce a normal signal.

Liquid-Cooled System

By far, the most popular and efficient method of engine cooling is the liquid-cooled system (Figure 4-32). In this system, heat is removed from around the combustion chambers by a heat-absorbing liquid (coolant) circulating inside the engine. This liquid is pumped through the engine and, after absorbing the heat of combustion, flows into the radiator where the heat is transferred to the atmosphere. The cooled liquid is then returned to the engine to repeat the cycle. These systems are designed to keep engine temperatures within a range where they provide peak performance.

The engine's thermostat controls the temperature and amount of coolant entering the radiator. While the engine is cold, the thermostat remains closed, allowing coolant to circulate only inside the engine. This allows the engine to warm up uniformly and eliminates hot spots. When the

Heat transfer by conduction means the heat transfers to a cooler object.

Heat is transferred by convection when heated atoms or molecules of fluid become less dense and lighter than the cooler atoms or molecules. The heated atoms or molecules rise upward, while the cooler portion of the fluid sinks downward.

Heat is transferred by radiation when heat waves travel through the atmosphere and strike another object causing a warming effect on that object.

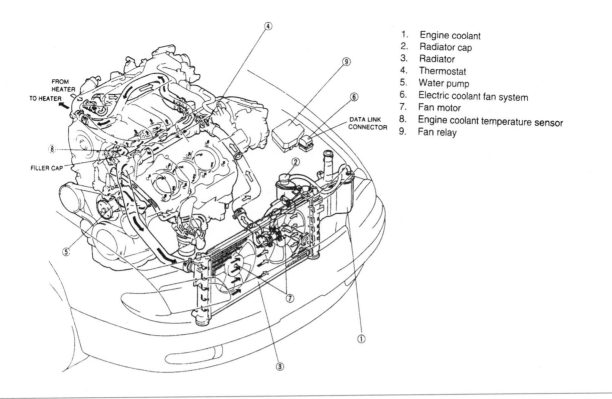

1. Engine coolant
2. Radiator cap
3. Radiator
4. Thermostat
5. Water pump
6. Electric coolant fan system
7. Fan motor
8. Engine coolant temperature sensor
9. Fan relay

DATA LINK CONNECTOR

FROM HEATER
TO HEATER

FILLER CAP

Figure 4-32 Major components of a typical liquid-cooling system. Arrows show coolant flow. *(Courtesy of Mazda)*

coolant reaches the opening temperature of the thermostat, the thermostat begins to open and allows the flow of coolant to the radiator. The hotter the coolant gets, the more the thermostat opens, allowing more coolant to flow through the radiator. Once the coolant has passed through the radiator and has given up its heat, it reenters the water pump. Here it is again pushed through the passages surrounding the combustion chambers to pick up heat and start the cycle once again.

The thermostat also maintains proper engine temperatures to keep the emission control systems working properly. Many vehicles fail the emissions test because they are operating at too low of a temperature.

Coolant

Coolant is a mixture of antifreeze and water. The antifreeze contains ethylene glycol and anticorrosion chemicals. If water is used alone in a cooling system, the water promotes rust, corrosion, and electrolysis. The water in the cooling system may become slightly acid because of minerals and metals in the cooling system. Metals such as brass, copper, aluminum, and cast iron alloy are contained in the cooling system. A small amount of electric current may flow from one of these metals to the other through the acid formed in the water. This electric current has a corrosive effect on cooling system metal components.

The antifreeze protects the coolant from freezing. Since water expands when it is frozen, the coolant jackets in the block and head may be cracked by frozen coolant. The radiator and heater core are also damaged by frozen coolant.

If the coolant boils, the liquid is no longer in contact with the cylinder walls and combustion chamber surfaces in the head. Under this condition, the heat is no longer transferred from these surfaces to the coolant, and the cylinder walls and combustion chamber surfaces become overheated. This action may score or collapse pistons and crack or warp cylinder heads.

A mixture of 50% antifreeze and water has a boiling point of approximately 230°F (103°C) at atmospheric pressure compared to water, which has a boiling point of 212°F (100°C). This mixture

Shop Manual
Chapter 4, page 129

Ethylene glycol is the basic liquid in automotive antifreeze.

of antifreeze and water provides extra protection from boiling. The same mixture of 50% antifreeze and water provides freezing protection to approximately -40°F (-35.5°C).

Most cooling systems today use an **expansion** or **recovery tank** (Figure 4-33). Cooling systems with expansion tanks are called closed-cooling systems. They are designed to catch and hold any coolant that passes through the pressure cap when the engine is hot. As the engine warms up, the coolant expands. This eventually causes the pressure cap to release. The coolant passes to an expansion tank. When the engine is shut down, the coolant begins to shrink. Eventually, the vacuum spring inside the pressure cap opens and the coolant in the expansion tank is drawn back into the cooling system.

The coolant recovery system is an emission control system that prevents spillage of coolant, which is a hazardous material.

⚠ **WARNING:** When working on the cooling system, remember that at operating temperature the coolant is extremely hot. Touching the coolant or spilling the coolant can cause serious body burns. Never remove the radiator cap when the engine is hot.

Water Pump

The heart of the cooling system is the water pump. Its job is to move the coolant through the cooling system. Typically the water pump is driven by the crankshaft through pulleys and a drive V-belt or ribbed V-belt. The pumps are centrifugal-type pumps (Figure 4-34), with a rotating paddle wheel type impeller to move the coolant. The shaft is mounted in the water pump housing and rotates on bearings. The pump contains a seal to keep the coolant from passing through it. At the drive end, the exposed end, a pulley is mounted to accept the belt. The pulley is driven by the crankshaft. The pump housing usually includes the mounting point for the lower radiator hose.

When the engine is started, the pump impeller pushes the water from its pumping cavity into the engine block. When the engine is cold, the thermostat is closed. This stops the coolant from reaching the top of the radiator. In order for the water pump to circulate the coolant through the engine during warmup, a bypass passage is added below the thermostat, which leads back to the water pump. This passage must be kept free to eliminate hot spots in the engine during warmup. It also allows hot coolant to pass through the valve, which will open the thermostat when it reaches the proper temperature.

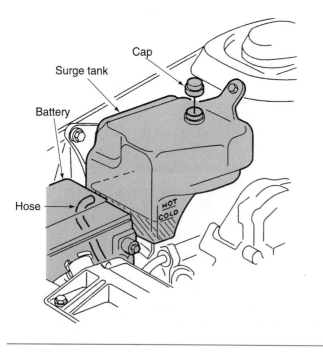

Figure 4-33 Coolant recovery system. *(Courtesy of Oldsmobile Division—GMC)*

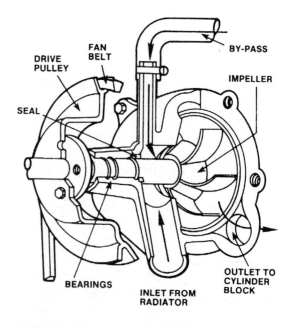

Figure 4-34 Impeller-type water pump.

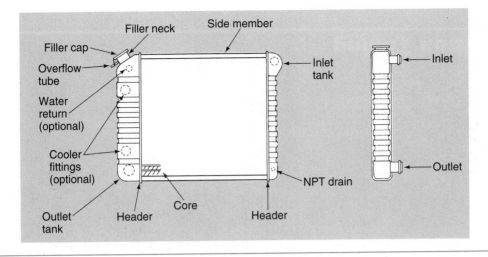

Figure 4-35 Cross flow radiator. *(Reprinted with permission from SAE 1983 Handbook, Volume 3, Society of Automotive Engineers, Inc.)*

Radiator

The radiator is basically a heat exchanger, transferring heat from the engine to the air passing through it. The radiator itself is a series of tubes and fins that expose the heat from the coolant to as much surface area as possible. This maximizes the potential of heat being transferred to the passing air. The inlet tank on the radiator contains a baffle to distribute and de-aerate the coolant.

Factors influencing the efficiency of the radiator are the basic design of the radiator, the area and thickness of the radiator core, the amount of coolant going through the radiator, and the temperature of the cooling air.

A radiator's efficiency can be greatly increased by increasing the difference between the temperature of the coolant and the outside air flowing through it. This can be done only by raising the temperature of the coolant. Doing this permits the use of a smaller radiator or the use of the same size radiator to cool a larger engine. This is the primary reason manufacturers have been specifying higher start-to-open temperatures for thermostats and higher pressure ratings for radiator pressure caps.

The radiator is usually based on one of these two designs: cross flow or down flow. In a cross flow radiator (Figure 4-35), coolant enters on one side, travels through tubes, and collects on the opposite side. In a down flow radiator (Figure 4-36), coolant enters the top of the radiator and is

Shop Manual
Chapter 4, page 131

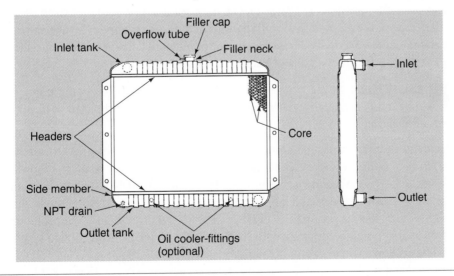

Figure 4-36 Down flow radiator. *(Reprinted with permission from SAE 1983 Handbook, Volume 3, Society of Automotive Engineers, Inc.)*

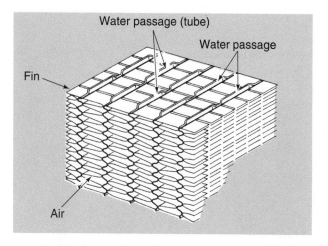

Figure 4-37 Cellular radiator core. *(Reprinted with permission from SAE 1983 Handbook, Volume 3, Society of Automotive Engineers, Inc.)*

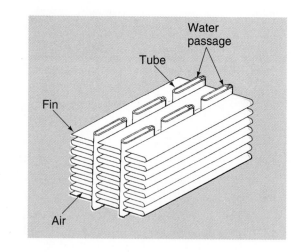

Figure 4-38 Tube and fin radiator core. *(Reprinted with permission from SAE 1983 Handbook, Volume 3, Society of Automotive Engineers, Inc.)*

drawn downward by gravity. Cross flow radiators are seen most often on large-engine or late-model cars because all the coolant flows through the fan air stream, which provides maximum cooling.

There are two types of radiator core construction: honeycomb or cellular type (Figure 4-37) and tube and fin (Figure 4-38). There are also two radiator core construction materials used—the copper/brass, soft solder coolers and the vacuum-brazed, aluminum cores cinched to nylon tanks.

The aluminum type is thin, lightweight, and less costly. (The initial price of aluminum is less than copper.) For these reasons, aluminum core radiators are being used more and more on newer model vehicles.

Most radiators feature petcocks or plugs that allow a technician to drain coolant from the system. Coolant is added to the system at the radiator cap or the recovery tank depending on the type of system being used.

Shop Manual
Chapter 4, page 133

Radiator Pressure Caps

Atmospheric pressure is 14.7 psi at sea level, and this pressure decreases as elevation increases above sea level. At elevations below sea level, atmospheric pressure increases. Water boils at 212°F (100°C) at sea level, and this boiling point decreases at higher elevations with reduced atmospheric pressure. The radiator cap contains a spring-loaded lower sealing gasket that contacts the lower sealing surface in the radiator filler neck (Figure 4-39).

> **CAUTION:** Never loosen the radiator cap on a warm or hot engine. This action may result in coolant boiling and severe burns to the technician or anyone standing near the vehicle.

On many radiator caps, the spring and sealing gasket are designed to maintain 15 psi (103 kPa) in the cooling system. While the cooling system pressure remains below 15 psi (103 kPa), the spring keeps the sealing gasket closed. For each 1 psi (6.8 kPa) of pressure increase, the boiling point of the water increases approximately 3°F (1.6°C). Therefore, the boiling point of water at 15 psi (103 kPa) is about 257°F (125°C). The 50% antifreeze and water solution increases the coolant boiling point to approximately 266°F (134°C). A pressurized cooling system provides additional protection against coolant boiling.

If the coolant overheats and the cooling system pressure exceeds 15 psi (103 kPa), this pressure forces the sealing gasket open against the spring tension. This action allows coolant to escape from the filler neck through the overflow tube into the coolant recovery system.

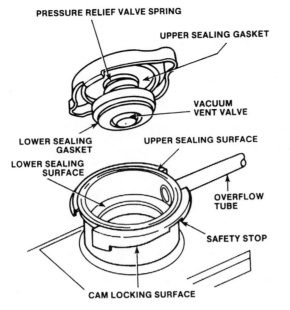

Figure 4-39 Radiator pressure cap assembly. *(Courtesy of Chrysler Corporation)*

Water Outlet

The water outlet is the connection between the engine and the upper radiator hose through which hot coolant from the engine is pumped back into the radiator. The water outlet has been called a gooseneck, elbow, inlet, outlet, or thermostat housing. Generally, it covers and seals the thermostat and, in some cases, includes the thermostat bypass.

Hoses

As already mentioned, the primary function of a cooling system hose is to carry the coolant between different elements of the system. Nearly all automobiles have an upper and lower radiator hose and two heater hoses. Some may also have a bypass hose. The majority of cooling system hoses are made of butyl or neoprene rubber. Many hoses are wire reinforced, and some are used to connect metal tubing between the engine and the radiator. Normally, radiator hoses are designed with expansion bends to protect radiator connections from excessive engine motion and vibration.

The upper radiator hose is subjected to the roughest service life of any hose in the cooling system. It must absorb more engine motion than any of the other hoses. It is exposed to the coolant at its hottest stage, plus it is insulated by the hood during hot soak periods. These conditions make the upper hose the most probable to fail.

Nearly all original equipment radiator hose is of the molded, curved design. Aftermarket products may be of this type or of a wire inserted flex type. This flex-type hose allows greater vehicle coverage per part number, but may not be designed for some cars that require radical bends and shapes. Lower radiator hoses are normally wire reinforced to prevent collapse due to the suction of the water pump. Hose clamps are used to secure the hoses to the inlets and outlets of the cooling system.

Shop Manual
Chapter 4, page 131

Thermostat

An automotive thermostat works somewhat like a typical home thermostat. It attempts to control the engine's operating temperature by routing the coolant either to the radiator or through the bypass or sometimes by a combination of both.

Shop Manual
Chapter 4, page 135

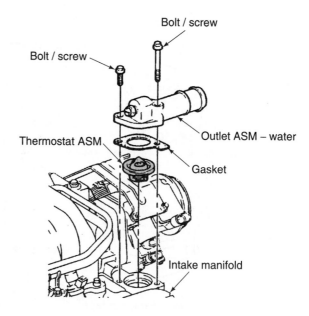

Figure 4-40 Typical thermostat location. *(Courtesy of Oldsmobile Division—GMC)*

Today's thermostat is composed of a specially formulated wax and powdered metal pellet, which is tightly contained in a heat-conducting copper cup equipped with a piston inside a rubber boot. Heat causes the wax pellet to expand, forcing the piston outward, which opens the valve of the thermostat. The pellet senses temperature changes and opens and closes the valve to control coolant temperature and flow. Today's thermostats are also designed to slow down coolant flow when they are open. This helps to prevent overheating, which can result from the coolant moving too quickly, through the engine, to absorb enough heat.

While the thermostat might be situated in several locations, the most common spot is at the front of the engine on top of the engine block (Figure 4-40). The heat element fits into a recess in the block where it will be exposed to hot coolant. The top of the thermostat is then covered by the water outlet housing, which holds it in place and provides a connection to the upper radiator hose.

There are two basic types of thermostats on the market today. Both function in the same manner, but each has distinct differences.

The reverse poppet thermostat (Figure 4-41A) opens against the flow of the coolant from the water pump. The coolant, under water pump pressure, is used to help the reverse poppet thermostat stay closed when it is cool. The valve is self-aligning and also self-cleaning.

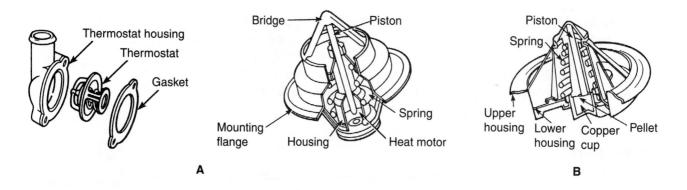

A   B

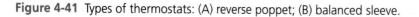

Figure 4-41 Types of thermostats: (A) reverse poppet; (B) balanced sleeve.

The balanced sleeve thermostat (Figure 4-41B) eliminates pressure shocks by allowing pressurized coolant to circulate around all of its working parts. By reducing pressure shocks, it provides coolant temperature stability during the most difficult operating conditions.

The thermostat permits fast warmup of the engine after it has been started. Slow warmup causes moisture condensation in the combustion chambers, which finds its way into the crankcase and causes sludge formation. The thermostat keeps the coolant above a designated minimum temperature required for efficient engine performance. Most engines are equipped with a coolant bypass, either outside the engine block or built into the casting. Some thermostat models are equipped with a bypass valve that shuts off the engine bypass after warmup, forcing all coolant to flow to the radiator.

Thermostats must start to open at a specified temperature, normally 3°F above or below its temperature rating. It must be fully opened at about 20°F above the "start to open" temperature. They must also permit the passage of a specified amount of coolant when fully open and leak no more than a specified amount when fully closed.

Belt Drives

Belt drives (Figure 4-42) have been used for many years. V-belts and serpentine belts are used to drive water pumps, power steering pumps, airconditioning compressors, alternators, and emission control pumps (Figure 4-43). The flexibility of belt drives has allowed many inexpensive improvements in the modern automobile.

The belt system is popular for a number of reasons. It is very inexpensive and quiet when compared to chains and gears. It is also very easy to repair. Because the belts are flexible, they will absorb some shock loads and cushion shaft bearings from excessive loads.

Serpentine belts are kept under constant tension by an idler pulley or tensioner. This pulley is designed to compensate for belt stretch and to keep a proper amount of tension on the belt.

Many drive units consist of two matched belts that run over parallel pulleys. If one pulley or belt should wear excessively, the belts will run at slightly different speeds, thus one will be constantly slipping. Matched belts should always be replaced in pairs so they wear together, thus maintaining the same length to prevent slippage and problems.

Fans and Fan Clutches

The efficiency of the cooling system is based on the amount of heat that can be removed from the system and transferred to the air. The system needs air. At highway speeds, the ram air through the

Shop Manual
Chapter 4, page 135

Serpentine belts are also called ribbed V-belts.

Shop Manual
Chapter 4, page 136

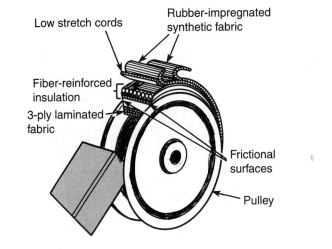

Figure 4-42 V-belt construction. *(Courtesy of Chrysler Corporation)*

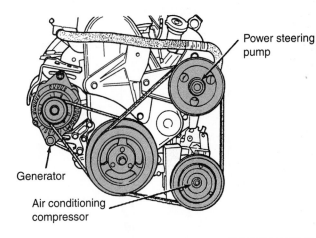

Figure 4-43 A single belt may drive many components. *(Courtesy of Chrysler Corporation)*

radiator should be sufficient to maintain proper cooling. At low speeds and idle, the system needs additional air. This air is delivered by a fan.

The design of fans varies with the air requirements of the engine's cooling system. Diameter, pitch, and the number of blades can be varied to attain the needed flow. Fans are usually shrouded to maintain efficiency by causing all of the flow to pass through the radiator. This shrouding also allows the fan to be placed relatively far from the radiator, thus allowing the engine to be set back in the car while the radiator is left near the front of the vehicle.

Five- or six-blade fans are found on some cars. Fans may be made of steel, nylon, or fiberglass, and are precisely balanced to prevent water pump bearing and seal damage. Since fan air is usually only necessary at idle and low speed operation, various design concepts are used to limit the fan's operation at higher speeds. Horsepower is required to turn the fan. Therefore, the operation of a cooling fan reduces the available horsepower to the drive wheels, as well as the fuel economy of the vehicle. Fans are also very noisy at high speeds, adding to driver fatigue and total vehicle noise.

In an effort to overcome these problems, some vehicle manufacturers use flexible blades or **flex-blades** (Figure 4-44) that bend or change pitch based on engine speed. That is, at slower speeds the blade pitch is at the maximum. As engine speed increases, the blade pitch decreases as does the horsepower losses and noise levels.

In most late-model applications, to save power and reduce the noise level, the conventional belt-driven, water-pump-mounted engine coolant fan has been replaced with an electrically driven fan (Figure 4-45). This fan and motor are mounted to the radiator shroud and are not connected mechanically or physically to the engine. The 12-volt, motor-driven fan is electrically controlled by either, or both, of two methods: an engine coolant temperature switch or sensor and the airconditioner switch.

Following the schematic in Figure 4-46, the cooling fan motor is connected to the 12-volt battery supply through a normally open (NO) set of contacts (points) in the cooling fan relay. Protection for this circuit is provided by a fusible link (F/L). During normal operation, with the air conditioner off and the engine coolant below a predetermined temperature of approximately 215°F, the relay contacts are open and the fan motor does not operate.

Should the engine coolant temperature exceed approximately 230°F, the engine coolant temperature switch closes. This energizes the fan relay coil, which in turn closes the relay contacts. The contacts provide 12 volts to the fan motor if the ignition switch is in the on position. The 12-volt supply for the relay coil circuit is independent of the 12-volt supply for the fan motor circuit. The

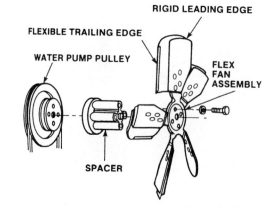

Figure 4-44 Flexible fan.

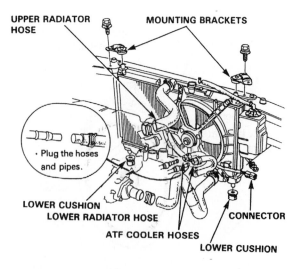

Figure 4-45 Electrically driven cooling fan. *(Courtesy of American Honda Motor Company, Inc.)*

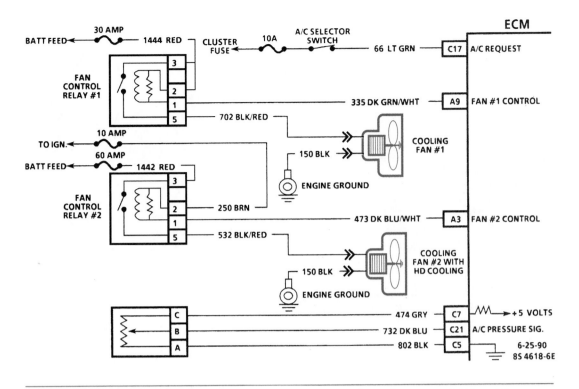

Figure 4-46 Electric cooling fan circuit with two cooling fans. *(Courtesy of Oldsmobile Division—GMC)*

coil circuit is from the on terminal of the ignition switch, through a fuse in the fuse panel, and to ground through the relay coil and temperature sensor.

Should the airconditioner select switch be turned to any cool position, regardless of engine temperature, a circuit is completed through the relay coil to ground through the select switch. This action closes the relay contacts to provide 12 volts to the fan motor. The fan then operates as long as the airconditioner and ignition switches are on.

There are many variations of electric cooling fan operation. Some provide a cool-down period whereby the fan continues to operate after the engine has been stopped and the ignition switch is turned off. The fan stops only when the engine coolant falls to a predetermined safe temperature, usually about 210°F. In some systems, the fan does not start when the airconditioner select switch is turned on unless the high side of the airconditioning (A/C) system is above a predetermined safe temperature.

Some late-model cars control the cooling fan by completing the ground through the engine control computer. Check the service manual to see how an electric cooling fan is controlled before working with it.

⚠ **WARNING:** The engine electric cooling fan can come on at any time without warning even if the engine is not running. For this reason, it is always wise to remove the negative terminal from the battery or the electric cooling fan connector while working around an electric fan. Make sure you reconnect the connector before giving the car back to the customer.

Another way of controlling fan noise and horsepower loss is by using a fan clutch (Figure 4-47). This unit connects the fan to the drive pulley (usually on the water pump shaft). The clutch slips at high speeds, therefore it is not turning at full engine speed.

Cooling fan clutches may be speed-controlled or temperature-controlled. The obvious advantage of the temperature-controlled unit is the fact that it knows when air is needed to cool the system. A silicone fluid couples the fan blade drive plate to a driven disc via a series of annular grooves

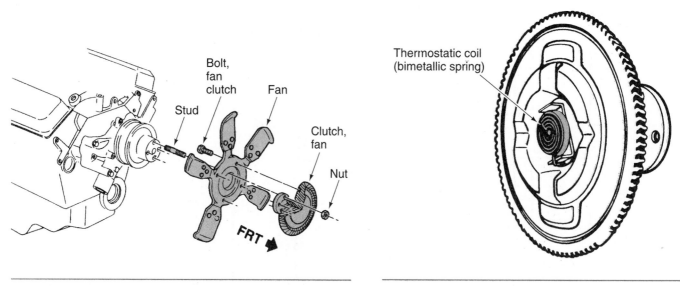

Figure 4-47 Viscous-drive fan clutch mounting. *(Courtesy of Chevrolet Motor Division—GMC)*

Figure 4-48 Thermostatic coil on a viscous-drive fan clutch. *(Courtesy of Chrysler Corporation)*

The bimetallic element is called a thermostatic coil.

in the two pieces. The fluid fills these grooves and drives the fan until the differential torque between the fan and the drive disc causes the fluid to shear or slip. The unit has a bimetallic element that senses the air temperature behind the radiator (Figure 4-48). This bimetal is calibrated to open and close a valve in the clutch that dispenses the silicone at particular temperatures. When the fluid is returned to its reservoir, the unit free wheels the fan until more cooling is required. Tests have proven that maximum cooling is necessary less than 10% of the time. Therefore, a temperature-controlled fan clutch saves fuel and reduces noise 90% of the time.

Water Jackets

Hollow passages in the block and cylinder heads surround the areas closest to the cylinders and combustion chambers (Figure 4-49). Coolant flow through the block and head can be in one of the following ways.

The most common engine using the **reverseflow** system is the GM 5.7L in Corvettes, Camaros, and Firebirds.

Series Flow System. In this system (Figure 4-50A), the coolant flows around all the cylinders on each bank as it flows to the rear of the block. Large main coolant passages at the rear of the block direct the coolant through the head gasket and to the head.

Parallel Flow System. In this system (Figure 4-50B), coolant flows into the block under pressure, then crosses the gasket to the head through main coolant passage openings beside each cylinder.

The warmer coolant at the bottom of the cylinders helps to reduce cylinder taper, which in turn reduces frictional horsepower loss. Lower frictional horsepower loss increases brake horsepower.

Series-Parallel Flow System. This system (Figure 4-50C) uses a combination of series and parallel flow systems. The cooling passages inside the engine are designed so that the whole system can be drained and that there are no pockets in which steam can form. Any steam that develops must be able to go directly to the top of the radiator, in order for the cooling system to function properly.

Reverse Flow System. This system is on a few high-performance engines. In it, the coolant flows through the cylinder heads first then it flows through the block. With the coolant flowing around the combustion chamber first, chamber temperatures are better maintained.

Included in the water jackets are soft (core) plugs and a block drain plug. The soft plugs and drain are usually removed during engine teardown. New ones are installed during reassembly. Core plugs are prone to rust and corrosion, and therefore will weep coolant or rust through completely. When this happens, the core plugs should be replaced.

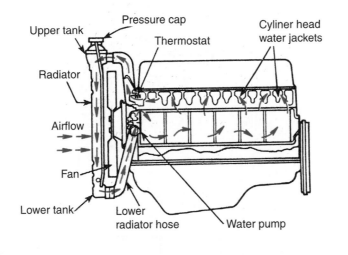

Figure 4-49 Coolant circulates through the engine.

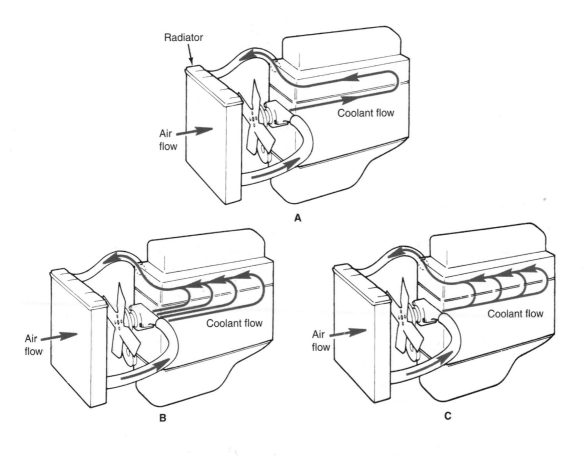

Figure 4-50 Different coolant flow designs: (A) series flow; (B) parallel flow; and (C) series-parallel flow.

Temperature Indicator

Coolant temperature indicators are mounted in the dashboard to alert the driver of an overheating condition. It consists of a temperature gauge and/or a light. A temperature sensor is screwed into a threaded hole in the water jacket. A schematic of an electric temperature gauge including an Instru-

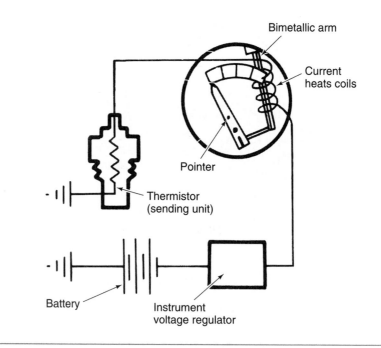

Figure 4-51 Temperature gauge circuit. *(Courtesy of Chrysler Corporation)*

ment Voltage Regulator is shown in Figure 4-51. Besides indicating coolant temperatures to the driver, temperature sensors supply some important information to today's computer-controlled engine control systems.

Heater System

A hot liquid passenger compartment heater is part of the engine's cooling system. Heated coolant flows from the engine through heater hoses and a heater control valve to a smaller heater core, or radiator, located in a hollow container on either side of the fire wall. Air is directed or blown over the hot heater core, and the heated air flows into the passenger compartment. Movable doors can be controlled to blend cool air with heated air for more or less heat.

Oil Cooler

Radiators used in vehicles with automatic transmissions have a sealed heat exchanger, or form of radiator, located in the coolant outlet tank of the regular radiator. Metal or rubber hoses carry hot automatic transmission fluid to the heat exchanger. The coolant passing over the sealed heat exchanger cools the fluid, which is then returned to the transmission. Cooling the transmission fluid is essential to the efficiency and durability of an automatic transmission.

Air-Cooled System

As mentioned at the beginning of this chapter, a few engines (such as the Porsche 911 and 930 models) use a cooling system that relies on air rather than a liquid to transfer heat from the engine to the atmosphere. Cylinders and heads have fins which are enclosed in a shroud to control airflow. The fins expose more of the surface area to airflow for better heat dissipation. Ducts and shrouds direct the airflow over the engine components, especially over the hotter cylinder head area. A belt-driven or electric blower provides for airflow. Fresh air is taken in and heated air expelled into the atmosphere. A thermostat connected to a control valve or door regulates airflow to control engine temperature.

Summary

❏ Automotive engines are classified by several different design features as operational cycles, number of cylinders, cylinder arrangement, valve train type, valve arrangement, ignition type, cooling system, and fuel system.

❏ The basis of automotive gasoline engine operation is the four-stroke cycle. This includes the intake stroke, compression stroke, power stroke, and exhaust stroke. The four strokes require two full crankshaft revolutions.

❏ The most popular engine designs are the in-line (in which all the cylinders are placed in a single row) and V-type (which features two rows of cylinders). The slant design is much like the in-line, but with the entire block placed at a slant. Opposed cylinder engines use two rows of cylinders located opposite the crankshaft.

❏ The two basic valve and camshaft placement configurations currently in use on four-stroke engines are the overhead valve and overhead cam. A third type, the flathead or side valve, was once popular but is no longer in use.

❏ Bore is the diameter of a cylinder and stroke is the length of piston travel between top dead center (TDC) and bottom dead center (BDC). Together these two measurements determine the displacement of the cylinder, which is the volume the cylinder holds between the TDC and BDC positions of the piston.

❏ Compression ratio is a measure of how much the air and fuel have been compressed. The higher the compression ratio is, the more power an engine can produce. The compression ratio of an engine must be suited to the fuel available. As compression ratio increases, the octane rating of the fuel must increase to prevent abnormal engine combustion.

❏ Horsepower is the rate at which torque is produced by an engine. The torque is then transmitted through the drivetrain to turn the driving wheels of the vehicle.

❏ Instead of relying on a spark for ignition, diesel engines use the heat produced by compression air in the combustion chamber to ignite the fuel. Three types of combustion chambers are used in diesel engines: open, precombustion, and turbulence.

❏ Features of both the gasoline and diesel engine are found in the stratified charge engine. Its major advantages are good part-load fuel economy, low exhaust emissions, and an ability to operate on low-octane fuel.

❏ In addition to the diesel and stratified charge, other automotive engines that may figure prominently in the future include the rotary or Wankel, Miller-cycle, and electric.

❏ An engine's lubrication system has several important purposes: hold an adequate supply of oil to cool, clean, lubricate, and seal the engine; remove contaminants from the oil; and deliver oil to all necessary areas of the engine.

❏ Engine oil additives include friction modifiers, antifoaming agents, corrosion and rust inhibitors, extreme pressure resistance, detergents and dispersants, and oxidation inhibitors.

❏ Society of Automotive Engineers (SAE) oil viscosity ratings compare oils in relation to their ability to flow.

❏ American Petroleum Institute (API) oil classification ratings indicate the type of service for which the oil is suitable.

❏ Synthetic oil is manufactured in a laboratory, and this oil has improved viscous friction qualities, but it is more expensive than refined oil.

❏ Recycled oil is used oil that has been recycled and restored to original specifications.

❏ The main components of a typical lubrication system are: an oil pump, oil pump pickup, oil pan, pressure relief valve, oil filter, engine oil passages, engine bearings, crankcase ventilation, oil pressure indicator, oil seals and gaskets, dipstick, and oil coolers.

Terms to Know

American Petroleum Institute (API)

Bore

Cam

Combustion chamber

Compression ratio

Coolant

Crank throw

Crude oil

Diesel engines

Displacement

Dual overhead camshaft (DOHC)

Efficiency

Energy-conserving oil

Expansion tank

Firing order

Flex-blade

Glow plug

Horsepower

Lobe

Multiviscosity

Overhead camshaft (OHC)

Overhead valve (OHV)

Oil pump pickup

Positive displacement pump

Pressure relief valve

Recovery tank

Reverse flow

Rotary

Society of Automotive Engineers (SAE)

Stroke

Torque

Viscosity

Wankel engine

- The purpose of the oil pump is to supply oil to the various moving parts in the engine.
- Because the faster the pump turns the greater the pressure becomes, a pressure-regulating valve is installed to control the maximum oil pressure.
- All automotive vehicles are equipped with either an oil pressure gauge or a low-pressure indicator light. The gauges are either mechanically or electrically operated.
- All oil leaving the oil pump is directed to the oil filter. The filter is a disposable metal container filled with a special type of treated paper or other filter substance that catches and holds the oil's impurities.
- The PCV system removes vapors from the crankcase and directs them into the intake manifold.
- Two basic types of cooling systems are utilized by automotive manufacturers: liquid-cooled systems and air-cooled systems. The most popular and efficient method of engine cooling is the liquid-cooled system.
- Heat may be transferred by conduction, convection, or radiation.
- A mixture of 50% ethylene glycol and water in the cooling system reduces acid formation, provides antifreeze protection, and increases the coolant boiling point.
- The function of the water pump is to move the coolant efficiently through the system. The radiator transfers heat from the engine to the air passing through it. The radiator is usually one of two designs: cross flow or down flow. The thermostat attempts to control the engine's operating temperature by routing the coolant either to the radiator or through the bypass or sometimes a combination of both.
- The radiator pressure cap pressurizes the cooling system and raises the coolant boiling point.
- V-belts and multiple-ribbed belts (called serpentine belts) are used to drive water pumps, power steering pumps, airconditioning compressors, alternators, and emission control pumps. The fan delivers additional air to the radiator to maintain proper cooling at low speeds and idle. Since fan air is usually only necessary at idle and low speed operation, various design concepts are used to limit the fan's operation at higher speeds.
- The hollow passages in the block and cylinder heads through which coolant flows may be arranged as a series flow system, a parallel flow system, a series-parallel, or reverseflow system.
- Flex-blade fans have blades that straighten out at higher speeds to reduce noise and the amount of power required to rotate the fan.
- Electric-drive cooling fans operate only when they are required to reduce the engine temperature.
- A viscous fan clutch is temperature operated, and it drives the fan faster when it is required to reduce coolant temperature.
- An engine temperature warning light is operated by a temperature switch in the cooling system.

Review Questions

Short Answer Essays

1. Explain how compression ratio is calculated.
2. Describe two different methods of camshaft drive in an overhead camshaft (OHC) engine.
3. Explain the advantage of a pressurized cooling system.
4. Describe thermostat purpose and operation.

5. Explain the reason for adding friction modifiers to engine oil.

6. Describe the difference between a 5W-40 oil and a 10W-30 oil.

7. How is the oil pump driven?

8. What is the name of the component in the lubrication system that prevents excessively high system pressures from occurring as engine speed increases?

9. Name the four strokes of a four-stroke cycle engine.

10. As an engine's compression ratio increases, what should happen to the octane rating of the gasoline?

Fill-in-the-Blanks

1. In most automotive applications, the water pump is driven by the _____.

2. The stroke of an engine is _____ the crank throw.

3. In a four-stroke engine, the exhaust valve opens _____ bottom dead center (BDC) on the power stroke.

4. In a four-stroke engine, the camshaft is rotating at _____ the crankshaft speed.

5. Lubricating oil cleans, cools, lubricates, and _____ engine components.

6. Engine displacement may be increased by increasing the bore diameter or the _____.

7. Higher octane gasoline may be required in an engine with a higher _____ _____.

8. For every pound per square inch (psi) of cooling system pressure, the boiling point of the coolant is increased about _____ degrees.

9. In a reverseflow cooling system, immediately after the coolant leaves the water pump, it flows through the _____ _____.

10. An automatic transmission cooler is usually mounted in the _____ _____.

ASE Style Review Questions

1. *Technician A* calculates compression ratio with the following formula:

$$\frac{\text{volume above the piston at BDC}}{\text{volume above the piston at TDC}}$$

Technician B calculates compression ratio with the following formula:

$$\frac{\text{total cylinder volume}}{\text{total combustion chamber volume}}$$

Who is correct?

A. A only C. Both A and B
B. B only D. Neither A nor B

2. While discussing ethylene glycol antifreeze:
Technician A says a solution of 50% ethylene glycol antifreeze and water provides protection against coolant freezing to -40°F (-35°C).
Technician B says a solution of 50% ethylene glycol antifreeze and water at atmospheric pressure has a boiling point of approximately 275°F (135°C).
Who is correct?

A. A only C. Both A and B
B. B only D. Neither A nor B

3. While discussing radiators:
   Technician A says the use of a cross flow radiator allows a more aerodynamic front body design.
   Technician B says some radiators have composite plastic tanks.
   Who is correct?
   A. A only   **C.** Both A and B
   B. B only   **D.** Neither A nor B

4. While discussing pressurized cooling systems:
   Technician A says the boiling point of a solution of 50% ethylene glycol and water is about 290°F (143°C) at 15 psi.
   Technician B says on many cooling systems if the radiator cap sealing gasket opens, coolant leaks out under the vehicle.
   Who is correct?
   A. A only   **C.** Both A and B
   B. B only   **D.** Neither A nor B

5. While discussing electric-drive cooling fans:
   Technician A says an electric-drive cooling fan requires more engine power for fan drive than a belt-driven cooling fan.
   Technician B says some electric-drive cooling fan circuits are designed to operate the fan continually when the air conditioning (A/C) is on.
   Who is correct?
   A. A only   **C.** Both A and B
   B. B only   **D.** Neither A nor B

6. While discussing viscous-drive fan clutches:
   Technician A says the viscous-drive fan clutch has more internal friction when the clutch is cold.
   Technician B says the viscous-drive fan clutch is connected between the water pump hub and the fan blade assembly.
   Who is correct?
   A. A only   **C.** Both A and B
   B. B only   **D.** Neither A nor B

7. While discussing oil additives:
   Technician A says friction modifiers in engine oils help to improve fuel economy.
   Technician B says oxidation inhibitors in engine oils help to prevent the formation of a sticky tar-like substance in the engine.
   Who is correct?
   A. A only   **C.** Both A and B
   B. B only   **D.** Neither A nor B

8. While discussing engine oils:
   Technician A says a 10W-30 oil pours more easily than a 10W oil at 0°F (-18°C).
   Technician B says a synthetic oil has improved stability over a wide temperature range compared to a refined oil.
   Who is correct?
   A. A only   **C.** Both A and B
   B. B only   **D.** Neither A nor B

9. *Technician A* says that the API classification S stands for standard passenger cars.
   Technician B says that the API classification C stands for vehicles with compression ignition.
   Who is correct?
   A. A only   **C.** Both A and B
   B. B only   **D.** Neither A nor B

10. *Technician A* says that the American Petroleum Institute has established an oil viscosity classification system.
    Technician B says that higher viscosity oils receive the higher rating numbers.
    Who is correct?
    A. A only   **C.** Both A and B
    B. B only   **D.** Neither A nor B

Ignition Systems

Upon completion and review of this chapter, you should be able to:

❑ Describe the three major functions of an ignition system.

❑ Name the operating conditions of an engine that affect ignition timing.

❑ Name the two major electrical circuits used in ignition systems and their common components.

❑ Define the purpose of the ignition coil.

❑ Define ignition dwell time, and explain its importance to ignition system operation.

❑ Explain how the high voltage is induced in the coil secondary winding.

❑ Explain the operation of a magnetic pick-up coil.

❑ Explain the operation of a Hall effect switch.

❑ Describe the various types of spark timing systems, including electronic switching

systems and their related engine position sensors.

❑ Describe spark plug internal design.

❑ Explain the difference between a hot spark plug and a cold spark plug.

❑ Describe spark plug wire internal design.

❑ Explain the operation of the centrifugal advance.

❑ Describe the operation of the vacuum advance.

❑ Describe the basic operation of distributor-based ignition systems.

❑ Describe the basic operation of distributorless ignition systems.

Introduction

One of the requirements for an efficient running engine is the correct amount of heat delivered into the cylinders at the right time. This requirement is the responsibility of the ignition system. The ignition system supplies properly timed, high-voltage surges to the spark plugs. These voltage surges cause combustion inside the cylinder.

The ignition system must create a spark, or current flow, across each pair of spark plug electrodes at the proper instant, under all engine operating conditions. This may sound relatively simple, but when one considers the number of spark plug firings required and the extreme variation in engine operating conditions, it is easy to understand why ignition systems are so complex.

If a six-cylinder engine is running at 4,000 revolutions per minute (rpm), the ignition system must supply 12,000 sparks per minute. Since the ignition system must fire three spark plugs per revolution, these plug firings must also occur at the correct time and generate the correct amount of heat. If the ignition system fails to do these things, fuel economy, engine performance, and emission levels will be adversely affected.

In order to have an efficient running engine there must be the correct amount of air mixed with the correct amount of fuel, in a sealed container, shocked by the correct amount of heat at the right time.

Purpose of the Ignition System

For each cylinder in an engine, the ignition system has three main jobs. First, it must generate an electrical spark that has enough heat to ignite the air-fuel mixture in the combustion chamber. Secondly, it must maintain that spark long enough to allow for the combustion of all the air and fuel in the cylinders. Lastly, it must deliver the spark to each cylinder so that combustion can begin at the right time during the compression stroke of each cylinder.

When the combustion process is completed, a very high pressure is exerted against the top of the piston. This pressure pushes the piston down on its power stroke. This pressure is the force that gives the engine power. In order for an engine to produce the maximum amount of power that

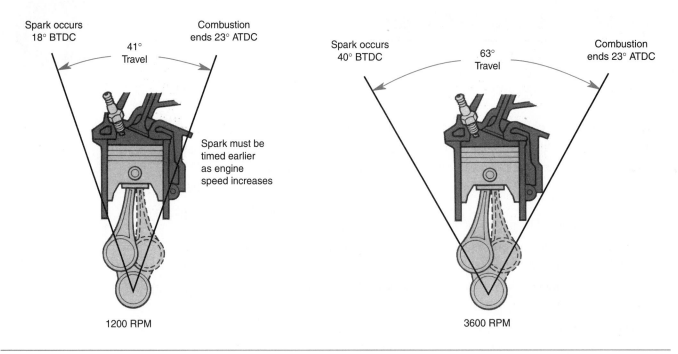

Spark occurs
18° BTDC

Combustion
ends 23° ATDC

41°
Travel

Spark must be
timed earlier
as engine
speed increases

1200 RPM

Spark occurs
40° BTDC

Combustion
ends 23° ATDC

63°
Travel

3600 RPM

Figure 5-1 With an increase in speed, ignition must begin earlier to end by 23 degrees ATDC. *(Reprinted with the permission of Ford Motor Company)*

it can, the maximum pressure from combustion should be present when the piston is at 10°–23° after top dead center (ATDC). Because combustion of the air-fuel mixture within a cylinder takes a short period of time, usually measured in thousandths of a second (milliseconds), the combustion process must begin before the piston is on its power stroke. Therefore, the delivery of the spark must be timed to arrive at some point before the piston reaches top dead center.

Determining how much before TDC the spark should begin gets complicated by the fact that as the speed of the piston as it moves from its compression stroke to its power stroke increases, the time needed for combustion stays about the same. This means that the spark should be delivered earlier as the engine's speed increases (Figure 5-1). However, as the engine has to provide more power to do more work, the load on the crankshaft tends to slow down the acceleration of the piston and the spark needs to be somewhat delayed.

Figuring out when the spark should begin gets more complicated with the fact that the rate of combustion varies according to certain factors. Higher compression pressures tend to speed up combustion. Higher octane gasolines ignite less easily and require more burning time. Increased vaporization and turbulence tend to decrease combustion times. Other factors, including intake air temperature, humidity, and barometric pressure, also affect combustion. Because of all of these complications, delivering the spark at the right time is a difficult task.

Ignition Timing

Shop Manual
Chapter 5, page 204

BTDC is the acronym for Before Top Dead Center.

TDC stands for top dead center or the uppermost position of the piston in its cylinder.

Ignition timing refers to the precise time that spark occurs. Ignition timing is specified by referring to the position of the number one piston in relationship to crankshaft rotation. Ignition timing reference timing marks can be located on engine parts and on a pulley or flywheel to indicate the position of the first piston (Figure 5-2). Vehicle manufacturers specify initial or **basic** ignition **timing**.

When the marks are aligned at TDC, or 0, the piston in cylinder one is at TDC of its compression stroke. Additional numbers on a scale indicate the number of degrees of crankshaft rotation **before TDC (BTDC)** or after TDC (**ATDC**). In a majority of engines, the initial timing is specified at a point between TDC and 20° BTDC.

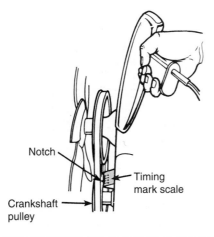

Figure 5-2 Reading ignition timing marks with a timing light.

If optimum engine performance is to be maintained, the ignition timing of the engine must change as the operating conditions of the engine change. All of the different operating conditions affect the speed of the engine and the load on the engine. All ignition timing changes are made in response to these primary factors.

Engine Speed

At higher engine speeds, the crankshaft turns through more degrees in a given period of time. If combustion is to be completed by 10° ATDC, ignition timing must occur sooner or be advanced.

However, air-fuel mixture turbulence increases with rpm. This causes the mixture inside the cylinder to turn faster. Increased turbulence requires that ignition must occur slightly later or be slightly retarded.

These two factors must be balanced for best engine performance. Therefore, while the ignition timing must be advanced as engine speed increases, the amount of advance must be decreased some to compensate for the increased turbulence.

Engine Load

The load on an engine is related to the work it must do. Driving up hills or pulling extra weight increases engine load. Under load, the pistons accelerate slower and the engine runs less efficiently. A good indication of engine load is the amount of vacuum formed during the intake stroke.

Under light loads and with the throttle plate(s) partially opened, a high vacuum exists in the intake manifold. The amount of air-fuel mixture drawn into the manifold and cylinders is small. On compression, this thin mixture produces less combustion pressure and combustion time is slow. To complete combustion by 10° ATDC, ignition timing must be advanced.

Under heavy loads, when the throttle is opened fully, a larger mass of air-fuel mixture can be drawn in and the vacuum in the manifold is low. High combustion pressure and rapid burning results. In such a case, the ignition timing must be retarded to prevent complete burning from occurring before 10° ATDC.

Engine vacuum is formed during the intake stroke of the cylinder. Engine vacuum is any pressure lower than atmospheric pressure.

Firing Order

Up to this point, the primary focus of discussion has been ignition timing as it relates to any one cylinder. However, the function of the ignition system extends beyond timing the arrival of a spark to a single cylinder. It must perform this task for each cylinder of the engine in a specific sequence.

Each cylinder of an engine produces power once every 720 degrees of crankshaft rotation. Each cylinder must have a power stroke at its own appropriate time during the rotation. To make

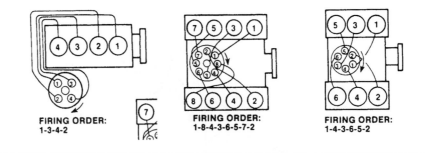

FIRING ORDER:
1-3-4-2

FIRING ORDER:
1-8-4-3-6-5-7-2

FIRING ORDER:
1-4-3-6-5-2

Figure 5-3 Cylinder numbering and firing orders.

this possible, the pistons and rods are arranged in a precise fashion. The **firing order** is arranged to reduce rocking and imbalance problems. Because the potential for this rocking is determined by the design and construction of the engine, the firing order varies from engine to engine. Vehicle manufacturers simplify identifying each cylinder identification by numbering them (Figure 5-3). Regardless of the particular firing order used, the number one cylinder always starts the firing order, with the rest of the cylinders following in a fixed sequence.

The ignition system must be able to monitor the rotation of the crankshaft and the relative position of each piston in order to determine which piston is on its compression stroke. It must also be able to deliver a high-voltage surge to each cylinder at the proper time during its compression stroke. How the ignition system does these things depends on the design of the system.

A BIT OF HISTORY

The man credited with the development of the ignition system as we know it today was Charles Kettering. He is also credited with the development of electric starters for Cadillacs in 1911, quick-drying paint, and ethyl gasoline.

Basic Ignition System

All ignition systems consist of two interconnected electrical circuits: a **primary** (low voltage) **circuit** and a **secondary** (high voltage) **circuit** (Figure 5-4).

Depending on the exact type of ignition system, components in the primary circuit include the following:

❏ battery

❏ ignition switch

❏ ballast resistor or resistance wire (some systems)

❏ starting bypass (some systems)

❏ ignition coil primary winding

❏ triggering device

❏ switching device or control module

The secondary circuit includes these components:

❏ ignition coil secondary winding

❏ distributor cap and rotor (some systems)

❏ ignition (spark plug) cables

❏ spark plugs

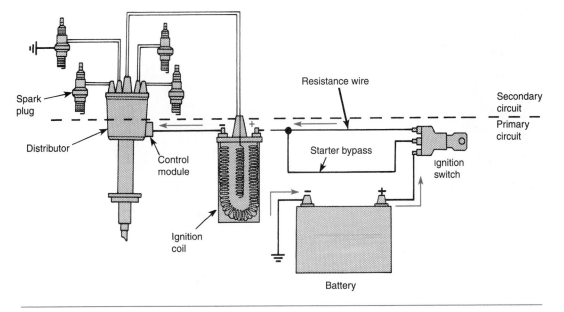

Figure 5-4 Basic primary and secondary ignition circuits.

Primary Circuit Operation

When the ignition switch is in the on position, current from the battery flows through the ignition switch and primary circuit resistor to the primary winding of the ignition coil. From here it passes through some type of switching device and back to ground. The switching device can be electronically or mechanically controlled by the triggering device. The current flow in the ignition coil's primary winding creates a magnetic field. The switching device or control module interrupts this current flow at predetermined times. When it does, the magnetic field in the primary winding collapses. This collapse generates a high-voltage surge in the secondary winding of the ignition coil. The secondary circuit of the system begins at this point.

Some ignition systems have a ballast resistor connected in series between the ignition switch and the coil positive terminal (Figure 5-5). This resistor supplies the correct amount of voltage and current to the coil. In some ignition systems, a calibrated resistance wire is used in place of the ballast resistor. Today, many ignition systems are not equipped with the resistor. These systems supply 12 volts directly to the coil.

Secondary Circuit Operation

The secondary circuit carries highvoltage to the spark plugs. The exact manner in which the secondary circuit delivers these high-voltage surges depends on the system design. Until 1984 all ignition systems used some type of **distributor** to accomplish this job. However, in an effort to reduce emissions, improve fuel economy, and boost component reliability, most auto manufacturers are now using distributorless or electronic ignition (EI) systems.

In a distributor ignition (DI) system, high voltage from the secondary winding passes through an ignition cable running from the coil to the distributor. The distributor then distributes the high

Shop Manual
Chapter 5, page 167

Distributorless ignition systems are commonly referred to as DIS.

In the SAE J1930 terminology, the term **distributor ignition (DI)** replaces all previous terms for electronically controlled distributor-type ignition systems.

In the SAE J1930 terminology, the term **electronic ignition (EI)** replaces all previous terms for distributorless ignition systems.

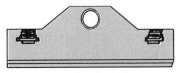

Figure 5-5 Ignition ballast resistor. *(Courtesy of Chrysler Corporation)*

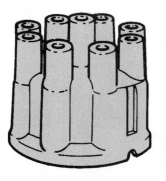

Figure 5-6 Distributor with alignment notches.
(Courtesy of Chrysler Corporation)

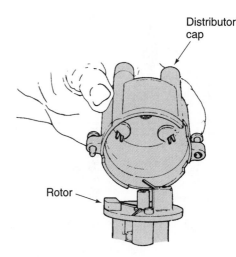

Figure 5-7 Distributor cap and rotor. *(Courtesy of Chrysler Corporation)*

voltage to the individual spark plugs through a set of ignition cables. The cables are arranged in the distributor cap according to the firing order of the engine. A **rotor** driven by the distributor shaft rotates and completes the electrical path from the secondary winding of the coil to the individual spark plugs. The distributor delivers the spark to match the compression stroke of the piston. The distributor assembly may also have the capability of advancing or retarding ignition timing.

The distributor cap is mounted on top of the distributor assembly and an alignment notch in the cap fits over a matching lug on the housing (Figure 5-6). Therefore the cap can only be installed in one position, thereby assuring the correct firing sequence.

The rotor is positioned on top of the distributor shaft, and a projection inside the rotor fits into a slot in the shaft. This allows the rotor to only be installed in one position. A metal strip on the top of the rotor makes contact with the center distributor cap terminal, and the outer end of the strip rotates past the cap terminals as it rotates (Figure 5-7). This action completes the circuit between the ignition coil and the individual spark plugs according to the firing order.

EI systems, have no distributor, rather spark distribution is controlled by an electronic control unit and/or the vehicle's computer (Figure 5-8). Instead of a single ignition coil for all cylinders, each cylinder may have its own ignition coil, or two cylinders may share one coil. The coils are wired directly to the spark plug they control. An ignition control module, tied into the vehicle's computer control system, controls the firing order and the spark timing and advance. This module is typically located under the coil assembly.

In many EI systems, a crank sensor located at the front of the crankshaft is used to trigger the ignition system. When a distributor is used in the ignition system, the distributor drive gear, shaft, and bushings are subject to wear. Worn distributor components cause erratic ignition timing and spark advance, which results in reduced fuel economy and performance plus increased exhaust emissions. Since the distributor is eliminated in EI systems, ignition timing remains more stable over the life of the engine, which means improved economy and performance with reduced emissions.

A specific amount of energy is available in a secondary ignition circuit. Electrical energy may be measured in watts, and watts are calculated by multiplying amperes and volts. In a secondary ignition circuit, the energy is normally produced in the form of voltage required to start firing the spark plug, and then a certain amount of current flow across the spark plug electrodes to maintain the spark. Distributorless ignition systems are capable of producing much higher energy than conventional ignition systems.

Since distributor ignition and electronic ignition systems are both firing spark plugs with approximately the same air gap across the electrodes, the voltage required to start firing the spark

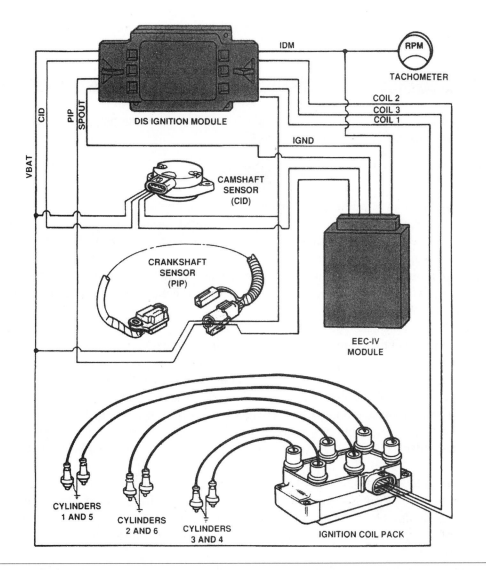

Figure 5-8 A six-cylinder distributorless ignition system. *(Reprinted with the permission of Ford Motor Company)*

plugs in both systems is similar. If the additional energy in the EI system is not released in the form of voltage, it will be released in the form of current flow. This results in longer spark plug firing times. The average firing time across the spark plug electrodes in an EI system is 1.5 milliseconds compared to approximately 1 millisecond in a DI system. This extra time may seem insignificant, but it is very important. Current emission standards demand leaner air-fuel ratios, and this additional spark duration on EI systems helps to prevent cylinder misfiring with leaner air-fuel ratios. For these reasons, many car manufacturers have equipped their engines with EI systems.

Ignition Coil Action

All ignition systems share a number of common components. Some, such as the battery and ignition switch perform simple functions. The battery supplies battery voltage to the ignition primary circuit. Current flows when the ignition switch is in the "start" or the "run" position.

To generate a spark to begin combustion, the ignition system must deliver high voltage to the spark plugs. Because the amount of voltage required to bridge the gap of the spark plug varies with the operating conditions, most late-model vehicles can easily supply 30,000 to 60,000 volts to

Shop Manual
Chapter 5, page 182

force a spark across the air gap. Since the battery delivers 12 volts, a method of stepping up the voltage must be used. Multiplying battery voltage is the job of a coil.

A turning point in the automotive industry came in 1902 when the Humber, an automobile from England, used a magneto electric ignition system.

The ignition coil is a **pulse transformer**. It transforms battery voltage into short bursts of high voltage. As explained previously, when a wire is moved through a magnetic field, voltage is induced in the wire. The inverse of this principle is also true—when a magnetic field moves across a wire, voltage is induced in the wire.

If a wire is bent into loops forming a coil and a magnetic field is passed through the coil, an equal amount of voltage is generated in each loop of wire. The more loops of wire in the coil, the greater the total voltage induced.

Also, the faster the magnetic field moves through the coil, even higher voltage is induced in the coil. If the speed of the magnetic field is doubled, the voltage output doubles.

An ignition coil uses these principles and has two coils of wire wrapped around an iron core. An iron or steel core is used because it has low **inductive reluctance**. The primary coil is normally composed of 100 to 200 turns of 20-gauge wire. This coil of wire conducts battery current. When a current is passing through the primary coil, it magnetizes the iron core. The strength of the magnet depends directly on the number of wire loops and the amount of current flowing through those loops. The secondary coil of wires may consist of 15,000 to 25,000, or more, turns of very fine copper wire.

Because of the effects of counter EMF (electromotive force) on the current flowing through the primary winding, it takes some time for the coil to become fully magnetized or saturated. Therefore, current flows in the primary winding for some time between firings of the spark plugs. The period of time that there is primary current flow is called **dwell**. The length of the dwell period is important.

When current flows through a conductor, it will immediately reach its maximum value as allowed by the resistance in the circuit. If a conductor is wound into a coil, maximum current will not be immediately present. As the magnetic field begins to form as the current begins to flow, the magnetic lines of force of one part of the winding pass over another part of the winding. This tends to cause an opposition to current flow. This occurrence is called reactance. Reactance causes a temporary resistance to current flow and delays the flow of current from reaching its maximum value. When maximum current flow is present in a winding, the winding is said to be saturated and the strength of its magnetic field will also be at a maximum.

Saturation can only occur if the dwell period is long enough to allow for maximum current flow through the primary windings. A less than saturated coil will not be able to produce the voltage it was designed to produce. If the energy from the coil is too low, the spark plugs may not fire long enough or not fire at all.

When the primary coil circuit is suddenly opened, the magnetic field instantly collapses. The sudden collapsing of the magnetic field produces a very high voltage in the secondary windings. This high voltage is used to push current across the gap of the spark plug. Figure 5-9 simplistically shows the coil's primary and secondary circuits.

Ignition Coil Construction

A laminated soft iron core is positioned at the center of the ignition coil and an insulated primary winding is wound around this core. The primary winding has approximately 200 turns of heavy

The low inductive reluctance of steel and iron means that it freely expands or strengthens the magnetic field around it.

Ignition dwell is the length of time there is current flow in the primary circuit prior to firing the spark plug.

Shop Manual
Chapter 5, page 193

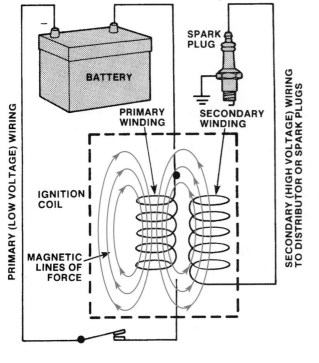

Figure 5-9 Primary and secondary ignition coil action.

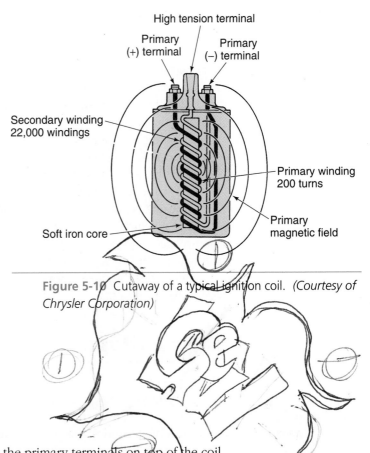

Figure 5-10 Cutaway of a typical ignition coil. (Courtesy of Chrysler Corporation)

wire and the two ends of this winding are connected to the primary terminals on top of the coil (Figure 5-10). These terminals are usually identified with positive and negative symbols. An enamel-type insulation prevents the primary windings from touching each other. Paper insulation is also placed between the layers of windings.

Secondary coil windings are made of very fine wire. These windings are on the inside of the primary winding. A similar insulation method is used on the secondary and primary windings. The ends of the secondary winding are usually connected to one of the primary terminals and to the high tension terminal in the coil tower. When the windings and core assembly is mounted in the coil container, the core rests on a ceramic insulating cup in the bottom of the container. Metal sheathing is placed around the outside of the coil windings in the container. This sheathing concentrates the magnetic field on the outside of the windings. A sealing washer is positioned between the tower and the container to seal the unit. The coil assembly is filled with oil through the screw hole in the high tension terminal. The unit is sealed to protect the windings and to keep the oil inside. The oil helps to cool the coil.

Some coils are not oil-cooled. Rather they are air-cooled. These coils are constructed in much the same way except the core is constructed from laminated sheets of iron. These sheets are shaped like the letter "E" and the primary and secondary windings are wound around the center of the E-core. This type of coil is commonly found on late-model engines. E-core coils cool more rapidly than oil-filled coils. It is also possible to have more windings around the core without drastically increasing the size of the ignition coil.

Secondary Voltage

The typical amount of secondary coil voltage required to jump the spark plug gap is 10,000 volts. Most coils have a maximum secondary voltage of 25,000 volts. The difference between the required voltage and the maximum available voltage is referred to as secondary reserve voltage. This reserve voltage is necessary to compensate for high cylinder pressures and increased sec-

The negative terminal of an ignition coil is commonly referred to as the tachometer (tach) terminal.

Shop Manual
Chapter 5, page 182

ondary resistances as the spark plug gap increases through use. The maximum available voltage must always exceed the required firing voltage or ignition misfire will occur. If there is an insufficient amount of voltage available to push current across the gap, the spark plug will not fire. It is important to note that a decrease in primary voltage will decrease the output of the secondary. Normally a decrease of one volt in the primary will cause a decrease of 5 kV in the secondary.

Since DI and EI systems are both firing spark plugs with approximately the same air gaps, the amount of voltage required to fire the spark is nearly the same in both systems. However, EI systems have higher voltage reserves. If the additional voltage in an EI system is not used to start the spark across the plug's gap, it is used to maintain the spark for a longer period of time. The average length of time an EI system can maintain the spark is 1.5 milliseconds. A typical DI system maintains the spark for 1 millisecond.

The number of ignition coils used in an ignition system varies with the type of ignition system found on a vehicle. In most ignition systems with a distributor, only one ignition coil is used. The high voltage of the secondary winding is directed by the distributor to the various spark plugs in the system. Therefore, there is one secondary circuit with a continually changing path.

While distributor systems have a single secondary circuit with a continually changing path, distributorless systems have several secondary circuits, each with an unchanging path.

Spark Plugs

Shop Manual
Chapter 5, page 198

The spark plug electrodes provide gaps inside each combustion chamber across which the secondary current flows to ignite the air-fuel mixture in the combustion chambers.

Every type of ignition system uses spark plugs. The spark plugs provide the crucial air gap across which the high-voltage current from the coil flows across in the form of an arc. The three main parts of a spark plug are the steel core, the ceramic core or insulator that acts as a heat conductor, and a pair of electrodes, one insulated in the core and the other grounded on the shell. The shell holds the ceramic core and electrodes in a gas-tight assembly and has the threads needed for plug installation in the engine (Figure 5-11). An ignition cable connects the secondary to the top of the plug. Current flows through the center of the plug and arcs from the tip of the center electrode to the ground electrode. The resulting spark ignites the air-fuel mixture in the combustion chamber. Most automotive spark plugs also have a resistor between the top terminal and the center electrode. This resistor reduces radio frequency interference (RFI), which prevents noise on stereo equipment. Voltage peaks from RFI could also interfere with, or damage, on-board computers. Therefore, when resistor-type spark plugs are used as original equipment, replacement spark plugs must also be resistor-type. The resistor, like all other resistances in the secondary, increases the voltage needed to jump the gap of the spark plug.

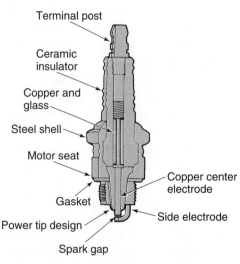

Figure 5-11 Construction of a typical spark plug. *(Courtesy of Chrysler Corporation)*

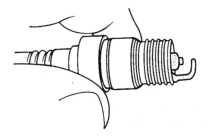

Figure 5-12 An air gap between the center and the ground electrodes is where the spark occurs.

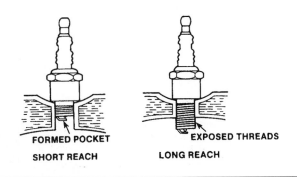

FORMED POCKET   EXPOSED THREADS
SHORT REACH   LONG REACH

Figure 5-13 Spark plug reach: short versus long.

Louis S. Clark of the Auto Car Company designed the porcelain spark plug insulation in 1902.

Spark plugs come in many different sizes and designs to accommodate different engines. To fit properly, spark plugs must be of the proper size and reach. Another design factor that determines the usefulness of a spark plug for a specific application is its **heat range**. The desired heat range depends on the design of the engine and on the type of driving conditions the vehicle is subject to.

A terminal post on top of the center electrode is the point of contact for the spark plug cable. The center electrode, commonly made of a copper alloy, is surrounded by a ceramic insulator and a copper and glass seal is located between the electrode and the insulator. These seals prevent combustion gases from leaking out of the cylinder. Ribs on the insulator increase the distance between the terminal and the shell to help prevent electric arcing on the outside of the insulator. The steel spark plug shell is crimped over the insulation and a ground electrode, on the lower end of the shell, is positioned directly below the center electrode. There is an air gap between these two electrodes and the width of this air gap is specified by the auto manufacturer (Figure 5-12).

A spark plug socket may be placed over a hex-shaped area near the top of the shell for plug removal and installation. Threads on the lower end of the shell allow the plug to be threaded into the engine's cylinder head. The thread length is referred to as reach (Figure 5-13). Various thread lengths are designed to match the cylinder head and combustion chamber design.

Some spark plugs have a metal gasket positioned between the seat on the steel shell and the cylinder head, whereas other shells have a tapered seat with a matching seat in the cylinder head (Figure 5-14). Some spark plugs have platinum-tipped electrodes, which greatly extend the life of the plug.

Shop Manual
Chapter 5, page 201

The average replacement interval for conventional spark plugs is often 30,000 miles (mi.), or 48,000 kilometers (km). Platinum-tipped spark plugs have a replacement interval of 60,000 mi. (96,000 km) or more.

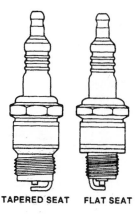

TAPERED SEAT   FLAT SEAT

Figure 5-14 Different spark plug seats: tapered versus flat.

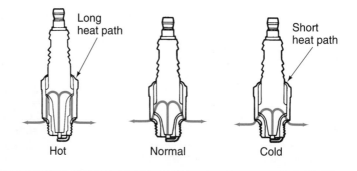

Figure 5-15 Various spark plug heat ranges. *(Courtesy of Chrysler Corporation)*

Shop Manual
Chapter 5, page 199

The **heat range** of a spark plug is a statement of how well a spark plug can conduct heat away from its tip.

Spark plugs are designed with different heat ranges to compensate for various operating conditions. When the engine is running, most of the plug's heat is concentrated on the center electrode. Heat is quickly dissipated from the ground electrode because it is threaded into the cylinder head. The spark plug heat path is from the center electrode through the insulator into the shell and to the cylinder head, where the heat is absorbed by engine coolant circulating through the cylinder head. Spark plug heat range is determined by the depth of the insulator before it contacts the shell. For example, in a cold-range spark plug, the depth of the insulator is short before it contacts the shell. This cold-type spark plug has a short heat path, which provides cooler electrode operation (Figure 5-15).

In a hot spark plug, the insulator depth is increased before it makes contact with the shell. This provides a longer heat path and increases electrode temperature. A spark plug needs to retain enough heat to clean itself between firings, but not so hot that it damages itself or causes premature ignition of the air-fuel mixture in the cylinder. If an engine is driven continually at low speeds, the spark plugs may become carbon fouled. Under this condition, a hotter range spark plug may be required. Severe high-speed driving over an extended time period may require a colder range spark plug to prevent electrode burning from excessive combustion chamber heat.

Ignition Cables

Shop Manual
Chapter 5, page 196

Spark plug, or ignition cables make up the secondary wiring. These cables carry the high-voltage current from the distributor or the multiple coils (EI systems) to the spark plugs. The cables are not solid wire, rather they contain fiber cores that act as a resistor in the secondary circuit (Figure 5-16). They cut down on radio and television interference, increase firing voltages, and reduce spark plug wear by decreasing current. Metal terminals on each end of the spark plug wires contact the spark plug and the distributor cap terminals. Insulated boots on the ends of the cables strengthen the connections as well as prevent dust and water infiltration and voltage loss.

Spark Timing Systems

Shop Manual
Chapter 5, page 204

To better understand the operation of current ignition systems, it is helpful to first review how older, fully mechanical distributor systems worked.

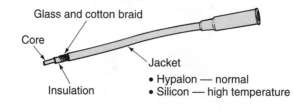

Figure 5-16 Construction of a typical ignition cable. *(Courtesy of Chrysler Corporation)*

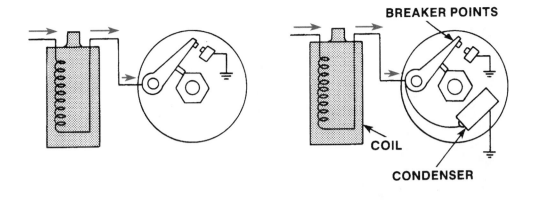

BREAKER POINTS

COIL

CONDENSER

Figure 5-17 A condenser prevents current from jumping across the open point gap.

Breaker Point Ignition

Breaker point ignition systems (Figure 5-17) were used on vehicles for more than sixty years but were abandoned many years ago as engineers looked for ways to decrease emissions and increase fuel efficiency.

The distributor assembly acted as a mechanical switch to turn the primary circuit on and off. The distributor's shaft, cam, breaker points, and condenser performed this function. **Contact points** were used as the primary circuit triggering and switching device. The breaker point assembly, which was mounted on the **breaker plate** inside the distributor, consisted of a fixed contact, movable contact, movable arm, rubbing block, pivot, and spring. The fixed contact was grounded through the distributor housing, and the movable contact was connected to the negative terminal of the coil's primary winding. As the cam turned by the camshaft, the movable arm opened and closed, which opened and closed the primary circuit in the coil. When the points were closed, primary current flow attempted to saturate the coil. When the points opened, primary current stopped and the magnetic field collapsed causing high voltage to be induced in the secondary. The firing of the plug was the result of opening the points.

Because voltage was still present at the movable arm when the breaker arms opened, current could continue to arc across the open point gap which could damage the points. To prevent this, a condenser was attached to the movable arm. In this way, the voltage at the movable arm was retained by the condenser instead of arcing across the gap.

A primary, or **ballast resistor** was located in series between the battery and the primary coil winding and was responsible for keeping the primary voltage at the desired level (about 9 or 10 volts). This prevented the contact points from burning due to high voltage. The ballast resistor could be either a separate unit or a specially made wire. During starting, the ballast resistor is by-passed to provide maximum current flow to the primary circuit.

The distributor also mechanically adjusted the time the spark arrived at the cylinder through the use of two mechanisms: the centrifugal advance and the vacuum advance units. This improved engine performance, fuel efficiency, and emission levels.

As its name implies, the distributor mechanically distributed the spark so that it arrived at the right time during the compression stroke of each cylinder. The distributor's shaft, rotor, and cap performed this function.

Electronic or Solid State Ignition

From the fully mechanical breaker point system, ignition technology progressed to basic electronic or solid state ignitions (Figure 5-18). Breaker points were replaced with an electronic triggering and

Although they are very much obsolete, breaker points are still mentioned in ASE's task list for Engine Performance Certification. Perhaps the feeling is that if you understand point-type ignitions, understanding the other systems will become easier.

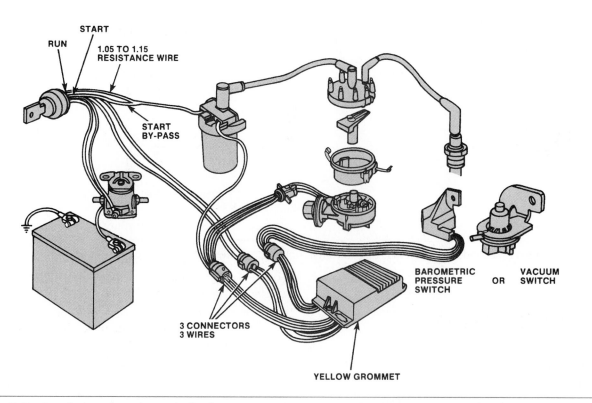

Figure 5-18 One design of an electronic distributor-type ignition system. *(Reprinted with the permission of Ford Motor Company)*

switching devices. The electronic switching components are normally inside a separate housing known as an **electronic control module** (ECU) or control unit (Figure 5-19). The original (solid-state) electronic ignitions still relied on mechanical and vacuum advance mechanisms in the distributor.

As technology advanced, many manufacturers expanded the ability of the ignition control modules. For example, by tying a manifold vacuum sensor into the ignition module circuitry, the module could now detect when the engine was under heavy load and retard the timing automatically. Similar add-on sensors and circuits were designed to control spark knock, start-up emissions, and altitude compensation.

Computer-Controlled Electronic Ignition

Computer-controlled ignition systems offer continuous spark timing control through a network of engine sensors and a central microprocessor. Based on the inputs it receives, the central micro-

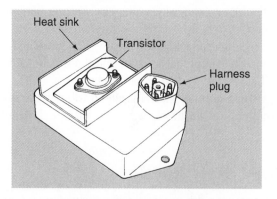

Figure 5-19 A typical electronic control module. *(Courtesy of Chrysler Corporation)*

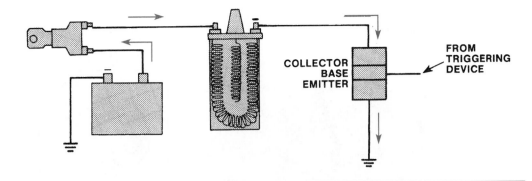

Figure 5-20 When the triggering device supplies a small amount of current to the transistor's base, the primary coil circuit is closed and current flows.

processor or computer makes decisions regarding spark timing and sends signals to the ignition module to fire the spark plugs according to those inputs and according to the programs in its memory.

Computer-controlled ignition systems may or may not use a distributor to distribute secondary voltage to the spark plugs. As mentioned earlier, distributorless systems use multiple coils and modules to provide and distribute high secondary voltages directly from the coil to the plug.

Electronic Switching Systems

Electronic ignition systems control the primary circuit, using an NPN transistor instead of breaker contact points. The transistor's emitter is connected to ground and takes the place of the fixed contact point. The collector is connected to the negative (-) terminal of the coil, taking the place of the movable contact point. When the triggering device supplies a small amount of current to the base of the switching transistor, the collector and emitter act as if they are closed contact points (a conductor), allowing current to build up in the coil primary circuit. When the current to the base is interrupted by the switching device, the collector and emitter act as an open contact (an insulator), interrupting the coil primary current. An example of how this works is shown in Figure 5-20, which is a simplified diagram of an electronic ignition system.

In a breaker-point ignition system, the air gap between the two contacts determines the dwell. As the rubbing block of the points wear down, point gap changes as does dwell. Since dwell is the length of time current flows through the primary windings of the coil, maintaining a proper dwell is important to having sufficient secondary coil voltage output. Electronic switching devices have no rubbing block and dwell tends to be maintained over long periods of time.

Engine Position Sensors

The time when the primary circuit must be opened and closed is related to the position of the pistons and the crankshaft. Therefore, the position of the crankshaft is used to control the flow of current to the base of the switching transistor.

A number of different types of sensors are used to monitor the position of the crankshaft and control the flow of current to the base of the transistor. These engine position sensors and generators serve as triggering devices and include magnetic pulse generators, metal detection sensors, Hall effect sensors, and photoelectric sensors.

The mounting location of these sensors depends on the design of the ignition system. All four types of sensors can be mounted in the distributor, which is turned by the camshaft.

Magnetic pulse generators and Hall effect sensors can also be located on the crankshaft (Figures 5-21 and Figure 5-22). These sensors are also commonly used on EI ignition systems. Both Hall effect sensors and magnetic pulse generators can also be used as camshaft reference sensors to identify which cylinder is the next one to fire.

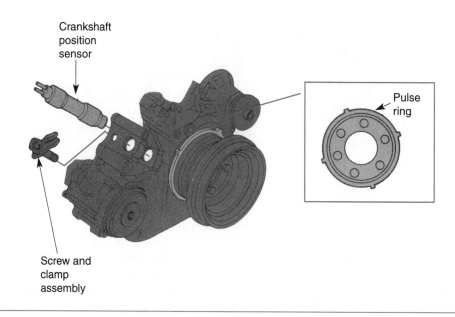

Figure 5-21 A typical crankshaft speed sensor. *(Reprinted with the permission of Ford Motor Company)*

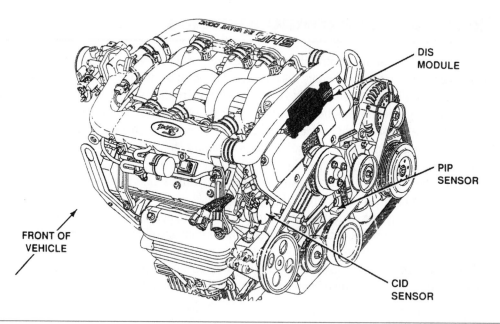

Figure 5-22 An engine with both a camshaft (CID) crankshaft (PIP) position sensor. *(Reprinted with the permission of Ford Motor Company)*

Shop Manual
Chapter 5, page 194

The pick-up coil may also be called a stator, sensor, or pole piece.

The timing disc may also be called a reluctor, trigger wheel, pulse ring, armature, or timing core.

Magnetic Pulse Generator

Basically, a magnetic pulse generator consists of two parts: a timing disc and a pick-up coil. The pick-up coil consists of a length of wire wound around a weak permanent magnet. Depending on the type of ignition system used, the timing disc may be mounted on the distributor shaft (Figure 5-23), at the rear of the crankshaft (Figure 5-24), or on the crankshaft vibration damper (Figure 5-25).

The magnetic pulse or PM generator operates on basic electromagnetic principles. Remember that a voltage can only be induced when a conductor moves through a magnetic field. The magnetic field is provided by the pick-up unit and the rotating timing disc provides the movement through the magnetic field needed to induce voltage.

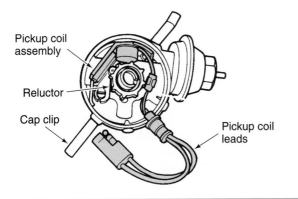

Figure 5-23 A pickup coil mounted in a distributor. *(Courtesy of Chrysler Corporation)*

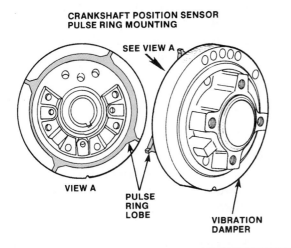

Figure 5-25 The pulse ring on a crankshaft pulley.

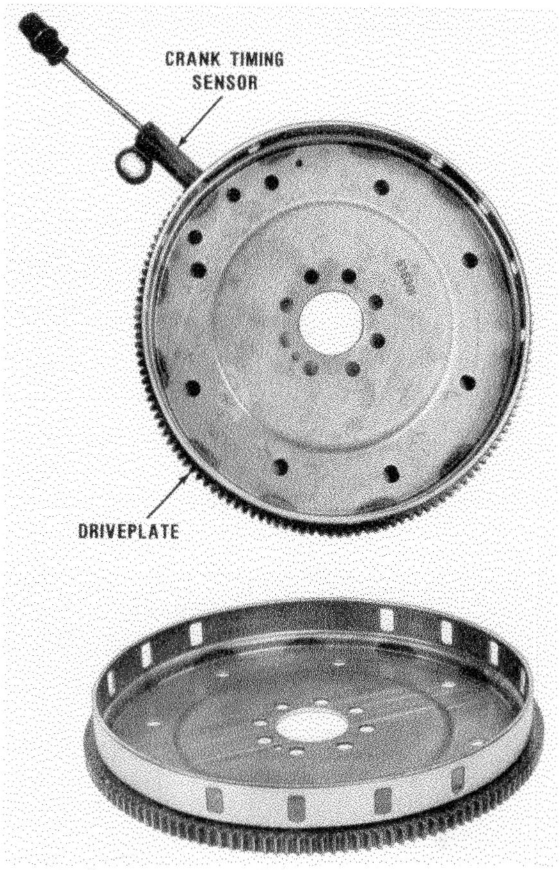

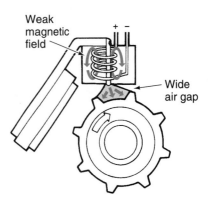

Figure 5-24 A flywheel crank timing sensor. *(Courtesy of Chrysler Corporation)*

As the disc teeth approach the pick-up coil, they repel the magnetic field, forcing it to concentrate around the pick-up coil (Figure 5-26). Once the tooth passes by the pick-up coil, the magnetic field is free to expand or unconcentrate (Figure 5-27), until the next tooth on the disc approaches. Approaching teeth concentrate the magnetic lines of force, while passing teeth allow

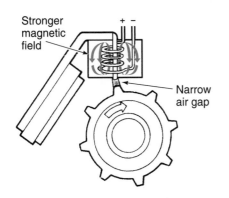

Figure 5-26 When the reluctor high point is aligned with the pick-up coil, the gap is narrower and the magnetic field is stronger. *(Courtesy of Chrysler Corporation)*

Figure 5-27 When the reluctor's high points are out of alignment with the pick-up coil, a wide gap and a weak magnetic field exist between the reluctor's high points and the pickup coil. *(Courtesy of Chrysler Corporation)*

them to expand. This pulsation of the magnetic field causes the lines of magnetic force to cut across the winding in the pick-up coil, inducing a small amount of AC voltage that is sent to the switching device in the primary circuit.

When a disc tooth is directly in line with the pick-up coil, the magnetic field is not expanding or contracting. Since there is no movement or change in the field, voltage at this precise movement drops to zero. At this point, the switching device inside the ignition module reacts to the zero voltage signal by turning the ignition's primary circuit current off. As explained earlier, this forces the magnetic field in the primary coil to collapse, discharging a secondary voltage to the distributor or directly to the spark plug.

As soon as the tooth rotates past the pick-up coil, the magnetic field expands again and another voltage signal is induced. The only difference is that the polarity of the charge is reversed. Negative becomes positive or positive becomes negative. Upon sensing this change in voltage, the switching device turns the primary circuit back on and the process begins all over.

The timing disc is mounted in a very precise manner. When the disc teeth align with the pick-up coil, this corresponds to the exact time certain pistons are nearing TDC. This means the zero voltage signal needed to trigger the secondary circuit occurs at precisely the correct time.

The pick-up coil might have only one pole as shown in Figure 5-28. Other magnetic pulse generators have pick-up coils with two or more poles. The one shown in Figure 5-29 has as many poles as it has trigger teeth.

Metal Detection Sensors

Metal detection sensors are found on many early electronic ignition systems. They work much like a magnetic pulse generator with one major difference.

A trigger wheel is pressed over the distributor shaft and a pick-up coil detects the passing of the trigger teeth as the distributor shaft rotates. However, unlike a magnetic pulse generator, the pickup coil of a metal detection sensor does not have a permanent magnet. Instead, the pick-up coil is an electromagnet. A low level of current is supplied to the coil by an electronic control unit, inducing a weak magnetic field around the coil. As the reluctor on the distributor shaft rotates, the trigger teeth pass very close to the coil (Figure 5-30). As the teeth pass in and out of the coil's magnetic field, the magnetic field builds and collapses, producing a corresponding change in the coil's voltage. The voltage changes are monitored by the control unit to determine crankshaft position.

Hall Effect Sensor

Introduced in early 1982, the Hall effect sensor or switch is now the most commonly used engine position sensor. There are several good reasons for this. Unlike a magnetic pulse generator, the

Shop Manual
Chapter 5, page 195

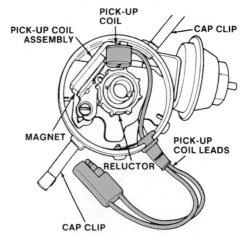

Figure 5-28 A single pole pick-up coil.

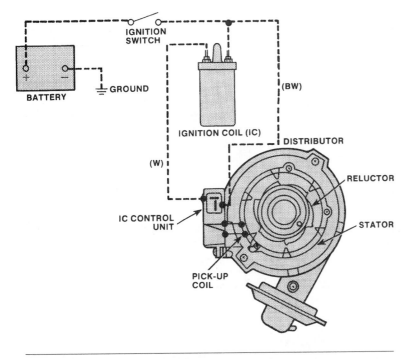

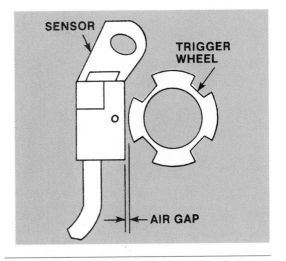

Figure 5-30 A metal detection sensor.

Figure 5-29 A pick-up coil with as many poles as there are cylinders. *(Courtesy of Nissan Motors)*

Hall effect sensor produces an accurate voltage signal throughout the entire rpm range of the engine. Furthermore, a Hall effect switch produces a square wave signal that is more compatible with the digital signals required by on-board computers.

Functionally, a Hall switch performs the same tasks as a magnetic pulse generator, but the Hall switch's method of generating voltage is quite unique. It is based on the Hall effect principle which states: If a current is allowed to flow through a thin conducting material, and that material is exposed to a magnetic field, voltage is produced.

The heart of the Hall generator is a thin semiconductor layer (Hall layer) derived from a gallium arsenate crystal. Attached to it are two terminals—one positive, and the other negative—that are used to provide the source current for the Hall transformation.

Directly across from this semiconductor element is a permanent magnet. It's positioned so that its lines of flux bisect the Hall layer at right angles to the direction of current flow. Two additional terminals, located on either side of the Hall layer, form the signal output circuit.

When a moving metallic shutter blocks the magnetic field from reaching the Hall layer or element, the Hall effect switch produces a voltage signal. When the shutter blade moves and allows the magnetic field to expand and reach the Hall element, the Hall effect switch does not generate voltage signal (Figure 5-31).

The Hall switch is described as being on any time the Hall layer is exposed to a magnetic field and a Hall voltage is being produced. However, before this signal voltage can be of any use, it must be modified. After leaving the Hall layer, the signal is routed to an amplifier where it is strengthened and inverted so that the signal reads high when it is actually coming in low and vice versa. Once it has been inverted, the signal goes through a Schmitt trigger where it is turned into a clean square wave signal. After conditioning, the signal is sent to the base of a switching transistor that is designed to turn on and off in response to the signals generated by the Hall effect switch assembly.

The shutter wheel is the last major component of the Hall switch. The shutter wheel consists of a series of alternating windows and vanes that pass between the Hall layer and magnet. The shutter wheel may be part of the distributor rotor (Figure 5-32) or separate from the rotor.

The shutter wheel performs the same function as the timing disc on magnetic pulse generators. The only difference is with a Hall effect switch there is no electromagnetic induction. Instead,

A Schmitt trigger is a pulse-shaping device.

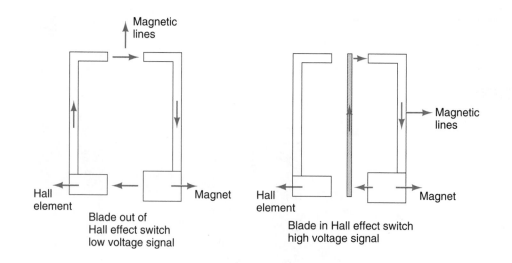

Figure 5-31 As the blade moves through the Hall effect switch, a high voltage signal is produced.

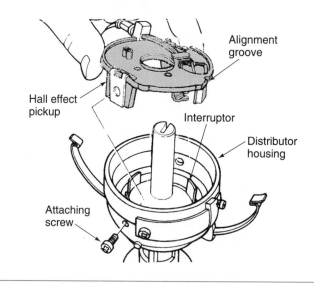

Figure 5-32 A distributor rotor with a shutter blade for the Hall effect switch. *(Courtesy of Chrysler Corporation)*

the shutter wheel creates a magnetic shunt that changes the field strength through the Hall element. When a vane of the shutter wheel is positioned between the magnet and Hall element, the metallic vane blocks the magnetic field and keeps it from permeating the Hall layer. As a result, only a few residual electrons are deflected in the layer and Hall output voltage is low. Conversely, when a window rotates into the air gap, the magnetic field is able to penetrate the Hall layer, which in turn pushes the Hall voltage to its maximum range.

The points where the shutter vane begins to enter and begins to leave the air gap are directly related to primary circuit control. As the leading edge of a vane enters the air gap, the magnetic field is deflected away from the Hall layer and Hall voltage decreases. When that happens, the modified Hall output signal increases abruptly and turns the switching transistor on. Once the transistor is turned on, the primary circuit closes and the coil's energy storage cycle begins.

Primary current continues to flow as long as the vane is in the air gap. As the vane starts to leave the gap, however, the reforming Hall voltage signal prompts a parallel decline in the modified output signal. When the output signal goes low, the bias of the transistor changes. Primary current flow stops.

In a Hall effect switch, the edges of the rotating blade that enter the switch are called the leading edges.

In a Hall effect switch, the edges of the rotating blade that exit the switch are referred to as the trailing edges.

In summary, the ignition module supplies current to the coil's primary winding as long as the shutter wheel's vane is in the air gap. As soon as the shutter wheel moves away and the Hall voltage is produced, the control unit stops primary circuit current, high secondary voltage is induced, and ignition occurs.

In addition to ignition control, a Hall effect switch can also be used to generate precise rpm signals (by determining the frequency at which the voltage rises and falls) and provide the sync pulse for sequential fuel ignition operation.

Photoelectric Sensor

A fifth type of crankshaft position sensor is the **photoelectric sensor**. The parts of this sensor include a light-emitting diode (LED), a light sensitive phototransistor, and a slotted disc called a light beam interrupter (Figure 5-33).

The slotted disc is attached to the distributor shaft. The LED and the photocell are situated over and under the disc opposite of each other. As the slotted disc rotates between the LED and photocell, light from the LED shines through the slots. The intermittent flashes of light are translated into voltage pulses by the photocell. When the voltage signal occurs, the control unit turns the primary system on. When the disc interrupts the light and the voltage signal ceases, the control unit turns the primary system off, causing the magnetic field in the coil to collapse and sending a surge of voltage to a spark plug.

The photoelectric sensor sends a very reliable signal to the control unit, especially at low engine speeds. These units have been primarily used on Chrysler and Mitsubishi engines. Some Nissan and General Motors products have used them as well.

> A light sensitive phototransistor is most often referred to as a photocell.

Timing Advance

As stated, early electronic ignition systems changed the timing mechanically just like breaker point systems. At idle, the firing of the spark plug usually occurs just before the piston reaches top dead center. At higher engine speeds however, the spark must be delivered to the cylinder much earlier in the cycle to achieve maximum power from the air-fuel mixture since the engine is moving through the cycle more quickly. To change the timing of the spark in relation to rpm, the **centrifugal advance** mechanism is used (Figure 5-34).

> Centrifugal advance systems are also called mechanical advance units.

Shop Manual
Chapter 5, page 206

A BIT OF HISTORY

The first use of automatic spark advance was in 1900. In spite of this development, many cars continued to have driver-operated timing advance controls.

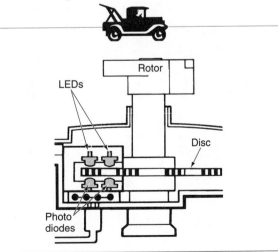

Figure 5-33 Distributor with an optical-type pickup. *(Courtesy of Chrysler Corporation)*

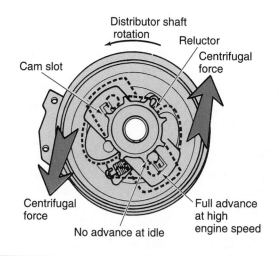

Figure 5-34 Action of a centrifugal advance assembly. *(Courtesy of Chrysler Corporation)*

This mechanism consists of a set of pivoted weights and springs connected to the distributor shaft and a distributor armature assembly. During idle speeds, the springs keep the weights in place and the armature and distributor shaft rotate as one assembly. When speed increases, centrifugal force causes the weights to slowly move out against the tension of the springs. This allows the armature assembly to move ahead in relation to the distributor shaft rotation. The ignition's triggering device is mounted to the armature assembly. Therefore, as the assembly moves ahead, ignition timing becomes more advanced.

During part-throttle engine operation, high vacuum is present in the intake manifold. To get the most power and the best fuel economy from the engine, the plugs must fire even earlier during the compression stroke than is provided by a centrifugal advance mechanism.

The heart of the **vacuum advance** mechanism (Figure 5-35) is the spring-loaded diaphragm, which fits inside a metal housing and connects to a movable plate on which the pick-up coil is mounted. Vacuum is applied to one side of the diaphragm in the housing chamber while the other side of the diaphragm is open to the atmosphere. Any increase in vacuum allows atmospheric pressure to push the diaphragm. In turn, this causes the movable plate to rotate. The more vacuum present on one side of the diaphragm, the more atmospheric pressure is able to cause a change in timing. The rotation of the movable plate moves the pick-up coil so the armature develops a signal earlier. These units are also equipped with a spring that returns the timing back to basic as the vacuum decreases.

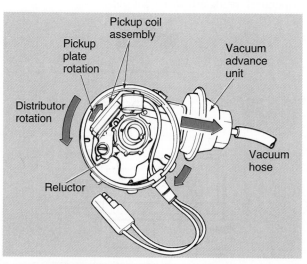

Figure 5-35 Action of a vacuum advance assembly. *(Courtesy of Chrysler Corporation)*

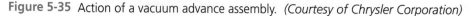

The vacuum advance unit is bolted to the side of the distributor housing. On some models, the vacuum hose for the vacuum advance is connected directly to the intake manifold, whereas on other models this hose is connected to a fitting right above the throttle plates. The diaphragm is pushed inward toward the distributor housing by a spring positioned between the diaphragm and the end of the sealed chamber. An arm is attached to the side of the diaphragm next to the distributor housing, and the inner end of the arm is mounted over a pin on the pick-up plate.

If the engine is operating at a moderate cruising speed, high intake manifold vacuum is applied to the vacuum advance diaphragm. Under this condition, the diaphragm is moved away from the distributor housing against the spring tension. This diaphragm movement rotates the pivoted pick-up plate in the opposite direction to distributor shaft rotation, thereby advancing the timing. When the engine is operating at a moderate cruising speed with a partially opened throttle, the amounts of cylinder air intake, compression pressure, and compression temperature are low. Under these conditions, the air-fuel mixture does not burn as fast and the additional spark advance provides more burn time which increases fuel economy and engine performance.

If the engine is operating at or near wide-open throttle, the amounts of cylinder air intake, compression pressure, and compression temperature are increased, which causes faster burning of the air-fuel mixture. During these operating conditions, the spark advance must be retarded to prevent engine detonation. Intake manifold vacuum is very low at wide throttle opening. Therefore little vacuum is present to work against the spring in the vacuum advance unit. When this action occurs, the pick-up coil is moved back toward the base timing position and the amount of spark advance is reduced.

If the vacuum advance hose is connected above the throttle plates, the vacuum advance does not provide any spark advance when the engine is idling. When the vacuum advance hose is connected directly into the intake manifold, the vacuum advance is fully advanced at idle speed, which provides lower engine temperatures.

Shop Manual
Chapter 5, page 208

Spark Distribution

The distributor cap and rotor receive the high voltage from the secondary winding via a high-tension wire. The voltage enters the distributor cap through the coil tower, or center terminal. The rotor then sends the voltage from the coil tower to the spark plug electrodes inside the distributor cap. The rotor mounts on the upper portion of the distributor shaft and rotates with it.

The distributor cap (Figure 5-36) is made from silicone plastic or similar material that offers protection from chemical attack. It is attached to the distributor housing with screws or spring-

Shop Manual
Chapter 5, page 167

Some call the distributor a "high voltage switch."

A distributor cap may be called a distributor head.

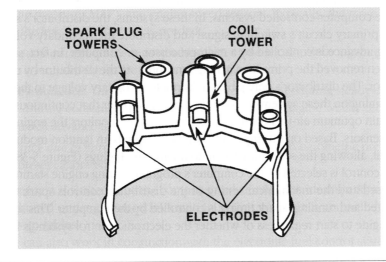

Figure 5-36 Cutaway of a distributor cap.

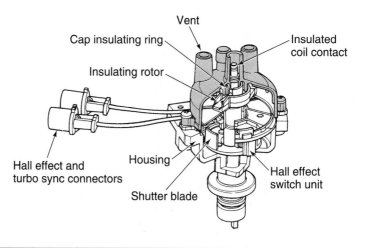

Vent
Cap insulating ring
Insulating rotor
Insulated coil contact
Hall effect and turbo sync connectors
Housing
Shutter blade
Hall effect switch unit

Figure 5-39 A Chrysler distributor with dual Hall effect switches, both located below the breaker plate. *(Courtesy of Chrysler Corporation)*

Detailed information on late-model DI and EI systems can be found in Chapters 12 and 13 of this book.

Chrysler's system has two Hall effect switches in the distributor when the engine is equipped with port fuel injection. In some units, the pickup unit used for ignition triggering is located above the pick-up plate in the distributor and is referred to as the reference pickup. The second pick-up unit is positioned below the plate. A ring with two notches is attached to the distributor shaft and rotates through the lower pick-up unit. This lower pickup is called the synchronizer (SYNC) pickup.

In other designs, the two pick-up units are mounted below the pick-up plate and one set of blades rotates through both Hall effect units (Figure 5-39). The shutter blade representing number one cylinder has a large opening in the center of the blade. When this blade rotates through the SYNC pickup, a different signal is produced compared to the other blades. This number one blade signal informs the PCM when to activate the injectors.

Summary

Terms to Know

ATDC

Ballast resistor

Basic timing

Before top dead center (BTDC)

Breaker plate

Breaker point

Centrifugal advance

Contact points

Distributor

Distrubutor ignition system (DI)

Dwell

Electronic control module

❑ The ignition system supplies high voltage to the spark plugs to ignite the air-fuel mixture in the combustion chambers.

❑ The arrival of the spark is timed to coincide with the compression stroke of the piston. This basic timing can be advanced or retarded under certain conditions, such as high engine rpm or extremely light or heavy engine loads.

❑ The ignition system has two interconnected electrical circuits: a primary circuit and a secondary circuit.

❑ The primary circuit supplies low voltage to the primary winding of the ignition coil. This creates a magnetic field in the coil.

❑ The primary ignition circuit resistor is connected in series between the ignition switch and the positive primary coil terminal and supplies the proper voltage and current flow to the coil.

❑ The electronic control unit (ECU), or module, opens and closes the primary circuit.

❑ A switching device interrupts primary current flow, collapsing the magnetic field and creating a high-voltage surge in the ignition coil secondary winding.

❑ The switching device used in electronic systems is an NPN transistor. Old ignitions use mechanical breaker point switching.

❑ The secondary circuit carries high voltage surges to the spark plugs. On some systems, the circuit runs from the ignition coil, through a distributor, to the spark plugs.

❏ Spark plugs provide an electrode gap in each combustion chamber, and the secondary current flows across this gap to ignite the air-fuel mixture in the combustion chamber.

❏ EI systems provide longer spark duration at the spark plug electrodes than conventional electronic ignition systems, and this helps to fire leaner air-fuel ratios in today's engines.

❏ The ignition coil primary winding contains a few hundred turns of heavy wire.

❏ The secondary coil winding contains thousands of turns of very fine wire.

❏ Dwell is the length of time that the primary circuit is turned on prior to each cylinder firing.

❏ Ignition timing is directly related to the position of the crankshaft. Magnetic pulse generators and Hall effect sensors are the most widely used engine position sensors. They generate an electrical signal at certain times during crankshaft rotation. This signal triggers the electronic switching device to control ignition timing.

❏ The distributor may house the switching device plus centrifugal or vacuum timing advance mechanisms. Some systems locate the switching device outside the distributor housing.

❏ The centrifugal advance controls spark advance in relation to engine rpm.

❏ When the centrifugal weights move outward in relation to engine speed, they turn the reluctor in the same direction as the distributor shaft is rotating.

❏ The vacuum advance controls spark advance in relation to engine load.

❏ The vacuum advance rotates the pick-up plate in the opposite direction to distributor shaft rotation.

❏ EI systems eliminate the distributor. Each spark plug, or in some cases, pair of spark plugs, has its (their) own ignition coil. Primary circuit switching and timing control is done using a special ignition module tied into the vehicle control computer.

❏ Computer-controlled ignition eliminates centrifugal and vacuum timing mechanisms. The computer receives input from numerous sensors. Based on this data, the computer determines the optimum firing time and signals an ignition module to activate the secondary circuit at the precise time needed.

Terms to Know (continued)

Electronic ignition system (EI)
Firing order
Heat range
Ignition timing
Inductive reluctance
Look-up tables
Photoelectric sensor
Primary circuit
Pulse transformer
Rotor
Saturation
Secondary circuit
Vacuum advance

Review Questions

Short Answer Essays

1. Describe the design of the primary and secondary coil windings.

2. Explain how the voltage is induced in the distributor pick-up coil as the reluctor high point approaches alignment with the pick-up coil.

3. Define spark plug reach.

4. Describe the type of driving conditions when a colder range spark plug may be required.

5. Explain how the high voltage is induced in the secondary coil winding in an electronic ignition system.

6. Explain why dwell time is important to ignition system operation.

7. Name the three major functions of an ignition system.

8. What is the basic difference between the primary and secondary ignition circuits?

9. Name the engine operating conditions that affect ignition timing requirements.

10. What primary role does a rotor have in the ignition system?

Fill-in-the-Blanks

1. Modern ignition cables contain fiber cores that act as a _____ in the secondary circuit to cut down on radio and television interference and reduce spark plug wear.

2. The ends of the secondary coil winding are usually connected to the terminal in the coil tower and one of the _____ terminals.

3. The arrival of the spark is timed to coincide with the _____ stroke of the piston.

4. Basic ignition timing is typically _____ with increases in engine speed and _____ with increases of engine load.

5. The ignition system has two interconnected electrical circuits: a _____ circuit and a _____ circuit.

6. A _____ device interrupts primary current flow, collapsing the magnetic field and creating a high-voltage surge in the ignition coil secondary winding.

7. The switching device used in electronic systems is a(n) _____.

8. _____ _____ _____ and _____-_____ sensors are the most widely used engine position sensors.

9. A cold-range spark plug has a _____ heat path compared to a hot-range spark plug.

10. Computer-controlled ignition systems rely on the inputs from various _____ to control ignition timing.

ASE Style Review Questions

1. While discussing the primary and secondary ignition circuits:
   Technician A says the module is part of the secondary ignition circuit.
   Technician B says the spark plugs are in the secondary ignition circuit.
   Who is correct?
   A. A only   **C.** Both A and B
   B. B only   **D.** Neither A nor B

2. While discussing secondary voltage:
   Technician A says the normal required secondary voltage is higher at idle speed than at wide-open throttle conditions.
   Technician B says the maximum available secondary voltage must always exceed the normally required secondary voltage.
   Who is correct?
   A. A only   **C.** Both A and B
   B. B only   **D.** Neither A nor B

3. While discussing distributor advances:
   Technician A says the centrifugal advance weights rotate the reluctor in the same direction as the distributor shaft rotation, as these weights fly outward.
   Technician B says the vacuum advance controls spark advance in relation to engine load.
   Who is correct?
   A. A only   **C.** Both A and B
   B. B only   **D.** Neither A nor B

4. *Technician A* says ignition systems equipped with Hall effect sensors do not require a ballast resistor or resistance wire to regulate primary circuit current.
   Technician B says these systems do require resistance control in their primary circuit.
   Who is correct?
   A. A only   **C.** Both A and B
   B. B only   **D.** Neither A nor B

5. While discussing what happens when the low-voltage current flow in the coil primary winding is interrupted by the switching device:
   Technician A says the magnetic field collapses.
   Technician B says a high-voltage surge is induced in the coil secondary winding.
   Who is correct?
   A. A only   **C.** Both A and B
   B. B only   **D.** Neither A nor B

6. *Technician A* says an ignition system must generate sufficient voltage to force a spark across the spark plug gap.
   Technician B says the ignition system must time the arrival of the spark to coincide with the movement of the engine's pistons and vary it according to the operating conditions of the engine.
   Who is correct?
   A. A only   **C.** Both A and B
   B. B only   **D.** Neither A nor B

7. While discussing electronic ignition systems:
   Technician A says a transistor actually controls primary current flow through the coil.
   Technician B says a reluctor controls the primary coil current.
   Who is correct?
   A. A only   **C.** Both A and B
   B. B only   **D.** Neither A nor B

8. *Technician A* says a magnetic pulse generator is equipped with a permanent magnet.
   Technician B says a Hall effect switch is equipped with a permanent magnet.
   Who is correct?
   A. A only   **C.** Both A and B
   B. B only   **D.** Neither A nor B

9. While discussing PM generators:
   Technician A says the pick-up coil does not produce a voltage signal when a reluctor tooth approaches the coil.
   Technician B says the pick-up coil does not produce a voltage signal when a reluctor tooth moves away from the coil.
   Who is correct?
   A. A only   **C.** Both A and B
   B. B only   **D.** Neither A nor B

10. While discussing engine position sensors:
    Technician A says a metal detection sensor needs to have its voltage signal amplified, inverted, and shaped into a clean square wave.
    Technician B says a Hall effect sensor needs to have its voltage signal amplified, inverted, and shaped into a clean square wave signal.
    Who is correct?
    A. A only   **C.** Both A and B
    B. B only   **D.** Neither A nor B

Fuel Systems

Upon completion and review of this chapter, you should be able to:

❏ Describe the four performance characteristics of gasoline.

❏ Describe the various types of gasoline and additives.

❏ Describe the different alternative fuels, including diesel.

❏ Describe the basic principles of carburetion.

❏ Explain the purpose and operation of the different carburetor circuits.

❏ Describe the various auxiliary carburetor controls.

❏ Describe the different types of carburetors.

❏ Explain the principles of operation of a fuel injection system. List the advantages of fuel injection.

❏ Explain the differences in point of injection in throttle body or port injection systems.

❏ Describe the difference between a sequential fuel injection (SFI) system and a multiport fuel injection (MFI) system.

❏ Explain the design and function of major electronic fuel injection (EFI) components.

❏ Explain the design and function of continuous injection systems-electronic (CIS-E) components.

❏ Describe fuel tank design and mounting on the vehicle.

❏ Describe a fuel tank filler and filler cap design.

❏ Describe three different types of fuel lines.

❏ Explain four different types of fuel line fittings.

❏ Describe the different fuel filter designs and mountings.

❏ Explain the operation of a mechanical fuel pump.

❏ Describe the operation of an electric fuel pump, including the operation of the relief valve and check valve.

❏ Describe the purpose of the inertia switch in a fuel pump circuit.

❏ Explain the purpose of the oil pressure switch that is connected in parallel with some fuel pump relays.

Introduction

This chapter focuses on an engine system that has been and is being modified year after year to meet the fuel economy and emissions demands of the government, in addition to the engine performance demands of the public. In order for an engine to be efficient, it must receive the correct amount of fuel mixed with the correct amount of air. Providing this is the purpose of the carburetion, fuel injection, and fuel delivery systems. The basics of these systems is covered in this chapter. More detailed coverage of computer-controlled carburetion and fuel injection systems is covered in later chapters of this book.

Before any discussion on fuel systems can take place, the properties of the most commonly used fuel in automobiles, gasoline, must be studied. This chapter takes a look at the composition and qualities of gasoline, as well as the other alternate fuels being used by some engines.

The carburetor is a device used to mix, or meter, fuel with air in proportions to meet the demands of the engine during all phases of operation. A carburetor is a very complex mechanism. Some of the larger two- and four-barrel carburetors can have over 200 parts. These parts make up the metering systems and subsystems that are necessary for matching air and fuel delivery with engine performance demands.

Although fuel injection systems have replaced carburetion in all late-model passenger cars and light trucks, there are many carbureted engines still on the road. Carburetors have a venturi and work under the principle of pressure differential. In other words, the amount of air and fuel

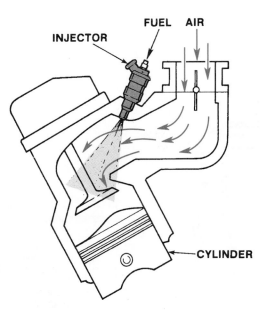

Figure 6-1 In a port fuel injection system, air and fuel are mixed in the intake manifold runners very close to the intake valve(s) of the combustion chamber.

that is delivered to the engine depends on the difference between the low pressure (vacuum) in the engine and the pressure of the outside air (atmospheric pressure).

The first use of fuel injection on a production American engine was a mechanical system installed by GM on some 1957 Corvettes and Pontiac Bonnevilles.

Fuel injection involves spraying or injecting fuel directly into the engine's intake manifold (Figure 6-1). Fuel injection, especially when it is electronically controlled, has several major advantages over carbureted systems. These include improved driveability under all conditions, improved fuel control and economy, decreased exhaust emissions, and an increase in engine efficiency and power.

The automobile's fuel system is both simple and complicated. It is simple in the systems that transfer fuel to the engine and complex in the carburetor or fuel injector system that mixes fuel with air in the correct amounts and proportion to meet all needs of the engine. The basic parts of the fuel delivery system include the fuel tank, fuel lines, filters, and pumps (Figure 6-2).

Gasoline Composition

Crude oil, as removed from the earth, is a mixture of hydrocarbon compounds ranging from gases to heavy tars and waxes. The crude oil can be refined into products such as lubricating oils, greases, asphalts, kerosene, diesel fuel, gasoline, and natural gas. Before its widespread use in the internal combustion engine of automobiles, gasoline was an unwanted byproduct of refining for oils and kerosene.

Gasoline contains hydrogen and carbon molecules. The chemical symbol for this liquid is C_8H_{15}, which indicates that each molecule of gasoline contains eight carbon atoms and fifteen hydrogen atoms. Gasoline is a colorless liquid with excellent vaporization capabilities.

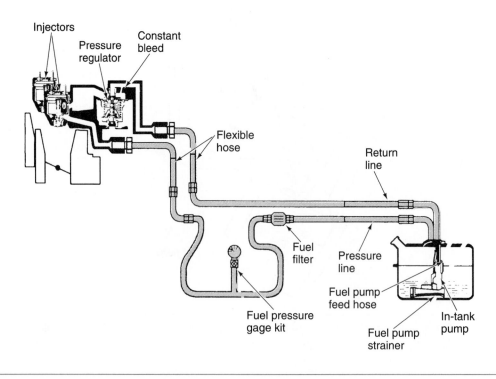

Figure 6-2 Fuel lines and location, TBI assembly. *(Courtesy of Chevrolet Motor Division—GMC)*

Gasoline Qualities

Many of the performance characteristics of gasoline can be controlled in refining and blending to provide proper engine function and vehicle driveability. The major factors affecting fuel performance are antiknock quality, volatility, sulfur content, and deposit control.

Antiknock Quality

Detonation is best described as abnormal combustion.

The flame front is the edge of the burning area inside the combustion chamber.

Two important factors affect the power and efficiency of a gasoline engine–compression ratio and **detonation**. The higher the compression ratio, the greater the engine's power output and efficiency. The better the efficiency, the less fuel consumed to produce a given power output. To have a high compression ratio requires an engine of greater structural integrity. Due to the use of low-octane unleaded gasoline in post-1975 models, compression ratios now generally range from 8:1 to 10:1. High performance cars may have higher compression ratios.

Normal combustion occurs gradually in each cylinder. The **flame front** advances smoothly across the combustion chamber until all the air-fuel mixture has been burned (Figure 6-3). Detonation occurs when the flame front fails to reach a pocket of air-fuel mixture before the temperature

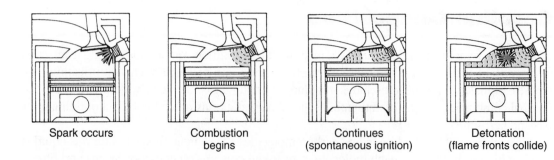

Spark occurs | Combustion begins | Continues (spontaneous ignition) | Detonation (flame fronts collide)

Figure 6-3 The stages of combustion that lead to detonation.

in that area reaches the point of self-ignition. Normal burning at the start of the combustion cycle raises the temperature and pressure of everything inside the cylinder. The last part of the air-fuel mixture is both heated and pressurized, and the combustion of those two factors can raise it to the self-ignition point. At that moment, the remaining mixture burns almost instantaneously. The two flame fronts create a pressure wave between them that can destroy cylinder head gaskets, break piston rings, and burn pistons and exhaust valves. When detonation occurs, a hammering, pinging, or knocking sound is heard. But, when the engine is operating at high speed, these sounds cannot be heard because of motor and road noise.

An **octane** number or rating was developed by the petroleum industry so the **antiknock** quality of a gasoline could be rated. The octane number is a measure of the fuel's tendency not to produce knock in an engine. The higher the octane number, the lesser the tendency to knock. By itself, the antiknock rating has nothing to do with fuel economy or engine efficiency.

Two commonly used methods for determining the octane number of motor gasoline are the motor octane number (MON) method and the research octane number (RON) method. Both use a laboratory single-cylinder engine equipped with a variable head and knock meter to indicate knock intensity. The test sample is used as fuel, and the engine compression ratio and air-fuel mixture are adjusted to develop a specified knock intensity. There are two primary standard reference fuels, isooctane and heptane. Isooctane does not knock in an engine, but is not used in gasoline because of its expense. Heptane knocks severely in an engine. Isooctane has an octane number of 100. Heptane has an octane number of zero.

A fuel of unknown octane value is run in the special test engine and the severity of knock is measured. Various proportions of heptane and isooctane are run in the test engine to duplicate the severity of the knock of the fuel being tested. When the knock caused by the heptane-isooctane mixture is identical to the test fuel, the octane number is established by the percentage of isooctane in the mixture. For example, if 85% isooctane and 15% heptane produce the same severity of knock as the fuel in question, the fuel is assigned an octane number of 85.

Factors that affect knock include:

❏ Lean Fuel Mixture. A lean mixture burns slower than a rich mixture. The heat of combustion is higher, which promotes the tendency for unburned fuel in front of the spark-ignition flame to detonate.

❏ Overadvanced Ignition Timing. Advancing the ignition timing induces knock. Slowing ignition timing suppresses knock.

❏ Compression Ratio. Compression ratio affects knock because cylinder pressures are increased with the increase in compression ratio.

❏ Valve Timing. Valve timing that fills the cylinder with more air-fuel mixture promotes higher cylinder pressures, increasing the chances for detonation.

❏ Turbocharging. Turbocharging or supercharging forces additional fuel and air into the cylinder. This induces higher cylinder pressures and promotes knock.

❏ Coolant Temperature. Hotspots in the cylinder or combustion chamber due to inefficient cooling or a damaged cooling system raise combustion chamber temperatures and promote knock.

❏ Cylinder-To-Cylinder Distribution. If an engine has poor distribution of the air-fuel mixture from cylinder to cylinder, the leaner cylinders could promote knock.

❏ Excessive Carbon Deposits. The accumulation of carbon deposits on the piston, valves, and combustion chamber causes poor heat transfer from the combustion chamber. Carbon accumulation also artificially increases the compression ratio. Both conditions cause knock.

❏ Air Inlet Temperature. The higher the air temperature when it enters the cylinder, the greater the tendency to knock.

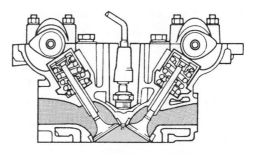

Figure 6-4 Cross section of a hemispherical combustion chamber.

❑ Combustion Chamber Shape. The optimum combustion chamber shape for reduced knocking is hemispherical with a spark plug located in the center (Figure 6-4). This hemi-head allows for faster combustion, allowing less time for detonation to occur ahead of the flame front.

❑ Octane Number. Only when an engine is designed and adjusted to take advantage of the higher octane gasoline can the value of the fuel be obtained. Most modern engines are designed to operate efficiently with regular grade gasoline and do not require a high-octane premium grade.

Volatility

The formation of gasoline vapors in the fuel system is called vapor lock.

Gasoline is very volatile. It readily evaporates so its vapor adequately mixes with air for combustion. Only vaporized fuel supports combustion. To ensure complete combustion, complete vaporization must occur. The volatility of gasoline affects the following performance characteristics or driving conditions.

Cold Starting and Warmup. A fuel can cause hard starting, hesitation, and stumbling during warmup if it does not vaporize readily. A fuel that vaporizes too easily in hot weather can form vapor bubbles in the fuel line and fuel pump, resulting in vapor lock or loss of performance.

Temperature. Because a highly volatile fuel vaporizes at a lower temperature than a less volatile fuel, winter grade gasoline is more volatile than summer grade gasoline.

Altitude. Gasoline vaporizes more easily at high altitudes, so volatility is controlled in blending according to the elevation of the place where fuel is sold.

Carburetor Icing Protection. Carburetor icing is not as common in modern engines as in older engines. It can occur when ambient temperatures reach between 28° to 55°F and the relative humidity rises above 65%. The humid air enters the carburetor and mixes with drops of fuel. When the fuel evaporates, it removes heat from the air and surrounding metal parts. When this occurs and when it is very cold outside, the throttle temperature is rapidly lowered to below 32°F, and the condensing water vapor forms ice.

Crankcase Oil Dilution. A fuel must vaporize well to prevent diluting the crankcase oil with liquid fuel. If parts of the gasoline do not vaporize, droplets of liquid break down the oil film on the cylinder wall, causing scuffing or scoring. The liquid eventually enters the crankcase oil and results in the formation of sludge, gum, and varnish accumulation as well as decreasing the lubrication properties of the oil.

Driveability. Poor vaporization can also affect the distribution of fuel from cylinder to cylinder since vaporized fuel travels farther and faster in the manifold.

Sulfur Content

Gasoline can contain some of the sulfur present in the crude oil. Sulfur content is reduced at the refinery to limit the amount of corrosion in the engine and exhaust system.

When the fuel is burned, one of the products of combustion is water. Water leaves the combustion chamber as steam but can condense back to water when passing through a cool exhaust system. Also when the engine is shut off and cools, steam condenses back to a liquid and forms water droplets. Steam present in crankcase blowby also condenses to water.

When the sulfur in the fuel is burned, it combines with the oxygen to form sulfur dioxide. This sulfur dioxide can then combine with water to form highly corrosive sulfuric acid. Corrosion caused by this acid is the leading cause of exhaust valve pitting and exhaust system deterioration. When the vehicle is equipped with a catalytic converter, the sulfur dioxide can cause an obnoxious odor of rotten eggs from the exhaust. To reduce corrosion caused by sulfuric acid, the sulfur content in gasoline is limited to less than 0.01%.

Deposit Control

Several additives are put in gasoline to control harmful deposits, including gum or oxidation inhibitors, detergents, metal deactivators, and rust inhibitors.

Basic Fuel Additives

At one time, all a gasoline-producing company had to do to produce their product was pump the crude from the ground, run it through the refinery to separate it, dump in a couple of grams of lead per gallon, and deliver the finished product to the service station. Of course, automobiles were much simpler then and what they burned was not very critical. As long as gasoline vaporized easily and did not cause the low compression engines to knock, everything was fine.

For many years, lead compounds such as tetraethyl lead (TEL) and tetramethyl lead (TML) were added to gasoline to improve its octane ratings. However, since the mid-1970s, vehicles have been designed to run on unleaded gasoline only. Leaded fuels are no longer available as automotive fuels. The main reason for the change to unleaded gasoline was to provide a fuel for cars with special antipollution devices—catalytic converters. These systems must have unleaded fuel in order to work properly.

Because of the deactivating or poisoning effect lead has on catalytic converters, gasolines are limited to a lead content of 0.06 gram per gallon. Since TEL or TML is not added to unleaded gasolines, the required octane number is obtained by blending compounds of the required octane quality. Methylcylopentadienyl manganese tricarbonyl (MMT) is a catalyst-compatible octane improver. Vehicles with catalytic converters are labeled at both the fuel gauge and fuel filler "Unleaded Fuel Only."

Times have changed. Today, refiners are under constant pressure to ensure that their product passes a series of rigorous tests for seasonal volatility, minimum octane, existent gum and oxidation stability as well as add the correct deicers and detergents to make the product competitive in a price-sensitive marketplace. Gasoline additives, primarily used in unleaded gasolines, have different properties and a variety of uses.

Anti-Icing or Deicer. Isopropyl alcohol is added seasonally to gasoline as an anti-icing agent to prevent gas line freeze-up in cold climates.

Metal Deactivators and Rust Inhibitors. These additives are used to inhibit reactions between the fuel and the metals in the fuel system that can form abrasive and filter-plugging substances.

Gum or Oxidation Inhibitors. Some gasolines contain aromatic amines and phenols to prevent formation of gum and varnish. During storage, harmful gum deposits can form due to the reaction of some gasoline components with each other and with oxygen. Oxidation inhibitors are added to promote gasoline stability. They help control gum, deposit formation, and staleness. Stale gasoline becomes cloudy and smells like paint thinner.

Detergents. The use of detergent additives in gasoline has been the subject of some public confusion. Detergent additives are designed to do only what their name implies—clean certain critical components inside the engine. They do not affect octane.

Shop Manual
Chapter 6, page 222

Gum content is influenced by the age of the gasoline and its exposure to oxygen and certain metals such as copper.

If gasoline is allowed to evaporate, the residue left can form gum and varnish.

Ethanol. By far the most widely used gasoline additive today is **ethanol**, or grain alcohol. Ethanol's value as an octane enhancer becomes more apparent when considered in the context of the government-mandated phasedown of tetraethyl lead. Blending 10% ethanol into gasoline is seen as a comparatively inexpensive octane booster that results in an increase of 2.5 to 3 road octane points. In addition to octane enhancement, ethanol blending keeps the carburetor or fuel injectors clean due to detergent additive packages found in most of the ethanol marketed. It also inhibits fuel system and injection corrosion due to additive packages and decreases carbon monoxide emissions at the tailpipe due to the higher oxygen content of blended fuel.

Methanol. Methanol is the lightest and simplest of the alcohols and is also known as wood alcohol. It can be distilled from coal, but most of what is used today is derived from natural gas. Many automakers continue to warn motorists about using a fuel that contains more than 10% methanol and co-solvents by volume. Methanol is recognized as being far more corrosive to fuel system components than ethanol, and it is this corrosion that has automakers concerned.

MTBE. Methyl tertiary butyl ether (**MTBE**) has been used as an octane enhancer and supply extender because of excellent compatibility with gasoline. Current EPA restrictions on oxygenates limit MTBE in unleaded gasoline to 11% of volume. At that level, it increases pump octane (RM/2) by 2.5 points. However, it is usually found in concentrations of 7–8% of volume. MTBE increases octane while reducing carbon monoxide emissions at the tail pipe and does it at a cost that makes it very attractive to gasoline marketers across the country.

Reformulated gasoline is also called "cleaner-burning" gasoline.

MTBEs and ethanol are the most commonly used oxygenates for producing reformulated gasoline (RFG). By blending oxygen into the gasoline, the fuel requires less ambient oxygen for complete burning. Therefore for the same carburetor or fuel injector settings, oxygenated gasoline produces a leaner air-fuel mixture and generates less carbon monoxide.

Alternative Fuels

Tighter federal, state, and local emissions regulations have led to a search for an alternative fuel. Many things are considered when determining the viability of an alternative fuel, including emissions, cost, fuel availability, fuel consumption, safety, engine life, fueling facilities, weight and space requirements of fuel tanks, and the range of a fully fueled vehicle. Currently, the major competing alternative fuels include ethanol, methanol, propane, and natural gas. Although diesel fuel has been in use for many years, its properties are included in this section. Diesel fuel has not proven, yet, to be a successful alternate fuel for automobiles.

Ethanol and methanol were presented earlier under other gasoline additives. Propane is a petroleum-based pressurized fuel used as a liquid. It is a constituent of natural gas. Natural gas comes in two forms: **compressed natural gas** (**CNG**) and liquefied natural gas (LNG).

LP-Gas

LP stands for liquefied petroleum.

LP-gas is a byproduct of crude oil refining and it is also found in natural gas wells. Fuel grade LP-gas is almost pure propane with a little butane and propylene usually present. Because of its high propane content, many people simply refer to LP-gas as propane.

Propane burns clean in the engine and can be precisely controlled. Because it vaporizes at atmospheric temperatures and pressures, it does not puddle in the intake manifold. This means it emits less hydrocarbons and carbon monoxide. Emission controls on the engine can be simpler. Cold starting is easy, down to much below zero. At normal cold temperatures, the propane engine fires easily and produces power without surge or stumble.

One of the most noticeable differences between propane and gasoline is that propane is a dry fuel. It enters the engine as vapor. Gasoline, on the other hand, enters the engine as tiny droplets of liquid, whether it flows through a carburetor or is sprayed in through a fuel injector.

The propane fuel system is a completely closed system that contains a supply of pressurized LP-gas. Since the fuel is already under pressure, no fuel pump is needed. From the pressurized fuel tanks, the fuel flows to a vacuum filter fuel lock. This serves as a filter and a control allowing fuel to flow to the engine. Fuel flows to a converter or heat exchanger where it changes from a liquid to a gas. When the propane flows through the converter, it expands as it changes into a gas. The carburetor mixes gaseous propane with the gaseous air. Airflow into the engine is controlled by a butterfly valve in the venturi. Mixture is controlled by a fuel metering valve operated by a diaphragm, which is controlled by intake manifold pressure. The idle system is an air bleed, similar to a gasoline engine. In fact, except for the fact that the propane carburetor does not require a fuel bowl, the two carburetor types are basically the same.

Natural Gas

Vehicles have been designed with gasoline/CNG, diesel/CNG, and dedicated CNG engine applications. Compressed natural gas (CNG) vehicles offer several advantages over gasoline.

> The designation "Dedicated Vehicle" means the vehicle is designed to use one type of fuel.

❏ The fuel costs less.

❏ It is the cleanest alternative fuel, generating up to 99% less carbon monoxide than gasoline, no particulates, almost no sulfur dioxide, and 85% less reactive hydrocarbons than gasoline.

❏ Natural gas vehicles are safer. The fuel tanks used for CNG are aluminum or steel cylinders with walls that are 1/2 to 3/4 inch thick. They can withstand severe crash tests, direct gunfire, dynamite explosions, and burning beyond any standard sheet metal gasoline tank. Because it is lighter than air, natural gas dissipates quickly. It also has a higher ignition temperature.

❏ It generally reduces vehicle maintenance since it burns cleanly. Oil changes may not be needed before 12,000 miles and spark plugs could last as long as 75,000 miles.

❏ Natural gas is abundant and readily available in the United States.

The chief disadvantage of CNG at present is its nonavailability to most users. Fuel facilities are needed in greater numbers than are currently in existence due to the relatively shorter range of CNG vehicles. The space taken by the CNG cylinders and their weight, about 300 pounds, also would be considered disadvantages in most applications.

Diesel Fuel

Diesel fuel is also made from petroleum. At the refinery, the petroleum is separated into three major components: gasoline, middle distillates, and all remaining substances. Diesel fuel comes from the middle distillate group, which has properties and characteristics different from gasoline.

The shape of the fuel spray, turbulence in the combustion chamber, beginning and duration of injection, and the chemical properties of the diesel fuel all affect the power output of the diesel engine. The significant chemical properties of diesel fuel are described briefly in the following.

Viscosity. The viscosity of diesel fuel varies with temperature. The viscosity of diesel fuel directly affects the spray pattern of the fuel into the combustion chamber. Fuel with a high viscosity produces large droplets that are hard to burn. Fuel with a low viscosity sprays in a fine, easily burned mist. If the viscosity is too low, however, it does not adequately lubricate and cool the injection pump and nozzles.

> Viscosity is a measure of a fluid's resistance to flow.

Wax Appearance Point and Pour Point. Temperature affects diesel fuel more than it affects gasoline. This is because diesel fuels contain paraffin. As temperatures drop past a certain point, wax crystals begin to form in the fuel. The point where the wax crystals appear is the wax appearance point (WAP) or cloud point. The better the quality of the fuel, the lower the WAP. As temper-

> Paraffin is a wax substance present in the middle distillates of crude oil.

atures drop, the wax crystals grow larger and restrict the flow of fuel through the filters and lines. Eventually, the fuel, which may still be liquid, stops flowing because the wax crystals plug a filter or line. As the temperature continues to drop, the fuel reaches a point where it solidifies and no longer flows. This is called the pour point. In cold climates it is recommended that a low-temperature pour point fuel be used.

Volatility. Gasoline is extremely volatile compared to diesel fuel. The amount of carbon residue left by diesel fuel depends on the quality and the volatility of that fuel. Fuel that has a low volatility is much more prone to leaving carbon residue. The small, high-speed diesels found in automobiles require a high-quality and highly volatile fuel because they cannot tolerate excessive carbon deposits. Large, low-speed industrial diesels are relatively unaffected by carbon deposits and can run on low-quality fuel.

Cetane Number or Rating. Diesel fuel's ignition quality is measured by the **cetane** rating. Much like the octane number, the cetane number is measured in a single-cylinder test engine with a variable compression ratio. The diesel fuel to be tested is compared to cetane, a colorless, liquid hydrocarbon that has excellent ignition qualities. Cetane is rated at 100. The higher the cetane number, the shorter the ignition lag time (delay time) from the point the fuel enters the combustion chamber until it ignites.

In fuels that are readily available, the cetane number ranges from 40 to 55 with values of 40 to 50 being most common. These cetane values are satisfactory for medium-speed engines whose rated speeds are from 500 to 1,200 rpm and for high-speed engines rated over 1,200 rpm. Low-speed engines rated below 500 rpm can use fuels in the above 30 cetane number range. The cetane number improves with the addition of certain compounds such as ethyl nitrate, acetone peroxide, and amyl nitrate. Amyl nitrate is commercially available for this purpose.

Diesel Fuel Grades. Minimum quality standards for diesel fuel grades have been set by the American Society for Testing Materials. Two grades of diesel fuels, Number 1 and Number 2, are available. Number 2 diesel fuel is the most popular and widely distributed. Number 1 diesel fuel is less dense than Number 2, with a lower heat content. Number 1 diesel fuel is blended with Number 2 to improve starting in cold weather. In the winter, diesel fuel is likely to be a mixture of Number 1 and 2 fuels. In moderately cold climates, the blend may be 90% Number 2 to 10% Number 1. In very cold climates, the ratio may be as high as 50-50. Diesel fuel economy can be expected to drop off during the winter months due to the use of Number 1 diesel in the fuel blend.

Basic Carburetor Design

A carburetor is a metering device that is used to mix air and fuel in the proper ratios to accommodate the engine's needs during a variety of operating conditions. The mixing of air and fuel takes place when the fuel in released in fine particles into the incoming air. The movement of the incoming air is caused by the pressure difference resulting from the vacuum formed by the engine and atmospheric pressure. A carburetor works on the principle of pressure differential. When two different pressures are exposed to each other, the higher of the two will move quickly toward the lower pressure in an attempt to balance the pressure (Figure 6-5). The more pressure differential there is, the faster the air will travel. In a carburetor, as airflow speeds increases, more fuel is mixed with the air. The carburetor has two basic functions: to regulate engine speed by controlling the amount of fuel and air the engine receives, and to deliver the correct ratio of air and fuel to the engine's cylinders.

Not only is a pressure differential formed by the pistons during their intake stroke, there is also one formed by the carburetor itself. In the throat or bore of a carburetor is a **venturi** (Figure 6-6). This restriction causes an increase in air velocity and lowers the pressure of the air passing through

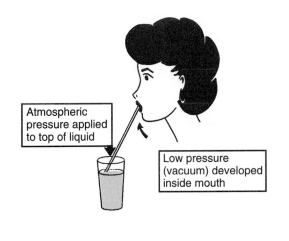

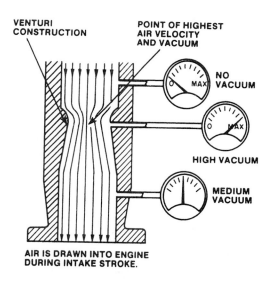

AIR IS DRAWN INTO ENGINE
DURING INTAKE STROKE.

Figure 6-5 Sucking on a straw creates a vacuum in your mouth; the atmospheric pressure above the soda pushes the soda up the straw into your mouth. *(Courtesy of Chevrolet Motor Division—GMC)*

Figure 6-6 A venturi is a restriction in the path of air flow. A vacuum is produced at the point of greatest restriction.

it. In many carburetors, a smaller boost venturi is positioned inside the main venturi to provide increased airflow speed and vacuum (Figure 6-7).

A carburetor controls engine speed and air-fuel ratios by controlling air velocity. Actually the carburetor's throttle plate assembly controls the velocity of the air. A throttle plate is a circular disc that is placed directly in the flow of air and fuel, below the venturi. It is connected to the driver's throttle pedal so that it opens to a vertical position as the pedal is depressed. When the throttle plate is all the way open, there is very little restriction of air. This is a maximum speed condition. As the driver's foot is removed, a spring closes the throttle plate. This restricts the amount of air going into the engine. This is a low speed condition.

The throttle plate is nearly closed at idle speed, preventing much airflow into the cylinders. As a result of this reduced airflow, there is not much pressure in the combustion chambers during the compression and power strokes, and this causes the engine to run at idle speed (Figure 6-8).

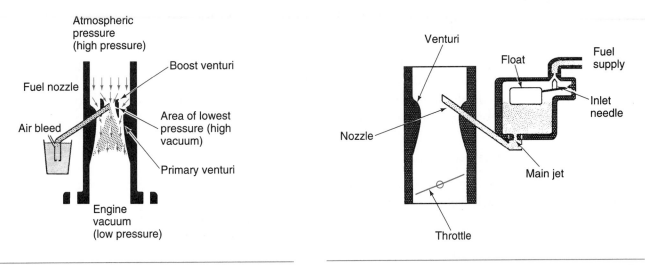

Figure 6-7 A boost venturi added to the center of the main venturi gives increased air flow speed. *(Courtesy of Chevrolet Motor Division—GMC)*

Figure 6-8 The throttle plate controls the amount of air entering the cylinders. *(Courtesy of Chevrolet Motor Division—GMC)*

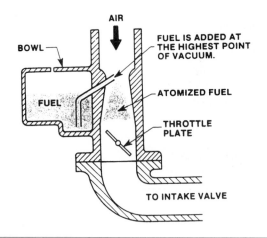

Figure 6-9 The vacuum that is produced at the venturi is used to draw in the fuel from the carburetor.

When the throttle is opened wider, the airflow into the cylinders increases, creating more pressure on the pistons during the compression and power strokes and causing the engine to run faster.

If air passes rapidly through a tube or cylinder that has a small hole in it, air will be drawn through the hole to combine with the moving air. Again this is the result of a higher pressure moving toward a lower pressure. If a column of air is moving quickly, there must be a large pressure differential. Therefore, outside air will enter the small hole and join with the moving air in attempt to balance the pressures. If a hose were connected to the small hole and to a container filled with a liquid, the liquid would be drawn out of the container and into the moving air, if the liquid in the container was exposed to atmospheric pressure. The greater the air velocity, the more fuel will be drawn into the moving column of air. This is another principle that must be understood before studying how a carburetor works.

As the throttle plate is opened, engine speed increases. This allows more air to be drawn into the carburetor. As a result, venturi vacuum increases because the greater the velocity of air passing through the venturi, the greater the vacuum. Venturi vacuum is used to draw in the correct amount of fuel through a discharge tube (Figure 6-9). As the air flows through the venturi, vacuum draws the fuel from the carburetor bowl into the stream of air going into the engine. More fuel is drawn in as venturi vacuum increases, and less as the vacuum decreases.

Venturi vacuum is created by the airflow through the venturi. This vacuum increases in relation to engine speed. Remember that an increase in vacuum is a decrease in pressure. Manifold vacuum is created by the downward piston movement on the intake strokes attempting to pull air past the restriction of the throttle. Manifold vacuum is highest with the throttle closed at idle and on deceleration. When the throttle is opened, manifold vacuum decreases. This vacuum is lowest at wide-open throttle.

Carburetion

The three general stages involved in carburetion are metering, atomization, and vaporization. **Metering** is another term for measuring. In the process of carburetion, fuel is metered into the air passing through the barrel of the carburetor. The ideal air-fuel ratio at which all the fuel blends with all the oxygen in the air is called the **stoichiometric ratio**. This ratio is about 14.7:1 (Figure 6-10). If there is more fuel in the mixture it is called a **rich** mix. If there is less fuel it is called a **lean** mix. The amount of fuel metered into the air is varied in relation to the amount of air passing through the carburetor. Additional factors that influence the amount of fuel metered into the air include engine temperature, load and speed requirements, and the amount of oxygen in the exhaust stream.

Shop Manual
Chapter 6, page 243

The mixture of air and fuel is called an **emulsion**.

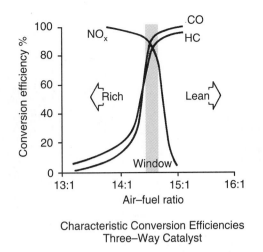

Characteristic Conversion Efficiencies
Three–Way Catalyst

Figure 6-10 The efficiency of a three-way catalytic converter with different air-fuel ratios. *(Courtesy of Chrysler Corporation)*

Atomization is the stage in which the metered fuel is drawn into the airstream in the form of tiny droplets. The droplets of fuel are drawn out of passages called discharge ports.

The surface area of an atomized droplet is in contact with a relatively large amount of surrounding air. In addition, the venturi is a low-pressure area. The boiling point of the fuel is greatly reduced by lower pressures, therefore liquid gasoline can become a vapor with low amounts of heat. These factors combine to create a fine mist of fuel below the venturi in the bore. This is called **vaporization**–the last stage of carburetion. It occurs below the venturi, in the intake manifold, and within the cylinder. Swirl, turbulence, and heat within the intake manifold and cylinder also enhance vaporization.

In order for the air and fuel mixture to burn efficiently, one gallon or approximately six pounds of gasoline must be mixed with 9,000 gallons or nearly 90 pounds of air! This is the ideal ratio.

Basic Carburetor Circuits

Variations in engine speed and load demand different amounts of air and fuel (often in differing proportions) for best performance, and present complex problems for the carburetor. At engine idle speeds, for example, there is insufficient air velocity to cause fuel to be drawn from the discharge nozzle and into the airstream. Also, with a sudden change in engine speed, such as rapid acceleration, the venturi effect (pressure differential) is momentarily lost. Therefore, the carburetor must have special circuits or systems to cope with these situations. There are seven basic circuits used on a typical carburetor.

1. Float
2. Idle
3. Off-idle
4. Main metering
5. Full power (or power enrichment)
6. Accelerator pump
7. Choke

Float Circuit

A **float circuit** or fuel inlet system (Figure 6-11) of a typical carburetor consists of the fuel bowl, fuel inlet fitting, fuel inlet needle valve and seat, and the float. A full screen or filter is usually installed at the fuel inlet to prevent dirty fuel from mixing in the carburetor and causing a problem.

Shop Manual
Chapter 6, page 257

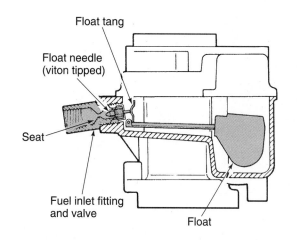

Figure 6-11 Float assembly. *(Courtesy of Chrysler Corporation)*

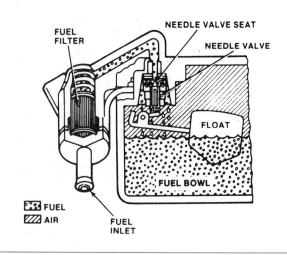

Figure 6-12 Fuel inlet system.

The float system stores fuel and holds it as a precise level as a starting point for uniform fuel flow. Fuel enters the carburetor through the inlet line and passes through an inlet filter to the inlet needle valve and seat. The incoming fuel is captured and stored in the reservoir or fuel bowl. The fuel bowl is normally an integral part of the main casting but can be a separate casting attached to the carburetor body with screws. Carburetors with primary and secondary venturis might have two separate fuel bowls.

The level of the fuel in the bowl is maintained at a specified height by the rising and falling of a pivoted float in the fuel bowl. Early floats were made of brass stampings soldered into an airtight lung. Floats made of nitrile rubber, a closed cell material made of thousands of tiny hollow spheres, are used most exclusively on domestic cars. Hollow plastic floats can also be found in carburetors. A tang near the pivot on the float arm contacts the float needle. This needle contains a tapered end with a Viton tip. When the float moves upward, the tang moves the needle down so the tapered tip fits tightly into its seat at the fuel inlet (Figure 6-12).

As fuel enters the bowl, the float opens and closes the inlet needle valve. With the needle valve closed, fuel is prevented from entering the carburetor. Fuel pressure against the inlet needle valve tends to force it open while the buoyancy of the float in the bowl tends to force it closed. This action establishes the precise fuel level for the carburetor. To prevent the float from bouncing and vibrating, a bumper spring is usually installed under the float or to a tang connected to the float.

The metering systems of a carburetor are designed to function properly only when the fuel level in the bowl is at specific level. The specific level is adjusted with the carburetor partially disassembled on most models. However, it can be adjusted externally on a few carburetors by turning a threaded inlet valve assembly. Since atmospheric pressure must be available above the fuel in the float bowl, this bowl is vented to the air horn area on top of the carburetor inside the air cleaner (Figure 6-13).

When the float bowl is vented to the air horn under the air cleaner, the carburetor may be referred to as a balanced carburetor. If the air cleaner element is partly restricted, the airflow into the carburetor is reduced, and the atmospheric pressure is also reduced inside the element. When this reduced pressure is supplied to the float bowl with a partially restricted air cleaner, a balance is maintained between the venturi vacuum and the float bowl pressure. This balance in pressure minimizes the effect of a partially restricted air cleaner on the air-fuel ratio. An antibackfire screen may be positioned in the air horn vent to prevent backfires into the intake manifold from entering the float bowl.

Prior to the introduction of emission controls, most primary fuel bowls were also vented to the atmosphere when the engine was at idle or turned off. Since the introduction of evaporative control systems, all fuel bowls are vented by a valve to a charcoal canister, which absorbs and stores fuel vapors. The vapors are returned to the engine when it is restarted.

The float needle is called the needle valve.

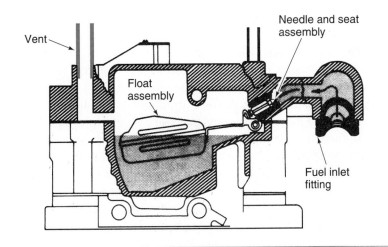

Figure 6-13 The float bowl is vented to the air horn area under the air cleaner. *(Courtesy of Chrysler Corporation)*

Idle and Off-idle Circuits

At idle, the engine requires a richer air-fuel mixture than during normal cruising conditions. This is because residual exhaust gases remain in the combustion chambers during low-engine rpm and dilute the air-fuel charge. The idle circuit supplies the richer air-fuel mixture to operate the engine at idle and low speeds.

During idle conditions, there is not enough air entering the venturi to cause a vacuum to move the fuel. The throttle plate is almost all the way closed as shown in Figure 6-14. During this condition, there is a large vacuum below the throttle plate. This vacuum causes fuel to be drawn from the carburetor float bowl through internal passages to the idle port below the throttle plate. As fuel is drawn from the float bowl to the idle port, air is drawn in through an air-bleed passageway near the top of the carburetor. Only a small amount of air passes by the throttle plate. The end result is the richer air-fuel mixture needed for idle operation.

As the throttle plate is opened during fast-idle or low-speed operation, a transfer slot located above the throttle plate is progressively exposed to vacuum and the air-fuel emulsion is also discharged from the **transfer slot**. The increased air-fuel mixture flow provides a smooth transition

Shop Manual
Chapter 6, page 243

The throttle plate is often called the throttle valve.

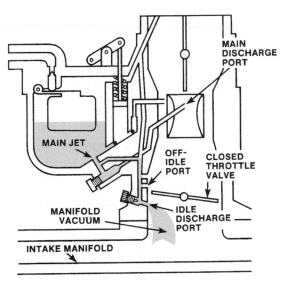

Figure 6-14 An idle discharge port draws fuel from the carburetor to allow the engine to run at idle.

between idle and cruising modes of operation. Some carburetors have a series of holes called off-idle air passages, instead of a transfer slot. Like the transfer slot, the holes permit increased fuel delivery as the throttle opens. This is called an off-idle system.

On older carburetors there is an idle mixture needle valve or screw. This valve is used to control or adjust the amount of air and fuel at idle. The idle mixture screw has a tapered tip on the inner end, and this screw is threaded into the carburetor base casting. A spring is positioned between the head of the idle mixture screw and the base casting to prevent vibration from moving the screw. The position of the idle mixture screw supplies the proper amount of air and fuel from the idle discharge port in relation to the amount of air flowing past the throttle to maintain the correct air-fuel ratio and emission levels at idle speed. Most carburetors, however, have limiting caps on the idle mixture screws, which limit the amount of adjustment available. On newer carburetors, the idle mixture screws are sealed with steel plugs to eliminate all adjustment.

To improve idle quality when meeting emission standards, a variable air bleed idle system is used on some carburetors (Figure 6-15). In this system there are two idle air bleeds. One idle air bleed is installed normally in the air horn. An auxiliary idle air bleed is drilled into the lower skirt of the venturi. The air entering through the auxiliary passage is adjusted by an idle air adjusting screw. The screw is turned clockwise to enrich the idle mixture and counterclockwise to lean out the idle air-fuel mixture.

Shop Manual
Chapter 6, page 251

Main Metering Circuit

The main metering circuit (Figure 6-16) comes into operation as engine speed increases. Opening the throttle plate past the idle position increases the air flowing through the venturi and creates enough vacuum to allow atmospheric pressure to force fuel through the main metering system and out the main fuel discharge nozzle, located in the center of the venturi. As engine speed is increased, the vacuum at the discharge nozzle increases. The main metering jet size, the size of the venturi, and the float bowl level are designed to provide the proper air-fuel ratio during main metering circuit operation.

This vacuum or pressure differential causes fuel to flow out of the fuel bowl, through the main metering jet, and into the main well. On most carburetors the main well is vented through a precisely sized opening called the main well air bleed. The main well air bleed allows air to enter at the top of the main well. Air entering the calibrated main air bleed prevents a vacuum from developing in the main well. The air also allows for aeration of the fuel as it leaves the main well

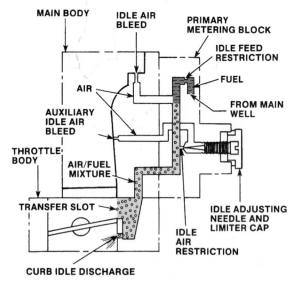

Figure 6-15 The idle discharge port and off-idle transfer slot supply fuel to the carburetor at low speeds.

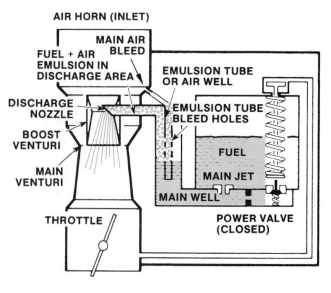

Figure 6-16 Air-fuel mixture routed to the venturi via the main metering circuit and discharge nozzle.

and travels up the well tube. This allows the fuel to be partially atomized as it travels toward the discharge nozzle.

Air flows through the bleeds because it draws air from the high-pressure area above the venturi and the main metering discharge nozzle is in the low-pressure venturi level.

As the air speed increases through the venturi, more fuel is drawn from the main well. This lowers the fuel level in the main well and exposes more air bleed openings in the main well tube. This causes extra air to enter the well tube, mix with the fuel, and dilute or lean the mixture. This action circumvents the richening effect caused by the increased carburetor airflow. If the fuel were not diluted as such, the air-fuel mixture would richen at high speed as the venturi vacuum increased faster than the engine's need for additional fuel.

Secondary Metering Systems

Some carburetors have more than one **barrel** or air horn. Each barrel has a throttle plate and main metering system (as well as other circuits). When all throttle plates open and close simultaneously, the carburetor is called a single stage carburetor.

Some carburetors have two stages, called primary and secondary stages. In the primary stage, one or two throttle plates operate normally as in a single stage carburetor. The secondary stage throttle plates, however, only open after the primary throttle plates have opened a certain amount. Thus, the primary stage controls off-idle and low cruising speeds and the secondary stage opens when high cruising speeds or loads require additional air and fuel. The added flow capacity raises the engine power output.

The secondary throttle plates can be opened mechanically or by a vacuum source. Mechanically actuated secondary throttle plates are opened by a tab on the primary throttle linkage. After the throttle primary plates open a set amount (usually 40°–45°), the tab engages the secondary throttle linkage, forcing the plates open. Vacuum-actuated secondaries have a spring-loaded diaphragm. Vacuum is supplied to the diaphragm from ports in the primary and secondary throttle bores. When the vacuum in the primary bore reaches a specific level, the vacuum supplied to the diaphragm overcomes the spring and opens the secondary throttle plates (Figure 6-17). The vacuum created in the secondary throttle bore increases the vacuum signal to the diaphragm, opening the secondary throttle valves still farther.

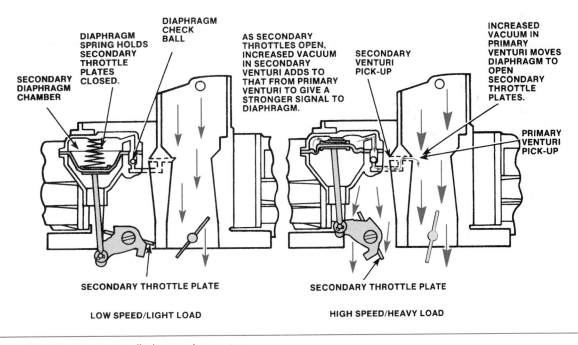

Figure 6-17 Vacuum-controlled secondary system.

Power Enrichment Circuit

At wide-open throttle, the engine needs a richer than normal air-fuel mixture. This mixture cannot be supplied by the main metering system. So, an additional fuel enrichment or full-power system is provided on most carburetors. The power enrichment system meters additional fuel into the mixture. This can be accomplished in several ways.

Metering Rods

In some carburetors, power enrichment is provided by **metering rods** placed in the main jets (Figure 6-18). The metering rods are actuated mechanically or by vacuum. When the throttle is not wide open, the throttle linkage keeps the rods in the jets, providing normal fuel flow. When the throttle is opened wide, either a mechanical link in the throttle linkage or vacuum-actuated lever lifts the rods out of the jets, enabling more fuel to be forced into the main well. The additional fuel flow richens the air-fuel mixture.

Normally, if the engine is operating at more than 55% throttle opening, the throttle shaft rotation blocks the vacuum passage to the power piston and vents this passage to the atmosphere. Under this condition, the spring pushes the power piston downward. When this action occurs, the lower end of the piston pushes the power valve open, allowing additional fuel through the power valve into the main well and main system. This additional fuel supply into the main system provides a richer air-fuel ratio and increased engine power when the engine is operating under high speed or heavy load conditions.

Power Valves

The **power valve** (Figure 6-19) is basically a vacuum-operated metering rod. It consists of a vacuum diaphragm or piston, a spring-loaded valve, and a metering rod inside an auxiliary fuel jet usually located in the bottom of the fuel bowl. A vacuum passageway machined into the main body casting supplies manifold vacuum to the diaphragm or piston. During idle and low cruising speeds, the vacuum holds the power valve closed. As engine speed and load increases and the vacuum signal drops to a specific level, the spring overcomes the vacuum and forces the power valve out of the jet. This increases the fuel flowing into the main well.

The power valve has been replaced or modified in today's feedback carburetor. In a feedback system, an electrical solenoid controls the metering fuel jets or idle air bleeds to regulate the air-fuel mixture. Feedback carburetors are discussed later in this chapter.

Accelerator Pump Circuit

The off-idle or transfer circuit discussed earlier allows the engine to be accelerated smoothly without hesitation or lags. However, during sudden acceleration, the engine experiences a momentary drop in power unless additional fuel is simultaneously introduced into the air charge.

During sudden acceleration the air flowing through the carburetor reacts almost immediately to each change in the throttle plate opening. However, since fuel is heavier than air, it has a slower response time. Fuel in the main metering system or idle system takes a fraction of a second to respond to the throttle opening. This lag in time creates a hesitation of fuel flow whenever the accelerator pedal is quickly depressed. The accelerator pump system solves this problem by mechanically supplying fuel until the other fuel metering systems are able to supply the proper mixture.

One type of accelerator pump (Figure 6-20) is the diaphragm type located in the bottom of the fuel bowl. Locating the pump in the bottom of the fuel bowl assures a more solid charge of fuel (fewer bubbles).

When the throttle is opened, the pump linkage, activated by a cam on the throttle lever, forces the pump diaphragm up. As the diaphragm moves up, the pressure forces the pump inlet check ball or valve onto its seat. This prevents the fuel from flowing back into the fuel bowl. At the

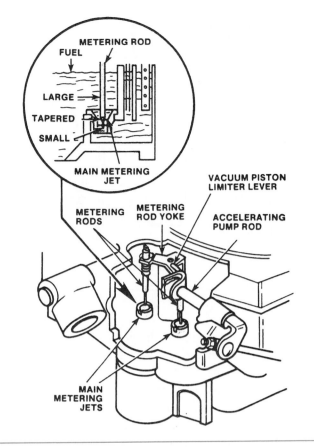

Figure 6-18 Some power systems consist of metering rods placed in the main jets.

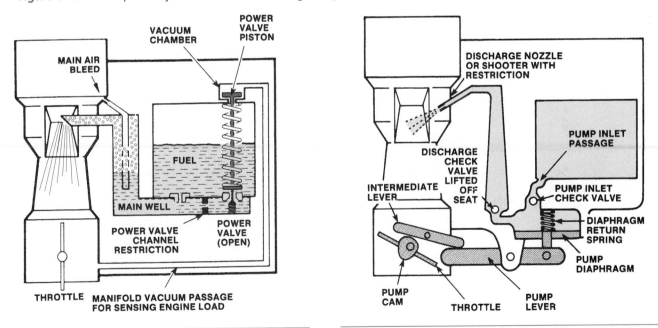

Figure 6-19 Power valve system.

Figure 6-20 Diaphragm accelerator pump.

same time, the pressure of the fuel causes the discharge check ball or valve to raise and fuel is then discharged into the venturi.

As the throttle returns toward the closed position, the linkage returns to its original position and the diaphragm return spring forces the diaphragm down. The pump inlet check valve is moved off its seat and the diaphragm chamber is refilled with fuel from the fuel bowl.

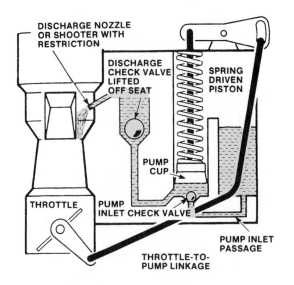

Figure 6-21 Plunger accelerator pump.

Another common type of accelerator pump is shown in Figure 6-21. The pump system contains a piston and stem assembly that is connected through a linkage to the throttle. A neoprene or rubber cup is positioned on the lower end of the accelerator pump piston, and this piston is mounted in a circular well. The piston and stem can move up and down inside the upper part of the piston, but the pump drive spring mounted on the piston keeps the piston extended. A passage is drilled through the carburetor casting from the bottom of the pump well to a discharge port in the venturi area. A check ball and weight are positioned in the pump discharge passage.

When the throttle is opened from the idle position, the linkage forces the top of the pump piston stem downward. Since the gasoline under the pump piston cannot be compressed, the upper part of the piston slides down on the piston, and this action compresses the pump drive spring. The pump drive spring slowly pushes the pump piston downward. This downward piston movement forces fuel through the pump discharge passage, past the check ball and weight, and out the pump discharge port into the venturi area.

When the throttle is about half open, the linkage has bottomed the plunger in the bottom of the pump well. Further opening of the throttle only collapses the two-piece plunger and compresses the pump drive spring. Since the plunger does not move when the throttle is opened past the half-way point, the accelerator pump does not discharge fuel under this condition. At half-throttle opening, the air velocity and fuel flow are already established in the venturi, and the accelerator pump action is not required.

When the engine is decelerated, the accelerator pump linkage pulls the pump piston upward in the well. Under this condition, fuel moves from the float bowl past the cup into the pump well. As the pump piston moves upward, the check ball seats in the discharge passage and prevents air intake from the pump discharge port into the discharge passage. The weight on top of the check ball prevents venturi vacuum from lifting the check ball and pulling fuel out of the discharge passage when the venturi vacuum is increased at high speed.

Choke Circuit

A cold engine needs a very rich air-fuel mixture during cranking and startup. Providing the rich mixture is the job of the **choke** circuit (Figure 6-22).

During a cold startup, the choke should be closed. This creates a very high vacuum level in the air horn below the choke plate. As the air pressure outside the carburetor forces its way into the low-pressure areas, it draws with it a rich air-fuel mixture. When the throttle plate is closed, the mixture is forced out through the idle port or ports below the throttle valve. If the throttle valve is

Shop Manual
Chapter 6, page 246

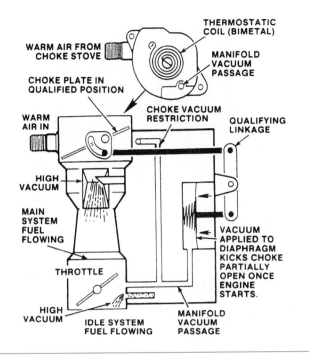

THERMOSTATIC COIL (BIMETAL)

WARM AIR FROM CHOKE STOVE

MANIFOLD VACUUM PASSAGE

CHOKE PLATE IN QUALIFIED POSITION

WARM AIR IN

CHOKE VACUUM RESTRICTION

QUALIFYING LINKAGE

HIGH VACUUM

MAIN SYSTEM FUEL FLOWING

VACUUM APPLIED TO DIAPHRAGM KICKS CHOKE PARTIALLY OPEN ONCE ENGINE STARTS.

THROTTLE

HIGH VACUUM

IDLE SYSTEM FUEL FLOWING

MANIFOLD VACUUM PASSAGE

Figure 6-22 Choke system.

opened to assist in starting the engine, additional ports are exposed to the low-pressure manifold pressure and additional fuel is forced into the air horn. After the engine starts, a leaner mixture can be used to keep the engine running. Therefore, the choke should be opened some to allow increased airflow. After the engine has warmed to normal operating temperatures, the choke should be opened completely to allow the throttle to control airflow and fuel metering.

Before the introduction of automatic chokes, the opening and closing of the choke plate was manually controlled by the driver. A choke cable was connected to a knob inside the passenger compartment on the dash. To close the choke, the driver simply pulled the knob out. As the engine warmed, the coke knob was gradually pushed in to open the choke.

Modern carbureted vehicles have an automatic choke that operates without any driver assistance. Being more sensitive to engine temperature and the demands of the engine, an automatic choke is more efficient. Many variations of automatic chokes have been used through the years, the following describes some of them.

Hot Air Choke System

The choke system contains a choke plate mounted on a shaft near the top of the carburetor. A linkage extends from the choke shaft to the choke housing, and a tang on this linkage extends into the housing. A thermostatic coil is mounted in the choke housing cover, and the inner end of the choke coil is attached to the cover. The outer end of the choke coil contacts the tang on the choke linkage.

A vacuum passage is connected from the carburetor base below the throttle to the choke housing, and a heat pipe is connected from a stove in the exhaust manifold to the choke housing.

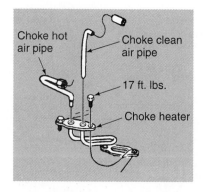

Figure 6-23 A choke heat stove. This type is located in the exhaust crossover passage in the intake manifold. *(Courtesy of Oldsmobile Division—GMC)*

On some V-type engines, this heat stove may be positioned in the exhaust crossover passage in the intake manifold (Figure 6-23).

A vacuum break diaphragm is linked to the choke plate, and a hose from this diaphragm chamber is connected to the intake manifold through a restriction. When the choke spring is cold, it closes the choke plate and provides a very rich air-fuel ratio while starting a cold engine. The closed choke plate also allows manifold vacuum to be available in the venturi area under the choke plate. This manifold vacuum pulls additional fuel out of the main discharge nozzle. Since a considerable amount of fuel condenses on the cold intake manifold before it enters the cylinders, a very rich air-fuel ratio is necessary while starting and running a cold engine.

A vacuum break diaphragm may be called a vacuum kick diaphragm.

After the engine is started, manifold vacuum gradually seats the vacuum break diaphragm which pulls the choke plate open a specific amount to supply the correct air-fuel ratio. The intake manifold vacuum in the choke housing pulls hot air from the manifold heat stove into the choke housing and intake manifold. This heat causes the bimetallic coil to slowly release its tension and allow the choke plate to open.

Electric Choke

The thermostatic coil may be called a bimetallic coil.

The choke, thermostatic coil, plate, housing, vacuum break, and linkage in an electric choke are similar to a hot air choke. In an electric choke, the bimetallic spring is heated electrically rather than by hot air from the exhaust manifold heat stove. In an electric choke, the housing does not have a heat pipe connection or an intake manifold vacuum connection. Many electric chokes have the cover riveted onto the choke housing to prevent improper choke adjustments.

A wire is connected from the ignition switch to a terminal on the choke cover. When the thermostatic coil is cold, its winding action closes the choke plate. If the engine is started, current flows through the choke cover terminal and ceramic resistor to ground on the choke housing. The ceramic resistor is heated by this current flow, and this heat causes the thermostatic spring to slowly lose its tension and open the choke plate.

A choke heater relay and an instrument panel warning indicator light are connected in some electric choke systems (Figure 6-24). The relay winding is grounded through the alternator field circuit, and the choke heater is connected to the relay contacts and to the warning light. The relay contacts are normally closed.

When the ignition switch is turned on, voltage is supplied through the warning light to the choke heater. Under this condition, the light is illuminated and the high resistance in this bulb provides a very low current flow to the choke heater so the choke spring is not heated if the driver leaves the ignition switch on with a cold engine. With the ignition switch on, current also flows from the ignition switch through the relay winding and the alternator field to ground. This current flow keeps the relay contacts open.

160

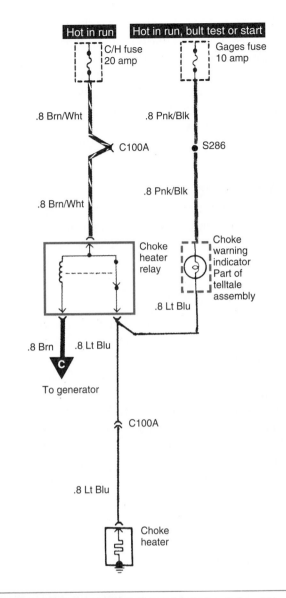

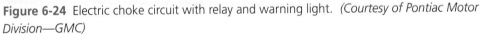

Figure 6-24 Electric choke circuit with relay and warning light. *(Courtesy of Pontiac Motor Division—GMC)*

Once the engine starts, the alternator supplies approximately 14 V to the entire electrical system. Under this condition, 14 V is supplied to both ends of the relay winding, and the current flow through the winding is stopped. Since this relay is normally closed, the contacts move to the closed position and supply full voltage to the choke heater to open the choke. When the choke relay contacts are closed, equal voltage is supplied to both sides of the choke warning light, and this light remains off. If a defect occurs in the system, such as a defective relay, current flows through the warning light and the choke heater to ground. This light is illuminated to inform the driver that a defect is present in the choke heater circuit.

Electric Assist Choke System

On some engines, the choke thermostatic spring is mounted in a hollow on top of the intake manifold crossover passage, and this spring is heated from the surface of this passage. A linkage is connected from the thermostatic spring to the choke plate. An electric heating element is positioned beside the thermostatic spring.

The choke heating element is connected through a control switch and oil pressure switch to the ignition switch. The oil pressure switch remains open until the engine starts, preventing the choke from opening if the driver leaves the ignition switch on with a cold engine.

The control switch may be a single- or dual-stage unit. The contacts in a single-stage unit are open below 55°F (13°C) and closed above 80°F (27°C). In a dual-stage control switch, a resistor in the control switch is connected in the circuit below 55°F (13°C), which supplies a reduced voltage and current to the choke heating coil. Above 80°F (27°C), the resistor in the control switch is bypassed, and full voltage is supplied to the choke heating coil to provide faster choke opening.

Carburetor Systems

Shop Manual
Chapter 6, page 247

To meet complex fuel economy and emission requirements, carburetors require the help of auxiliary controls. The following describes some of the more common assist devices you are likely to encounter when servicing a carburetor.

Choke Qualifier

Once the engine has started, a leaner mixture is needed. If the choke stays shut, the rich mixture floods the engine and causes stalling. Therefore, the automatic choke has a choke qualifying mechanism to open the choke plate slightly after the engine has started.

A qualifying diaphragm is also called a choke pull-off diaphragm or vacuum break.

Many integral chokes have a vacuum piston in the choke housing that opens the choke slightly when manifold vacuum reaches a certain level (immediately after the engine starts). Some integral chokes and all divorced chokes have a qualifying diaphragm instead of a vacuum piston (Figure 6-25). The diaphragm is connected to manifold vacuum, and when the engine starts, the diaphragm retracts, pulling the choke open. The amount of opening, or the distance between the upper edge of the choke plate and the side of the air horn, is called the qualifying dimension or setting. In some carburetors, the pull-off diaphragm has a modulator spring that varies the qualifying setting based on ambient temperatures.

Dashpot

The **dashpot** (Figure 6-26) is used during rapid deceleration to retard the closing of the throttle. This allows a smooth transition from the main metering system to the idle system and prevents stalling due to the overly rich air-fuel mixture. It also controls the level of hydrocarbons (HC) in the exhaust during deceleration.

The dashpot consists of a small chamber with a spring-loaded diaphragm and a plunger. A link from the throttle comes in contact with the dashpot plunger as the throttle closes. As the throttle linkage exerts force on the plunger, air slowly bleeds out of the diaphragm chamber through a small hole. This slows the closing action of the throttle plate.

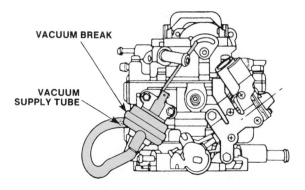

Figure 6-25 Vacuum break.

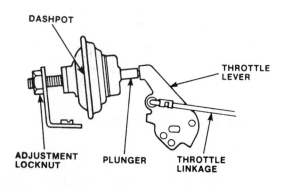

Figure 6-26 Typical carburetor dashpot installation.

Hot-Idle Compensator (HIC) Valve

When the engine is overheated, a hot-idle compensator opens an air passage to lean the mixture slightly. This increases idle speed to help cool the engine (by increased coolant flow) and to prevent excess fuel vaporization within the carburetor. The *hot-idle compensator* is a bimetal, thermostatically controlled air bleed valve.

Dual Vacuum Break

Some carburetors are equipped with a fuel vacuum break system, which includes a primary diaphragm and a secondary diaphragm. The primary vacuum diaphragm opens the choke valve slightly as soon as the engine starts to keep the engine from overchoking and stalling. The secondary vacuum diaphragm, which is also attached to the choke lever, opens the choke valve slightly wider in warm weather or when a warm engine is being started.

Vacuum to the secondary diaphragm is controlled by a thermal vacuum switch or valve (TVV). The TVV releases vacuum to the secondary vacuum break when the engine reaches a certain temperature. This prevents a rich fuel mixture and the high emissions that result from starting a cold engine in warm weather or when a warm engine is started.

Choke Unloader

To be able to start a cold engine that has been flooded with gasoline, a **choke unloader** is required. The choke unloader (Figure 6-27) is throttle linkage actuated and opens the choke whenever the accelerator pedal is floored. At wide-open throttle, the partially opened choke allows additional air to lean out the mixture and reduce fuel flow.

Deceleration Valve

This valve (Figure 6-28) is designed to prevent backfire during deceleration as the fuel mixture becomes richer. The valve, which operates only during deceleration, is usually located between the intake manifold and the air cleaner. A typical valve has a cam-shaped diaphragm housing on one

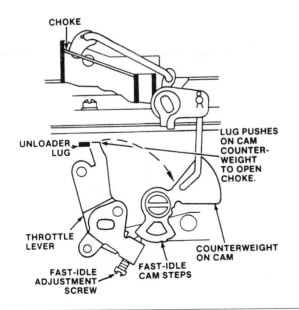

Figure 6-27 The choke unloader opens the choke any time the gas pedal is floored. Lug pushed on cam counterweight to open choke.

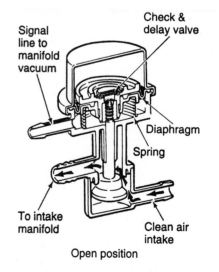

Figure 6-28 A vacuum-operated decel valve. *(Courtesy of Pontiac Motor Division—GMC)*

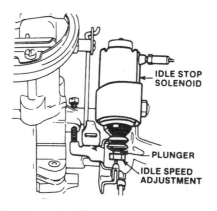

Figure 6-29 An idle stop solenoid is used to control the position of the idle setting when the ignition is turned off and on. *(Courtesy of Echlin Manufacturing Co.)*

end. A control manifold-vacuum line is attached to a port under the diaphragm housing. The other end of the valve is connected by hoses to the intake manifold and air cleaner. When deceleration causes an increase in manifold vacuum, the diaphragm opens the deceleration valve and allows air to pass from the air cleaner into the intake manifold, leaning the fuel mixture and preventing exhaust system backfire.

Throttle Position Solenoid

The throttle position solenoid is used to control the position of the throttle plate (Figure 6-29). It can have several functions, depending on its application. When the basic function is to prevent dieseling, the solenoid is called a **throttle stop solenoid** or an **idle stop solenoid**. When the engine is started, the solenoid is energized and the plunger extends, pushing against the throttle linkage. This forces the throttle plates open slightly to the curb idle position. When the ignition switch is turned off, the solenoid is de-energized and the plunger retracts. This allows the throttle plate to close completely, and it shuts off the air-fuel supply to prevent dieseling or run-on.

The throttle position solenoid is also used to increase the curb idle speed to compensate for extra loads on the engine. When this is its primary function, the solenoid might also be called an idle speed-up solenoid or a **throttle kicker** (Figure 6-30). This feature is most often used on cars with air conditioners. When the air conditioning is turned on, a relay energizes the solenoid so that the plunger extends farther, raising the idle rpm. This keeps the engine running at a higher speed, which is required to maintain a smooth idle speed and to ensure adequate emission control.

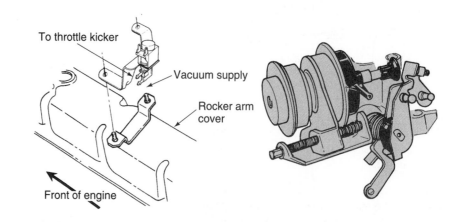

Figure 6-30 A throttle kicker assembly. *(Reprinted with the permission of Ford Motor Company)*

The throttle position solenoid is also used to control idle speed when the transmission is engaged. A relay in the park/neutral switch signals the solenoid to extend when the transmission is shifted into gear. This opens the throttle slightly to compensate for the increased load on the engine.

Types of Carburetors

Many types of carburetors have been built in order to accommodate different load conditions, engine designs, and air-fuel requirements. Different carburetors feature different drafts, different numbers of barrels, different types of venturi, and different flow rates.

Carburetor Draft

Draft is defined as the act of pulling or drawing air. A carburetor's direction of draft is one way in which carburetors are classified. Most engines have a downdraft carburetor that has air flowing vertically down into the engine. In the sidedraft carburetor, air flows through the carburetor in a horizontal direction. Many early sports cars used a sidedraft carburetor. An updraft carburetor brings the air and fuel into the engine in an upward direction. Not many automobiles use this type, but they are used in forklifts and other industrial engine applications.

Carburetor Barrels

A carburetor barrel is a passageway or bore used to mix the air and fuel. It consists of the throttle plate, venturi, and air horn. A one-barrel carburetor is used on small engines that do not require large quantities of air and fuel.

A two-barrel carburetor has two throttle plates and two venturis. The area where the air comes into the carburetor is common on both barrels. A two-barrel carburetor may have one barrel that is smaller in diameter than the other one.

Figure 6-31 shows a four-barrel carburetor. It has four barrels to mix the air and fuel. The engine operates on two barrels during most driving conditions. When more power is needed, the other two barrels add fuel to increase the amount of horsepower and torque produced by the engine.

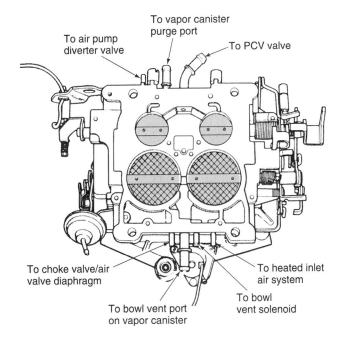

Figure 6-31 Throttles of a four-barrel carburetor. *(Courtesy of Chrysler Corporation)*

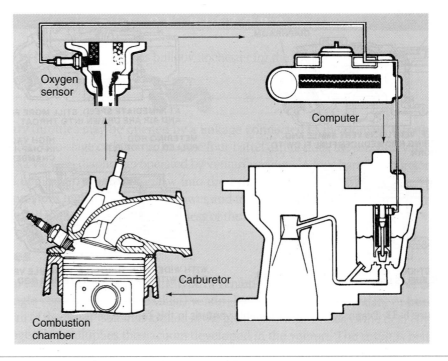

Figure 6-34 Oxygen feedback solenoid. *(Courtesy of Chrysler Corporation)*

ate output signals to a mixture control device that meters more or less fuel into the air charge as it is needed to maintain the 14.7 to 1 ratio.

Open loop means the control cycle has been broken.

Whenever these components are working to control the air-fuel ratio, the carburetor is said to be operating in closed loop. Closed loop is illustrated in the schematic shown in Figure 6-35. The oxygen sensor is constantly monitoring the oxygen in the exhaust, and the control module is constantly making adjustments to the air-fuel mixture based on the fluctuations in the sensor's voltage output. However, there are certain conditions under which the control module ignores the signals from the oxygen sensor and does not regulate the ratio of fuel to air. During these times, the carburetor is functioning in a conventional manner and is said to be operating in *open loop*.

More detail on feedback carburetors is given in Chapter 10 of this book.

The carburetor operates in open loop until the oxygen sensor reaches a certain temperature (approximately 600°F). The carburetor also goes into open loop when a richer than normal air-fuel mixture is required–such as during warmup and heavy throttle application. Several other sensors are needed to alert the electronic control module of these conditions. A coolant sensor provides input relating to engine temperature. A vacuum sensor and a throttle position sensor indicate a wide open throttle.

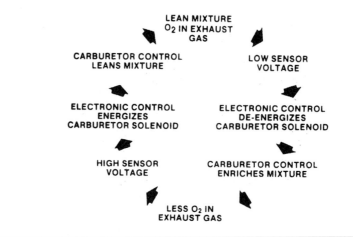

Figure 6-35 Closed loop operation.

Fuel Injection

Most electronic fuel injection systems only inject fuel during part of the engine's combustion cycle. The engine fuel needs are measured by intake airflow past a sensor or by intake manifold pressure (vacuum). The airflow or manifold vacuum sensor converts its reading to an electrical signal and sends it to the engine control computer. The computer processes this signal (and others) and calculates the fuel needs of the engine. The computer then sends an electrical signal to the fuel injector or injectors. This signal determines the amount of time the injector opens and sprays fuel. This interval is known as the *injector pulse width*.

Electronic fuel injection (EFI) has proven to be the most precise, reliable, and cost effective method of delivering fuel to the combustion chambers of today's vehicles. EFI systems must provide the correct air-fuel ratio for all engine loads, speeds, and temperature conditions. To accomplish this, an EFI system uses a fuel delivery system, air induction system, input sensors, control computer, fuel injectors, and some sort of idle speed control.

Many of the same sensors are used in both, computer-controlled carburetor systems and electronic fuel injection systems. These sensors include:

1. Oxygen (O_2) sensor
2. Engine coolant temperature (ECT) sensor
3. Air charge temperature (ACT) sensor
4. Throttle position sensor (TPS)
5. Manifold absolute pressure (MAP) sensor
6. Mass air flow (MAF) sensor
7. Knock sensor
8. Exhaust gas recirculation valve position (EVP) sensor
9. Vehicle speed sensor (VSS)
10. Neutral drive switch (NDS)

Types of Fuel Injection

Although fuel injection technology has been around since the 1920s, it wasn't until the 1980s that manufacturers began to replace carburetors with **electronic fuel injection** (**EFI**) systems. Many of the early EFI systems were **throttle body injection** (**TBI**) systems in which the fuel was injected above the throttle plates. Recently a similar system, **central port injection** (**CPI**), was introduced. In these systems the injector assembly is located in the lower half of the intake manifold. Engines equipped with TBI have gradually become equipped with **port fuel injection** (**PFI**), which has injectors located in the intake ports of the cylinders. Since the 1995 model-year, all new cars are equipped with an EFI system.

Throttle body injection systems have a throttle body assembly mounted on the intake manifold in the position usually occupied by a carburetor. The throttle body assembly usually contains one or two injectors.

On port fuel injection systems, fuel injectors are mounted at the back of each intake valve. Aside from the differences in injector location and number of injectors, operation of throttle body and port systems is quite similar with regard to fuel and air metering, sensors, and computer operation.

Port-type **continuous injection systems** (**CIS**) have been used on many import vehicles. These systems deliver a steady stream of pressurized fuel into the intake manifold. The injectors do not pulse on and off as in port and throttle body systems. In CIS, the amount of fuel delivered is controlled by the rate of airflow entering the engine. An airflow sensor controls movement of a plunger that alters fuel flow to the injectors. When introduced, CIS was a mechanically controlled system. However, oxygen sensor feedback circuits and other electronic controls have been added to the system. CIS systems that have electronic controls are commonly referred to as **CIS-E** systems.

Diesel engines, for quite some time, have been equipped with fuel injection systems. The two basic differences between gasoline injection and diesel injection are: diesel fuel is injected directly into the cylinders and diesel fuel injection systems are operated mechanically rather than electronically. Although late-model diesel systems use electronic fuel controls, the fuel injection system is a mechanical system that is controlled mechanically.

Continuous Injection Systems

Continuous injection systems (Figure 6-36) are used almost exclusively on import vehicles. The basic technology for CIS was introduced in the early 1970s and has been continuously updated and refined.

In a CIS-equipped engine, the amount of fuel delivered to the cylinders is not varied by pulsing the injectors on and off. Instead, CIS injectors spray fuel continuously. What does vary is the amount of fuel contained in the spray. CIS systems do this by maintaining a constant relative fuel system pressure and metering the amount of fuel to the injectors.

Basic Operation. Metering is done through a *mixture control unit*. This unit consists of an airflow sensor and a special *fuel distributor* with fuel lines running to all injectors. A control plunger in the fuel distributor is mechanically linked to the airflow sensor plate by means of a lever. As the airflow sensor measures the volume of engine intake air, its plate moves. The lever transfers this motion to the control plunger in the fuel distributor. The plunger moves up or down changing the size of the fuel metering openings in the fuel lines. This increases or decreases the volume of fuel flowing to the injectors.

Air Delivery System. Air enters the system through the air filter and is measured by the airflow sensor. The amount of airflow is controlled by the throttle. The sensor plate is located in an air venturi or funnel-shaped passage in the mixture control unit. Because of the shape of this venturi or

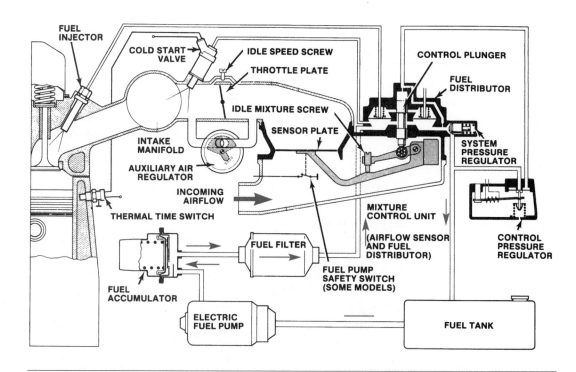

Figure 6-36 Components of a continuous injection system (CIS) that meters fuel mechanically. *(Courtesy of Robert Bosch Corporation)*

funnel, the airflow sensor moves more when more air flows into the engine. Any air that enters the intake without passing the sensor plate interferes with the proper air-fuel mixture, causing the engine to run lean.

Fuel Delivery System. Fuel is drawn from the tank by an electric fuel pump. It passes through the fuel accumulator and filter before reaching the fuel distributor in the metering control unit. Some models use a *prefeed pump* to supply the main pump. This prefeed pump helps prevent vapor lock in hot driving conditions.

The *fuel accumulator* is needed to prevent a sudden rapid rise in fuel pressure inside the fuel distributor when the vehicle is being started. Besides stabilizing the pressure, the accumulator also maintains a rest pressure within the fuel system when the engine is off. This helps eliminate vapor lock in the fuel lines.

The fuel distributor consists of a fuel control unit, pressure regulating valves for each cylinder, and a system pressure regulator. The fuel control unit consists of a slotted metering cylinder. This cylinder contains the fuel control plunger. Part of the control plunger protrudes past the fuel distributor and rests on the airflow sensor lever.

Fuel flows through the slots in the fuel metering cylinder. There is one metering slot for each engine cylinder. Control plunger movement within the metering cylinder determines the amount of fuel released to the fuel injectors. Each cylinder has its own pressure regulating valve. These valves maintain a constant pressure differential of approximately 1.5 psi on either side of the fuel metering slot. This pressure differential remains the same, regardless of the size of the slot opening. Without pressure regulating valves, the amount of fuel injected would not remain proportional to the size of the metering slot opening.

The fuel distributor also contains a pressure relief valve that regulates system pressure. This regulator maintains a constant system pressure by allowing excess fuel to return to the fuel tank via a return line.

Control Pressure Regulator. A control pressure regulator, or *warm-up regulator*, is also used to provide correct fuel pressure on top of the fuel control plunger. This helps regulate the engine air-fuel needs. A dampening restriction over the fuel control plunger also eliminates any fluctuations that may occur in the airflow sensor lever.

Fuel Injectors. CIS fuel injectors (Figure 6-37) open at a set fuel pressure. Once the engine is started, each injector continuously sprays finely atomized fuel into the intake port of the cylinder.

A vibrator pin or needle inside each injector helps break up and atomize the fuel. This vibrating action also helps keep the injectors from clogging. When the engine is stopped, the pin and spring assembly seal off the injector to retain fuel pressure in the lines. This helps assure quick starting. CIS systems are normally equipped with a cold start injector and auxiliary air valve system to control cold starting and engine idling. These systems operate similarly to the EFI systems discussed earlier in the chapter.

Oxygen Control Feedback System. Continuous injection systems can be fitted with an oxygen sensor for feedback control. The sensor is mounted in the exhaust manifold so it heats up rapidly when the engine is started.

An oxygen sensor is sometimes called a lambda sensor.

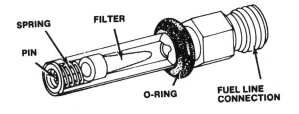

SPRING FILTER

PIN

O-RING

FUEL LINE
CONNECTION

Figure 6-37 Typical CIS injector. *(Courtesy of Robert Bosch Corporation)*

Signals from the oxygen sensor are sent to the oxygen control unit. The control unit modifies the fuel flow in the mixture control unit so the engine operates with the proper ratio. The changing exhaust gas affects the oxygen sensor and it sends a signal in a closed loop through the mixture control unit to the engine.

The oxygen control valve operates on signals from the oxygen control unit. It opens and closes to allow more or less fuel to return to the tank through the fuel return. This is called dwell time. By reducing the pressure in the lower part of the differential pressure valve, fuel flow to the injector can be increased, enriching the mixture. Shortening the time that the oxygen control valve is open increases the pressure beneath the differential pressure valve diaphragm. This lessens the amount of fuel injected, leaning the mixture.

The oxygen control unit switches to open loop during conditions when the oxygen sensor is cold or when the engine is cold. This open loop operation holds the oxygen control valve open for a fixed amount of time.

The oxygen control valve is sometimes called a timing or frequency valve.

Electronic Continuous Injection System

The CIS-E system is used on many European imported vehicles. Robert Bosch Corporation manufactures a variety of SFI and MFI systems that have components that are used or are similar to the ones used by other original equipment manufacturers (OEMs). In the CIS-E system, the injectors in each intake port are injecting fuel continually while the engine is running. These injectors do not open and close while the engine is running. Fuel is supplied from a central fuel distributor to all the injectors. An electric fuel pump supplies fuel through an accumulator and filter to the fuel distributor. The accumulator prevents fuel pressure pulses. Fuel is returned from the fuel distributor and the pressure regulator to the fuel tank.

A pivoted airflow sensor plate is positioned in the air intake. With the engine not running, the airflow sensor plate closes the air passage in the air intake. A control plunger in the fuel distributor rests against the airflow sensor lever (Figure 6-38).

When engine speed increases, the air velocity in the air intake gradually opens the airflow sensor plate. This plate movement lifts the control plunger, which precisely meters the fuel to the

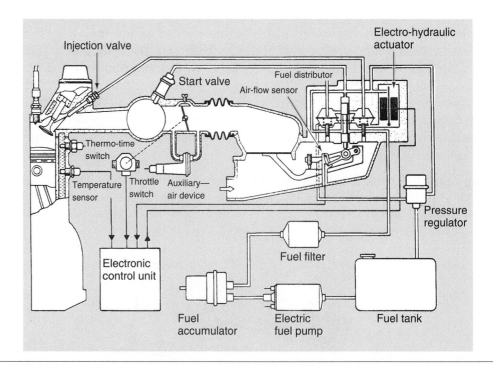

Figure 6-38 CIS-E components. *(Reprinted with permission from SAE ©1982, Society of Automotive Engineers, Inc.)*

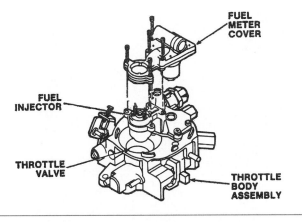

Figure 6-39 Parts of a throttle body injection unit.

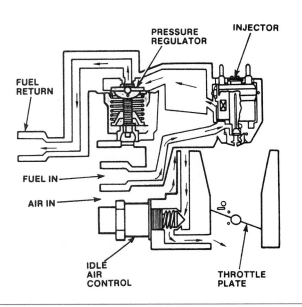

Figure 6-40 Fuel and airflow in a TBI injection system. The injector is a bottom fuel feed design. The idle air control allows air to bypass the throttle plate to regulate engine idle in all operating conditions.

injectors. The differential pressure regulator winding is energized by the electronic control unit. The differential pressure regulator winding is cycled on and off by the electronic control unit. This action moves the plunger up and down in the differential pressure regulator, which controls fuel pressure in the lower chambers of the fuel distributor and provides precise control of the fuel delivery to the injectors and the air-fuel ratio.

The input sensors connected to the electronic control unit vary depending on the vehicle, but these sensors include an oxygen sensor and coolant temperature sensor. Some CIS-E systems are complete engine management systems in which the electronic control unit provides fully integrated control of the air-fuel ratio, spark advance, emission control devices, and idle speed.

Throttle Body Fuel Injection

For some auto manufacturers, TBI served as a stepping stone from carburetors to more advanced port fuel injection systems. TBI units were used on many engines during the 1980s and are still used on some engines. The throttle body unit is similar in size and shape to a carburetor, and like a carburetor, mount on the intake manifold (Figure 6-39). The injector(s) spray fuel down into a throttle body chamber leading to the intake manifold. The intake manifold feeds the air-fuel mixture to all cylinders.

Shop Manual
Chapter 6, page 260

TBI Operation. The basic TBI assembly consists of two major castings: a throttle body with a valve to control airflow and a fuel body to supply the required fuel. A fuel pressure regulator and fuel injector are integral parts of the fuel body (Figure 6-40). Also included as part of the assembly is a device to control idle speed and one to provide throttle valve positioning data.

The throttle body casting has ports that can be located above, below, or at the throttle valve depending on the manufacturer's design. These ports generate vacuum signals for the manifold absolute pressure sensor and for devices in the emission control system, such as the exhaust gas recirculation (EGR) valve, the canister purge system, and so on.

The **fuel pressure regulator** used on the throttle body assembly is similar to a diaphragm-operated relief valve (Figure 6-41). Fuel pressure is on one side of the diaphragm and atmospheric pressure is on the other side. The regulator is designed to provide a constant pressure on the fuel

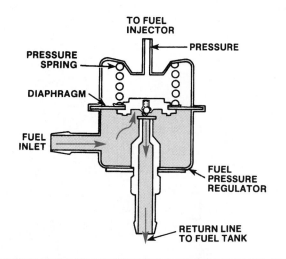

Figure 6-41 A fuel pressure regulator for a TBI system.

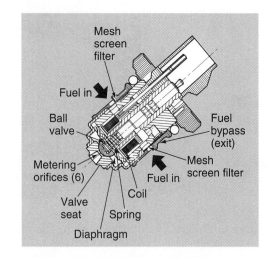

Figure 6-42 Typical TBI injector. *(Reprinted with the permission of Ford Motor Company)*

injector throughout the range of engine loads and speeds. If regulator pressure is too high, a strong fuel odor is emitted and the engine runs too rich. On the other hand, regulator pressure that is too low results in poor engine performance or detonation can take place, due to the lean mixture.

The fuel injector is solenoid operated and pulsed on and off by the vehicle's engine control computer. Surrounding the injector inlet is a fine screen filter where the incoming fuel is directed. When the injector's solenoid is energized, a normally closed ball valve is lifted (Figure 6-42). Fuel under pressure is then injected at the walls of the throttle body bore just above the throttle plate.

A fuel injector has a movable armature in the center of the injector, and a pintle with a tapered tip is positioned at the lower end of the armature. A spring pushes the armature and pintle downward so the pintle tip seats in the discharge orifice. The injector coil surrounds the armature, and the two ends of the winding are connected to the terminals on the side of the injector. An integral filter is located inside the top of the injector. When the ignition switch is turned on, voltage is supplied to one injector terminal and the other terminal is connected through the computer. Each time the control unit completes the circuit from the injector winding to ground, current flows through the injector coil, and the coil magnetism moves the plunger and pintle upward. Under this condition, the pintle tip is unseated from the injector orifice, and fuel sprays out this orifice.

TBI Advantages and Disadvantages. Throttle body systems provide improved fuel metering when compared to carburetors. They are also less expensive and simpler to service. TBI units also have some advantages over port injection. They are less expensive to manufacture, simpler to diagnose and service, and don't have injector balance problems to the extent that port injection systems do when the injectors begin to clog.

However, throttle body units are not as efficient as port systems. The disadvantages are primarily manifold related. Like a carburetor system, fuel is still not distributed equally to all cylinders, and a cold manifold may cause fuel to condense and puddle in the manifold. Like a carburetor, throttle body injection systems must be mounted above the combustion chamber level, which eliminates the possibility of tuning the manifold design for more efficient operation.

Port Fuel Injection

PFI systems use one injector at each cylinder. They are mounted in the intake manifold near the cylinder head where they can inject a fine, atomized fuel mist as close as possible to the intake valve (Figure 6-43). Fuel lines run to each cylinder from a fuel manifold, usually referred to as a **fuel rail**. The fuel rail assembly on a PFI system of V6 and V8 engines usually consists of a left- and right-hand rail assembly. The two rails can be connected either by crossover and return fuel tubes

The term pulse width is applied to the length of time the injector plunger is lifted and the injector is discharging fuel.

Shop Manual
Chapter 6, page 261

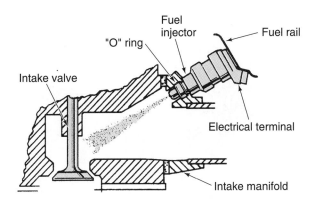

Figure 6-43 Port injection sprays fuel into the intake port and fills the port with fuel vapor before the valve opens. *(Courtesy of Oldsmobile Division—GMC)*

or by a mechanical bracket arrangement. A typical fuel rail arrangement is shown in Figure 6-44. Fuel tubes crisscross between the two rails. Figure 6-45 shows a fuel rail on a transverse four-cylinder engine. Since each cylinder has its own injector, fuel distribution is exactly equal. With little or no fuel to wet the manifold walls, there is no need for manifold heat or any early fuel evaporation system. Fuel does not collect in puddles at the base of the manifold. This means that the intake manifold passages can be tuned or designed for better low-speed power availability (Figure 6-46).

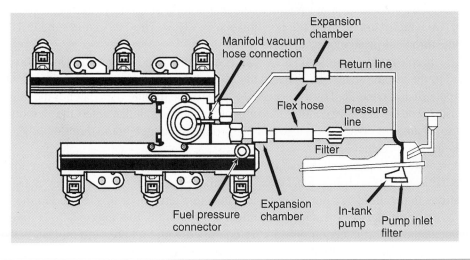

Figure 6-44 A one-piece fuel rail for a V-6 engine. *(Courtesy of Oldsmobile Division—GMC)*

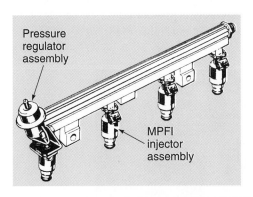

Figure 6-45 Fuel rail and injectors for a 4-cylinder engine. *(Courtesy of Oldsmobile Division—GMC)*

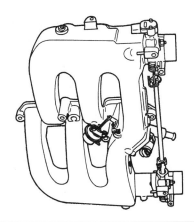

Figure 6-46 Typical tuned intake manifold for a port injection engine. *(Courtesy of Chrysler Corporation)*

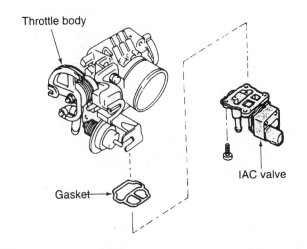

Figure 6-47 A throttle body assembly of a port fuel injection system.

The port type systems provide a more accurate and efficient delivery of fuel. Some engines are now equipped with variable induction intake manifolds that have separate runners for low and high speeds. This technology is only possible with port injection.

The throttle body in a port fuel injection system controls the amount of air that enters the engine as well as the amount of vacuum in the manifold. It also houses and controls the idle air control (IAC) motor and the throttle position sensor (TPS). The TPS enables the ECU to know where the throttle is positioned at all times.

The throttle body (Figure 6-47) is a single cast aluminum housing with a single throttle blade attached to the throttle shaft. The TPS and the IAC valve/motor are also attached to the housing. The throttle shaft is controlled by the accelerator pedal. The throttle shaft extends the full length of the housing. The throttle bore controls the amount of incoming air that enters the air induction system. A small amount of coolant is also routed through a passage in the throttle body to prevent icing during cold weather.

Port systems require an additional control system that throttle body injection units do not require. While throttle body injectors are mounted above the throttle plates and are not affected by fluctuations in manifold vacuum, port system injectors have their tips located in the manifold where constant changes in vacuum would affect the amount of fuel injected. To compensate for these fluctuations, port injection systems are equipped with fuel pressure regulators that sense manifold vacuum and continually adjust the fuel pressure to maintain a constant pressure drop across the injector tips at all times.

Pressure Regulators. The pressure regulator in port injection systems is similar to the regulator used in TBI systems. A diaphragm and valve assembly is positioned on the center of the regulator, and a diaphragm spring seats the valve on the fuel outlet (Figure 6-48).

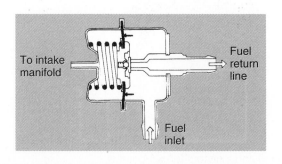

Figure 6-48 Internal pressure regulator. *(Courtesy of Chrysler Corporation)*

When fuel pressure reaches the setting of the regulator, the diaphragm moves against the spring tension and the valve opens. This action allows fuel to flow through the return line to the fuel tank. The fuel pressure drops slightly when the pressure regulator valve opens, and the spring closes the regulator valve. In many systems, the regulator maintains fuel pressure at 39 psi (269 kPa).

A vacuum hose is connected from the intake manifold to the vacuum inlet on the pressure regulator. This hose supplies vacuum to the area where the diaphragm spring is located. This vacuum works with the fuel pressure to move the diaphragm and open the valve. When the engine is running at idle speed, high manifold vacuum is supplied to the regulator. Under this condition, the specified fuel pressure opens the regulator valve. When the engine is running under heavy load and/or wide-open throttle, a very low vacuum is supplied to the regulator. During these times, the vacuum does not help open the regulator valve and a higher fuel pressure is required to open the valve.

The change in fuel pressure allows the fuel to be sprayed into the manifold with the same effect, regardless of the pressure present in the manifold. When there is a high vacuum in the manifold, a very low pressure exists and the pressure difference between the fuel spray and the vacuum is the same as when there is a higher pressure in the manifold (low vacuum) and a higher fuel pressure.

Port Firing Control. While all port injection systems operate using an injector at each cylinder, they do not fire the injectors in the same manner. This one statement best defines the difference between typical **multiport injection** systems (**MPI**) and **sequential fuel injection** systems (**SFI**).

SFI systems control each injector individually so that it is opened just before the intake valve opens. This means that the mixture is never static in the intake manifold and that adjustments to the mixture can be made almost instantaneously between the firing of one injector and the next. Sequential firing is the most accurate and desirable method of regulating port injection.

In MPI systems, the injectors are grouped together in pairs or groups, and these pairs or groups of injectors are turned on at the same time. When the injectors are split into two equal groups, the groups are fired alternatively, with one group firing each engine revolution.

Since only two injectors can be fired relatively close to the time when the intake valve is about to open, the fuel charge for the remaining cylinders must stand in the intake manifold for varying periods of time. These periods of time are very short, therefore the standing of fuel in the intake manifold is not that great a disadvantage of MPI systems. At idle speeds this wait is about 150 milliseconds, and at higher speeds the time is much less. The primary advantage of SFI is the ability to make instantaneous changes to the mixture.

In SFI systems, each injector is connected individually into the computer, and the computer completes the ground for each injector, one at a time. In MPI systems, the injectors are grouped and all injectors within the group share the same common ground wire.

Some injection systems fire all of the injectors at the same time for every engine revolution. This type of system offers easy programming and relatively fast adjustments to the air-fuel mixture. The injectors are connected in parallel so the ECU sends out just one signal for all injectors. They all open and close at the same time. It simplifies the electronics without compromising injection efficiency. The amount of fuel required for each four-stroke cycle is divided in half and delivered in two injections, one for every 360° of crankshaft rotation. The fact that the intake charge must still wait in the manifold for varying periods of time is the system's major drawback.

Central Port Injection

In a central port injection (CPI) system, a central injector assembly is mounted in the lower half of the intake manifold. Fuel inlet and return lines are connected from the rear of the intake to the CPI assembly. A retaining clip attaches these lines to the CPI assembly. Small poppet nozzles are positioned in each intake port in the lower half of the intake. Nylon fuel lines connect these nozzles to the CPI assembly (Figure 6-49).

The pressure regulator is mounted with the central injector. Since this regulator is mounted inside the intake manifold, vacuum from the intake is supplied through an opening in the regulator cover to the regulator diaphragm. The regulator spring pushes downward on the diaphragm and

Shop Manual
Chapter 6, page 263

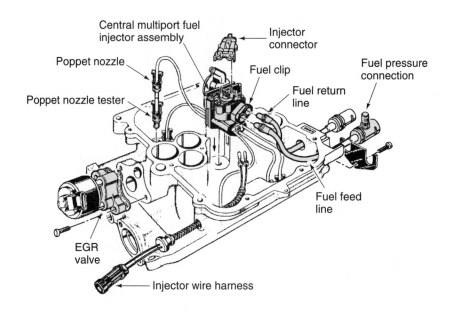

Figure 6-49 CPI components shown in the lower half of the intake manifold. *(Courtesy of Chevrolet Motor Division—GMC)*

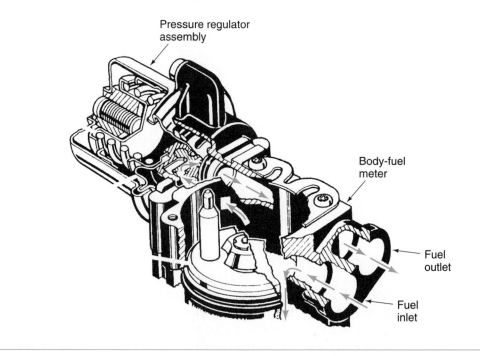

Figure 6-50 The pressure regulator system of a CPI. *(Courtesy of Chevrolet Motor Division—GMC)*

closes the valve. Fuel pressure from the in-tank fuel pump pushes the diaphragm upward and opens the valve, which allows fuel to flow through this valve and the return line to the fuel tank (Figure 6-50).

The pressure regulator is designed to regulate fuel pressure from 54 to 64 psi (370 to 440 kPa), which is higher than many port fuel injection systems. Higher pressure is required in the CPI system to prevent fuel vaporization from the extra heat encountered with the CPI assembly, poppet nozzles, and lines mounted inside the intake manifold. The pressure regulator operates the same as the regulators explained previously in this chapter.

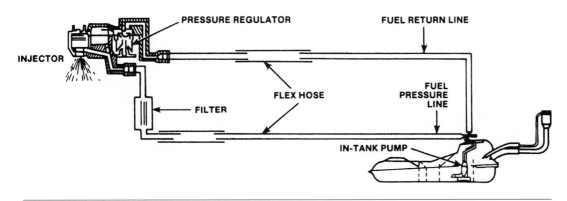

Figure 6-51 Components of a typical fuel circuit used in a throttle body fuel injection system.

Fuel Delivery System

The components of a typical gasoline delivery system are fuel tanks, fuel lines, fuel filters, and fuel pumps. Figure 6-51 shows a typical EFI fuel delivery system, Fuel is drawn from the fuel tank by an in-tank or chassis-mounted electric fuel pump. Before it reaches the injectors, the fuel passes through a filter that removes dirt and impurities. A fuel line pressure regulator maintains a constant fuel line pressure that may be as high as 50 psi in some systems. This fuel pressure generates the spraying force needed to inject the fuel. Excess fuel not required by the engine returns to the fuel tank through a fuel return line.

Fuel Tanks

Modern fuel tanks include devices that prevent vapors from leaving the tank. For example, to contain vapors and allow for expansion, contraction, and overflow that result from changes in the temperature, the fuel tank has a separate air chamber dome at the top. Another way to contain vapors is to use a separate internal expansion tank within the main tank (Figure 6-52). All fuel tank designs provide some control of fuel height when the tank is filled. Frequently, this is achieved by using vent lines within the filler tube or tank. With tank designs such as this, only 90% of the tank is ever filled, leaving 10% for expansion in hot weather. Some vehicles have an overfill limiting valve to prevent overfilling of the tank. If a tank is filled to capacity, it overflows whenever the temperature of the fuel increases.

Fuel tanks can be constructed of pressed corrosion-resistant steel, aluminum, or molded reinforced polyethylene plastic. Aluminum and molded plastic fuel tanks are becoming more common as manufacturers attempt to reduce the overall weight of the vehicle. Metal tanks are usually ribbed to provide added strength. Seams are welded, and heavier gauge steel is often used on exposed

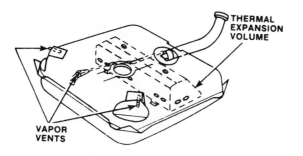

Figure 6-52 An internal expansion tank helps contain vapors in the fuel tank.

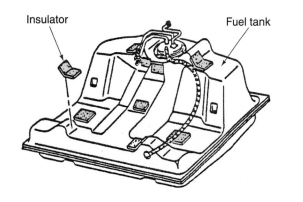

Insulator   Fuel tank

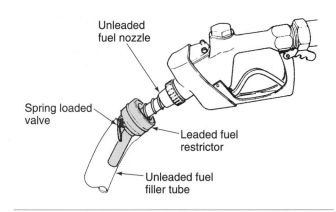

Unleaded fuel nozzle

Spring loaded valve

Leaded fuel restrictor

Unleaded fuel filler tube

Figure 6-53 Insulators positioned between the top of the fuel tank and the chassis. *(Courtesy of Oldsmobile Division—GMC)*

Figure 6-54 Filler pipe with restriction for unleaded fuel nozzles. *(Courtesy of Chrysler Corporation)*

sections for added strength. Fuel tanks are normally mounted with insulators between the top of the tank and the chassis to protect the tank and prevent noise from transferring into the passenger compartment (Figure 6-53).

Most tanks have slosh baffles or surge plates to prevent the fuel from splashing around inside the unit. In addition to slowing down fuel movement, the plates tend to keep the fuel pick-up or sending assembly immersed in the fuel during hard braking and acceleration. The plates or baffles also have holes or slots in them to permit the fuel to move from one end of the tank to the other. Except for rear engine vehicles, the fuel tank in a passenger car is located in the rear of the vehicle for improved safety.

The fuel tank is provided with an inlet filler tube and cap. The filler pipe is usually supported by a bracket that is bolted to the chassis. The location of the fuel inlet filler tube depends on the tank design and tube placement. It is usually positioned behind the filler cap or a hinged door in the center of the rear panel or in the outer side of either rear fender panel. Vehicles designed for unleaded fuel use have a restrictor in the filler tube that prevents the entry of the larger leaded fuel delivery nozzle at the gas pumps (Figure 6-54). The filler pipe can be a rigid one-piece tube soldered to the tank or a three-piece unit. The three-piece unit has a lower neck soldered to the tank and an upper neck fastened to the inside of the body sheet metal panel.

Filler tube caps are non-venting and usually have some type of pressure-vacuum relief valve arrangement (Figure 6-55). Under normal operating conditions the valve is closed, but whenever pressure or vacuum is more than the calibration of the cap, the valve opens. Once the pressure or

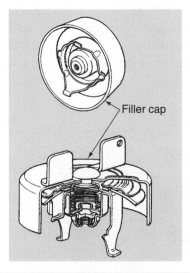

Filler cap

Figure 6-55 Pressure-vacuum gasoline tank filler cap. *(Courtesy of Chrysler Corporation)*

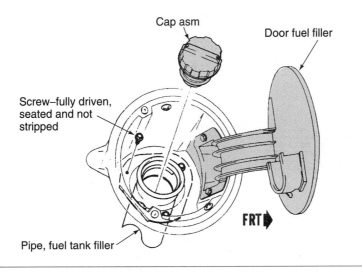

Cap asm

Door fuel filler

Screw–fully driven, seated and not stripped

Pipe, fuel tank filler

FRT►

Figure 6-56 Filler door and threaded filler cap. *(Courtesy of Oldsmobile Division—GMC)*

vacuum has been relieved, the valve closes. Many pressure caps have four anti-surge tangs that lock onto the filler neck to prevent the delivery system's pressure from pushing fuel out of the tank. By turning such a cap one-half turn, the tank pressure is relieved. Then, with another quarter turn, the cap can be removed.

A threaded filler cap may be threaded into the upper end of the filler pipe. These caps require several turns counterclockwise to remove (Figure 6-56). The long threaded area on the cap is designed to allow any remaining fuel tank pressure to escape during cap removal. The cap and filler neck have a ratchet-type, torque-limiting design to prevent overtightening. When the cap is installed, tighten the cap until a clicking noise is heard. This noise indicates the cap is properly tightened and fully seated.

> ⚠ **WARNING:** When a gasoline tank filler cap is replaced, the replacement cap must be exactly the same as the original cap or a malfunction may occur in the filling and venting system, resulting in higher emission levels and the escape of dangerous hydrocarbons.

Starting with the 1976 model year, a Federal Motor Vehicle Safety Standard (FMVSS 301) required a control on gasoline leakage from passenger cars, certain light trucks, and buses after they were subjected to barrier impacts and rolled over. Tests conducted under these severe conditions showed the most common gasoline leak path was the gasoline supply line from the fuel tank to the carburetor.

Most rollover leakage protection devices used on carburetor-equipped engines are variations of a basic one-way check valve. These protective check valves are usually installed in the fuel vapor vent line between the tank and the vapor canister and at the carburetor fuel feed or return line fitting (Figure 6-57). In some systems the check valve is part of the carburetor inlet fuel filter (Figure 6-58).

Under normal operation, fuel pump pressure is sufficient to open the check valve and supply fuel to the engine. However, if the vehicle is involved in a rollover accident, fuel spills out of the carburetor, the engine stalls, and the fuel pump ceases to operate. This decreases fuel system pressure to the point where the check valve closes. This prevents fuel from reaching the carburetor where it would leak out.

A check valve might also be fitted in the fuel tank filler cap, and most caps' pressure-vacuum relief valve settings have been increased so that fuel pressure cannot open them in a rollover.

Many electric fuel pumps found on vehicles with fuel injection systems have an **inertia switch** that shuts off the pump if the vehicle is involved in a collision or rolls over. The Ford inertia switch consists of a permanent magnet, a steel ball inside a conical ramp, a target plate, and a

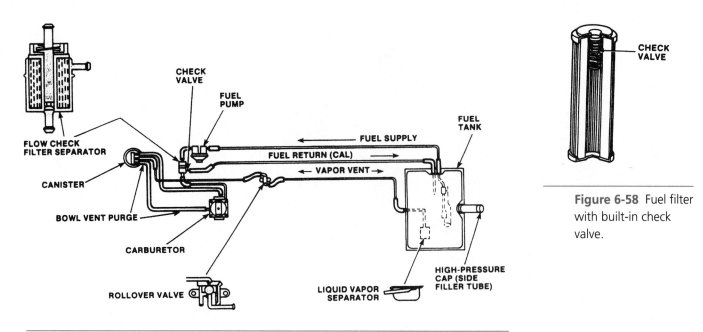

Figure 6-57 Rollover leakage protection devices.

Figure 6-58 Fuel filter with built-in check valve.

set of electrical contacts. The magnet holds the steel ball in the bottom of the conical ramp. In the event of a collision, the inertia of the ball causes it to break away from the magnet, roll up the conical ramp, and strike the target plate. The force of the ball striking the plate causes the electrical contacts in the inertia switch to open, cutting off power to the fuel pump. The switch has a reset button that must be depressed to close the contacts before the pump operates again.

Some fuel line systems contain a fuel return arrangement that aids in keeping the gasoline cool, thus reducing chances of vapor lock. The return system consists of a special fuel pump equipped with an extra outlet fitting and necessary fuel line. The fuel return line generally runs next to the conventional fuel line. The fuel return system allows a metered amount of cool fuel to circulate through the tank and fuel pump, thus reducing vapor bubbles caused by overheated fuel in the tank.

Some form of liquid vapor separator is incorporated into most modern vehicles to stop liquid fuel or bubbles from reaching the vapor storage canister or the engine crankcase. It can be located inside the tank, on the tank (Figure 6-59), in fuel vent lines, or near the fuel pump.

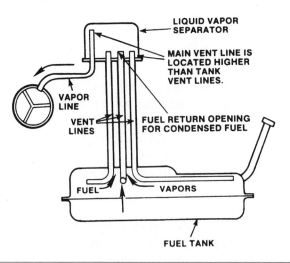

Figure 6-59 A fuel tank vapor separator allows some of the fuel vapors to condense back into liquid and return to the tank.

182

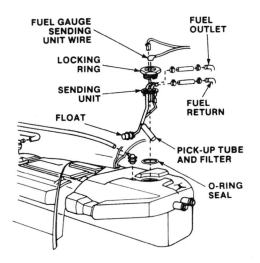

Figure 6-60 Fuel tank sending unit assembly.

Inside the fuel tank there is also a sending unit that includes a pick-up tube and float-operated fuel gauge (Figure 6-60). The fuel tank pick-up tube is connected to the fuel pump by fuel line. Some electric fuel pumps are combined with the sending unit. The pick-up tube extends nearly, but not completely, all the way to the bottom of the tank. This prevents rust, dirt, sediment, and water from being drawn into the fuel tank filter, which can cause it to clog. A ground wire is often attached to the fuel tank unit.

Fuel Lines and Fittings

Fuel lines can be made of either metal tubing or flexible nylon or synthetic rubber hose. The latter must be able to resist gasoline. It must also be nonpermeable, so gas and gas vapors cannot evaporate through the hose. Ordinary rubber hose, such as that used for vacuum lines, deteriorates when exposed to gasoline. Only hoses made for fuel systems should be used for replacement. Similarly, vapor vent lines must be made of material that resist attack by fuel vapors. Replacement vent hoses are usually marked with the designation EVAP to indicate their intended use. The inside diameter of a fuel delivery hose is generally larger (5/16 to 3/8 inch) than that of a fuel return hose (1/4 inch).

Shop Manual
Chapter 6, page 226

> ⚠ **WARNING:** Never substitute any other type of tubing such as copper or aluminum for a steel, nylon, or reinforced rubber fuel line.

The fuel lines carry fuel from the fuel tank to the fuel pump, fuel filter, and carburetor or fuel injection assembly. These lines are usually made of rigid metal, although some sections are constructed of rubber hose to allow for car vibrations. Clips retain the fuel lines to the chassis to prevent line movement and damage (Figure 6-61). This fuel line, unlike filler neck or vent hoses, must work under pressure or vacuum. Because of this, the flexible synthetic hoses must be stronger. This is especially true for the hoses on fuel injection systems, where pressures reach 50 psi or more. For this reason, flexible fuel line hose must also have special resistance properties. Many auto manufacturers recommend that flexible hose only be used as a delivery hose to the fuel metering unit in fuel injection system. It should not be used on the pressure side of the injector systems. This application requires a special high-pressure hose.

Fuel supply lines from the tank to the carburetor or injectors are routed to follow the frame along the underchassis of vehicles. Generally, rigid lines are used extending from near the tank to a point near the fuel pump. To absorb engine vibrations, the gaps between the frame and tank or fuel pump are joined by short lengths of flexible hose.

Sections of fuel line are assembled together by fittings. Some of these fittings are a threaded-type fitting, while others are a quick-release design. Quick-disconnect fuel line fittings may be

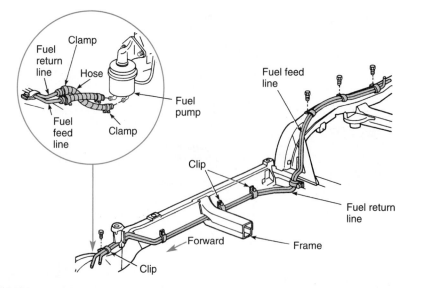

Figure 6-61 Gaps between the frame and tank or fuel pump are joined with flexible hose and the tubing is held to the chassis by clips.

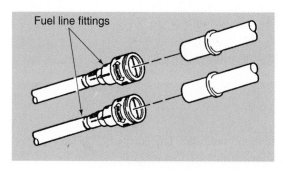

Figure 6-62 Quick-disconnect hand releasable fuel line fittings. *(Courtesy of Oldsmobile Division—GMC)*

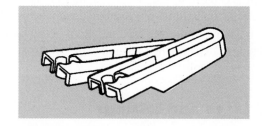

Figure 6-63 Fuel line quick-disconnect separator tools. *(Courtesy of Cadillac Motor Division—GMC)*

hand releasable, or a special tool may be required to release these fittings. Many fuel lines have quick-disconnect fittings with a unique female socket and a compatible male connector (Figure 6-62). These quick-disconnect fittings are sealed by an O-ring inside the female connector. Some of these quick-disconnect fittings have hand-releasable locking tabs, while others require a special tool to release the fitting (Figure 6-63).

▲ **WARNING:** Other types of O-rings should not be substituted for a Viton O-ring.

The two most common threaded-type tube fittings are the *compression* and the *double-flare* (Figure 6-64). The double-flare, which is the most common, is made with a special tool that has an anvil and a cone (Figure 6-65). The double-flaring process is performed in two steps. First, the anvil begins to fold over the end of the tubing. Then, the cone is used to finish the flare by folding the tubing back on itself, doubling the thickness, and creating two sealing surfaces.

▲ **WARNING:** Fuel line fittings must be tightened to the specified torque.

Some fuel lines have threaded fittings with an O-ring seal to prevent fuel leaks (Figure 6-66). These O-ring seals are usually made from Viton, which resists deterioration from gasoline. On some other fuel lines, the fuel hose is clamped to the steel line and the hose and clamp must be properly positioned on the steel line (Figure 6-67).

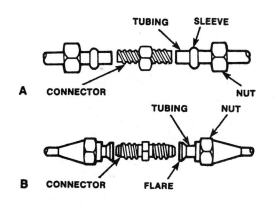

Figure 6-64 (A) Compression and (B) double-flare tubing fittings.

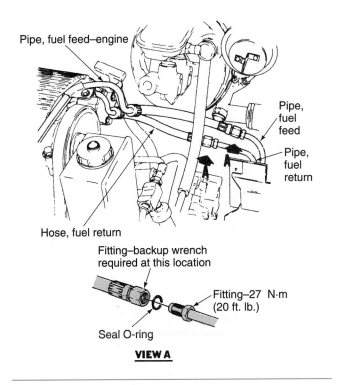

Fitting—backup wrench required at this location

Fitting—27 N·m (20 ft. lb.)

Seal O-ring

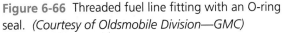
VIEW A

Figure 6-66 Threaded fuel line fitting with an O-ring seal. *(Courtesy of Oldsmobile Division—GMC)*

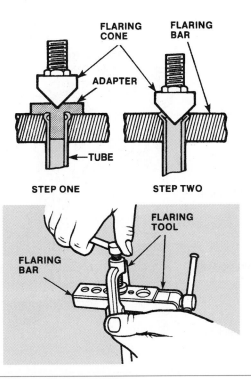

Figure 6-65 The double-flare tubing fitting is made with an anvil and cone.

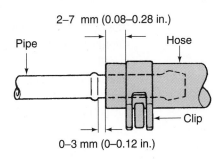

Figure 6-67 Fuel hose clamped to the steel tubing.

To control the rate of vapor flow from the fuel tank to the vapor storage tank, a plastic or metal restrictor may be placed in either the end of the vent pipe or in the vapor-vent hose itself.

Fuel Filters

Automobiles and light trucks usually have an in-tank strainer and a gasoline filter (Figure 6-68). The strainer, located in the gasoline tank, is made of a finely woven fabric. The purpose of this strainer is to prevent large contaminant particles from entering the fuel system where they could cause excessive fuel pump wear or plug fuel metering devices. It also helps to prevent passage of any water that might be present in the tank. Servicing of the fuel tank strainer is seldom required.

On fuel injected vehicles, the fuel filter is connected in the fuel line between the fuel tank and the engine. Many of these filters are mounted under the vehicle (Figure 6-69), and others are

Shop Manual
Chapter 6, page 231

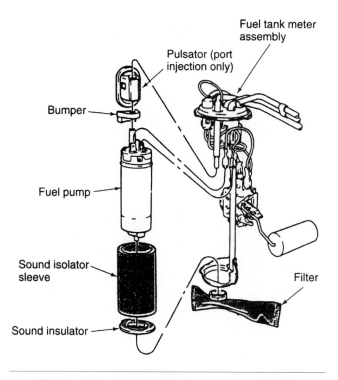

Figure 6-68 An in-tank fuel pump assembly with a fuel filter at the pump inlet. *(Courtesy of Oldsmobile Division—GMC)*

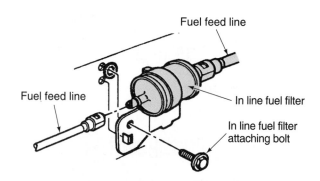
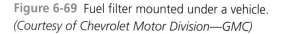

Figure 6-69 Fuel filter mounted under a vehicle. *(Courtesy of Chevrolet Motor Division—GMC)*

mounted in the engine compartment. Most fuel filters contain a pleated paper element mounted in the filter housing, which may be made from metal or plastic. Fuel filters on fuel injected vehicles usually have a metal case. On many fuel filters, the inlet and outlet fittings are identified, and the filter must be installed properly. An arrow on some filter housings indicates the direction of fuel flow through the filter.

On carbureted engines, the fuel filter is often mounted in the fuel line between the fuel pump and the carburetor (Figure 6-70). Some fuel filters on carbureted engines have the usual inlet and outlet fittings and a vapor return tube. A restricted orifice is positioned in the return tube on the filter,

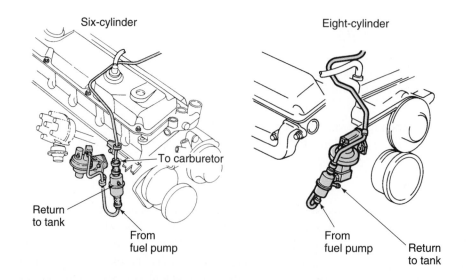

Figure 6-70 Fuel filter mounted between fuel pump and carburetor. *(Courtesy of Chrysler Corporation)*

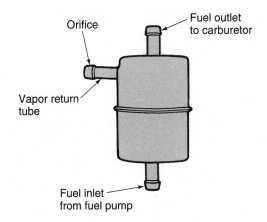

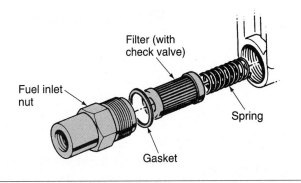

Figure 6-72 Fuel filter mounted in the carburetor inlet nut. *(Courtesy of Chevrolet Motor Division—GMC)*

Figure 6-71 Fuel filter with a vapor return line. *(Courtesy of Chrysler Corporation)*

and a fuel line is connected from this tube to the fuel tank (Figure 6-71). During hot engine operation, fuel vapors are returned from the filter through the return line to the fuel tank rather than flowing to the carburetor. This action helps to maintain lower fuel temperatures in the carburetor and prevents vapor lock.

On some carbureted engines, the fuel filter is mounted in the carburetor inlet nut (Figure 6-72). Some of these filters contain a one-way check valve that allows fuel into the carburetor but does not allow fuel to flow out of the carburetor. There are three basic types of in-carburetor gasoline filters. Pleated paper filters use pleated paper as the filtering medium. Paper elements are more efficient than screen-type elements, such as nylon or wire mesh, in removing and trapping small particles, as well as large size contaminants. Sintered bronze filters are often referred to as a stone or ceramic filter. Screw-in filters are designed to screw into the carburetor fuel inlet. The fuel line attaches to a fitting on the filter. This filter has a magnetic element to remove metallic contamination before it reaches the carburetor (Figure 6-73).

If this type of filter becomes severely restricted, the fuel pump pressure pushes the filter away from the inlet nut against the spring tension. This action allows fuel to continue flowing into the carburetor with a plugged filter, which prevents the engine from stalling. However, some dirt may enter the carburetor with the fuel.

Mechanical Fuel Pumps

The fuel pump is the device that draws the fuel from the fuel tank through the fuel lines to the engine's carburetor or injectors. Basically, there are two types of fuel pumps: mechanical and electrical. The latter is the most commonly used today.

Shop Manual
Chapter 6, page 235

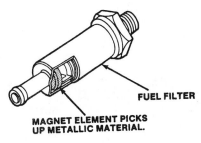

Figure 6-73 Screw-in fuel filter with magnetic element.

The first successful mechanical fuel pump was introduced in 1927 by the AC Spark Plug Company.

The return spring behind the fuel pump rocker arm may be called an antirattle spring.

Many carbureted engines have a mechanical fuel pump. The mechanical fuel pump is bolted on the side of the engine block or on the side of the timing gear cover. A gasket between the fuel pump and the mounting surface prevents engine oil leaks. A rocker arm pivots on a pin in the fuel pump casting near the pump mounting surface. The outer end of the rocker arm contacts a camshaft lobe or an eccentric attached to the front of the camshaft. On some engines, a pushrod is connected between the camshaft lobe and the rocker arm. The inner end of the rocker arm is linked to the diaphragm pull rod in the fuel pump (Figure 6-74). A small return spring between the rocker arm and the housing keeps the rocker arm in contact with the camshaft lobe to prevent rattling.

When the high part of the camshaft lobe contacts the rocker arm, the camshaft end of the rocker arm is forced downward. Since the rocker arm is pivoted, this action pulls the diaphragm end of the rocker arm upward against the spring tension. Upward diaphragm movement creates a vacuum between the diaphragm and the pump valves, and this vacuum opens the inlet valve and closes the outlet valve. Since there is atmospheric pressure above the fuel in the tank, fuel moves from the tank through the fuel line and the pump inlet valve into the diaphragm chamber.

When the high point on the camshaft lobe moves past the rocker arm, the camshaft end of the rocker arm moves upward. This action allows the diaphragm spring to force the diaphragm downward. This diaphragm movement creates a pressure in the diaphragm chamber that closes the inlet valve and opens the outlet valve. The pressure in the diaphragm chamber forces fuel through the outlet valve, fuel line, and filter to the carburetor. The amount of fuel pump pressure is determined by the diaphragm spring and the diaphragm stroke. A vent opening in the upper half of the pump casting prevents pressure buildup between the diaphragm and the seal.

When the carburetor float bowl has the proper level of fuel, the float movement closes the inlet needle and seat. This needle and seat action causes pressure to build up in the fuel line, filter, and fuel pump diaphragm chamber. This increase in fuel pressure forces the diaphragm upward against the spring tension, and the pin on the lower end of the pull rod is moved upward away from the rocker arm. Under this condition, the rocker arm continues to move up and down, but the

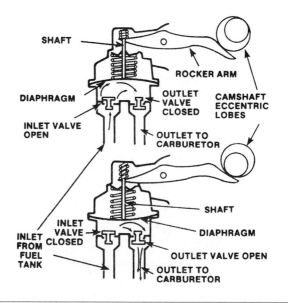

Figure 6-74 Mechanical fuel pump assembly.

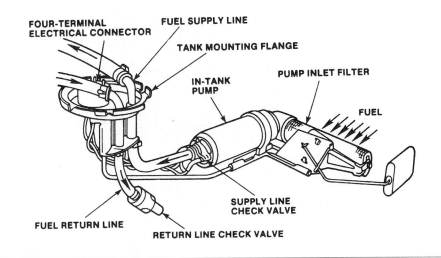

Figure 6-75 An in-tank electric fuel pump.

rocker arm no longer moves the pull rod. This diaphragm and pull rod action limits the fuel pump pressure. The pull rod seal prevents crankcase oil from entering the diaphragm area of the pump.

Electric Fuel Pumps

Electric fuel pumps offer important advantages over mechanical fuel pumps. Because electric fuel pumps maintain constant fuel pressure, they aid in starting and reduce vapor lock problems.

An electric fuel pump can be located inside (Figure 6-75) or outside the fuel tank. The electric fuel pump may also be combined with the fuel gauge's sending unit (Figure 6-76). There are four basic types of electric fuel pumps: the diaphragm, plunger, bellows, and impeller or rotary pump.

Shop Manual
Chapter 6, page 225

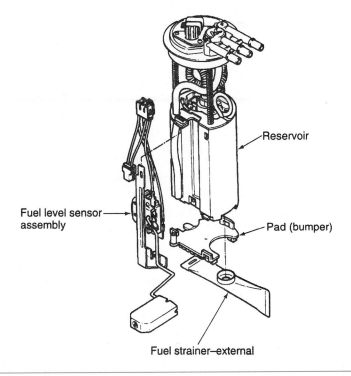

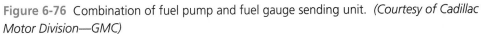

Figure 6-76 Combination of fuel pump and fuel gauge sending unit. *(Courtesy of Cadillac Motor Division—GMC)*

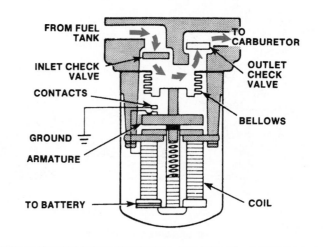

Figure 6-77 Bellows fuel pump.

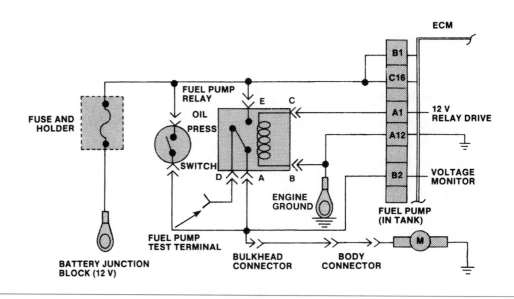

Figure 6-78 Typical wiring diagram for an electric fuel pump.

Although it is dangerous to have a spark near gasoline, the in-tank fuel pump is safe because there is no oxygen to support combustion in the tank.

The in-tank electric pump is usually a rotary type. The diaphragm, plunger, and bellows type (Figure 6-77) are usually the demand style. That is, when the ignition is turned on, the pump begins operation. They shut off automatically when the fuel line is pressurized. When there is a demand for more fuel, the pump provides more. When demand is lower, it pumps less, so proper fuel flow and pressure are constantly maintained. A typical wiring diagram for an electric fuel pump is shown in Figure 6-78.

Vehicles equipped with EFI usually have electric fuel pumps mounted in the fuel tank. A few carbureted engines also have this type of fuel pump. Some carbureted engines have an electric fuel pump mounted under the vehicle. Some EFI equipped vehicles have an in-tank electric pump and a second electric fuel pump mounted under the vehicle. The electric fuel pump in the fuel tank contains a small direct current (DC) electric motor with an impeller mounted on the end of the motor shaft. A pump cover is mounted over the impeller, and this cover contains inlet and discharge ports. When the armature and impeller rotate, fuel is moved from the tank to the inlet port, and the impeller grooves pick up the fuel and force it around the impeller cover and out the discharge port (Figure 6-79).

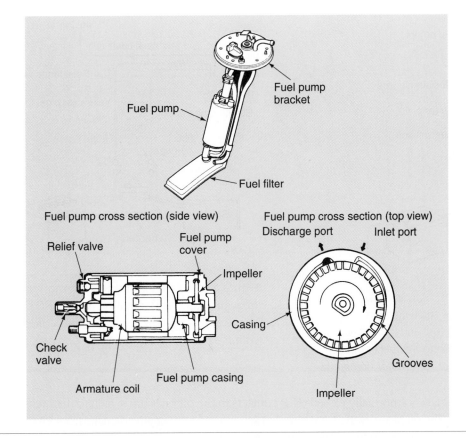

Figure 6-79 Electric fuel pump. *(Courtesy of American Honda Motor Company, Inc.)*

Fuel moves from the discharge port through the inside of the motor and out the check valve and outlet connection, which is connected via the fuel line to the fuel filter and underhood fuel system components. A pressure relief valve near the check valve opens if the fuel supply line is restricted and pump pressure becomes very high. When the relief valve opens, fuel is returned through this valve to the pump inlet. This action protects fuel system components from high fuel pressure. Each time the engine is shut off, the check valve prevents fuel from draining out of the underhood fuel system components into the fuel tank. A fuel filter is attached to the pump inlet. This filter prevents dirt or water from entering the pump.

The main reasons for the fuel pump being located in the fuel tank are: (1) to keep the fuel pump cool while it is operating and (2) to keep the entire fuel line pressurized to prevent premature fuel evaporation.

Electric Fuel Pump Circuits

Electric fuel pump circuits vary depending on the vehicle make and year. In some electric fuel pump circuits for carbureted engines, an oil pressure switch is connected in series with the fuel pump. When the engine is cranking, current flows from the starter side of the starter relay through

Shop Manual
Chapter 6, page 238

191

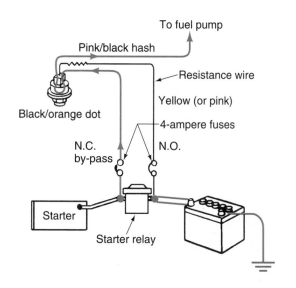

Figure 6-80 While cranking the engine, current flows from the starter relay through a fuse and a set of NC contacts in the oil pressure switch to the fuel pump. *(Reprinted with the permission of Ford Motor Company)*

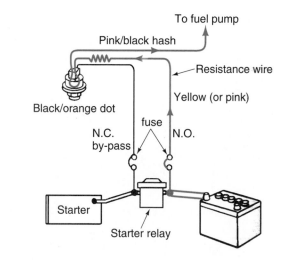

Figure 6-81 While the engine is running, current flows through a fuse, resistance wire, and the NO contacts of the oil pressure switch to the fuel pump. *(Reprinted with the permission of Ford Motor Company)*

a fuse and a set of normally closed (NC) contacts in the oil pressure switch to the fuel pump (Figure 6-80). The oil pressure switch is wired in series to prevent engine damage from low oil pressure by turning off the fuel pump making the engine stall.

Once the engine starts and there is oil pressure, the normally open (NO) contacts in the oil pressure switch close. The current now flows through another fuse, resistance wire, and the NO contacts in the oil pressure switch to the fuel pump (Figure 6-81). In this type of circuit, a defective oil pressure switch or low engine oil pressure results in an inoperative fuel pump and engine stalling. The resistance wire lowers the voltage supplied to the fuel pump and slows the pump to reduce fuel pressure while the engine is running. Since electric fuel pumps on electronic fuel injection systems are computer-controlled, these circuits are explained in later chapters.

A BIT OF HISTORY

In the early part of this century, many cars did not have a fuel pump. On some cars, like the Model A Ford, the fuel tank was mounted in the front of the instrument panel. The filler cap was positioned in front of the windshield. This was a very dangerous position for a fuel tank if the vehicle was involved in a collision. Since the gasoline tank was mounted higher than the carburetor, the fuel flowed by gravity from the tank to the carburetor. Other cars in those years had a vacuum tank, which used engine vacuum to move the fuel from the tank to the carburetor.

Inertia Switch. On Ford products, an inertia switch is connected in series in the fuel pump circuit. A steel ball inside the inertia switch is held in place by a permanent magnet. If the vehicle is involved in a collision, this ball pulls away from the magnet and strikes a target plate to open the points in the switch (Figure 6-82). This inertia switch action opens the fuel pump circuit and stops the fuel pump.

A reset button on top of the inertia switch must be pressed to close the switch and restore fuel pump operation. The inertia switch is located in the trunk area on most Ford cars. A decal in

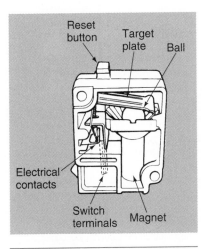

Figure 6-82 The internal design of an inertia switch. *(Reprinted with the permission of Ford Motor Company)*

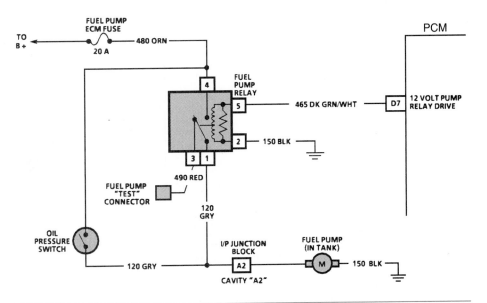

Figure 6-83 Typical GM fuel pump circuit. *(Courtesy of Oldsmobile Division— GMC)*

the trunk of later model cars indicates the inertia switch location. On EFI-equipped trucks, the inertia switch is located under the dash.

Oil Pressure Switch. In a late-model General Motors fuel pump circuit, the powertrain control module (PCM) supplies voltage to the winding of the fuel pump relay when the ignition switch is turned on. This action closes the relay points and voltage is supplied through the points to the in-tank fuel pump (Figure 6-83). The fuel pump remains on while the engine is cranking or running. If the ignition switch is on for two seconds and the engine is not cranked, the PCM shuts off the voltage to the fuel pump relay and the relay points open to stop the pump.

If the ignition switch is on and the fuel line is broken during an accident, PCM and fuel pump relay action is a safety feature that prevents the fuel pump from pumping gasoline from the ruptured fuel line. An oil pressure switch is connected parallel to the fuel pump relay points. If the relay becomes defective, voltage is supplied through the oil pressure switch points to the fuel pump. This action keeps the fuel pump operating and the engine running, even though the fuel pump relay is defective. When the engine is cold, oil pressure is not available immediately and the engine may be slow to start if the fuel pump relay is defective.

⚠️ **WARNING:** Never turn on the ignition switch or crank the engine with a fuel line disconnected. This action may result in gasoline discharge from the fuel line and a fire, which could cause personal injury and/or property damage.

ASD Relay. The fuel pump relay in Chrysler EFI systems is referred to as an automatic shutdown (ASD) relay. If the ignition switch is turned on, the PCM grounds the windings of the fuel pump relay and the relay points close. The ASD relay points supply voltage to the fuel pump, positive primary coil terminal, oxygen sensor heater, and the fuel injectors in some systems (Figure 6-84).

On Chrysler products with a power module and logic module, the engine must be cranked before the power module grounds the ASD relay winding. The later model PCM grounds the ASD relay winding when the ignition switch is turned on, and the relay remains closed while the engine is cranking or running. If the ignition switch is on for 1/2 second and the engine is not cranked, the PCM opens the circuit from the ASD relay winding to ground. Under this condition, the ASD relay points open and voltage is no longer supplied to the fuel pump, positive primary coil terminal, injectors, and oxygen sensor heater.

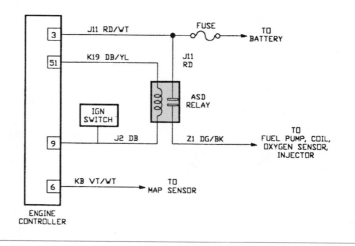

Figure 6-84 Chrysler fuel pump circuit with an ASD relay. *(Courtesy of Chrysler Corporation)*

Later model Chrysler fuel pump circuits have a separate ASD relay and a fuel pump relay. In these circuits, the fuel pump relay supplies voltage to the fuel pump, and the ASD relay powers the positive primary coil terminal, injectors, and oxygen sensor heater. The ASD relay and the fuel pump relay operate the same as the previous ASD relay. The PCM grounds both relay windings through the same wire.

Circuit Opening Relay. On many Toyota vehicles, the fuel pump relay is called a circuit opening relay. This relay has dual windings and it is mounted on the firewall (Figure 6-85). One of the windings of the circuit opening relay is connected between the starter relay points and ground, and the second relay winding is connected from the battery positive terminal to the PCM. When the engine is cranking and the starter relay points are closed, current flows through the starter relay points and the circuit opening relay winding to ground. This current flow creates a magnetic field

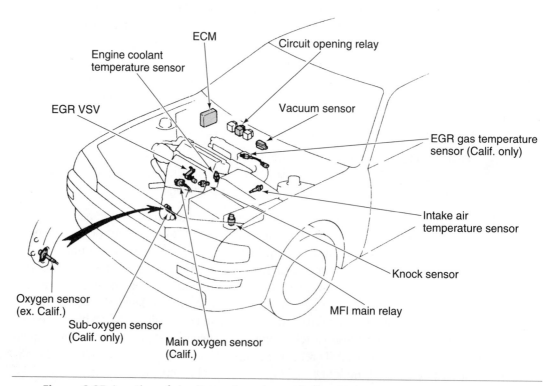

Figure 6-85 Location of circuit opening relay and other EFI components.

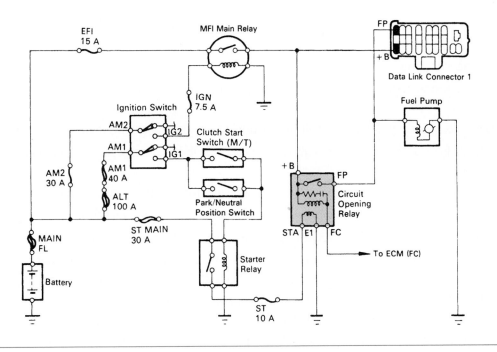

Figure 6-86 Wiring diagram for the circuit opening relay.

around the circuit opening relay winding that closes the relay points. When these points close, current flows through the points to the fuel pump (Figure 6-86).

Once the engine starts, the starter relay is no longer energized, and current stops flowing through these relay points and the winding of the circuit opening relay. However, the PCM grounds the other winding of the circuit opening relay when the engine starts. This action keeps the relay points closed while the engine is running.

Summary

❏ Gasoline is made from crude oil.

❏ The major factors affecting fuel performance are antiknock quality, volatility, sulfur content, and deposit control.

❏ An octane number is a measure of a gasoline's burning quality.

❏ The main reason unleaded gasoline was introduced was to provide a fuel for cars with catalytic converters.

❏ The significant chemical properties of diesel fuel are its wax appearance point, pour point, viscosity volatility, and cetane number.

❏ Because of its high propane content, LP-gas is often referred to as propane. Propane is a dry fuel, meaning that it enters the engine as a vapor. Gasoline, on the other hand, enters as tiny droplets of liquid.

❏ Natural gas costs less than gasoline. It is the cleanest alternative fuel. Therefore, it generally reduces vehicle maintenance. It is safe and abundant and readily available in the United States.

❏ The ideal air-fuel ratio for a gasoline engine is 14.7 pounds of air to 1 pound of fuel.

❏ Carburetion means enriching a gas by combining it with a carbon-containing compound. Three general stages of carburetion are metering, atomization, and vaporization.

❏ A venturi is a streamlined restriction that partly closes the carburetor bore. This restriction causes an increase in vacuum.

Terms to Know

Antiknock

Atomization

Barrel

Boost venturi

Cetane

Choke

Choke unloader

Continuous injection system (CIS)

Compressed natural gas (CNG)

Central port injection (CPI)

Dashpot

Detonation

Electronic fuel injection (EFI)

❏ The flow of air and fuel through the carburetor is controlled by the throttle plate.

❏ There are seven basic carburetor circuits: float, idle, off-idle, main metering, full power, accelerator pump, and choke.

❏ The float system stores fuel and holds it at a precise level as a starting point for uniform fuel flow.

❏ The idle circuit supplies a richer air-fuel mixture to operate the engine at idle and low speeds.

❏ The main metering circuit comes into operation when the engine speed reaches about 1,500 rpm or higher.

❏ The power system provides the engine with a richer air-fuel mixture at wide-open throttle.

❏ The accelerator pump mechanically supplies extra fuel during sudden acceleration.

❏ The choke circuit provides a very rich mixture during cranking and startup of a cold engine.

❏ Additional carburetor controls include the choke qualifier, dashpot, hot-idle compensator, dual vacuum break, choke unloader, deceleration valve, and throttle position solenoid.

❏ The latest type of carburetor system is the electronic feedback design, which provides better combustion by control of the air-fuel mixture.

❏ There are three types of electronic fuel injection systems: throttle body, central port, and port injection.

❏ While some electronic control elements are being added to the basic system, continuous injection systems (CIS) meter fuel delivery mechanically, not electronically.

❏ CIS injectors spray fuel constantly. They do not pulse on and off. The proper air-fuel mixture is attained by varying the amount of fuel delivered to the injectors.

❏ Modern fuel tanks include devices that prevent vapors from leaving the tank. They have slash baffles or surge plates to prevent the fuel from splashing around. Each tank has an inlet filler tube and a non-vented cap. Rollover protection devices prevent gasoline from leaking out of the fuel tank in an accident. A liquid vapor separator stops liquid fuel or bubbles from reaching the vapor storage canister or the engine crankcase.

❏ The fuel lines carry fuel from the tank to the fuel pump, fuel filter, and carburetor or fuel injection metering pump. They are made of either metal tubing or flexible nylon or synthetic rubber hose.

❏ Basically, there are two types of fuel pumps: mechanical and electrical. There are four basic types of electric fuel pumps: the diaphragm, plunger, bellows, and impeller (or rotary). Incorrect fuel pressure and low volume are the two most likely fuel pump troubles that affect engine performance.

❏ An inertia switch in the fuel pump circuit opens the fuel pump circuit immediately if the vehicle is involved in a collision.

❏ The oil pressure switch connected in parallel with the fuel pump relay operates the fuel pump if the fuel pump relay is defective.

Review Questions

Short Answer Essays

1. Explain the major differences between throttle body fuel injection and port fuel injection systems.

2. What is meant by sequential firing of fuel injectors?

3. Name the two forms of natural gas commonly used as an automotive fuel.

4. Name the components of a typical fuel delivery system.

5. What is the passageway in a carburetor that speeds up the flow of intake air?

6. Name the three general stages involved in carburetion.

7. Describe the difference between venturi vacuum and intake manifold vacuum.

8. Describe how a mechanical fuel pump limits fuel pressure.

9. Explain the purpose of the relief valve and one-way check valve in an electric fuel pump.

10. Explain the purpose of the inertia switch in a fuel pump circuit, including the switch reset procedure.

Fill-in-the-Blanks

1. Some fuel tank filler caps contain a pressure valve and a _____ _____ valve.

2. Some fuel filters in the carburetor inlet fitting have a one-way check valve that does not allow fuel flow out of the _____.

3. Fuel flows from the fuel tank into the mechanical fuel pump because there is _____ in the diaphragm chamber and _____ above the fuel in the tank.

4. In an electric fuel pump, the relief valve opens if the fuel line becomes _____.

5. Compared to TBI systems, MFI and SFI systems require _____ fuel pressure.

6. In TBI, MFI, and SFI systems, the fuel pressure must be high enough to prevent _____ _____.

7. On an SFI system, each injector has an individual _____ _____ connected to the computer.

8. If the injector pulse width is increased on TBI, MFI, or SFI systems, the air-fuel ratio becomes _____.

9. Fuel is discharged from the idle transfer slot when the throttle is _____ _____.

10. In a gasoline engine, _____ occurs when the flame front fails to reach a pocket of air-fuel mixture before the temperature in that area reaches the point of self-ignition.

ASE Style Review Questions

1. While discussing fuel tank filler pipes and caps:
   Technician A says the threaded filler cap should be tightened until it clicks.
   Technician B says the vent pipe is connected from the top of the filler pipe to the bottom of the fuel tank.
   Who is correct?
   A. A only   **C.** Both A and B
   B. B only   **D.** Neither A nor B

2. While discussing electric fuel pumps:
   Technician A says the one-way check valve prevents fuel flow from the underhood fuel system components into the fuel pump and tank when the engine is shut off.
   Technician B says the one-way check valve prevents fuel flow from the pump to the fuel filter and fuel system if the engine stalls and the ignition switch is on.
   Who is correct?
   A. A only   **C.** Both A and B
   B. B only   **D.** Neither A nor B

3. While discussing electric fuel pumps:
   Technician A says some electric fuel pumps are combined in one unit with the gauge sending unit.
   Technician B says on an engine with an electric fuel pump, low engine oil pressure may cause the engine to stop running.
   Who is correct?
   A. A only   **C.** Both A and B
   B. B only   **D.** Neither A nor B

4. While discussing basic carburetor principles:
   Technician A says the fuel is pulled out the main discharge nozzle by the intake manifold vacuum.
   Technician B says the intake manifold vacuum increases as the throttle is opened.
   Who is correct?
   A. A only   **C.** Both A and B
   B. B only   **D.** Neither A nor B

5. While discussing the float system:
   Technician A says many float bowls are vented from the float bowl to the air horn area.
   Technician B says the other carburetor systems depend on the proper level of fuel in the float bowl for proper operation.
   Who is correct?
   A. A only   **C.** Both A and B
   B. B only   **D.** Neither A nor B

6. While discussing the accelerator pump system:
   Technician A says the accelerator pump system is operated by the throttle linkage.
   Technician B says the accelerator pump system prevents a hesitation when accelerating from half-throttle to wide-open throttle.
   Who is correct?
   A. A only   **C.** Both A and B
   B. B only   **D.** Neither A nor B

7. While discussing TBI, MFI, and SFI systems:
   Technician A says the PCM provides the proper air-fuel ratio by controlling fuel pressure.
   Technician B says the PCM provides the proper air-fuel ratio by controlling injector pulse width.
   Who is correct?
   A. A only   **C.** Both A and B
   B. B only   **D.** Neither A nor B

8. While discussing central port injection (CPI):
   Technician A says the poppet nozzles are opened by the computer.
   Technician B says the poppet nozzles are opened by fuel pressure.
   Who is correct?
   A. A only   **C.** Both A and B
   B. B only   **D.** Neither A nor B

9. *Technician A* says that a richer air-fuel mixture is needed for idle operation.
   Technician B says that some carburetors have a transfer slot instead of an idle port.
   Who is correct?
   A. A only   **C.** Both A and B
   B. B only   **D.** Neither A nor B

10. *Technician A* says that a very rich air-fuel mixture is needed during cranking and start-up.
    Technician B says that a rich mixture is needed at wide-open throttle.
    Who is correct?
    A. A only   **C.** Both A and B
    B. B only   **D.** Neither A nor B

Intake and Exhaust Systems

Upon completion and review of this chapter, you should be able to:

❏ Describe air cleaner purpose and operation.

❏ Describe different types of air cleaner elements.

❏ Explain the operation of a heated air inlet system in the hot air, modulated air, and cold air positions.

❏ Explain the purpose of a heated air inlet system.

❏ Explain the purpose of the exhaust crossover passage in the intake manifold on V-type engines.

❏ Describe the advantage of port fuel injection (PFI) in relation to intake manifold design.

❏ Explain the purpose and operation of an electrically heated intake manifold grid.

❏ Describe how intake manifold vacuum is created, and explain the relation of throttle opening to intake manifold vacuum.

❏ Explain the difference between tuned exhaust headers and a conventional exhaust manifold.

❏ Describe the operation of the heat riser valve.

❏ Explain how carbon monoxide (CO), unburned hydrocarbons (HC), and oxides of

nitrogen (NO_x) are formed during the combustion process.

❏ Explain the difference between a two-way and a three-way catalytic converter.

❏ Define diesel particulate emissions.

❏ Briefly explain the operating principle of an electronic muffler.

❏ Explain basic turbocharger operation.

❏ Describe how the turbocharger boost pressure is controlled.

❏ Explain three ways in which the turbocharger bearings are cooled.

❏ Describe the results of inadequate turbocharger bearing cooling.

❏ List three items that cause premature turbocharger failure.

❏ Explain basic supercharger operation.

❏ Describe the difference in compression ratio in a turbocharged or supercharged engine compared to a normally aspirated engine.

❏ List nine components that are strengthened in a supercharged engine compared to a normally aspirated engine.

Introduction

The intake and exhaust systems are extremely important to engine operation. An air intake system must provide an adequate supply of clean air to the engine cylinders to maintain volumetric efficiency. If any component in the air intake system offers excessive air restriction, volumetric efficiency is reduced. The air cleaner element must remove all dust and abrasives from the incoming air. If the air cleaner allows any dust and abrasives to enter the engine, scoring of the cylinder walls, pistons, and piston rings will occur. In many engines, the air intake system is responsible for maintaining air intake temperature to provide improved fuel vaporization and engine performance.

The exhaust system must deliver exhaust from the cylinder head exhaust ports to the atmosphere with a minimum amount of restriction and noise. If any exhaust system component offers excessive restriction to exhaust flow, volumetric efficiency is reduced. Exhaust leaks in any of the exhaust system components provide objectionable noise. Catalytic converters in the exhaust system are responsible for reducing tail pipe emission levels.

The Corporate Average Fuel Economy (CAFE) regulations required vehicle manufacturers to produce a greater number of small, fuel-efficient cars with small engines. In many vehicles, this shift to smaller engines resulted in a reduction in engine power and acceleration compared to the large engines of the past. However, a turbocharger or supercharger provides the best of two worlds. A turbocharger or supercharger may be used on a relatively small engine that provides adequate fuel economy when driven at normal cruising speeds. However, the turbocharger or supercharger increases engine power to provide the faster acceleration desired by the driver.

Importance of Intake and Exhaust Systems

An internal combustion engine requires air to operate. This air supply is drawn into the engine by the vacuum created during the intake stroke of the pistons. The air is mixed with fuel and delivered to the combustion chambers. Controlling the flow of air and the air-fuel mixture is the job of the induction system.

Prior to the introduction of emission control devices, the induction system was quite simple. It consisted of an air cleaner housing mounted on top of the engine with a filter inside the housing. Its function was to filter dust and grit from the air being drawn into the carburetor.

Modern air induction systems filter the air and do much more. The introduction of emission standards and fuel economy standards encouraged the development of intake air temperature controls. These components draw air warmed by the exhaust manifold into the engine during cold startup and warmup.

The air intake system on a modern fuel injected engine is much more complicated. Ducts channel cool air from outside the engine compartment to the throttle plate assembly. The air filter has been moved to a position below the top of the engine to allow for aerodynamic body designs. Electronic meters measure airflow, temperature, and density. Heat riser tubes warm the intake air during cold weather warmup. Pulse air systems provide fresh air to the exhaust stream to oxidize unburned hydrocarbons in the exhaust. These components allow the air induction system to perform the following functions.

❏ Provide the air that the engine needs to operate

❏ Filter the air to protect the engine from wear

❏ Monitor airflow temperature and density for more efficient combustion and a reduction of hydrocarbon (HC) and carbon monoxide (CO) emissions

❏ Operate with the positive crankcase ventilation (PCV) system to burn the crankcase fumes in the engine

❏ Provide air for some air injection systems

Before each cylinder can be refilled with a fresh charge of air-fuel mixture, the cylinders must be free of the burnt gases resulting from the last four-stroke cycle. This is the primary purpose of the exhaust system. The exhaust system collects these gases, quiets exhaust noise, and carries the gases out at the rear of the vehicle.

Air Cleaners

The primary function of the air filter (Figure 7-1) is to prevent airborne contaminants and abrasives from entering into the engine with the air-fuel mixture. Without proper filtration, these contaminants can cause serious damage and appreciably shorten engine life. All incoming air should pass through the filter element before entering the engine.

Most air filter assemblies are located inside the air cleaner housing (Figure 7-2), which are mounted directly to a flange on the carburetor air horn, the fuel injection throttle plate assembly, the intake manifold (on diesel engines), or on the engine or engine compartment. In such cases, the air cleaner unit is connected to the air horn, throttle plate assembly, or manifold by an air transfer tube or duct.

The air cleaner assembly also helps muffle noise caused by the airflow through the throttle plates. The air cleaner also provides filtered air to the PCV system and provides engine compartment fire protection in the event of backfire.

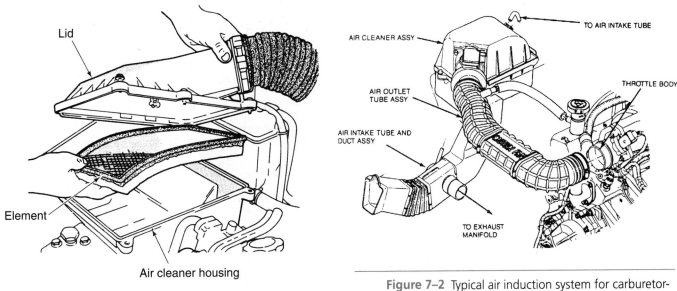

Lid

Element

Air cleaner housing

AIR CLEANER ASSY

TO AIR INTAKE TUBE

AIR OUTLET
TUBE ASSY

THROTTLE BODY

AIR INTAKE TUBE AND
DUCT ASSY

TO EXHAUST
MANIFOLD

Figure 7–1 Typical air cleaner assembly. (*Courtesy of Chrysler Corporation*)

Figure 7–2 Typical air induction system for carburetor-equipped engines.

Modern air cleaners provide many other important functions besides filtering the intake air. Many air cleaners have a heated air inlet system to heat the intake air on carbureted and throttle body injected engines. This system improves engine performance and emissions during warmup.

Some air cleaners on fuel injected engines contain an air charge temperature sensor that sends a signal to the computer in relation to air intake temperature. The computer uses this signal and other inputs to control the air-fuel mixture.

A small filter in the air cleaner, on some engines, is responsible for filtering the air entering the PCV system. Some air cleaners have an outlet connected from inside the air filter to the pulse air injection system. Other air cleaners have a hose connected to the air injection pump.

Air Cleaner Assembly and Filter Design

⚠️ **WARNING:** Even a small hole in an air cleaner element allows dirt particles to enter the engine, and these particles will cause cylinder scoring. Always check and service the air cleaner at the vehicle manufacturer's recommended service intervals.

The lower half of the air cleaner body is mounted on top of the carburetor or throttle body on fuel injected engines. On a diesel engine, the lower part of the air cleaner is mounted directly on the intake manifold. Regardless of the air cleaner mounting position, a gasket must be positioned between the air cleaner and the mounting flange (Figure 7–3). All the intake air must pass through the air filter. If the gasket is leaking between the air cleaner body and the mounting flange, air and abrasives are moved through this leak into the engine.

The air filter element is mounted inside the lower part of the air cleaner body. A large sealing surface is located on the top and bottom surfaces of the filter element, and the lower sealing surface must fit properly against a matching surface on the lower half of the air cleaner body. The air cleaner cover fits on top of the air cleaner body, and a sealing surface on the underside of the cover is designed to fit against the filter element. A sealing gasket is located around the outside of the cover. This gasket seals against the lower air cleaner body. A stud is threaded into the air cleaner mounting area, and this stud extends through the center of the air cleaner cover. A wingnut on top of the air cleaner stud retains the cover on the air cleaner. When this wingnut is tightened, the sealing surfaces on the air filter element contact the matching surfaces on the cover and the lower half of the air cleaner body.

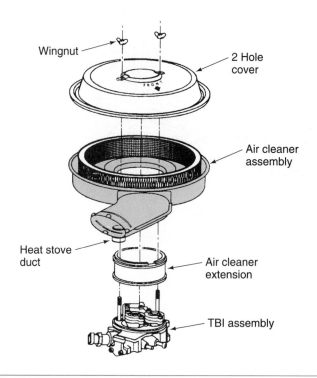

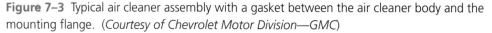

Figure 7–3 Typical air cleaner assembly with a gasket between the air cleaner body and the mounting flange. (*Courtesy of Chevrolet Motor Division—GMC*)

Many air filters have a pleated paper element with a wire or expanded metal screen on the outside of the element to provide protection against accidental paper damage while shipping and handling. A fine mesh screen is mounted on the inside of the pleated paper. The paper element and wire screens are molded into a heat-resistant plastisol on the top and bottom of the element. This heat-resistant plastisol has large sealing beads on both sides of the element. Air filter elements are available in many different shapes and sizes to fit various air cleaners. Some pleated-paper air filter elements contain oil-wetted resin-impregnated paper to provide longer element life and improved element efficiency.

Some heavy-duty air cleaners have an oil-wetted polyurethane cover placed over the paper element. The polyurethane traps larger dirt particles, and the smaller particles pass through the polyurethane where they are caught in the paper element. The polyurethane cover may be removed, cleaned, oil-wetted, and reinstalled.

The crankcase or PCV filter (Figure 7-4) inside some air cleaner assemblies is designed to filter the air that enters the crankcase through the PCV system. The filter also helps keep the engine's

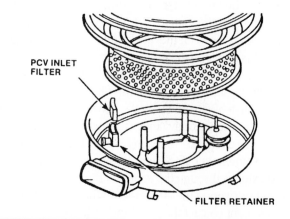

Figure 7–4 Typical location of a PCV filter. (*Reprinted with Permission of Ford Motor Company*)

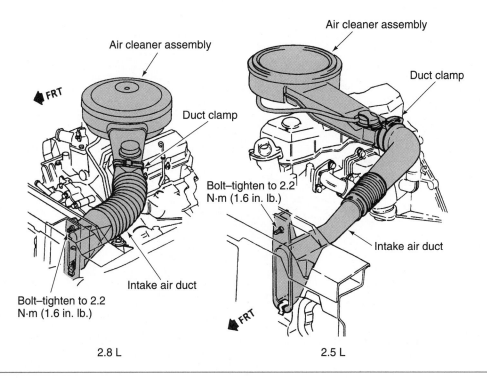

Figure 7–5 Air cleaner and snorkel with duct and cool air intake. (*Courtesy of Chevrolet Motor Division—GMC*)

air filter clean. When the engine is running under heavy loads, its manifold vacuum is very low. This means atmospheric pressure is pushing the crankcase fumes (blowby) out through the PCV valve and into the intake manifold. Rather, the pressure of the crankcase blowby pushes the fumes out of the crankcase filter where they are mixed with the incoming air at air filter. The oil vapor in the blowby gases gets trapped in the crankcase filter rather than entering the engine or clogging the engine's air filter.

Air Filter Ducts

There are many different air cleaner designs depending on the type of fuel system and the available underhood space. Many air cleaner assemblies have a snorkel attached to the air cleaner body, and the air intake at the end of the snorkel. On some engines, the outer end of the air cleaner snorkel is exposed to underhood air. Other air cleaner assemblies have a plastic duct connected to the snorkel, and the outer end of this duct is positioned in a cool air location, such as in front of or beside the radiator (Figure 7-5). A flexible coupling is positioned between the plastic duct and the snorkel. Some **air cleaner ducts** contain a large specially shaped resonator chamber which silences the airflow into the air cleaner (Figure 7-6).

Some air cleaner assemblies are mounted remotely from the engine. In these applications, a steel duct is attached to the carburetor or throttle body, and a flexible hose is connected from this duct to the air cleaner. Another flexible hose and duct are positioned to move cool air into the air cleaner (Figure 7-7). All air cleaner ducting must be leak-free, properly clamped, and secured. Steel air cleaner ducts are used in hot areas, and plastic ducts are usually mounted in cooler locations.

Heated Air Inlet Systems

Heated air inlet systems control the air intake temperature in the air cleaner. These systems are used on carbureted engines and some throttle body injected applications. A vacuum-operated **air door** is located in the air cleaner snorkel, and vacuum is supplied from the intake manifold through a **bimetal sensor** mounted inside the air cleaner to the air door vacuum motor (Figure 7-8).

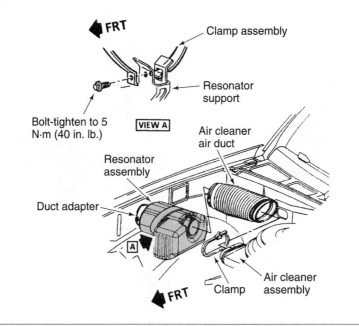

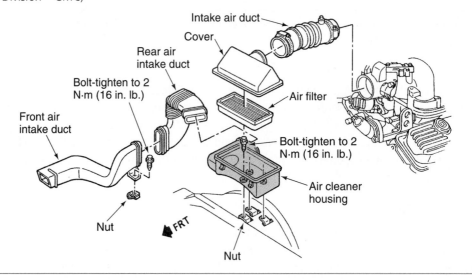

Figure 7–6 Resonator connected in an air cleaner duct. (*Courtesy of Chevrolet Motor Division—GMC*)

Figure 7–7 Remote air cleaner mounting. (*Courtesy of Chevrolet Motor Division—GMC*)

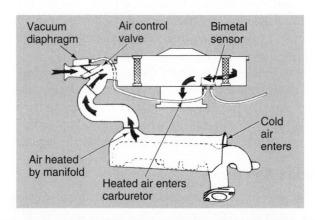

Figure 7–8 Heated air inlet system with vacuum-operated door in the air cleaner snorkel and bimetal sensor in the air cleaner. (*Courtesy of Chrysler Corporation*)

When the temperature inside the air cleaner is below approximately 125°F (52°C), the bimetal sensor closes the bleed port in the sensor. Under this condition, vacuum is supplied through the bimetal sensor to the air door vacuum motor. This vacuum lifts the diaphragm and air door upward, and the door closes the outside air passage into the air cleaner snorkel (Figure 7-9). In this door position, hot air is pulled into the air cleaner through the hot air inlet in the bottom of the snorkel. This hot air inlet is connected by a flexible hose to a metal heat stove surrounding the exhaust manifold.

The hot airflow into the air cleaner reduces fuel condensation on the cold surfaces of the intake manifold and provides more complete vaporization of the fuel, which results in improved engine performance, especially when the engine is cold.

When the air cleaner begins to warm up, the bimetal sensor starts to bleed off some of the vacuum supplied to the vacuum motor. Under this condition, the vacuum motor diaphragm and air door move downward until the door is partially open. In this position, the air flowing into the air cleaner is a blend of hot air from the hot air inlet and cold air from the snorkel. This action maintains air cleaner temperature and engine performance.

When the air cleaner temperature increases above 125°F (52°C), the bimetal sensor bleeds off all the vacuum through the sensor. Under this condition, there is no vacuum supplied to the air

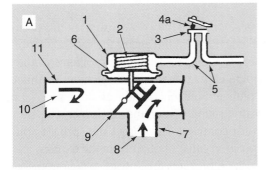

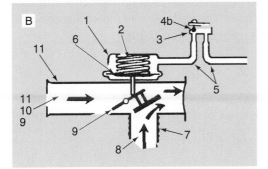

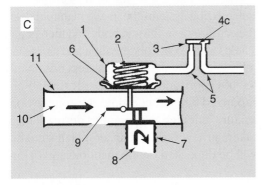

1	Vacuum diaphragm motor
2	Diaphragm spring
3	Temperature sensor
4a	Air bleed valve - closed
4b	Air bleed valve - partially open
4c	Air bleed valve - open
5	Vacuum hoses
6	Diaphragm
7	Heat stove
8	Hot air (exhaust manifold)
9	Damper door
10	Outside inlet air
11	Snorkel

A - Hot air delivery mode
B - Regulating mode
C - Outside air delivery mode

Figure 7-9 Heated air inlet system at various air cleaner temperatures. (*Courtesy of Chevrolet Motor Division—GMC*)

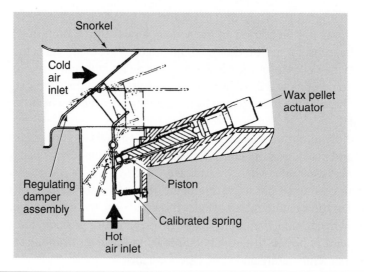

Figure 7–10 Heated air inlet with thermostatic element controlling the air valve door position. (*Courtesy of Chevrolet Motor Division—GMC*)

motor diaphragm, and the diaphragm and air door move fully downward. In this position, the air door closes the hot air inlet, and all the intake air flows in the end of the snorkel. If hot air continued to enter the air cleaner when the engine is hot, engine detonation problems would occur.

Some air cleaners have an air door operated by a thermostatic element mounted in the air cleaner snorkel. The thermostatic element plunger is attached to the air door, and a spring on this door pulls the door upward toward the hot air position. When the air cleaner temperature is cold, the wax pellet in the **thermostatic element** contracts and allows the spring to pull the door upward to the hot air position. This door position causes hot air from the exhaust manifold heat stove to enter the air cleaner, but the door position prevents the flow of cold air from the end of the snorkel. As the air cleaner temperature increases, the thermostatic element expands and pushes the air door to the downward position, where it closes the hot air intake into the air cleaner and opens the cold air intake from the end of the snorkel (Figure 7-10).

Intake Manifolds

The intake manifold distributes the clean air or air-fuel mixture as evenly as possible to each cylinder of the engine.

Intake Manifold Design

On carbureted and throttle body (central) fuel injection systems the intake manifold delivers an air-fuel mixture. The air and fuel are mixed in the carburetor or throttle body and then enter the intake manifold. The design of the intake manifold helps to prevent condensation and assists in the vaporization of the air-fuel mixture. Smooth and efficient engine performance depends on mixtures that are uniform in strength, quality, and degree of vaporization. This is partly the job of the intake manifold. Ideally, the fuel is completely vaporized when it goes into the combustion chamber. Complete vaporization requires high temperature, but high temperatures decrease the volumetric efficiency of the engine. Therefore the best alternative is to deliver an air-fuel mixture that is vaporized above the point where fuel particles will be deposited on the manifold and below the point where excess heat results in power losses.

On port fuel injection and diesel injection systems the intake manifold delivers only clean air to the cylinders. Fuel is introduced into the combustion chamber by the individual fuel injectors. There are many new intake manifold designs for ported gasoline fuel injection. They are often

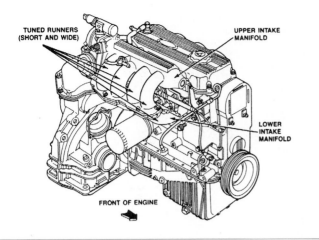

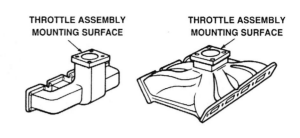

Figure 7–12 (A) Typical intake manifold design used on in-line engines; (B) typical design used on V-type engines.

Figure 7–11 This type of tuned intake manifold, used with ported fuel injection, gives a more equal amount of air to each cylinder. (*Reprinted with permission of Ford Motor Company*)

referred to as tuned intake manifolds because they have been redesigned to deliver equal amounts of airflow to each cylinder (Figure 7-11).

Intake manifolds are made of cast-iron or die-cast aluminum. Aluminum manifolds reduce engine weight. A few are cast integrally with the cylinder head.

Manifolds can be either wet or dry. **Wet intake manifolds** have coolant passages cast directly into them. **Dry intake manifolds** do not have cooling passages. Some intake manifolds have an integrally cast water crossover passage. This allows coolant to flow through a passage below the carburetor and warm the incoming air-fuel mixture. The intake manifold has a throttle body or carburetor mounting surface (Figure 7-12), and another mounting surface that fits against the intake port area on the cylinder head. Some intake manifolds have an upper and a lower half. The lower half of the intake manifold is bolted to the cylinder heads, and the upper half is bolted to the lower half (Figure 7-13).

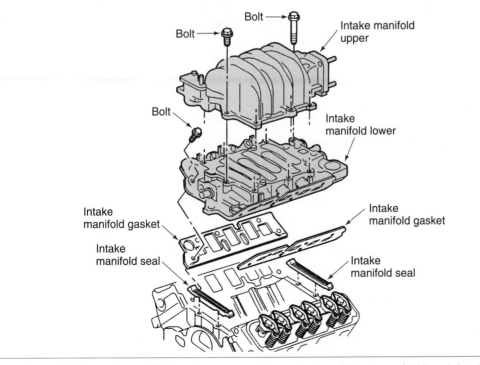

Figure 7–13 Intake manifold—the upper and lower halves. (*Courtesy of Oldsmobile Division— GMC*)

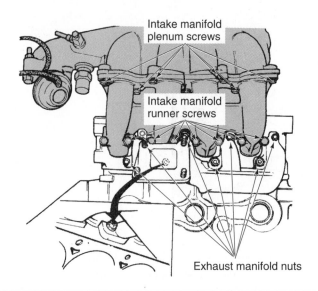

Intake manifold plenum screws

Intake manifold runner screws

Exhaust manifold nuts

Figure 7–14 Intake manifold from a 4-cylinder engine with curved runners in the lower half and a plenum area in the upper half. (*Courtesy of Chrysler Corporation*)

On an in-line engine, the intake manifold-to-cylinder head area is a straight machined surface, whereas the intake manifold-to-cylinder head mounting surface on a V-type engine fits in the V between the cylinder heads. The intake manifold must distribute air and fuel, or air, as equally as possible to all the cylinders. Since the distance from the carburetor or TBI assembly to each cylinder is not equal, the intake manifold air passages may be designed to make this distance as equal as possible. For example, in the manifold from a four-cylinder engine, four curved air passages split off from the upper plenum and connect with passages in the lower half of the intake, which is bolted to the cylinder head (Figure 7-14). The curved passages in the upper plenum and lower half of the intake are designed with similar lengths.

A series of bolts retains the intake manifold to the cylinder head, or heads, and a gasket is positioned between these two components. The throttle body or carburetor is bolted to the intake manifold, with a mounting gasket between the two components. All intake manifold gaskets and mounting surfaces must be in satisfactory condition to prevent air leaks into the intake manifold, which cause a lean air-fuel mixture, rough idle operation, and reduced engine performance at low speeds. A few intake manifolds are cast integrally with the cylinder head. The air passages in an intake manifold must be smooth and curved to provide unrestricted airflow and improved volumetric efficiency.

Intake Manifold Temperature

Intake manifold temperature is critical. If the intake manifold is too cold on TBI and carbureted engines, the air-fuel mixture condenses on the manifold surfaces. Since the gasoline must be vaporized and mixed with oxygen in the air to burn, the gasoline is wasted if it condenses on the intake manifold surface. Therefore, the intake manifold surface must be heated sufficiently to improve vaporization and prevent fuel condensation inside the manifold air passages. If the intake manifold is heated more than necessary, the heat expands the air-fuel mixture and reduces the amount of mixture entering the cylinders. This action decreases volumetric efficiency.

Most intake manifolds on TBI and carbureted V-type engines have an exhaust crossover passage through the intake manifold under the TBI assembly or carburetor mounting surface (Figure 7-15). This exhaust passage is connected from the exhaust system in one cylinder head through the intake manifold to the exhaust system in the opposite cylinder head. Exhaust flow through the crossover passage heats the intake manifold and improves fuel vaporization, especially when the engine is warming up. Heat from the crossover passage also helps to prevent frost formation around the throttles.

Carburetor icing is frost forming around the throttles or in the carburetor bores.

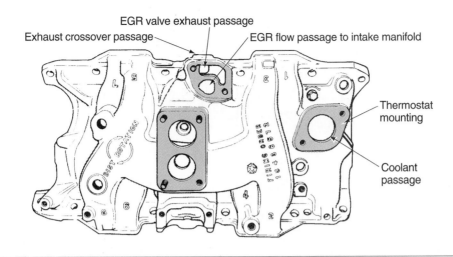

EGR valve exhaust passage
Exhaust crossover passage
EGR flow passage to intake manifold
Thermostat mounting
Coolant passage

Figure 7–15 Intake manifold from a V-8 engine with an exhaust crossover passage and an automatic choke well. (*Courtesy of Chrysler Corporation*)

In some intake manifolds on V-type engines, the automatic choke spring is located in a well above the crossover passage, and the heat from this passage is applied to the choke spring. Some intake manifolds have coolant circulated through them to provide heating.

Some intake manifolds have an electric heating grid mounted between the carburetor and the intake manifold to heat the air-fuel mixture (Figure 7-16). Voltage is supplied to this heating grid through a temperature switch mounted in the cylinder head. In some applications, the temperature switch senses coolant temperature, whereas in other vehicles this switch senses metal temperature in the cylinder head. If the engine is cold, the temperature switch is closed, and voltage is supplied to the heating grid to warm the air-fuel mixture. When the engine reaches a preset temperature, the temperature switch opens the circuit, and voltage is no longer supplied to the grid. The heating grid improves fuel vaporization and engine performance while the engine is cold.

On PFI engines, the intake manifold does not require as much heat because the fuel is injected at the intake ports, and the manifold distributes air to the intake ports. Since manifold heating requirements are reduced on these engines, the manifold may be designed with longer curved air passages to improve airflow and volumetric efficiency (Figure 7-17). This type of intake

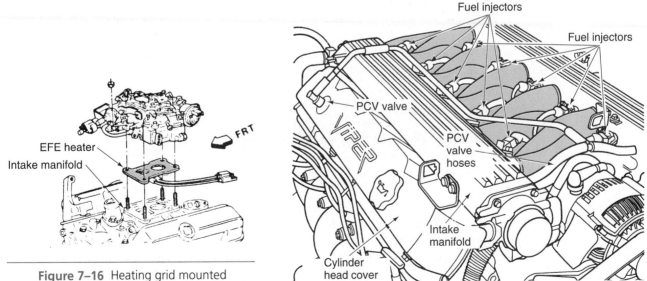

EFE heater
Intake manifold
FRT

Fuel injectors
Fuel injectors
PCV valve
PCV valve hoses
Intake manifold
Cylinder head cover

Figure 7–16 Heating grid mounted between the carburetor and the intake manifold. (*Courtesy of Chevrolet Motor Division—GMC*)

Figure 7–17 Tuned intake manifold with long curved intake runners on a port injected engine. (*Courtesy of Chrysler Corporation*)

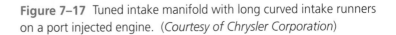

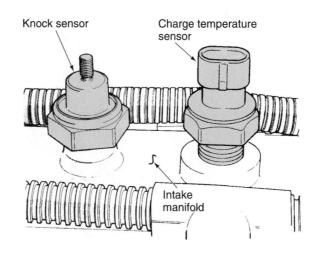

Figure 7–18 Air charge temperature sensor mounted in the air flow passage in an intake manifold. (*Courtesy of Chrysler Corporation*)

may be referred to as a tuned intake. Heated air inlet systems in the air cleaner are not required on PFI engines. Some PFI engines have small heater hoses connected from the cooling system to the throttle body assembly to prevent frost formation around the throttles.

Some high performance engines have two intake runners for each cylinder. There is a short set and a long set. Each of these sets are designed for different operating conditions—one set for low-speed, high torque operation and the other for high-speed operation. The switching from one set of intake runners to the next is normally controlled by the PCM. The PCM relies on input from a variety of sensors to determine the optimum time to switch from a set of runners to the other.

Intake manifolds and related intake systems contain a variety of computer input sensors on PFI, TBI, and computer-controlled carburetor systems. These sensors include the engine coolant temperature sensor (ECT), air charge temperature (ACT) sensor, mass air flow (MAF) sensor, and knock sensor. The ECT sensor may be mounted in a coolant passage in the intake manifold. This sensor sends a signal to the computer in relation to coolant temperature. An ACT sensor is mounted in one of the airflow passages in the intake manifold. This sensor sends a signal to the computer in relation to air temperature in the intake manifold (Figure 7-18). On some engines, the ACT sensor is mounted in the air cleaner.

The MAF sensor transmits a signal to the computer in relation to the total volume of air entering the engine. This sensor is located in the air intake hose between the air cleaner and the throttle body assembly, or directly in the throttle body assembly (Figure 7-19). A detailed explanation of these sensors is included later in the appropriate chapters.

Vacuum Systems

The vacuum in the intake manifold is used to operate many systems, such as emission controls, brake boosters, parking brake releases, headlight doors, heater/air conditioners, and cruise controls. Vacuum is applied to these systems through a system of hoses and tubes, which can become quite elaborate.

Vacuum is measured in relation to atmospheric pressure. Atmospheric pressure is the pressure exerted on every object on earth and is caused by the weight of the surrounding air. At sea level, the pressure exerted by the atmosphere is 14.7 psi. The normal measure of vacuum is in inches of mercury (in. Hg) instead of psi. Other units of measurement for vacuum are kilopascals and bars. Normal atmospheric pressure at sea level is about 1 bar or 100 kilopascals.

Vacuum in any four-stroke engine is created by the downward movement of the piston during the intake stroke. With the intake valve open and the piston moving downward, a partial vacuum is created within the cylinder and intake manifold. The air passing the intake valve does not

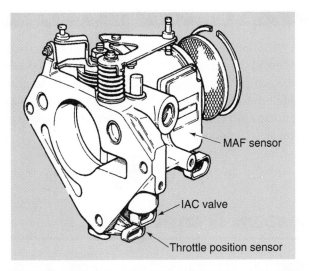

Figure 7–19 Mass air flow sensor mounted in the throttle body assembly. *(Courtesy of Oldsmobile Division—GMC)*

move fast enough to fill the cylinder, thereby causing the lower pressure. This partial vacuum is continuous in a multi-cylinder engine, since at least one cylinder is always at some stage of its intake stroke.

The amount of vacuum created is partially related to the positioning of the choke (on carbureted vehicles) and to the throttle plates. The throttle plate not only admits air or air-fuel into the intake manifold, it also helps to control the amount of vacuum available during engine operation. At closed-throttle idle, the vacuum available is usually between 15 to 22 inches. At a wide-open throttle acceleration, the vacuum can drop to zero.

Vacuum Controls

The intake manifold on early engines simply connected the carburetor to the engine. It did not have any auxiliary vacuum plugs or connections.

From the 1920s to the early 1970s, a single vacuum line was normally found. It either operated the vacuum advance of the distributor or the windshield wipers.

Since the 1970s, many systems rely on vacuum controls. The use of electronic engine controls have further complicated the vacuum system. Where once a vacuum line may have run from a carburetor to the EGR valve, it now goes from the carburetor to a computer-controlled vacuum relay solenoid and then to the valve.

In addition to being used to draw filtered air into the engine, vacuum is typically used to control these systems.

Fuel Induction System. Certain vacuum-operated devices are added to carburetors and some central fuel-injection throttle bodies to ease engine startup, warm-and-cold engine driveaway, and to compensate for air conditioner load on the engine.

Emission Control System. While some emission control output devices are solenoid or linkage controlled, many operate on vacuum. This vacuum is usually controlled by solenoids that are opened or closed, depending on electrical signals received from the **Powertrain Control Module (PCM)**. Other systems use switches that are controlled by engine coolant temperature, such as a **ported vacuum switch (PVS)** or by ambient air such as a **temperature vacuum switch (TVS)**.

Accessory Controls. Engine vacuum is used to control operation of certain accessories, such as air conditioner/heater systems, power brake boosters, speed-control components, automatic transmission vacuum modulators, and so on.

From the 1920s to the 1960s, intake manifold vacuum was supplied to two components: the distributor vacuum advance and the windshield wipers. From the 1970s to the present time, a wide variety of vacuum and electric/vacuum emission and computer system components have been added to the average automobile. Intake manifold vacuum is now responsible for such items as brake boosting, cruise control, air conditioning, computer input sensor signals, computer output control devices, and emission components.

Exhaust System Components

Shop Manual
Chapter 7, page 290

CAUTION: Exhaust system components are extremely hot if the engine has been running. Wear protective gloves to avoid burns when servicing these components.

The exhaust system is responsible for collecting the exhaust gas from each cylinder and discharging this gas at the rear of the vehicle. While performing this function, the exhaust system must silence the exhaust flow to an acceptable level outside and inside the vehicle. Catalytic converters in the exhaust system reduce emission levels. The main components in a typical exhaust system are these:

    1. Exhaust manifolds
    2. Exhaust pipe and seal
    3. Catalytic converter
    4. Muffler
    5. Resonator
    6. Tailpipe
    7. Heat shields
    8. Hangers, brackets, and clamps

All the parts of the system are designed to confirm to the available space of the vehicle's undercarriage and yet be a safe distance above the road.

WARNING: When inspecting or working on the exhaust system, remember that its components get very hot when the engine is running and contact with them could cause a severe burn. Also, always wear safety glasses or goggles when working under a vehicle.

Exhaust Manifolds

Many exhaust manifolds are made from cast iron or nodular iron. Some exhaust manifolds are made from stainless steel or heavy-gauge steel. The exhaust manifold contains an exhaust port for each exhaust port in the cylinder head, and a flat machined surface on this manifold fits against a matching surface on the exhaust port area in the cylinder head. Some exhaust manifolds have a gasket between the manifold and the cylinder head (Figure 7-20). In other applications, the machined surface fits directly against the matching surface on the cylinder head. The exhaust passages from each port in the manifold join into a common single passage before they reach the manifold flange. An exhaust pipe is connected to the exhaust manifold flange. On a V-type engine an exhaust manifold is bolted to each cylinder head.

A positive pressure may be defined as a pressure higher than atmospheric pressure.

Exhaust system components are designed for a specific engine. The pipe diameter, component length, **catalytic converter** size, muffler size, and exhaust manifold design are engineered to provide proper exhaust flow, silencing, and emission levels on a particular engine. Exhaust headers are used in place of exhaust manifolds on some engines (Figure 7-21). Each time a power stroke occurs and an exhaust valve opens, a positive pressure occurs in the exhaust manifold. A

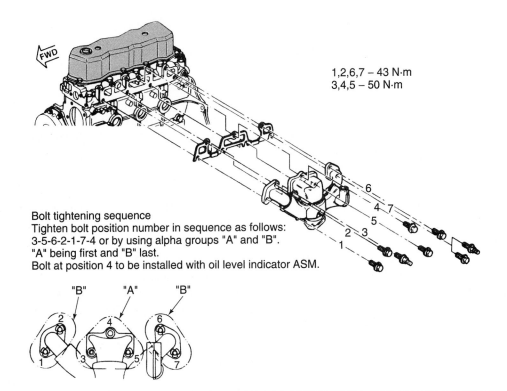

1,2,6,7 – 43 N·m
3,4,5 – 50 N·m

Bolt tightening sequence
Tighten bolt position number in sequence as follows:
3-5-6-2-1-7-4 or by using alpha groups "A" and "B".
"A" being first and "B" last.
Bolt at position 4 to be installed with oil level indicator ASM.

Figure 7–20 Exhaust manifold for an in-line engine. (*Courtesy of Chevrolet Motor Division—GMC*)

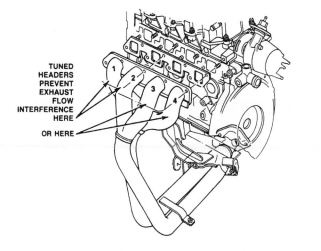

Figure 7–21 Efficiency can be improved with tuned exhaust headers. (*Reprinted with permission of Ford Motor Company*)

A negative pressure is a pressure less than atmospheric pressure.

A tuned exhaust header is a special exhaust manifold designed to prevent interference between the exhaust pulses from the cylinders.

negative pressure occurs in the exhaust manifolds between the positive pressure pulses, especially at lower engine speeds.

Some **exhaust headers** are tuned so the exhaust pulses enter the exhaust manifold between the exhaust pulses from other cylinders, preventing interference between the exhaust pulses. If the exhaust pressure pulses interfere with each other, the exhaust flow is slowed, causing a decrease in volumetric efficiency. Proper exhaust manifold tuning actually creates a vacuum, which helps to draw exhaust out of the cylinders and improve volumetric efficiency. The interference between exhaust pulses is most noticeable on a V8 engine compared to a V6, or four-cylinder engine.

Exhaust Pipe and Seal

The exhaust pipe is connected from the exhaust manifold to the catalytic converter. On in-line engines the exhaust pipe is a single pipe, but on V-type engines the exhaust pipe is connected to each manifold flange, and these two pipes are connected into a single pipe under the rear of the engine. This single pipe is then attached to the catalytic converter. Exhaust pipes may be made from stainless steel or zinc-plated steel, and some exhaust pipes are double-walled. In some exhaust systems, an **intermediate pipe** is connected between the exhaust pipe and the catalytic converter. Some exhaust systems have a heavy tapered steel or steel composition sealing washer positioned between the exhaust pipe flange and the exhaust manifold flange. Other exhaust pipes have a tapered end that fits against a ball-shaped surface on the exhaust manifold flange. Bolts or studs and nuts retain the exhaust pipe to the exhaust manifold (Figure 7-22). Some V-type engines have dual exhaust systems with separate exhaust pipes and exhaust systems connected to each exhaust manifold.

On some carbureted engines, a **heat riser valve** is positioned between the exhaust manifold and the exhaust pipe (Figure 7-23). The heat riser valve is positioned at one exhaust pipe flange on V-type engines. A heat riser valve may be mounted in the exhaust manifold on some engines.

When the engine is cold, a bimetallic spring holds the heat riser valve in the closed position. This valve position forces more exhaust gas through the crossover passage in the intake manifold to the exhaust system on the opposite side of the engine. This action heats the intake manifold and helps to improve fuel vaporization and prevent carburetor icing. The heat from the crossover passage also warms the choke spring, if this spring is located on the crossover passage. The bimetallic spring tension is reduced as the spring is heated. Under this condition, the exhaust flow forces the heat riser valve open.

If the heat riser valve is stuck open, the engine may hesitate on acceleration during warmup and experience carburetor icing problems. If the choke spring is located on the crossover passage, the choke may be slow to open if the heat riser valve is stuck open.

On some engines, the heat riser valve is vacuum operated, and a vacuum diaphragm is connected by a link to the heat riser valve. When the engine is cold, vacuum is supplied through a thermal vacuum switch (TVS) in the cooling system to the heat riser valve. Under this condition,

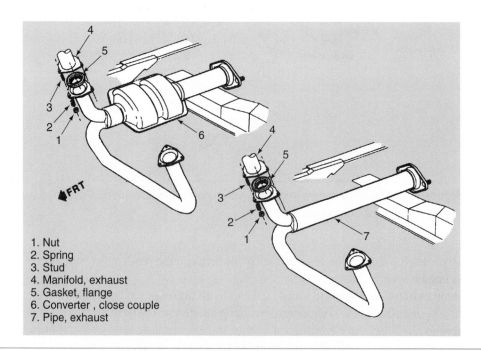

1. Nut
2. Spring
3. Stud
4. Manifold, exhaust
5. Gasket, flange
6. Converter , close couple
7. Pipe, exhaust

Figure 7–22 Bolts and nuts retain these pipes to the manifold flange. (*Courtesy of Chevrolet Motor Division—GMC*)

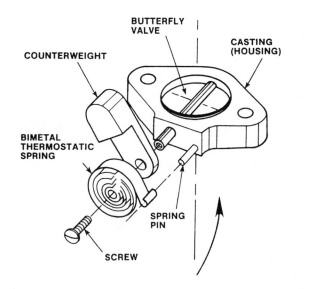

Figure 7–23 Mechanical heat riser valve.

the heat riser valve is pulled closed by the vacuum supplied to the diaphragm. At a preset coolant temperature, the TVS closes and shuts off the vacuum to the heat riser valve diaphragm, allowing the heat riser valve to move to the open position. Other emission devices such as the EGR valve may have an exhaust connection to the exhaust manifold.

On PFI, TBI, or computer-controlled carburetor systems, the oxygen (O_2) sensor is threaded into the exhaust pipe or exhaust manifold (Figure 7-24). This sensor informs the computer if the air-fuel mixture is rich or lean.

Automotive Pollutants and Catalytic Converters

Three major automotive pollutants are carbon monoxide (CO), unburned hydrocarbons (HC), and oxides of nitrogen (NO_x). Gasoline is a hydrocarbon fuel containing hydrogen and carbon. Since the combustion process in the cylinders is never 100% complete, some unburned HC are left over in the exhaust. Some HC emissions occur from evaporative sources, such as gasoline tanks and carburetors.

When air and gasoline are mixed and burned in the combustion chambers, the by products of combustion are carbon, carbon dioxide (CO_2), CO, and water vapor. Therefore, CO is a by-product of the combustion process.

Shop Manual
Chapter 7, page 291

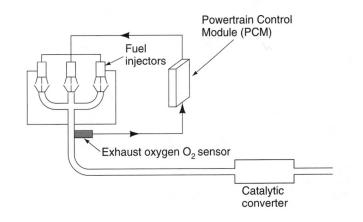

Figure 7–24 Simplified oxygen sensor circuit. *(Courtesy of Oldsmobile Division—GMC)*

Oxides of nitrogen (NO_x) are caused by high cylinder temperature. Nitrogen and oxygen are both present in air. If the combustion chamber temperatures are above 2,500°F (1,371°C), some of the oxygen and nitrogen combine to form NO_x. In the presence of sunlight, HC and NO_x join to form smog.

Catalytic converters may be pellet-type or monolithic-type. A **pellet-type converter** contains a bed made from hundreds of small beads, and the exhaust gas passes over this bed (Figure 7-25). In a **monolithic-type converter**, the exhaust gas passes through a honeycomb ceramic block (Figure 7-26). The converter beads, or ceramic block are coated with a thin coating of platinum, palladium, and rhodium, and mounted in a stainless steel container.

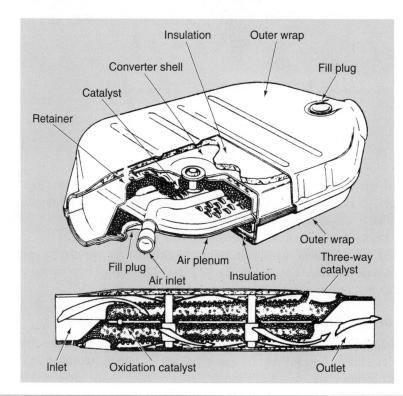

Figure 7–25 Pellet-type catalytic converter. (*Courtesy of Chevrolet Motor Division—GMC*)

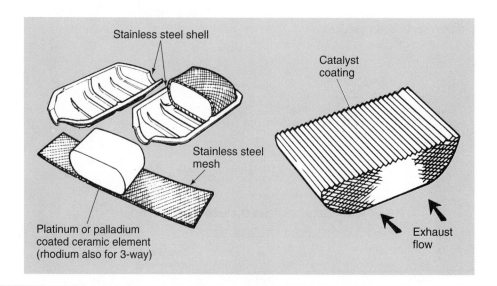

Figure 7–26 Monolithic-type catalytic converter. (*Courtesy of Chrysler Corporation*)

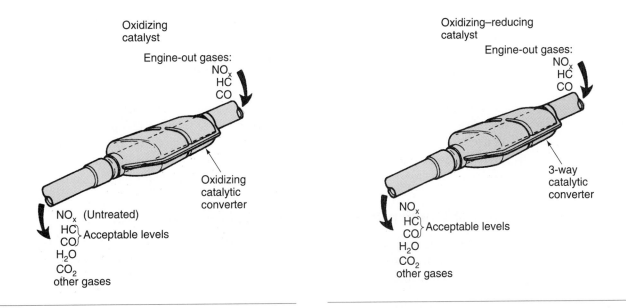

Figure 7–27 An oxidation catalyst. (*Courtesy of Chrysler Corporation*)

Figure 7–28 A three-way catalyst. (*Courtesy of Chrysler Corporation*)

An oxidation catalyst changes HC and CO to CO_2 and water vapor (H_2O). The oxidation catalyst may be referred to as a two-way catalytic converter (Figure 7-27).

In a three-way catalytic converter, the three-way converter is positioned in front of the oxidation catalyst. A three-way catalytic converter reduces NO_x emissions as well as CO and HC. The three-way catalyst reduces NO_x into nitrogen and oxygen (Figure 7-28). Some catalytic converters contain a thermo-sensor that illuminates a light on the instrument panel if the converter begins to overheat. Unleaded gasoline must be used in engines with catalytic converters. If leaded gasoline is used, the lead in the gasoline coats the catalyst and makes it ineffective. Under this condition, tail pipe emissions become very high. An engine that is improperly tuned causes severe overheating of the catalytic converter. Examples of improper tuning would be a rich air-fuel mixture or cylinder misfiring.

Most vehicles are equipped with a mini-catalytic converter, which provides a close coupled converter that is either built in the exhaust manifold or located next to it (Figure 7-29). It is primarily used to clean the exhaust during engine warmup. These converters are commonly called warmup converters or preheaters.

Many catalytic converters have an air hose connected from the belt-driven air pump to the oxidation catalyst. This converter must have a supply of oxygen to operate efficiently. On some engines, a mini-catalytic converter is built into the exhaust manifold or bolted to the manifold flange.

During oxidation catalyst operation, small amounts of sulfur in the gasoline combine with oxygen in the air to form oxides of sulfur (SO_x). Sulphur dioxide (SO_2) gas is also formed during

An oxidation catalyst may be referred to as a two-way catalytic converter.

An oxidation catalyst in a catalytic converter changes hydrocarbons (HC) and carbon monoxide (CO) to carbon dioxide (CO_2) and water vapor (H_2O).

A three-way catalytic converter performs the same function as a two-way catalytic converter, but it also reduces oxides of nitrogen (NO_x) to oxygen and nitrogen.

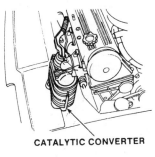

CATALYTIC CONVERTER

Figure 7–29 Mini- or pre-catalytic converter.

oxidation converter operation. This SO_2 gas is the same gas produced by rotting eggs. Although this gas has an unpleasant smell, it is not considered a major pollutant at present.

The SO_X combines with the water vapor in the converter to form small amounts of sulfuric acid (H_2SO_4). There is some concern among environmental agencies about these small amounts of H_2SO_4 from many vehicles contributing to acid rain.

OBD II which went into effect in 1997 (with a phase-in starting in 1995) had a clause that mandates the manufacturers to provide a way to warn the driver of a problem with the catalytic converter. This is accomplished by placing a oxygen sensor upstream and downstream of the converter. The computer will compare both sensor voltages. If the readings are the same, there is a problem with the converter or with the engine. This will light the malfunction indicator light (MIL).

Diesel Particulate Oxidizer Catalytic Converter

Particulate emissions on a diesel engine may be called soot.

Tailpipe emissions on a diesel engine include HC, CO, and NO_X emissions, but a diesel engine also produces **particulate** emissions, which are small carbon particles. A ceramic monolith inside the diesel particulate converter reduces particulate emissions (Figure 7-30).

Mufflers

The **muffler** is an oval-shaped, or cylindrical, component made from coated and aluminized steel or stainless steel. Inlet and outlet pipes extend from the ends of the muffler. Inside the muffler, the exhaust gas flows through a series of perforated tubes and a tuning chamber to silence the exhaust. The perforated tubes inside the muffler cancel out and silence the pressure pulsations in the exhaust each time an exhaust valve opens. The muffler is located behind the catalytic converter in the exhaust system. On many vehicles, the muffler is positioned just behind the center of the vehicle, but space requirements on some vehicles demand muffler installation near the rear of the car. When the muffler is positioned near the rear of the vehicle, it runs cooler and may experience more internal condensation. Mufflers rust on the inside if excessive internal condensation occurs.

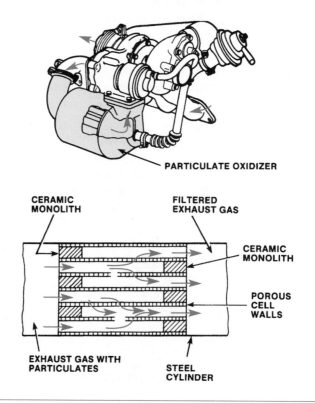

Figure 7–30 Particulate catalytic converter.

218

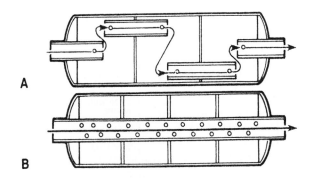

Figure 7–31 (A) Reverse-flow muffler; (B) straight-through muffler.

The most common type of muffler is the **reverse-flow** design (Figure 7-31), which changes the direction of exhaust flow inside the muffler. Some mufflers are a straight-through design in which the exhaust passes through a single perforated tube.

There have been several important changes in recent years in the design of mufflers. Most of these changes have been centered at reducing weight and emissions, improving fuel economy, and simplifying assembly. These changes include the following.

New Materials. More and more mufflers are being made of aluminized and stainless steel. Using these materials reduces the weight of the units as well as extends their life.

Double-Wall Design. Retarded engine ignition timing that is used on many small cars tends to make the exhaust pulses sharper. Many small cars now use a double-wall exhaust pipe to better contain the sound and reduce pipe ring.

Rear-Mounted Mufflers. More and more often, the only space left under the car for the muffler is at the very rear. This means that the muffler runs cooler than before and is more easily damaged by condensation in the exhaust system. This moisture, combined with nitrogen and sulfur oxides in the exhaust gas, forms acids that rot the muffler from the inside out. Many mufflers are being produced with drain holes drilled into them.

Backpressure. Even a well-designed muffler will produce some **backpressure** in the system. Backpressure reduces an engine's volumetric efficiency, or ability to "breathe." Excessive backpressure caused by defects in a muffler or other exhaust system part can slow or stop the engine. However, a small amount of backpressure can be used intentionally to allow a slower passage of exhaust gases through the catalytic converter. This slower passage results in more complete conversion to less harmful gases.

Electronic Mufflers. Basically, sensors and microphones pick up the pattern of the pressure waves an engine emits from its exhaust pipe (Figure 7-32). This data is analyzed by a computer. A mirror-image pattern of pulses is instantly produced and sent to speakers mounted near the exhaust outlet. Contrawaves are created that cancel out the noise. Noise is removed without creating backpressure in the muffler. Electronic mufflers can be designed to emit certain sounds, or no sound at all.

Resonator

On some vehicles, there is an additional muffler, known as a **resonator** or silencer. This unit is designed to further reduce the sound level of the exhaust. It is located toward the end of the system and generally looks like a smaller, rounder version of a muffler. The resonator is constructed like a straight-through muffler and is connected to the muffler by an intermediate pipe. The resonator on some cars is an integral part of the tailpipe, forming a one-piece unit. Resonators have been eliminated on nearly all late-model cars as manufacturers attempt to reduce production costs and weight.

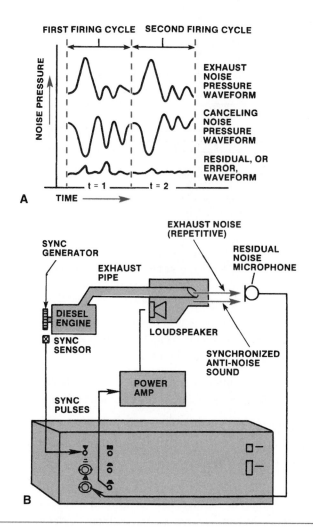

Figure 7–32 (A) An electronic muffler might seem far-fetched, but the production of a mirror image, out-of-phase waveform can cancel out sound. (B) Schematic shows how sensors, microphones, and speakers are teamed up to alternate noise.

Tailpipe

Pipes from the engine to the muffler are called exhaust pipes.

Pipes from the muffler to the rear of the vehicle are called tailpipes.

The tailpipe carries the flow of exhaust from the muffler to the rear of the vehicle. Some vehicles have an integral resonator in the tailpipe. This resonator is similar to a small muffler, and it provides additional exhaust silencing. In some exhaust systems, the resonator is clamped into the tailpipe. Tailpipes have many different bends to fit around the chassis and driveline components. All exhaust system components must be positioned away from the chassis and driveline to prevent rattling. The tailpipe usually extends under the rear bumper, and the end of this pipe is cut at an angle to deflect the exhaust downward. Chrome tailpipe extensions are available in auto parts stores. These extensions are attached to the tailpipe with lock screws.

Heat Shields, Clamps, Gaskets, and Hangers

Since exhaust system components become extremely hot, many vehicles have heat shields between some of these components and the chassis (Figure 7-33). Without the heat shields, some chassis components such as the floor pan may become hot enough to burn the padding under the floor mat.

Hangers are used to secure the exhaust system components to the chassis without transferring engine vibration to the chassis. The hanger is clamped to an exhaust system component, and a hanger bracket is bolted to the chassis (Figure 7-34). A piece of fabric-reinforced rubber between the hanger clamp and bracket prevents vibration transfer from the exhaust system to the chassis.

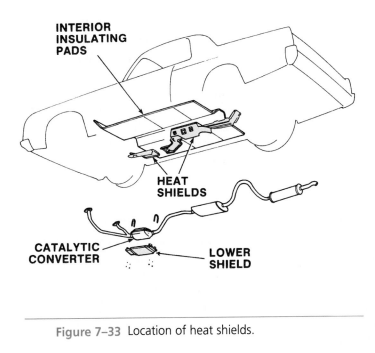

Figure 7–33 Location of heat shields.

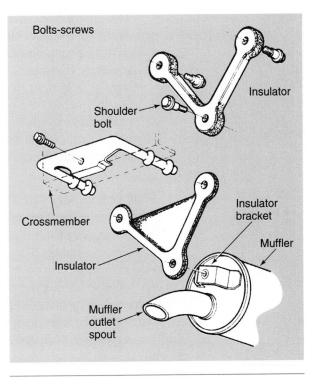

Figure 7–34 Exhaust system hangers and clamps. (*Courtesy of Chrysler Corporation*)

Exhaust system clamps are bolts curved into a U-shape with two nuts that retain a bar across the U-shaped bolt.

Exhaust system components such as the muffler and exhaust pipe are designed to slide over each other. These exhaust system components should overlap 1.5 in. and the clamp should be installed on the center of this overlap (Figure 7-35). When the clamp is tightened, the two pipes are squeezed together to prevent component movement and provide a seal between the pipes.

Forced Induction

The power generated by the internal combustion engine is directly related to the amount of air that is compressed in the cylinders. In other words, the greater the compression (within reason), the greater the output of the engine.

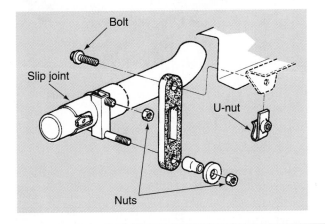

Figure 7–35 Exhaust pipe clamp installation. (*Courtesy of Chrysler Corporation*)

Two approaches can be used to increase engine compression. One is to modify the engine to increase the compression ratio. This has been done in many ways including the use of such things as domes or high top pistons, altered crankshaft strokes, or changes in the shape and structure of the combustion chamber.

Another, less expensive way to increase compression (and engine power) without physically changing the shape of the combustion chamber is to simply increase the intake charge. By pressurizing the intake mixture before it enters the cylinder, more air and fuel molecules can be packed into the combustion chamber. The two processes of artificially increasing the amount of airflow into the engine are known as turbocharging and supercharging.

Turbochargers

Turbochargers are used to increase engine power by compressing the air that goes into the engine's combustion chambers. They do not require a mechanical connection between the engine and the pressurizing pump to compress the intake gases. Instead, it relies on the rapid expansion of hot exhaust gases exiting the cylinders. These gases spin the turbine blades (hence the name turbocharger) of the pump. Because exhaust gas is a waste product, the energy developed by the turbine is said to be free since it theoretically does not use any of the engine's power it helps to produce.

A typical turbocharger, usually called a turbo, consists of the following components (Figure 7-36):

- ❏ Turbine or hot wheel
- ❏ Shaft
- ❏ Compressor or cold wheel
- ❏ Wastegate valve
- ❏ Actuator
- ❏ Center housing and rotating assembly (CHRA). This component contains the bearings, shaft, turbine seal assembly, and compressor seal assembly.

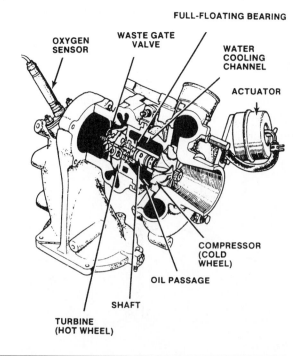

Figure 7–36 Cross section of a turbocharger shows the turbine wheel, the compressor wheel, and their connecting shaft.

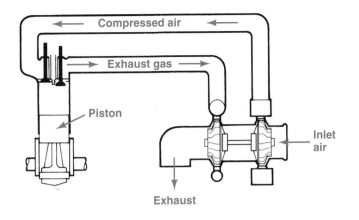

Figure 7–37 Basic turbocharger with turbine and compressor wheels. (*Courtesy of Chrysler Corporation*)

The turbocharger is normally located close to the exhaust manifold. Since exhaust gas expansion also helps spin the impeller faster, placing the turbocharger close to the exhaust manifold helps its efficiency. An exhaust pipe runs between the exhaust manifold and the turbine housing to carry the exhaust flow to the turbine wheel. Another pipe connects the compressor housing intake to an injector throttle plate assembly or a carburetor.

Basic Operation

A turbocharger contains a turbine wheel and a compressor wheel mounted on a common shaft. This shaft is supported on bearings in the turbocharger housing, and both wheels contain blades. The exhaust gas from the cylinders is directed past the turbine wheel, and the force of the exhaust gas against the turbine wheel blades causes the turbine wheel and shaft to rotate. Since the compressor wheel is positioned on the opposite end of this shaft, the compressor wheel must rotate with the shaft (Figure 7-37).

The compressor wheel is mounted in the air intake, and as the compressor wheel rotates, it forces air into the intake manifold. Since most turbocharged engines are port injected, the fuel is injected into the intake ports. The rotation of the compressor wheel compresses the air and fuel in the intake manifold, creating a denser air-fuel mixture. This increased intake manifold pressure forces more air-fuel mixture into the cylinders to provide increased engine power.

Turbocharger wheels rotate at very high speeds in excess of 100,000 rpm. Therefore, turbocharger wheel balance and bearing lubrication is very important. The turbocharger shaft must reach a certain rpm before it begins to pressurize the intake manifold. Some turbochargers begin to pressurize the intake manifold at 1,250 engine rpm and reach full boost pressure in the intake manifold at 2,250 rpm.

Air is typically drawn into the cylinders by the difference in pressure between the atmosphere and engine vacuum. A turbocharger, however, is capable of pressurizing the intake charge above normal atmospheric pressure. Turbo boost is the term used to describe the positive pressure increase created by a turbocharger. For example, 10 psi of boost means the air is being fed into the engine at 24.7 psi (14.7 psi atmospheric plus 10 pounds of boost).

A BIT OF HISTORY

Turbochargers have been common in heavy-duty applications for many years, but they were not widely used in the automotive industry until the 1980s. Turbochargers had two traditional problems that prevented their wide acceptance in automotive applications. Older turbochargers had a lag, or hesitation, on low-speed acceleration, and there was the problem of bearing cooling. Engineers greatly reduced the low-speed lag by designing lighter turbine and compressor wheels with improved blade design. Water cooling combined with oil cooling provided improved bearing life. These changes made the turbocharger more suitable for automotive applications.

Boost Pressure Control

A turbocharger is often called a turbo.

A wastegate diaphragm may be referred to as a bypass valve controller.

If the turbocharger boost pressure is not limited, excessive intake manifold and combustion pressure may damage engine components. Many turbochargers have a **wastegate** diaphragm mounted on the turbocharger. A linkage is connected from this diaphragm to a wastegate valve in the turbine wheel housing (Figure 7-38).

The diaphragm spring holds the wastegate valve closed. Boost pressure from the intake manifold is supplied to the wastegate diaphragm (Figure 7-39). When the boost pressure in the intake manifold reaches the maximum safe limit, the boost pressure pushes the wastegate diaphragm and opens the wastegate valve. This action allows some exhaust to bypass the turbine wheel, which limits turbocharger shaft rpm and boost pressure.

On some engines, the boost pressure supplied to the wastegate diaphragm is controlled by a computer-operated solenoid. In many systems, the PCM pulses the wastegate solenoid on and off to control boost pressure. Some computers are programmed to momentarily allow a higher boost pressure on sudden acceleration to improve engine performance.

VG30ET engine (with turbocharger)

Vacuum control valve
F.I.C.D. solenoid valve
A.A.C. valve
Air regulator
Emergency relief valve
Throttle valve
By-pass valve controller
By-pass valve
A.I.V. control solenoid valve
Air cleaner
Muffler
Catalyst
Turbine
Compressor
Turbocharger unit
Air flow meter
A.I.V. unit

⇦ Intake air flow
◀ Exhaust gas flow

Figure 7-38 Wastegate diaphragm mounted to a turbocharger. (*Courtesy of Chrysler Corporaton*)

Figure 7-39 A boost pressure hose is connected from the intake manifold to the wastegate diaphragm. (*Courtesy of Nissan Motors*)

Turbocharger Cooling

Exhaust flow past the turbine wheel creates very high turbocharger temperature, especially under high engine load conditions. Many turbochargers have coolant lines connected from the turbocharger housing to the cooling system (Figure 7-40). Coolant circulation through the turbocharger housing helps to cool the bearings and shaft. Full oil pressure is supplied from the main oil gallery to the turbocharger bearings and shaft to lubricate and cool the bearings. This oil is drained from the turbocharger housing back into the crankcase. Seals on the turbocharger shaft prevent oil leaks into the compressor or turbine wheel housings. Worn turbocharger seals allow oil into the compressor or turbine wheel housings, resulting in blue smoke in the exhaust and oil consumption. Some heat is also dissipated from the turbocharger to the surrounding air.

Some turbochargers do not have coolant lines connected to the turbocharger housing. These units depend on oil and air cooling. On these units, if the engine is shut off immediately after heavy-load or high-speed operation, the oil may burn to some extent in the turbocharger bearings. When this action occurs, hard carbon particles, which destroy the turbocharger bearings, are created. The coolant circulation through the turbocharger housing lowers the bearing temperature to help prevent this problem. When turbochargers that depend on oil and air cooling have been operating at heavy load or high speed, idle the engine for at least one minute before shutting it off. This action will help to prevent turbocharger bearing failure.

CAUTION: Lack of lubrication and oil lag are major causes of turbocharger failure.

Retarding spark timing is an often used method of controlling detonation on turbocharged engine systems. Unfortunately, any time the ignition is permanently retarded to prevent detonation, power is lost, fuel economy suffers, and the engine tends to run hotter. Because of these trade-offs, most systems use knock-sensing devices to retard timing only when detonation is detected.

Computer-controlled devices limit the amount of boost to prevent detonation and engine damage. In fact, some turbocharging systems on computer-controlled vehicles use an electronic control unit to operate the wastegate control valve through sensor signals.

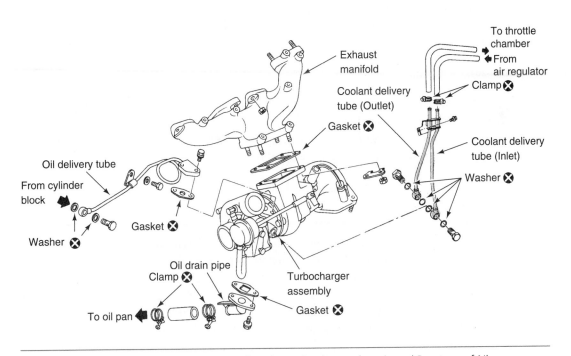

Figure 7-40 Coolant lines connected to the turbocharger housing. (*Courtesy of Nissan Motors*)

Turbo-Lag

Increases in horsepower are normally evidenced by an engine's response to a quick opening of the throttle. The lack of throttle response is felt with some turbocharged systems. This delay or turbo-lag occurs because exhaust gas requires a little time to build enough energy to spin the blower up to speed.

The variable nozzle turbine (**VNT**) has greatly improved the turbo-lag traditionally associated with turbochargers. **Turbo-lag** occurs when the turbocharger is unable to meet the immediate demands of the engine. This causes the power from the engine to temporarily lag behind the need. VNT units allow the turbine to accelerate quicker than the conventional turbos thereby reducing the lag time.

Scheduled Maintenance

▲ **WARNING:** When the compressor wheel is connected between the throttle body and the intake manifold, do not start the engine with an air intake hose disconnected between the turbocharger and the throttle body. This action may result in high engine rpm and engine damage.

Three main turbocharger killers are:

1. Lack of oil
2. Contaminants in the oil
3. Ingestion of foreign material through the air intake

To prevent these turbocharger killers from causing premature turbocharger failure, engine oil and filters should be changed at the vehicle manufacturer's recommended intervals. The engine oil level must be maintained at the specified level on the dipstick. The air cleaner element and the air intake system must be maintained in satisfactory condition. Dirt entering the engine through an air cleaner will damage the compressor wheel blades. When coolant lines are connected to the turbocharger housing, the cooling system must be maintained according to the vehicle manufacturer's maintenance schedule to provide normal turbocharger life.

Turbocharged engines have a lower compression ratio than a normally aspirated engine, and many parts are strengthened in a turbocharged engine because of the higher cylinder pressure. Therefore, many components in a turbocharged engine are not interchangeable with the parts in a normally aspirated engine.

▲ **WARNING:** When performing an oil change on a turbocharged engine, the oil should be drained in the normal way. However, after the new oil is put in the engine, the ignition should be disabled and the engine cranked until there is oil pressure. Then the ignition can be enabled and the engine started.

A nonturbocharged engine may be called a normally aspirated engine.

Shop Manual
Chapter 7, page 301

Superchargers

Supercharging fascinated auto engineers even before they decided to steer with a wheel instead of a tiller. The 1906 American Chadwick had a **supercharger**. Since then many manufacturers have equipped engines with superchargers. Supercharged Dusenbergs, Hispano-Suizas, and Mercedes-Benzs were giants among luxury-car marques, as well as winners on the race tracks in the 1920s and '30s. Then, after World War II, supercharging started to fade, although both Ford and American Motors sold supercharged passenger cars into the late 1950s. However, after being displaced first by larger V8 engines, then by turbochargers, the supercharger started to make a comeback with 1989 models. Some automobile manufacturers are offering superchargers, on performance cars, as an alternative to the turbo (Figure 7-41).

Figure 7–41 A supercharger installed on a V6 engine.

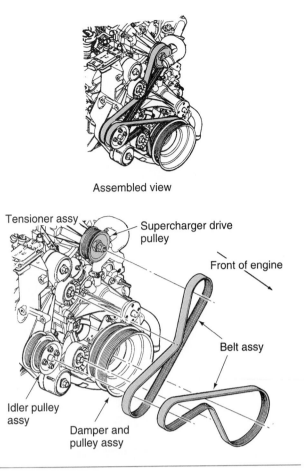

Assembled view

Tensioner assy

Supercharger drive pulley

Front of engine

Belt assy

Idler pulley assy

Damper and pulley assy

Figure 7–42 Supercharger pulley and drive belts. (*Reprinted with permission of Ford Motor Company*)

Basic Design

The supercharger is belt-driven from the crankshaft by a ribbed V-belt (Figure 7-42). A shaft is connected from the pulley to one of the drive gears in the front supercharger housing, and the driven gear is meshed with the drive gear. The rotors inside the supercharger are attached to the two drive gears (Figure 7-43).

The drive gear design prevents the rotors from touching. However, there is a very small clearance between the drive gears. In some superchargers, the rotor shafts are supported by roller bear-

A supercharger may be called a blower.

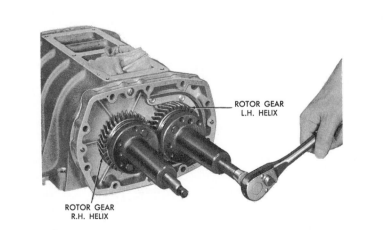

ROTOR GEAR L.H. HELIX

ROTOR GEAR R.H. HELIX

Figure 7–43 Supercharger gears. (*Courtesy of Detroit Diesel Allison*)

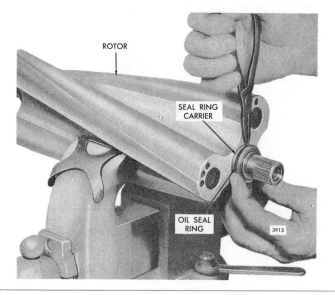

Figure 7–44 Supercharger rotor design. (*Courtesy of Detroit Diesel Allison*)

ings on the front and needle bearings on the back. During the manufacturing process, the needle bearings are permanently lubricated. The ball bearings are lubricated by a synthetic base high-speed gear oil. A plug is provided for periodic checks of the front bearing lubricant. Front bearing seals prevent lubricant loss into the supercharger housing.

Supercharger Operation

CAUTION: The supercharger and related components are very hot if the engine has been running. Wear protective gloves when handling these components.

Many superchargers have three lobe helical-cut rotors for quieter operation and improved performance (Figure 7-44). Intake air enters the inlet plenum at the back of the supercharger, and the rotating blades pick up the air and force it out the top of the supercharger (Figure 7-45).

The blades rotate in opposite directions, and they act like a pump as they rotate. This pumping action pulls air through the supercharger inlet and forces the air from the outlet. There is a very small clearance between the meshed rotor lobes and between the rotor lobes and the housing (Figure 7-46).

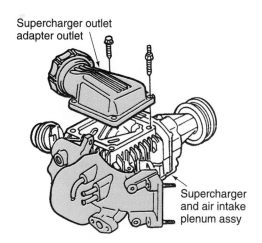

Figure 7–45 Supercharger air inlet and outlet housings. (*Reprinted with permission of Ford Motor Company*)

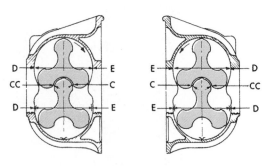

Figure 7–46 Rotor pumping action. (*Courtesy of Detroit Diesel Allison*)

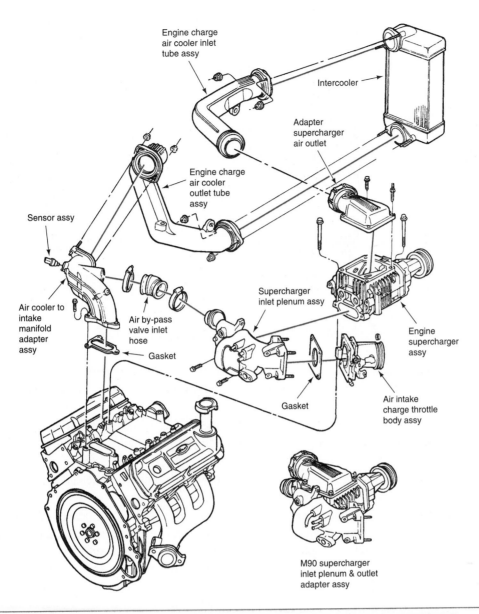

Engine charge
air cooler inlet
tube assy

Intercooler

Adapter
supercharger
air outlet

Engine charge
air cooler
outlet tube
assy

Sensor assy

Air cooler to
intake
manifold
adapter
assy

Air by-pass
valve inlet
hose

Gasket

Supercharger
inlet plenum assy

Engine
supercharger
assy

Gasket

Air intake
charge throttle
body assy

M90 supercharger
inlet plenum & outlet
adapter assy

Figure 7–47 Supercharger airflow. (*Reprinted with permission of Ford Motor Company*)

Air flows through the supercharger system components in the following order:

1. Air flows through the air cleaner and mass air flow sensor into the throttle body. The air cleaner and mass air flow sensor are not shown.

2. Airflow enters the supercharger intake plenum.

3. From the intake plenum the air flows into the rear of the supercharger housing (Figure 7-47).

4. The compressed air flows from the supercharger to the intercooler inlet.

5. Air leaves the intercooler and flows into the intercooler outlet tube.

6. Air flows from the intercooler outlet tube into the intake manifold adapter.

7. Compressed, cooled air flows through the intake manifold into the engine cylinders.

8. If the engine is operating at idle or very low speeds, the supercharger is not required. Under this condition, airflow is bypassed from the intake manifold adapter through a butterfly valve to the supercharger inlet plenum (Figure 7-48).

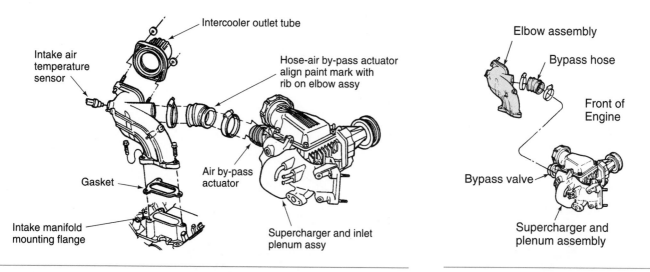

Figure 7–48 Supercharger air bypass actuator. (*Reprinted with permission of Ford Motor Company*)

Figure 7–49 Supercharger bypass hose and intake elbow.

The bypass butterfly valve (Figure 7-49) is operated by an air bypass actuator diaphragm as follows:

1. When manifold vacuum is 7 in. Hg or higher, the bypass butterfly valve is completely open and a high percentage of the supercharger air is bypassed to the supercharger inlet.

2. If the manifold vacuum is 3 to 7 in. Hg, the bypass butterfly valve is partially open and some supercharger air is bypassed to the supercharger inlet, while the remaining air flow is forced into the engine cylinders.

3. When the vacuum is less than 3 in. Hg, the bypass butterfly valve is closed and all the supercharger airflow is forced into the engine cylinders.

On some superchargers, the pulley size causes the rotors to turn at 2.6 times the engine speed. Since supercharger speed is limited by engine speed, a supercharger wastegate is not required.

Belt-driven superchargers provide instant low-speed action compared to exhaust-driven turbochargers, which may have a low-speed lag because of the brief time interval required to accelerate the turbocharger shaft. Compared to a turbocharger, a supercharger turns at much lower speed.

Friction between the air and the rotors heats the air as it flows through the supercharger. The intercooler dissipates heat from the air in the supercharger system to the atmosphere, creating a denser air charge. When the supercharger and the intercooler supply cooled, compressed air to the cylinders, engine power and performance are improved.

A Ford supercharged 3.8-L engine provides 210 hp at 4,000 rpm and 315 ft. lb. of torque at 2,600 rpm. A Ford turbocharged 3.8-L engine provides 190 hp at 4,600 rpm and 240 ft. lb. torque at 3,400 rpm. Therefore, the supercharged engine produces more horsepower and torque at a lower rpm.

The compression ratio in the supercharged 3.8-L engine is 8.2:1, compared to a normally aspirated 3.8-L engine, which has a 9.0:1 compression ratio. The following components are reinforced in the supercharged engine because of the higher cylinder pressure:

Engine block
Main bearings
Crankshaft bearing caps
Crankshaft
Steel crankshaft sprocket
Timing chain
Cylinder head
Head bolts
Rocker arms

Shop Manual
Chapter 7, page 309

Supercharger Designs

While there have been a number of supercharger designs on the market over the years, the most popular is the **Roots** type. The pair of three lobed rotor vanes in the Roots supercharger is driven by the crankshaft. The lobes force air into the intake manifold.

The key to the supercharger's operation, of course, is primarily the design of the rotors. Some Roots-type superchargers use straight-lobe rotors that result in uneven pressure pulses and, consequently, relatively high noise levels. Therefore, the supercharger used with most of today's engines uses a helical design for the two rotors. The helical design evens out the pressure pulses in the blower and reduces noise. It was found that a 60-degree helical twist works best for equalizing the inlet and outlet volumes.

Another benefit of the helical rotor design is it reduces carryback volumes—air that is carried back to the inlet side of the supercharger because of the unavoidable spaces between the meshing rotors—which represents a loss of efficiency.

The rotors are held in a proper relationship to each other by timing gears. They are supported by ball bearings in the front and needle roller bearings in the back. The needle bearings are greased for life. The ball bearings are generally lubricated by gear oil and a plug is provided for periodic checks of the oil level at the ball bearings. The oil is a synthetic base specifically produced for high-speed use. When servicing the gear case, it is very important to keep out dirt and moisture.

To handle the higher operating temperatures imposed by supercharging, an engine oil cooler is usually built into the engine system. This water-to-oil cooler is generally mounted between the engine front cover and oil filter.

Superchargers can be enhanced with electrically operated clutches and bypass valves. These allow the same computer that controls fuel and ignition to kick the boost on and off precisely as needed. This results in far greater efficiency than a full-time supercharger.

Another popular supercharger design, especially in Europe, is the **G-Lader** supercharger, which is based on a 1905 French design. As shown in Figure 7-50, spiral ramps in both sides of the rotor intermesh with similar ramps in the housing. Unlike most superchargers, the rotor of the G-Lader does not spin on its axis, rather it moves around an eccentric shaft. This motion draws in air, squeezes it inward through the spiral, which compresses it, then forces it through ducts in the center into the engine. Airflow is essentially constant, so intake noise is lower than that of a Roots blower. Because there is only a slight wiping motion between the spiral and housing, wear is minimal.

A comparison of supercharger and turbocharger operations is given in Table 7-1.

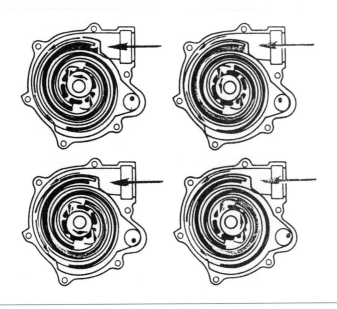

Figure 7–50 Operation of a G-Lader type supercharger.

Table 7-1 COMPARISON OF SUPERCHARGER VS TURBOCHARGER

	Supercharger	Turbocharger
Efficiency loss	Mechanical friction and power to rotate	Increased exhaust backpressure
Packaging	Belt driven	Major revisions required to package—new exhaust manifold
Lube system	Self-contained	Uses engine oil and coolant
System noise	Pressure pulsation and gear noise	High frequency whine and wastegate
Durability	In use by OEMs, seems to be durable	In use by OEMs, improvements have been made and durability has increased
Performance	30% to 40% power increase over naturally aspirated Better low end torque No lag—dramatic improvement in startup and passing acceleration, particularly with automatic transmission	30% to 40% power increase over naturally aspirated No loss of horsepower to operate
Temperatures	No change from naturally aspirated	Increased underhood temperatures

Intercoolers

The **intercooler** (Figure 7-51) cools the air from a supercharger or turbocharger before it reaches the combustion chamber. Cooling the air makes it denser, therefore its oxygen content increases and increases the engine's power from a given boost pressure. It also lowers the temperature produced in the combustion chamber. These factors help reduce engine knock and increase engine output. Intercoolers are like radiators in that heat from the air passing through it is removed and dissipated to the atmosphere. An air-to-air intercooler is normally located next to the conventional engine cooling radiator. Intercoolers can be air or water cooled.

Without an intercooler, the pressurized air is very hot. When this hot air is compressed during the engine's compression stroke, it will get even hotter. The hot air-fuel mixture in the cylinder may ignite on its own and at the wrong time, resulting in detonation.

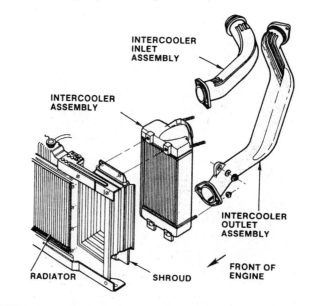

Figure 7–51 A typical intercooler system.

A turbocharger or supercharger changes the effective compression ratio of an engine, simply by packing in air that has a pressure greater than atmospheric pressure. For example, an engine that has a compression ratio of 8:1 and receives 10 pounds of boost will have an effective compression ratio of 10.5:1. This is why these boost devices increase power.

As air pressure increases, the temperature of the air also increases. The idea behind an intercooler is simply to let the turbocharger or supercharger increase the pressure of air. Then let's remove heat from the air. This allows cool, dense, high pressure air to enter the cylinders.

Summary

- ❑ If the air cleaner allows dust and abrasives to enter the engine, cylinder walls, pistons, and piston rings are scored.
- ❑ The heat-resistant plastisol seal on the top and bottom of the air cleaner element must contact the air cleaner body to provide proper sealing.
- ❑ Some air cleaner elements contain oil-wetted, resin-impregnated pleated paper for longer life.
- ❑ Some heavy-duty air cleaners have an oil-wetted polyurethane cover over the pleated paper element.
- ❑ In a heated air inlet system, a bimetal sensor controls the vacuum supplied to the air door vacuum motor.
- ❑ When the air cleaner is cold, the heated air inlet system supplies warm air from a manifold stove to the air cleaner.
- ❑ When the air cleaner is partially warmed up, the heated air inlet system supplies a blend of warm and cold air to the air cleaner.
- ❑ Once the air cleaner is hot, the heated air inlet system supplies cooler air from the snorkel to the air cleaner.
- ❑ The heated air inlet system improves fuel vaporization and engine performance during engine warmup.
- ❑ The intake manifold conducts clean air from the throttle body or carburetor to the intake ports.
- ❑ In some engines, the exhaust crossover passage in the intake manifold heats the intake to improve fuel vaporization and engine performance.
- ❑ The heat from the exhaust crossover passage in the intake manifold also helps to prevent carburetor icing, and the crossover passage supplies heat to the choke spring on some engines.
- ❑ Since the fuel is injected at the intake ports in a port fuel injected (PFI) engine, the intake manifold heating requirements are reduced, and the manifold may be designed to improve air flow and increase engine performance.
- ❑ On some engines, the heat riser valve supplies more exhaust flow through the exhaust crossover passage in the intake manifold during engine warmup.
- ❑ Intake manifold vacuum is higher with the throttle closed, and this vacuum decreases as the throttle is opened.
- ❑ Vehicles manufactured in recent years have many vacuum-operated components such as air conditioning, cruise control, emission devices, and computer system parts.
- ❑ All domestic and import vehicles must have a vacuum schematic decal in the underhood area.
- ❑ Exhaust manifolds are made from cast iron, nodular iron, stainless steel, or heavy gauge steel.
- ❑ Exhaust manifolds are connected to the exhaust port area on the cylinder head.
- ❑ Some exhaust manifolds have a gasket between the manifold and the cylinder head, but other exhaust manifolds have a tight fit between the machined surfaces on the manifold and cylinder head.

Terms to Know

Air cleaner ducts

Air door

Backpressure

Bimetal sensor

Carburetor icing

Catalytic converter

Dry intake manifold

Exhaust headers

G-Lader

Heat riser valve

Heated air inlet

Intercooler

Intermediate pipe

Monolithic-type converter

Muffler

Particulates

Pellet-type converter

Ported vacuum switch (PVS)

Powertrain control module (PCM)

Resonator

Reverse-flow muffler

Roots

Supercharger

Tailpipe

Temperature vacuum switch (TVS)

Thermostatic element

Turbo-lag

Turbocharger

❏ Tuned exhaust headers are designed so the exhaust pulses from each cylinder occur between the exhaust pulses from the other cylinders, thereby reducing interference between these pulses to provide improved exhaust flow.

❏ The exhaust pipe is connected between the exhaust manifolds and the catalytic converter.

❏ Automotive pollutants include carbon monoxide (CO), unburned hydrocarbons (HC), and oxides of nitrogen (NO_X).

❏ A catalytic converter may be pellet-type or monolithic-type.

❏ The pellets or ceramic honeycomb in a catalytic converter have a thin coating of platinum, palladium, or rhodium.

❏ A two-way catalyst oxidizes CO and HC into carbon dioxide (CO_2) and water vapor (H_2O).

❏ A three-way catalyst provides the same function as the two-way catalyst, and also reduces NO_X to nitrogen and oxygen.

❏ Diesel engines produce particulate emissions as well as the other emissions produced by a gasoline engine.

❏ Mufflers may be reverse-flow or straight-through design.

❏ Electronic mufflers electronically produce sound waves that oppose and cancel the original exhaust sound waves.

❏ In a turbocharger, the exhaust gas from the cylinders flows past the turbine wheel causing it to rotate at high speed.

❏ The turbine wheel and compressor wheel in a turbocharger are mounted on a common shaft.

❏ The compressor wheel in a turbocharger forces more air into the intake manifold. Since the intake manifold is pressurized, more air flows into the engine cylinders, resulting in increased engine power.

❏ The wastegate valve in a turbocharger is controlled by a wastegate diaphragm, and this diaphragm is operated by boost pressure.

❏ When the wastegate diaphragm opens the wastegate valve, some exhaust bypasses the turbine wheel, which limits boost pressure.

❏ On many applications, the PCM operates a solenoid connected in the hose between the intake manifold and the wastegate diaphragm. The PCM operates this solenoid to control boost pressure supplied to the wastegate diaphragm and limit boost pressure.

❏ Turbocharger bearings are cooled by oil, coolant, and the surrounding air.

❏ If turbocharger bearings are not properly cooled, the oil in the bearings may burn to some extent after a hot engine is shut off. This action results in hard carbon formations in the bearings, which cause premature bearing failure.

❏ Proper maintenance of the lubrication and cooling systems is very important on a turbocharged engine.

❏ A supercharger is belt-driven from the engine.

❏ Supercharger rotors usually contain three rotors, which force air into the intake manifold.

❏ The supercharger system has a bypass butterfly valve connected to a vacuum diaphragm, and intake manifold vacuum is supplied to this diaphragm.

❏ When the engine is operating at idle or very low speed, the vacuum diaphragm opens the by-pass butterfly valve, which allows a high percentage of the supercharger airflow to be by-passed to the supercharger inlet.

❏ When intake manifold vacuum drops below 3 in. Hg, the vacuum diaphragm closes the bypass butterfly valve, which forces all the supercharger air to flow into the intake manifold.

❏ Since the supercharger speed is limited by the pulley size, a wastegate for high-speed operation is not required.

❏ The compression ratio is lower in a supercharged or turbocharged engine compared to a normally aspirated engine.

❏ Many components are strengthened in a supercharged or turbocharged engine because of the higher cylinder pressures.

❏ An intercooler removes heat from the pressurized air from a turbocharger or supercharger thereby making it more dense.

Review Questions

Short Answer Essays

1. List the purposes of the air cleaner on various engines.

2. Explain the differences between an intake manifold design on a carbureted engine and the intake manifold design on a port fuel injected engine.

3. Explain why intake manifold heating is necessary on a carbureted engine.

4. Describe three methods of intake manifold and air-fuel mixture heating.

5. Explain the operation of a two-way catalyst.

6. Explain how a turbocharger or supercharger supplies more engine power.

7. Describe basic turbocharger operation.

8. Describe how the PCM controls turbocharger boost pressure.

9. Describe how the turbocharger bearings are cooled.

10. Explain basic supercharger operation.

Fill-in-the-Blanks

1. Ten psi of turbo boost means that air is being fed into the engine at _____ when the engine is operating at sea level.

2. A wet-type intake manifold has _____ circulated through the manifold.

3. Voltage is supplied through a _____ _____ to the electric heating grid between the carburetor and the intake manifold.

4. Intake manifold vacuum is _____ at wide-open throttle compared to the intake manifold vacuum at idle speed.

5. Pellets or the honeycomb ceramic block in a catalytic converter may be coated with _____ _____ _____ or _____.

6. The exhaust flows past the _____ wheel in a turbocharger.

7. The intake airflow is forced into the intake manifold by the _____ wheel in a turbocharger.

8. The wastegate diaphragm is moved by _____ pressure from the _____ _____ .

9. A supercharger is _____-driven from the engine.

10. Supercharger rotor speed is limited by _____ _____.

1. While discussing heated air inlet systems:
   Technician A says the intake manifold vacuum is supplied to the bimetal sensor.
   Technician B says the hot air inlet on the air cleaner is connected to a heat stove on the exhaust manifold.
   Who is correct?
   A. A only   **C.** Both A and B
   B. B only   **D.** Neither A nor B

2. While discussing intake manifolds:
   Technician A says the intake manifold on a port fuel injected (PFI) engine requires heating to prevent fuel condensation on the surfaces of the air passages in the manifold.
   Technician B says an exhaust crossover passage through the intake manifold provides manifold heating on carbureted engines.
   Who is correct?
   A. A only   **C.** Both A and B
   B. B only   **D.** Neither A nor B

3. While discussing intake manifold heating:
   Technician A says the heat from the exhaust crossover passage through the intake manifold helps to prevent carburetor icing.
   Technician B says carburetor icing may be defined as frost formation around the throttles.
   Who is correct?
   A. A only   **C.** Both A and B
   B. B only   **D.** Neither A nor B

4. While discussing catalytic converters:
   Technician A says small amounts of hydrochloric acid are formed in the catalytic converter.
   Technician B says small amounts of sulfur dioxide (SO_2), and sulfuric acid (H_2SO_4) are formed in the catalytic converter.
   Who is correct?
   A. A only   **C.** Both A and B
   B. B only   **D.** Neither A nor B

5. While discussing turbocharger operation:
   Technician A says the turbocharger shaft may turn at speeds in excess of 100,000 rpm.
   Technician B says the exhaust is routed past the compressor wheel.
   Who is correct?
   A. A only   **C.** Both A and B
   B. B only   **D.** Neither A nor B

6. When discussing turbocharger boost pressure control:
   Technician A says when the wastegate valve is open, intake air is bypassed around the compressor wheel.
   Technician B says a hose is connected from the intake manifold to the wastegate diaphragm.
   Who is correct?
   A. A only   **C.** Both A and B
   B. B only   **D.** Neither A nor B

7. While discussing turbocharger bearing cooling:
   Technician A says inadequate bearing cooling may result in the formation of hard carbon particles in the bearings.
   Technician B says if turbocharger bearing cooling is inadequate, the oil may burn to some extent in the bearings after a hot engine is shut off.
   Who is correct?
   A. A only   **C.** Both A and B
   B. B only   **D.** Neither A nor B

8. While discussing supercharger operation:
   Technician A says a supercharger requires a wastegate to prevent excessive intake manifold pressure at high engine speeds.
   Technician B says a typical supercharger speed would be 2.6 times the engine speed.
   Who is correct?
   A. A only   **C.** Both A and B
   B. B only   **D.** Neither A nor B

9. While discussing supercharger control:
   Technician A says the vacuum diaphragm opens the bypass butterfly valve at low engine speeds, and this valve bypasses a high percentage of the supercharger airflow to the supercharger inlet.
   Technician B says the vacuum diaphragm opens the bypass butterfly valve under hard acceleration, and this valve bypasses some of the supercharger airflow to the atmosphere.
   Who is correct?
   A. A only   **C.** Both A and B
   B. B only   **D.** Neither A nor B

10. While discussing turbochargers and superchargers:
    Technician A says a supercharger turns at a lower speed than a turbocharger.
    Technician B says a supercharger provides faster low-speed response in pressurizing the intake manifold compared to a turbocharger.
    Who is correct?
    A. A only   **C.** Both A and B
    B. B only   **D.** Neither A nor B

Emission Control Systems

Upon completion and review of this chapter, you should be able to:

❑ Explain why hydrocarbon (HC) emissions are released from an engine's exhaust.

❑ Explain how carbon monoxide (CO) emissions are formed in the combustion chamber.

❑ Describe oxygen (O_2) emissions in relation to air-fuel ratio.

❑ Describe how carbon dioxide (CO_2) is formed in the combustion chamber.

❑ Describe how oxides of nitrogen (NO_X) are formed in the combustion chamber.

❑ Describe the operation of an evaporative control system during the canister purge and nonpurge modes.

❑ Explain the purpose of the positive crankcase ventilation (PCV) system.

❑ Describe the operation of the PCV system at idle speed, part throttle, and heavy load conditions.

❑ Describe the operation of the thermal vacuum valve connected in the distributor vacuum advance hose.

❑ Explain the operation and purpose of a vacuum delay valve in the distributor vacuum advance hose.

❑ Describe the operation of the detonation sensor and electronic spark control module.

❑ Describe the operation of a port exhaust gas recirculation (EGR) valve.

❑ Explain the design and operation of a positive backpressure EGR valve.

❑ Describe the design and operation of a negative backpressure EGR valve.

❑ Explain the operation of a digital EGR valve.

❑ Explain the operation of a linear EGR valve.

❑ List and describe the various controls commonly used in EGR systems.

❑ Explain the operation and purpose of a mechanically controlled mixture heater system.

❑ Describe the operation of a computer-controlled mixture heater system.

❑ Explain the operation and purpose of a computer-controlled heat riser valve.

❑ Define the purpose of a catalytic converter.

❑ Explain how a catalytic converter works.

❑ Describe the operation of a pulsed secondary air system.

❑ Describe the operation of a secondary air injection system.

Introduction

Emission controls on cars and trucks have one purpose: to reduce the amount of pollutants and environmentally damaging substances released by the vehicles. The consequences of the pollutants are grievous. The air we breathe and the water we drink have become contaminated with chemicals that adversely affect our health. It took many years for the public and the industry to address the problem of these pollutants. Not until smog became an issue did anyone in power really care and do something about these pollutants.

Smog not only appears as dirty air, it is also an irritant to your eyes, nose, and throat. The things necessary to form photochemical smog are HC and NO_X exposed to sunlight in stagnant air. When there is enough HC in the air, it reacts with the NO_X in the air. The energy of sunlight causes these two chemicals to react and form photochemical smog.

The name "smog" comes from a combination of the words "smoke" and "fog," which pretty much tells what smog looks like.

Legislative History

The first Clean Air Act prompted Californians to create the **California Air Research Board** (**ARB**). California ARB's purpose was to implement strict air standards. These became the standard for federal mandates. One of the approaches to clean the air by the ARB was to start **periodic motor**

The California Air Research Board is commonly called the California ARB.

PMVI is basically the same thing a I/M (Inspection and Maintenance).

Shop Manual
Chapter 8, page 323

vehicle inspection (**PMVI**). The purpose for the PMVI is to inspect a vehicle's emission controls once a year. This inspection includes a tailpipe emissions test and an underhood inspection. The tailpipe test certifies that the vehicle's exhaust emissions are within the limits set by law. The underhood and/or vehicle inspection verifies that the pollution control equipment has not been tampered with or disconnected.

California is not the only state that requires annual emissions testing. Many states have incorporated an emissions test with their annual vehicle registration procedures. Most states have or are planning to implement the I/M 240 program that was first implemented in California. The I/M 240 tests the emissions of a vehicle while it is operating under a variety of load conditions. This is an improvement over exhaust testing during idle and high speed with no load.

The I/M 240 test requires the use of a chassis (road) dynonameter, commonly called a *dyno*. While on the dyno, the vehicle is operated for 240 seconds and under different load conditions. The test drive on the dyno simulates both in-traffic and highway driving and stopping. The emissions tester tracks the exhaust quality through these conditions.

The I/M 240 program also includes a functional test of the evaporative emission control devices and a visual inspection of the total emission control system. If the vehicle fails the test, it must be repaired and certified before it can be registered.

According to the Environmental Protection Agency (EPA) National Air Pollutant Emission Estimates 1940-1990, November 1991, the total air pollution in the United States during 1990 included:

1. 18.7 million metric tons of volatile organic compounds, mainly hydrocarbons
2. 60.1 million metric tons of carbon monoxide
3. 19.6 million metric tons of oxides of nitrogen

This same source also states that passenger cars are responsible for 17.8% of the total hydrocarbon emissions, 30.9% of the total carbon monoxide emissions, and 11.1% of the oxides of nitrogen emissions. After 30 years of emission regulations, these figures remain staggering! Imagine what these figures would be if automotive and industrial emissions had remained unregulated during the last 30 years!

There are three main automotive pollutants: hydrocarbons (HC), carbon monoxide (CO), and oxides of nitrogen (NO_X). Particulate emissions are also present in diesel engine exhaust. HC emissions are caused largely by unburned fuel from the combustion chambers. HC emissions can also originate from evaporative sources such as the gasoline tank. CO emissions are a byproduct of the combustion process, resulting from incorrect air-fuel mixtures. NO_X emissions are caused by nitrogen and oxygen uniting at cylinder temperatures above 2,500°F (1,371°C).

Emission standards have been one of the driving forces behind many of the technological changes in the automotive industry. Catalytic converters and other emission systems were installed to meet emission standards. Computer-controlled carburetors and fuel injection systems were installed to provide more accurate control of the air-fuel ratio to reduce emission levels and allow the catalytic converter to operate efficiently.

During the 1990s, emission standards in the United States have become increasingly stringent (Table 8-1). In 1994, an ambitious emission program began in California. This program specified emission standards for **transitional low emission vehicles** (**TLEV**). In 1997, the California emission program specified a further reduction in emission levels for **low emission vehicles** (**LEV**), and in the year 2000, emission levels are specified for **ultra low emission vehicles** (**ULEV**).

In 1998, California emission regulations require 2 percent of the cars sold in the state to be **zero emission vehicles** (**ZEV**). The regulators also call for a higher percentage in the years 2003 and 2005. With present technology, only electric cars will meet the ZEV standards. Other states have adopted or are considering adopting the California emission standards.

Present government goals call for a 98 percent reduction of unburned hydrocarbons, a 97 percent reduction of carbon monoxide, and a 90 percent reduction of oxides of nitrogen compared to precontrolled cars.

Shop Manual
Chapter 8, page 322

Table 8-1. EMISSION STANDARDS, U.S. AND CALIFORNIA 1990 TO 2000

	HC	CO	NO$_x$
1990 – U.S.	0.41	3.4	1.0
1994 – U.S.	0.25*	3.4	0.4
1993 – Calif.	0.25	3.4	0.40
1994 – TLEV	0.125	3.4	0.40
1997 – LEV	0.075	3.4	0.20
2000 – ULEV	0.040	1.7	0.20
*non-methane HC			

(Courtesy of Johnson Matthey, Catalytic Systems Divisions)

Many changes will be necessary in vehicles, engines, and emission systems to meet the emission standards of the 1990s. Since approximately 90 percent of hydrocarbon emissions occur before the catalytic converter is hot enough to provide proper HC oxidation, engineers are designing catalytic converters that are heated from the vehicle electrical system. Since these heaters have high current flow requirements, some charging system modifications may be necessary. This is just one of the many changes in emission systems that we will likely see in the near future.

The federal government has set standards for these pollutants. The exceptions to these standards are a few high-altitude western states and California. Because there is less oxygen at high altitudes to promote combustion, emission standards at high altitudes are slightly less strict. Because of the dense population of the state, California's standards allow less pollution than federal standards. Many states have followed California's mandate and have instituted emission standards of the same levels as California.

Automobile manufacturers have been working toward reduction of automotive air pollutants since the early 1950s, when auto emissions were found to be part of the cause of smog in Los Angeles. Governmental interest in controlling emissions developed around the same time.

Development of Emission Control Devices

In late 1959, California established the first standards for automotive emissions. In 1967, the Federal Clean Air Act was amended to provide for federal standards to apply to motor vehicles.

The first source of emissions to be brought under control was the crankcase. Positive crankcase ventilation systems that route crankcase vapors back to the engine's intake manifold were developed and incorporated into 1961 cars and light trucks sold in California. These systems were installed on all cars nationwide beginning with the 1963 models.

Control of unburned hydrocarbons and carbon monoxide in the engine's exhaust was the next major development. An **air injection reactor** (**AIR**) system was built into cars and light trucks sold in California in 1966. Other systems, including the controlled combustion system, were developed and used nationwide in 1968. Further improvements in the following years improved combustion to reduce hydrocarbon and carbon monoxide emissions.

Fuel vapors from the gasoline tank and the carburetor float bowl were brought under control with the introduction of evaporation control systems. These systems were first installed in 1970 model cars sold in California and in most domestic-made cars beginning with 1971 models.

Most vehicle manufacturers started to provide emission control systems that reduced oxides of nitrogen as early as 1970. The exhaust gas recirculation system used on some 1972 models was used extensively for 1973 models when federal standards for oxides of nitrogen took effect.

One of the most important developments for lowering emission levels has been the availability and use of unleaded gasolines. Beginning with 1971, engines have been designed to operate on unleaded fuels.

Removing lead from gasoline brings some immediate benefits. It eliminates the emission of lead particles from automobile's exhaust and increases spark plug life, important from an emission standpoint. It avoids formation of lead deposits in the combustion chambers that tend to increase hydrocarbon emissions.

The catalytic converter, a later development, provided a means for oxidizing the carbon monoxide and hydrocarbon emissions in the engine exhaust. Beginning with the 1975 model year, passenger cars and light trucks have been equipped with converters.

Three basic types of emission control systems are used in modern vehicles: evaporative control systems, precombustion, and post-combustion.

The **evaporative control** system is a sealed system. It traps the fuel vapors (HC) that would normally escape from the fuel tank and carburetor into the air.

Most of the pollution control systems used today prevent emissions from being created in the engine either during or before the combustion cycle. The common **precombustion control** systems are as follows:

❏ Positive Crankcase Ventilation (PCV). The PCV system removes pollutants that blow by the pistons into the crankcase and recirculates them into the induction system.

❏ Engine Modification Systems. These systems improve combustion and reduce HC and CO in the exhaust. They include a heated primary air system, air-fuel control changes, engine breathing refinements, and some spark timing controls.

❏ Exhaust Gas Recirculating (EGR) Systems. EGR reduces NO_x by diluting the air-fuel mixture with some exhaust gas, which does not burn.

Post-combustion control systems clean up the exhaust gases after the fuel has been burned. Secondary air or air injector systems put fresh air into the exhaust to reduce HC and CO to harmless water vapor, and carbon dioxide by chemical (thermal) reaction with oxygen in the air. Catalytic converters help this process. Most catalysts now reduce NO_x as well as HC and CO.

Pollutants

The gases that are of most concern to environmentalists, engineers, and technicians are HC, CO, NO_x, CO_2, and O_2. The latter two are not really pollutants but are monitored because they are indicators of combustion efficiency.

Hydrocarbons

Hydrocarbon (HC) emissions are unburned gas and are caused by incomplete combustion. Even an engine in good condition with satisfactory ignition and fuel systems produces some HC. When the flame front in the combustion chamber approaches the cooler cylinder wall, the flame front quenches leaving some unburned HC.

An excessively lean air-fuel ratio also results in cylinder misfiring and high HC emissions. A very rich air-fuel ratio also causes higher-than-normal HC emissions. At the stoichiometric air-fuel ratio, HC emissions are low. Evaporative emissions from fuel tanks, carburetor float bowls, and evaporative systems are also a source of HC emissions.

On cars with catalytic converters, the HC reading at the tailpipe is very low if the engine, ignition, fuel, and emission systems are in normal working condition. When a cylinder misfires, all the

Shop Manual
Chapter 8, page 322

The **evaporative emission control system** is often referred to as the **EVAP** system.

240

unburned gas in the cylinder is delivered to the exhaust system, resulting in a high HC reading. Cylinder misfiring and high HC readings may be caused by ignition defects, such as defective spark plugs, spark plug wires, coil, and distributor cap or rotor. Low cylinder compression also causes high HC emissions.

Excessive HC emissions may be caused by:

- ❏ Ignition system misfiring.
- ❏ Improper ignition timing.
- ❏ Excessively lean or rich air-fuel ratio.
- ❏ Low cylinder compression.
- ❏ Defective valves, guides, or lifters.
- ❏ Defective rings, pistons, or cylinders.
- ❏ Vacuum leaks.
- ❏ Plugged PCV system.
- ❏ Excessively rich air-fuel ratio.
- ❏ Stuck open heat riser valve.
- ❏ AIR pump inoperative or disconnected.
- ❏ Engine oil diluted with gasoline.

Carbon Monoxide

Carbon monoxide (CO) is a byproduct of combustion. CO is a poisonous chemical compound of carbon and oxygen. It forms in the engine when there is not enough oxygen to combine with the carbon during combustion. When there is enough oxygen in the mixture, carbon dioxide (CO_2) is formed. CO_2 is not a pollutant and is the gas used by plants to manufacture oxygen. CO is primarily found in the exhaust, but can also be in the crankcase. CO is odorless and tasteless, but in concentrated form it is toxic.

CO emissions are caused by a lack of air or too much fuel in the air-fuel mixture. CO will not occur if combustion does not take place in the cylinders, therefore the presence of CO means

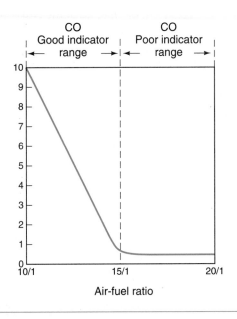

Figure 8-1 CO emissions in relationship to air-fuel ratio. (*Courtesy of Sun Electric Corporation*)

combustion is taking place. As the air-fuel ratio becomes richer, the CO levels increase (Figure 8-1). At the stoichiometric air-fuel ratio, the CO emissions are very low. If the air-fuel ratio is leaner than stoichiometric, the CO emissions remain very low. Therefore, CO emissions are a good indicator of a rich air-fuel ratio, but are not an accurate indication of a lean air-fuel ratio. When cylinder misfiring occurs, there is less total combustion in the engine cylinders. Since CO is a byproduct of combustion, cylinder misfiring does not increase CO emissions. If a cylinder is misfiring, CO emissions may decrease slightly.

Excessive CO emissions may be caused by:

- ❏ Rich air-fuel mixtures.
- ❏ Dirty air filter.
- ❏ Faulty injectors.
- ❏ Higher than normal fuel pressures.
- ❏ Defective system input sensor.
- ❏ Plugged PCV system.
- ❏ Excessively rich air-fuel ratio.
- ❏ Stuck open heat riser valve.
- ❏ AIR pump inoperative or disconnected.
- ❏ Engine oil diluted with gasoline.

Oxides of Nitrogen

NO$_x$ is pronounced "knocks".

This pollutant is actually various compounds of nitrogen and oxygen. Both of these gases are present in the air used for combustion. The formation of **oxides of nitrogen** (**NO$_x$**) is the result of high combustion temperatures. When combustion temperature reaches more than 2500°F (1,370°C), the N and the O$_2$ in the air combine to form these oxides of nitrogen. NO$_x$ emissions are a major contributor to the problem of smog. Since outside air is 78 percent N, the gas cannot be prevented from entering the combustion chamber. The key to controlling NO$_x$ is to prevent N from joining with oxygen during the combustion process.

The "x" in NO$_x$ stands for the proportion of oxygen mixed with a nitrogen atom. The "x" is a variable, which means it could be the number 1, 2, 3, etc., therefore, the term NO$_x$ refers to many different oxides of nitrogen (NO, NO$_2$, NO$_3$, etc.). Most of the NO$_x$ emissions from an engine are NO or nitrous oxide. When NO is released in the air, NO seeks, finds, and combines with an oxygen atom to form nitrous dioxide (NO$_2$). NO$_2$ is a very toxic gas and contributes to the formation of smog, ozone, and acid rain.

Higher-than-normal NO$_x$ emissions may be caused by:

- ❏ An overheated engine.
- ❏ Lean air-fuel mixtures.
- ❏ Vacuum leaks.
- ❏ Overadvanced ignition timing.
- ❏ Defective EGR system.

Oxygen

Acid rain is NO$_2$ mixed with water. This forms nitric acid.

O$_2$ is not a pollutant, therefore its presence in the exhaust does not pose any threat to our environment. However, too much oxygen in the exhaust does indicate that an improper mixture was in the cylinders or poor combustion has occurred in the engine. An improper air-fuel mixture will also cause a high reading of oxygen.

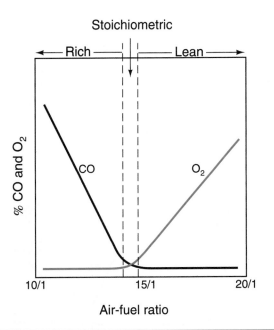

Stoichiometric

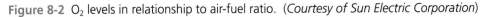

Figure 8-2 O_2 levels in relationship to air-fuel ratio. (*Courtesy of Sun Electric Corporation*)

If the air-fuel ratio is rich, all the oxygen in the air is mixed with fuel and the O_2 levels in the exhaust are very low. When the air-fuel ratio is lean, there is not enough fuel to mix with all the air entering the engine and O_2 levels in the exhaust are higher (Figure 8-2). Therefore, O_2 levels are a good indicator of a lean air-fuel ratio, and they are not affected by catalytic converter operation.

Lower-than-normal O_2 emissions may be caused by:

- ❏ Rich air-fuel mixture.
- ❏ Dirty air filter.
- ❏ Faulty injectors.
- ❏ Higher than normal fuel pressures.
- ❏ Defective system input sensor.
- ❏ Restricted PCV system.
- ❏ Charcoal canister purging at idle and low speeds.

Higher-than-normal O_2 emissions may be caused by:

- ❏ An engine misfire.
- ❏ Lean air-fuel mixtures.
- ❏ Vacuum leaks.
- ❏ Lower than specified fuel pressures.
- ❏ Defective fuel injectors.
- ❏ Defective system input sensor.

Carbon Dioxide

CO_2 is also not a pollutant and its presence in the exhaust does not pose a great threat to our environment. An ideal byproduct of combustion is CO_2, therefore large amounts of this gas are desired. If the air-fuel ratio goes from 9:1 to 14.7:1, the CO_2 levels gradually increase from approximately 6 percent to 13.5 percent (Figure 8-3). CO_2 levels are highest when the air-fuel ratio is slightly leaner than stoichiometric. At the stoichiometric air-fuel ratio, CO_2 levels begin to decrease.

CO_2 is a greenhouse gas and may be one of the causes of global warming.

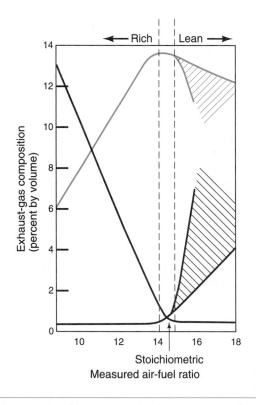

Figure 8-3 CO_2 levels in relationship to air-fuel ratio. (*Courtesy of Sun Electric Corporation*)

Lower-than-normal CO_2 emissions may be caused by:

❑ Leaking exhaust system.

❑ Rich air-fuel mixture.

Evaporative Emission Control Systems

The fuel evaporative emission control system reduces the amount of raw fuel vapors that are emitted into the air from the fuel tank and carburetor (if so equipped). These vapors must not be allowed to escape from the fuel system to the atmosphere. Since the first systems were used nationwide in the early 1970s, several refinements have been added. Current systems include the following components.

❑ A special filler design to limit the amount of fuel that can be put in the tank.

❑ A pressure/vacuum relief fuel tank cap instead of a plain vented cap (Figure 8-4).

❑ A vapor separator in the top of the fuel tank (Figure 8-5). This device collects droplets of liquid fuel and directs them back into the tank.

❑ A domed fuel tank in which its upper portion is raised. Fuel vapors rise to this upper portion and collect.

❑ Check, or one-way, valves keep vapors confined. When an engine runs, vacuum- or electrically operated valves open.

❑ Hoses and tubes connect parts of a vapor-recovery system. Special fuel/vapor rubber tubing must be used.

The most obvious part of the evaporative emission control system is the **charcoal (carbon) canister** in the engine compartment. This canister is located in or near the engine compartment.

Shop Manual
Chapter 8, page 328

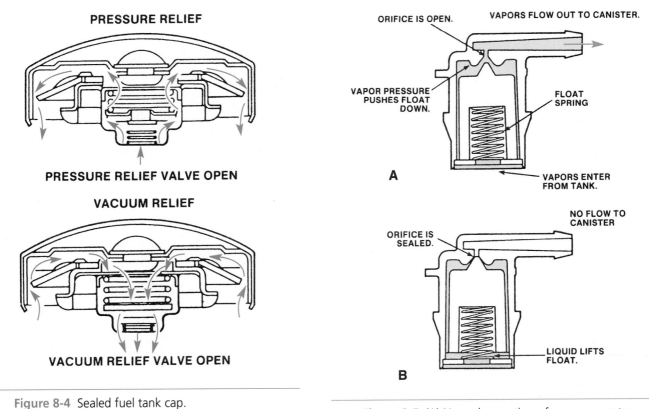

Figure 8-4 Sealed fuel tank cap.

Figure 8-5 (A) Normal operation of vapor separator; (B) with liquid in separator.

Fuel vapors from the gas tank and carburetor float bowl are routed to and absorbed onto the surfaces of the canister's charcoal granules. When the vehicle is restarted, vapors are drawn by the vacuum into the carburetor or intake manifold to be burned in the engine. Canister purging varies widely with make and model. In some instances a fixed restriction allows constant purging whenever there is manifold vacuum. In others, a staged valve provides purging only at speeds above idle. Generally, the **canister purge** valve is normally closed. It opens the inlet to the purge outlet when vacuum is applied. Some units incorporate a thermal-delay valve so the canister is not purged until the engine reaches operating temperature. Purging at idle or with a cold engine creates other problems, such as rough running and increased emissions because of the additional vapor added to the intake manifold. Typical purge valve mountings are shown in Figure 8-6.

When the engine is running, intake manifold vacuum is supplied to the tank pressure control valve (TPCV). This vacuum opens the valve and allows vapors to flow through the valve into the

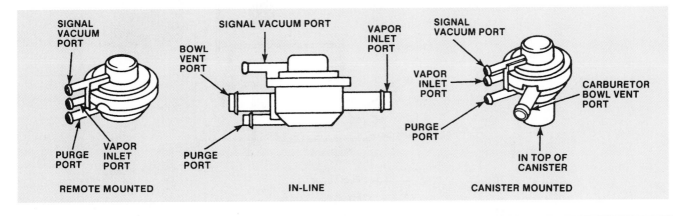

Figure 8-6 Typical purge valve mountings.

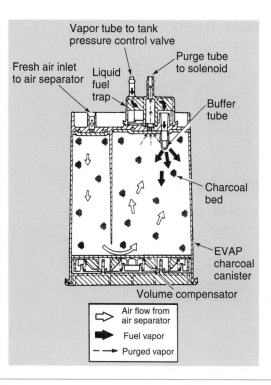

Figure 8-7 Typical charcoal canister. (*Courtesy of Cadillac Motor Division—GMC*)

canister. When the engine is not running, the TPCV valve closes and fuel vapors are contained in the fuel tank. If the tank pressure exceeds a specified amount with the engine not running, this pressure forces the TPCV valve open and allows vapor flow to the canister.

The canister contains a liquid fuel trap that collects any liquid fuel entering the canister (Figure 8-7). Condensed fuel vapor forms liquid fuel. This liquid is returned from the canister to the tank when a vacuum is present in the tank. This liquid fuel trap prevents liquid fuel from contaminating the charcoal in the canister. The EVAP system reduces the escape of HC evaporative emissions from the gasoline tank to the atmosphere.

In some EVAP systems, the purge hose between the charcoal canister and the intake manifold is opened and closed by a **thermal vacuum valve (TVV)** that is mounted in the cooling system (Figure 8-8). The TVV contains a thermowax element and a plunger. When the thermowax is heated, it expands and moves the plunger. If the engine coolant temperature is below 95°F (35°C), the plunger

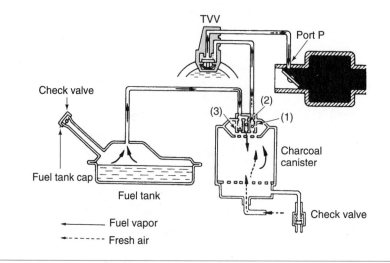

Figure 8-8 EVAP system with a TVV.

To reduce HC emissions, evaporated fuel from the fuel tank is routed through the charcoal canister to the intake manifold for combustion in the cylinders.

Engine Coolant Temp.	TVV	Throttle Valve Opening	Canister Check Valve			Check Valve in Cap	Evaporated Fuel (HC)
			(1)	(2)	(3)		
Below 35°C (95°F)	CLOSED	–	–	–	–	–	HC from tank is absorbed into the canister.
Above 54°C ((129°F)	OPEN	Positioned below port P	CLOSED	–	–	–	
		Positioned above port P	OPEN	–	–	–	HC from canister is led into air intake chamber.
High pressure in tank	–	–	–	OPEN	CLOSED	CLOSED	HC from tank is absorbed into the canister.
High vacuum in tank	–	–	–	CLOSED	OPEN	OPEN	Air is led into the fuel tank.

Figure 8-9 Operation of a TVV equipped EVAP system.

in the TVV closes the purge hose between the intake manifold and the canister. Above this temperature, the TVV plunger opens the purge hose. Three check valves are located in the top of the charcoal canister. When the throttle is open enough so the edge of the throttle uncovers the purge port and the TVV is open, check valve one in the canister is opened by vacuum. Under this condition fuel vapors are purged from the canister through the TVV into the intake manifold (Figure 8-9).

Check valve two in the canister is opened with pressure in the fuel tank and closed with vacuum in the tank. Check valve three operates in the opposite way to check valve two. A vacuum valve is located in the fuel tank filler cap. This valve opens and allows air into the tank if a specific amount of vacuum develops in the tank.

WARNING: Gasoline vapors are extremely explosive! Do not smoke or allow sources of ignition near any component in the EVAP system. Explosion of gasoline vapors may result in property damage or personal injury.

Canister purging may also be electronically controlled. The on-board computer enables a purge solenoid to initiate the purge cycle. A canister purge solenoid is connected in the purge hose from the canister to the intake port near the edge of the throttle (Figure 8-10). The PCM provides a ground for the canister purge solenoid winding to operate the solenoid. When the solenoid is energized by the computer, the purge valve opens and allows the intake manifold vacuum to draw the trapped fuel vapors from the canister.

The PCM energizes the canister purge solenoid and allows vacuum to purge vapors from the canister under these conditions:

1. 150 seconds have elapsed since the PCM entered closed loop.

2. Coolant temperature is above 176°F (80°C).

3. When the PCM is not enabling injector shut-off, such as on vehicles with traction control while one drive wheel is spinning.

4. The idle contact switch in the idle air control motor is open.

5. Vehicle speed is above 20 mph (32 kph).

6. Engine speed is above 1,100 rpm.

7. The engine metal temperature sensor is not indicating excessive temperature.

8. Low coolant level is not indicated.

Shop Manual
Chapter 8, page 329

The idle air control motor is often called the IAC motor.

247

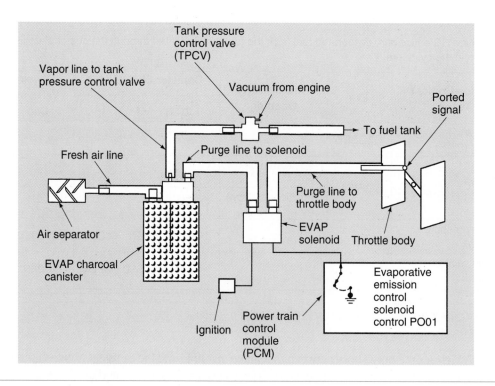

Figure 8-10 EVAP system components. (*Courtesy of Cadillac Motor Division—GMC*)

If any of these conditions are not present, the PCM does not energize the canister purge solenoid, and the gasoline vapors from the fuel tank are stored in the canister.

Reports published in the early 1950s by Professor A. J. Haagen-Smit of California were the beginning of modern automotive emission standards. Haagen-Smit discovered that two invisible automotive emissions, hydrocarbons and oxides of nitrogen, react in the presence of sunlight to form oxidants such as ozone, a main ingredient in smog. Professor Haagen-Smit also stated that automobiles, along with such industries as petroleum and rubber, were the main sources of these emissions. The professor's reports also warned about the danger of allowing these emissions to go unregulated. Forty years later, the wisdom of Professor Haagen-Smit's conclusions is very evident.

Precombustion Systems

Systems designed to prevent or limit the amount of pollutants produced by an engine are called precombustion emission control devices. Although there are specific systems and engine designs that are classified this way, anything that makes an engine more efficient can be categorized as a precombustion emission control. The specific systems discussed in this chapter are the:

Engine design changes

PCV system

Spark control systems

EGR systems

Intake heat control systems

Engine Design Changes

In recent years, the basic engine has seen many technological changes. For the most part, the basics of operation have not been changed. Engineers have worked overtime in an attempt to squeeze as much out of small engines as they can. Many of these changes not only have increased the efficiency of the engines, but have decreased the pollutants released by the engines. Also, some of the changes were only necessary or brought about to accommodate changes in the engine's fuel and/or ignition systems.

Better Sealing Pistons

One of the first areas of concern for engineers and emission control devices was the crankcase of an engine. PCV systems relieve the crankcase of unwanted gases and pressure. The PCV did not solve the problem of these pressurized gases blowing by the piston rings. Not only do these blowby gases represent a problem for the engine's lubrication, but they are an indication of wasted energy as well. The blowby gases start out to be fuel-air mixture. The mixture leaks past the piston rings during the compression and power strokes of the engine. Instead of being used to produce energy, this sampling of mixture is used only to dilute the oil. Blowby gases can be and are being reduced in some engines through the use of better sealing piston rings and improved cylinder wall surfaces. Many of these better sealing piston rings also have frictional qualities that make them less of a drag. This increases fuel economy and engine power.

Poor sealing pistons also allow engine oil to enter the combustion chamber while the piston is on its intake stroke.

Combustion Chamber Designs

Combustion chamber design has seen many changes. The aim of these changes is to reduce or eliminate the quench area of the combustion chamber. The quench area (Figure 8-11) is any place in the chamber where the flame front of combustion is cooled as it tries to move into a small area. By removing the quench areas in a combustion chamber, HC and CO emissions can be reduced. Another change in the design of combustion chambers has been the locating of the spark closer to the center of the chamber. This provides for a more even burn and allows for leaner air-fuel mixtures. Manufacturers have also worked with designs that cause controlled turbulence in the chamber (Figure 8-12). This turbulence improves the mixing of the fuel with the air which results in improved combustion.

Lower Compression

Combustion chamber designs have also affected the compression ratios of engines. By keeping the compression ratio low, combustion temperatures can be kept below the point where NO_x is

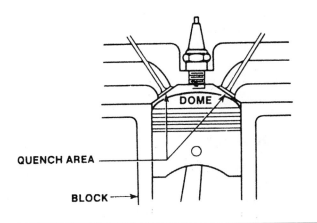

Figure 8-11 Quench areas in a combustion chamber.

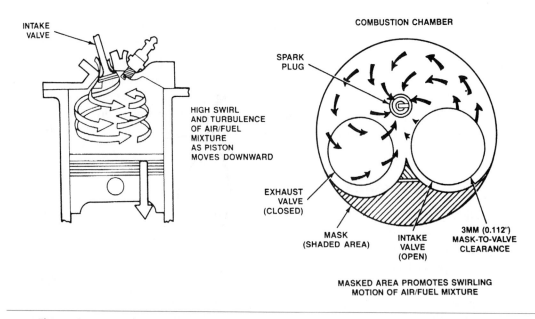

INTAKE VALVE

COMBUSTION CHAMBER

SPARK PLUG

HIGH SWIRL AND TURBULENCE OF AIR/FUEL MIXTURE AS PISTON MOVES DOWNWARD

EXHAUST VALVE (CLOSED)

MASK (SHADED AREA)

INTAKE VALVE (OPEN)

3MM (0.112") MASK-TO-VALVE CLEARANCE

MASKED AREA PROMOTES SWIRLING MOTION OF AIR/FUEL MIXTURE

Figure 8-12 Combustion chamber designed to control turbulence. (*Reprinted with the permission of Ford Motor Company*)

formed. Initially when compression ratios dropped, engine performance and fuel economy also dropped. With the application of new technology, engine performance and fuel economy have greatly improved in spite of the lower compression ratios. These other technological changes have also allowed for compression ratios to come back up on some high performance engines.

Decreased Friction

Power losses through friction are a major contributor to the overall inefficiency of a gasoline engine. The friction of all of the moving parts of an engine results in a large loss of useable power and energy. By reducing the friction at key points within the engine, engineers have reduced the amount of power lost. By regaining the power, fuel economy, and engine performance is increased. Emissions are also reduced because less fuel is burned to produce the same amount of useable power. Improved engine oils, new component materials, and weight reductions have had the biggest impact on reducing friction.

Unleaded Fuel

Leaded fuels are no longer used in automobiles because they poison catalytic converters and lead emitted in the exhaust poisons the atmosphere. Lead was added to fuels to increase its octane rating. Lead also served as a cushion between the engine's valve and valve seats. Unleaded fuel does not offer this cushioning, therefore engines are built with hardened valve seats. This change has been in place for many years, however, engineers are still working to design valves and valve seats that maintain their seal after hundreds of thousands of miles. By maintaining a proper seal at the valves, engines can run more efficiently for many miles of use.

Intake Manifold Designs

Intake manifolds have been designed to distribute equal amounts of air to each cylinder. Most of the design changes to intake manifolds have been afforded by the use of port fuel injection (Figure 8-13). The use of plastics in the manufacture of intake manifolds have allowed for smoother runners and better heat control of the air. Intake manifolds can be designed to be more efficient at low or high engine speeds, or both. Again, increased efficiency results in decreased emissions.

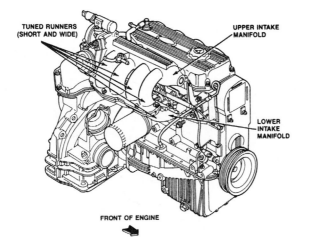

TUNED RUNNERS
(SHORT AND WIDE)

UPPER INTAKE
MANIFOLD

LOWER
INTAKE
MANIFOLD

FRONT OF ENGINE

Figure 8-13 A tuned intake manifold. (*Reprinted with the permission of Ford Motor Company*)

Improved Cooling Systems

Today's engines have a higher normal operating temperature than older engines. The higher engine temperature reduces HC and CO emissions, however, the higher temperature also makes the formation of NO_x harder to control. Most engine cooling systems have been designed to run at high temperatures but are prevented from getting too hot. This prevention reduces NO_x. The high engine temperature results in increased efficiency because the engine parts around the combustion chamber are hotter. This means less heat from combustion is absorbed by the metal engine parts leaving more heat for power conversion. Today's engine control systems incorporate many features that change air-fuel mixture, ignition timing, and idle speed to control the temperature of the engine.

PCV Systems

During the last part of the engine's combustion stroke, some unburned fuel and products of combustion—water vapor, for instance—leak past the engine's piston rings into the crankcase. This leakage into the engine crankcase is called **blowby**. Blowby must be removed from the engine before it condenses in the crankcase and reacts with the oil to form sludge. Sludge, if allowed to circulate with engine oil, corrodes and accelerates the wear of pistons, piston rings, valves, bearings, and other internal working parts of the engine. Blowby gases must also be removed from the crankcase to prevent premature oil leaks. Because these gases enter the crankcase by the pressure formed during combustion, they pressurize the crankcase. The gases exert pressure on the oil pan gasket and crankshaft seals. If the pressure is not relieved, oil is eventually forced out of these seals.

Because the air-fuel mixture in an engine never completely burns, blowby also carries some unburned fuel into the crankcase. If not removed, the unburned fuel dilutes the crankcase oil. When oil is diluted with gasoline it does not lubricate the engine properly, causing excessive wear.

Combustion gases that enter the crankcase are removed by a positive crankcase ventilation system that uses engine vacuum to draw fresh air through the crankcase. This fresh air dissipates the harmful gases and enters through the air filter or through a separate PCV breather filter located on the inside of the air filter housing.

Because the vacuum supply for the PCV system is from the engine's intake manifold, the airflow through this system must be controlled in such a way that it varies in proportion to the regular air-fuel ratio being drawn into the intake manifold. Otherwise, the additional air that is drawn into the system would cause the air-fuel mixture to become too lean for efficient engine operation.

Shop Manual
Chapter 8, page 331

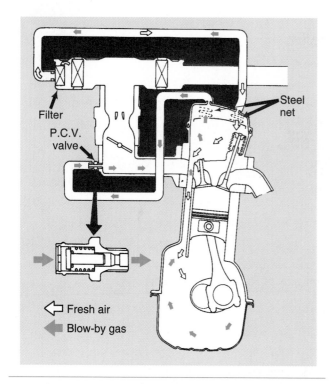

Figure 8-14 Typical PCV system. (*Courtesy of Nissan Motors*)

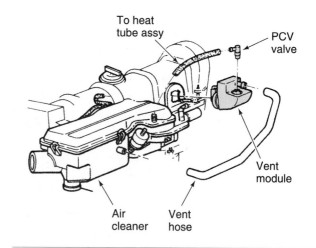

Figure 8-15 PCV valve mounted in a vent module. (*Courtesy of Chrysler Corporation*)

The filter mounted in the PCV valve clean air hose inside the air cleaner may be called an antibackfire filter.

Therefore, a PCV valve is placed in the flow just before the intake manifold to regulate the flow according to vacuum.

The positive crankcase ventilation system has two major functions. It prevents the emission of blowby gases from the engine crankcase to the atmosphere. These gases were once vented through a road draft tube. Now they are recirculated to the engine intake and burned during combustion. It also scavenges the crankcase of vapors that could dilute the oil and cause it to deteriorate or that could build undesirable pressure in the crankcase. Fresh air from the air cleaner mixes with these vapors and makes them flow to the intake.

The PCV system benefits the vehicle's driveability by eliminating harmful crankcase gases, reducing air pollution, and promoting fuel economy. An inoperative PCV system could shorten the life of the engine by allowing harmful blowby gases to remain in the engine, causing corrosion and accelerating wear.

The PCV valve is usually mounted in a rubber grommet in one of the valve covers. A hose is connected from the PCV valve to the intake manifold. A clean air hose is connected from the air cleaner to the opposite rocker arm cover. A filter is positioned in the air cleaner end of the clean air hose (Figure 8-14). On some systems, the PCV valve is mounted in a vent module and the clean air filter is located in this module (Figure 8-15).

When the engine is running, intake manifold vacuum is supplied to the PCV valve. This vacuum moves air through the clean air hose into the rocker arm cover. From this location, air flows through cylinder head openings into the crankcase where it mixes with blowby gases that escape from the combustion chamber past the piston rings. The mixture of blowby gases and air flows up through cylinder head openings to the rocker arm cover and PCV valve. Intake manifold vacuum moves the blowby gas mixture through the PCV valve into the intake manifold (Figure 8-16). The blowby gases are then moved through the intake valves into the combustion chambers where they are burned.

On many engines, the PCV system delivers blowby gases to one location in the intake manifold. This type of system may not deliver these gases equally to all the cylinders. This action may result in an air-fuel ratio variation between the cylinders, which results in rougher idle operation.

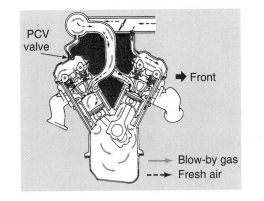

PCV valve

→ Front

→ Blow-by gas
---→ Fresh air

Figure 8-16 Operation of PCV system.

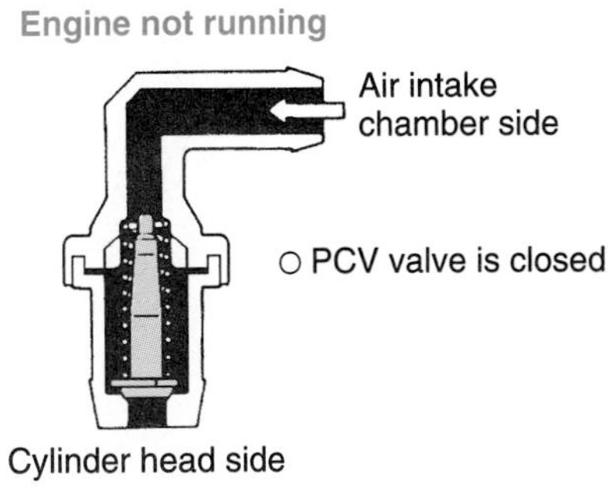

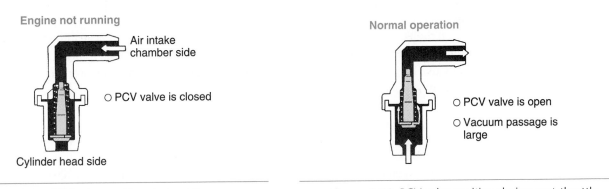

Engine not running

Air intake chamber side

○ PCV valve is closed

Cylinder head side

Figure 8-17 PCV valve position with the engine not running.

Engine not running

Air intake chamber side

○ PCV valve is closed

Cylinder head side

Figure 8-18 PCV valve position during idle or deceleration.

Normal operation

○ PCV valve is open

○ Vacuum passage is large

Figure 8-19 PCV valve position during part-throttle operation.

Some engines, such as the Ford 4.6 L V8, have passages from the PCV valve through the intake manifold to supply blowby gases equally to each cylinder, resulting in smoother idle operation.

A PCV valve contains a tapered valve. When the engine is not running, a spring keeps the tapered valve seated against the valve housing (Figure 8-17). During idle or deceleration, the high intake manifold vacuum moves the tapered valve upward against the spring tension. Under this condition there is a small opening between the tapered valve and the PCV valve housing (Figure 8-18). Since the engine is not under heavy load during idle or deceleration operation, blowby gases are minimal and the small PCV valve opening is adequate to move the blowby gases out of the crankcase.

Intake manifold vacuum is lower during part-throttle operation than during idle operation. Under this condition the spring moves the tapered valve downward to increase the opening between this valve and the PCV valve housing (Figure 8-19). Since engine load is higher at part-throttle operation than at idle operation, blowby gases are increased. The larger opening between the tapered valve and the PCV valve housing allows all the blowby gases to be drawn into the intake manifold.

When the engine is operating under heavy load conditions with a wide throttle opening, the decrease in intake manifold vacuum allows the spring to move the tapered valve further downward in the PCV valve (Figure 8-20). This action provides a larger opening between the tapered valve and the PCV valve housing. Since higher engine load results in more blowby gases, the larger PCV valve opening is necessary to allow these gases to flow through the valve into the intake manifold.

When worn rings or scored cylinders allow excessive blowby gases into the crankcase, the PCV valve opening may not be large enough to allow these gases to flow into the intake manifold. Under this condition, the blowby gases create a pressure in the crankcase and some of these gases

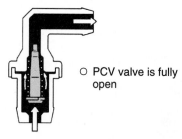

Acceleration or high load

○ PCV valve is fully
  open

Figure 8-20 PCV valve position during hard acceleration or heavy load.

are forced through the clean air hose and filter into the air cleaner. When this action occurs there is oil in the PCV filter and air cleaner. This same action occurs if the PCV valve is restricted or plugged.

If the PCV valve sticks in the wide-open position, excessive air flow through the valve causes rough idle operation. If a backfire occurs in the intake manifold, the tapered valve is seated in the PCV valve as if the engine is not running. This action prevents the backfire from entering the engine where it could cause an explosion.

Fixed Orifice Tube PCV System

Some engines are equipped with a PCV system that does not use a PCV valve. Rather the blowby gases are routed into the intake manifold through a fixed orifice tube. The basic system works the same as if it had a valve except that the system is regulated only by the vacuum on the orifice. The size of the orifice limits the amount of blowby flow into the intake. The engine's air-fuel system is calibrated for this calibrated air leak. Since the action of the PCV allows unmetered air into the intake, the air-fuel system must be set for this amount of extra air.

Spark Control Systems

Shop Manual
Chapter 8, page 334

Spark control systems have been in use since the earliest gasoline engines. It was discovered that the proper timing of the ignition spark helped to reduce exhaust emissions and develop more power output. Incorrect timing affects the combustion process. Incomplete combustion results in HC emissions. High CO emissions can also result from incorrect ignition timing. Advanced timing can also increase the production of NO_x. When timing is too far advanced, combustion temperatures rise. For every one degree of overadvance, the temperature increases by 125°F. Throughout the years each car manufacturer developed slightly different spark timing controls according to engine requirements and emission standards for each model year, but the systems and devices all operate on the same principles.

Thermal Vacuum Valve

Carbureted engines have a variety of spark control systems. The TVV valve is mounted in the cooling system and three vacuum hoses are connected to this valve. Two of these hoses are connected to the EGR valve and the third hose is connected to the distributor vacuum advance hose (Figure 8-21). A thermowax element is positioned in the lower end of the TVV and a plunger is mounted above this element. As the thermowax element is heated, it expands and moves the plunger.

If the coolant temperature is between 50°to 122°F (10–50°C), the plunger is positioned so it vents the vacuum supplied to the distributor vacuum advance. At all other temperatures, the

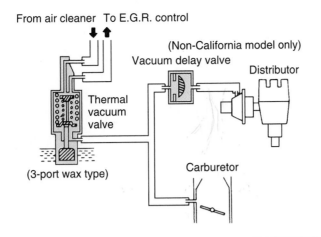

From air cleaner   To E.G.R. control

(Non-California model only)
Vacuum delay valve

Distributor

Thermal vacuum valve

(3-port wax type)

Carburetor

Figure 8-21 Spark control system with a TVV and delay valve. (*Courtesy of Nissan Motors*)

plunger closes the vent port connected to the distributor advance hose. Venting the vacuum supplied to the distributor vacuum advance retards the vacuum advance. With this retard in spark advance, combustion temperatures are increased and hydrocarbon emissions are reduced. However, retarding the spark advance decreases engine performance and economy to some extent.

Vacuum Delay Valve

The vacuum delay valve is connected in series in the distributor advance hose. The vacuum port connected to the distributor vacuum advance is located above the edge of the throttle when the throttle is in the idle position. Therefore, no vacuum is supplied to the distributor advance when the engine is idling.

When the throttle is opened slightly, the edge of the throttle exposes the vacuum advance port to manifold vacuum. Under this condition, vacuum is supplied to the vacuum delay valve and the check valve in the vacuum delay valve closes to prevent vacuum advance. However, this check valve is designed with a slow leak, so the vacuum gradually leaks past the valve in approximately 20 seconds to supply normal vacuum advance. Vacuum delay valves were designed to delay the vacuum advance for different times depending on the engine. The retarded spark advance increased combustion chamber temperatures and decreased HC emissions.

One side of the vacuum delay valve is usually marked carburetor, and the opposite side is identified as distributor. This valve must be installed according to the valve markings. If the vacuum delay valve is reversed during installation, valve operation is improper.

Knock Sensor and Knock Sensor Module

WARNING: Operating an engine for a period of time with a severe detonation problem may result in engine damage.

Many engines with an EFI have a knock sensor or sensors. The knock sensors may be mounted in the block, cylinder head, or intake manifold. A piezoelectric sensing element is mounted in the knock sensor, and a resistor is connected parallel to this sensing element (Figure 8-22). When the engine detonates, a vibration occurs in the engine. The piezoelectric sensing element changes this vibration to an analog voltage, and this signal is sent to the knock sensor module.

The knock sensor module changes the analog voltage signal to a digital voltage signal and sends this signal to the PCM (Figure 8-23). When the PCM receives this signal, it reduces the spark advance to prevent detonation.

Shop Manual
Chapter 8, page 334

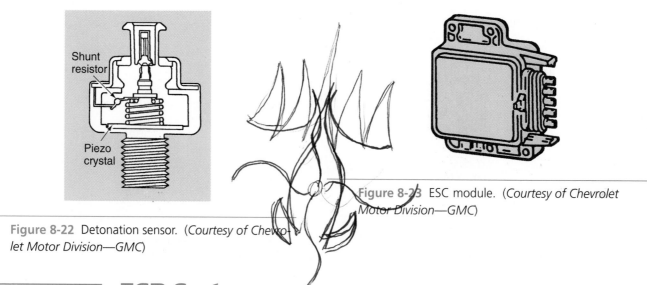

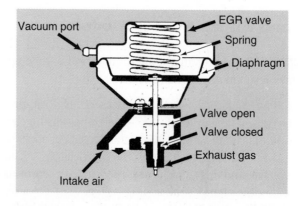

Figure 8-22 Detonation sensor. (*Courtesy of Chevrolet Motor Division—GMC*)

Figure 8-23 ESC module. (*Courtesy of Chevrolet Motor Division—GMC*)

The term driveability is commonly used to describe the ability of an engine to move a vehicle under the different driving conditions.

Shop Manual
Chapter 8, page 338

EGR Systems

Exhaust gas recirculating systems reduce the amount of oxides of nitrogen emitted. The EGR system dilutes the air-fuel mixture with controlled amounts of exhaust gas. Since exhaust gas does not burn, this reduces the peak combustion temperatures. At lower combustion temperatures very little of the nitrogen in the air combines with oxygen to form NO_X. Most of the nitrogen is simply carried out with the exhaust gases. For driveability it is desirable to have the EGR valve opening (and the amount of gas flow) proportional to the throttle opening. Driveability is also improved by shutting off the EGR when the engine is started up cold, at idle, and at full throttle. Since the NO_X control requirements vary on different engines, there are several different systems, with various controls to provide these functions.

Most of these systems use a vacuum-operated EGR valve to regulate the exhaust gas flow into the intake manifold. Exhaust crossover passages under the intake manifold channel the exhaust gas to the valve. (Some in-line engines route the exhaust gas to the valve through an external tube.) Typical mounting of the EGR valve is either on a plate under the carburetor or directly on the manifold.

Figure 8-24 illustrates the basic valve design. The EGR valve is a vacuum-operated, flow control valve. On most systems, it is attached to a carburetor spacer. The carburetor spacer is sandwiched between the carburetor and intake manifold. Gaskets are used above and below the spacer to seal the EGR system and the carburetor-to-manifold air-fuel flow. A small exhaust crossover passage in the intake manifold admits exhaust gases to the spacer. These gases flow through the spacer to the inlet port of the EGR valve. Opening the EGR valve by control vacuum at the

Figure 8-24 Typical design of an EGR valve. (*Courtesy of Chevrolet Motor Division—GMC*)

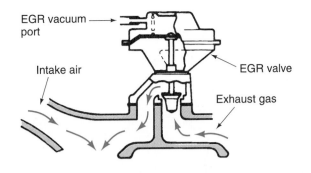
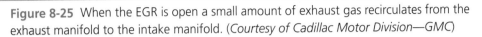

Figure 8-25 When the EGR is open a small amount of exhaust gas recirculates from the exhaust manifold to the intake manifold. (*Courtesy of Cadillac Motor Division—GMC*)

diaphragm allows exhaust gases to flow through the valve and back to another port of the spacer (Figure 8-25). Here, the exhaust gas mixes with the air-fuel mixture leaving the carburetor and then entering the intake manifold. The effect is to dilute or lean-out the mixture so that it still burns completely but with a reduction in combustion chamber temperatures.

On some engines, such as the General Motors Northstar 4.6L V8, the exhaust gas from the EGR system is distributed through passages in the cylinder heads and distribution plates to each intake port (Figure 8-26). The distribution plates are positioned between the cylinder heads and the intake manifold.

Since the exhaust gas from the EGR system is distributed equally to each cylinder, smoother engine operation results. Exhaust gases in the Northstar 4.6L V8 engine flow through passages in the crossover water pump housing where they are cooled by engine coolant. This cooling action lowers the exhaust gases below their carbon-forming temperature prior to routing these gases into the cylinder distribution passages, which reduces carbon deposits in the EGR system.

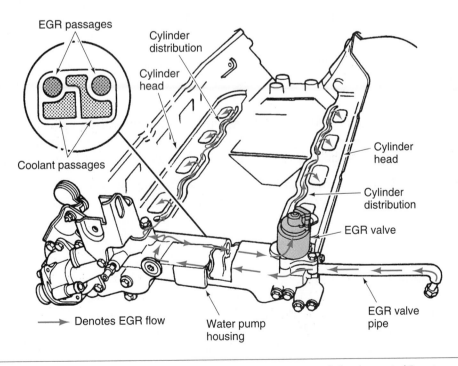

Figure 8-26 A GM Northstar engine with EGR passages to each intake port. (*Courtesy of Cadillac Motor Division—GMC*)

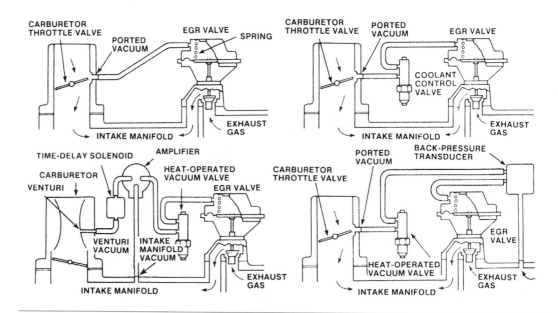

Figure 8-27 Different ways of controlling an EGR valve with vacuum.

EGR valve control vacuum is from either of two sources (Figure 8-27). In the port vacuum system, a vacuum line connects the EGR valve to a slot port in the carburetor throttle body above the throttle plate. When the throttle plate is closed no vacuum is transmitted. The port is exposed to increasing manifold vacuum as the throttle plate opens. The exhaust gas flow rate depends on manifold vacuum, throttle position, and exhaust gas back pressure.

The EGR system works when the engine reaches operating temperature or when the engine is operating under conditions other than idle or wide-open throttle, EGR systems include various functions that control the operation of the EGR valve. Some applications use cold engine EGR lockout and wide-open throttle EGR lockout. Cold EGR lockout is necessary to keep the EGR valve closed during cold engine operation. Wide-open throttle EGR lockout might be required to keep the EGR valve closed when the engine is operating under maximum load. The following are various controls that relate directly to the EGR system.

Shop Manual
Chapter 8, page 345

> **CAUTION:** In the United States it is against the law for an automotive technician to tamper with or alter emission systems on a vehicle. Such action may result in fines and/or imprisonment.

Thermal Vacuum Switch (TVS). The **TVS** senses the air temperature in the carburetor air cleaner to control vacuum to the EGR valve. When the engine reaches operating temperature, the TVS opens to supply vacuum to the EGR valve. This opens the EGR valve for exhaust gas recirculation.

Ported Vacuum Switch (PVS). The **PVS** senses the coolant temperature to control vacuum to the EGR valve (Figure 8-28). The PVS operates in the same manner as the TVS, except it senses the coolant temperature instead of the air temperature. That is, the PVS function is to cut off vacuum to the EGR valve when the engine is cold and connect the vacuum to the EGR valve when the engine is warm.

Venturi Vacuum Amplifier (VVA). Some EGR systems use the VVA so that the carburetor venturi vacuum can control the EGR valve operation (Figure 8-29). Venturi vacuum is more desirable because it is in proportion to the airflow through the carburetor. Since the venturi vacuum is a relatively weak vacuum signal, the VVA converts it to a strong enough signal to operate the EGR valve. The VVA system uses manifold vacuum for strength and venturi vacuum for the control signal.

EGR Delay Timer Control. Some vehicles have an EGR delay system, which consists of an electrical timer that connects to an engine-mounted solenoid. The purpose of the delay timer and solenoid

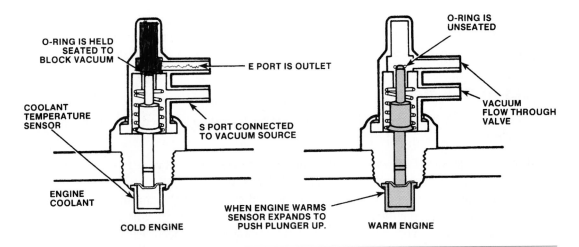

Figure 8-28 How the two-port PVS switch works.

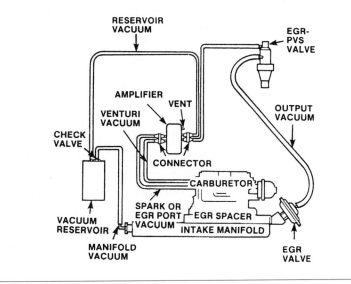

Figure 8-29 Venturi vacuum amplifier circuit

is to prevent EGR operation for a predetermined amount of time after arm engine startup. On cold engine startups, the TVS and PVS valves override the delay timer.

Early Fuel Evaporation/Thermal Vacuum Switch (EFE/TVS). In most common applications the EFE system uses a valve that increases the exhaust gas flow under the intake manifold during cold engine operation through a crossover passage to heat up the incoming air-fuel charge. The EFE is vacuum-operated and controlled by a TVS that applies vacuum to the EFE valve when the coolant temperature is low. Once the engine reaches operating temperature, the TVS blocks off vacuum to the EFE and directs it to the EGR valve for EGR operation.

Wide-Open Throttle Valve (WOT). Some applications use the WOT valve where it is desirable to cut off EGR flow at wide-open throttle. This dump valve compares venturi vacuum to manifold vacuum to determine when the throttle is all of the way open.

Backpressure Transducer. This device can be used to modulate or change the amount the EGR valve opens. It controls the amount of air bleed in the EGR vacuum line according to the level of exhaust gas pressure, which is dependent on engine rpm. The EGR valve can be closed or partially opened at different engine speeds. Air bleed is stopped completely when the exhaust backpressure is high. Thus maximum EGR occurs during acceleration, when backpressure is high. When backpressure decreases the vacuum line bleed is reopened. This decreases the vacuum at the EGR valve, which then reduces the amount of exhaust gas recirculated. In the past, the backpressure transducer was a separate unit. Now it is incorporated into the design of the EGR valve itself.

Types of EGR Valves

The design of the EGR valve may change depending on the system it is used in. Often these design changes incorporate some of the system controls.

Shop Manual
Chapter 8, page 338

Positive Backpressure EGR Valve. The **positive backpressure** EGR valve has a bleed port and valve positioned in the center of the diaphragm. A light spring holds this bleed valve open and an exhaust passage is connected from the lower end of the tapered valve through the stem to the bleed valve. The area under the diaphragm is vented to the atmosphere. When the engine is running, exhaust pressure is applied to the bleed valve. At low engine speeds exhaust pressure is not high enough to close the bleed valve. If control vacuum is supplied to the diaphragm chamber, the vacuum is bled off through the bleed port and the valve remains closed (Figure 8-30).

As engine and vehicle speed increase, the exhaust pressure also increases. At a preset throttle opening, the exhaust pressure closes the EGR valve bleed port. When control vacuum is supplied to the diaphragm, the diaphragm and valve are lifted upward and the valve is open. If vacuum from an external source is supplied to a positive backpressure EGR valve with the engine not running, the valve will not open because the vacuum is bled off through the bleed port.

Negative Backpressure EGR Valve. In a **negative backpressure** EGR valve, a normally closed bleed port is positioned in the center of the diaphragm. An exhaust passage is connected from the lower end of the tapered valve through the stem to the bleed valve (Figure 8-31). When the engine is running at lower speeds, each time a cylinder fires and an exhaust valve opens there is a high-pressure pulse in the exhaust system. Between these high-pressure pulses there are low-pressure pulses. As the engine speed increases, more cylinder firings occur in a given time and the high-pressure pulses become closer together in the exhaust system. At low speed the negative exhaust pulses are more predominant compared to higher engine speeds.

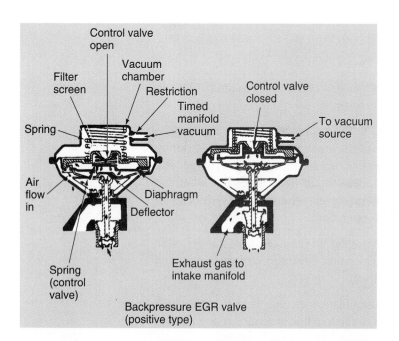

Figure 8-30 A positive backpressure EGR valve. (*Courtesy of Cadillac Motor Division—GMC*)

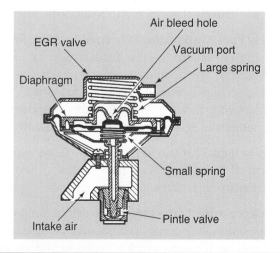

Figure 8-31 A negative backpressure EGR valve. (*Courtesy of Chevrolet Motor Division—GMC*)

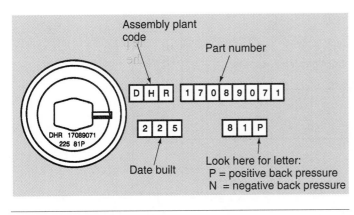

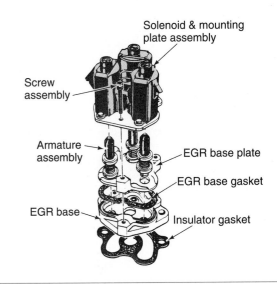

Figure 8-32 A "N" or "P" stamped on the top of an EGR valve identifies the type of backpressure valve it is. (*Courtesy of Cadillac Motor Division—GMC*)

Figure 8-33 A digital EGR valve with three solenoids. (*Courtesy of Oldsmobile Division—GMC*)

At lower engine and vehicle speeds, the negative pulses in the exhaust system hold the bleed valve open. When the engine and vehicle speed increase to a preset value, the negative exhaust pressure pulses decrease and the bleed valve closes. Under this condition, if control vacuum is supplied to the diaphragm chamber, the EGR valve is opened. When vacuum from an external source is supplied to a negative backpressure EGR valve with the engine not running, the bleed port is closed and the vacuum should open the valve.

Digital EGR Valve. A digital EGR valve contains up to three electric solenoids that are operated directly by the PCM (Figure 8-33). Each solenoid contains a movable plunger with a tapered tip that seats in an orifice. When any solenoid is energized, the plunger is lifted and exhaust gas is allowed to recirculate through the orifice into the intake manifold. The solenoids and orifices are different sizes. The PCM can operate one, two, or three solenoids to supply the amount of exhaust recirculation required to provide optimum control of NO_x emissions.

Linear EGR Valve. The linear EGR valve contains a single electric solenoid that is operated by the PCM. A tapered pintle is positioned on the end of the solenoid plunger. When the solenoid is energized, the plunger and tapered valve are lifted and exhaust gas is allowed to recirculate into the intake manifold (Figure 8-34). The EGR valve contains an EGR valve position (EVP) sensor,

Negative or positive backpressure EGR valves may be identified by an N or a P stamped on top of the valve with the part number and plant code identification (Figure 8-32).

The PCM is the Powertrain Control Module.

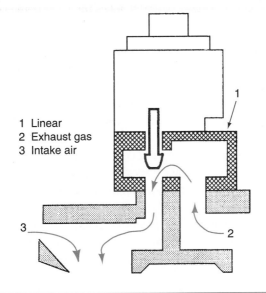

1 Linear
2 Exhaust gas
3 Intake air

Figure 8-34 Linear EGR valve operation. (*Courtesy of Chevrolet Motor Division—GMC*)

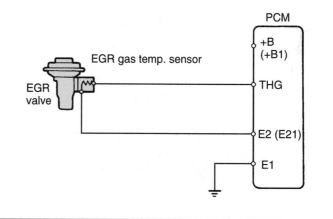

Figure 8-35 EGR exhaust gas temperature sensor.

which is a linear potentiometer. The signal from this sensor varies from approximately 1 V with the EGR valve closed to 4.5 V with the valve wide open.

The PCM pulses the EGR solenoid winding on and off with a pulse width modulation principle to provide accurate control of the plunger and EGR flow. The EVP sensor acts as a feedback signal to the PCM to inform the PCM if the commanded valve position was achieved.

EGR Valve with Exhaust Gas Temperature Sensor. Some EGR valves, particularly on vehicles sold in California, contain an exhaust gas temperature sensor. This sensor contains a thermistor that changes resistance in relation to temperature. An increase in exhaust temperature decreases the sensor resistance. Two wires are connected from the exhaust gas temperature sensor to the PCM (Figure 8-35). The PCM senses the voltage drop across this sensor. Cool exhaust temperature and higher sensor resistance cause a high-voltage signal to the PCM, whereas hot exhaust temperature and low sensor resistance result in a low-voltage signal to the PCM.

Electronic EGR Controls

Shop Manual
Chapter 8, page 339

These various EGR system controls represent some of the common controls currently used by automobile manufacturers. Control devices used in the various systems might have different labels but actually complete the same function within the EGR system. Engines with electronic engine control systems control EGR action in many different ways, the most common of these follow.

Twin Solenoid EGR System. The EGR valve within this system resembles and is operated in a manner similar to the conventional EGR valves. This system uses sensors, solenoids, and an electronic control assembly to modulate and control EGR system components as shown in Figure 8-36. A pintle valve is often used in the valve to better control the flow rate of exhaust gases. A sensor mounted on the valve stem sends an electronic signal to the on-board computer, which in turn tells how far the EGR valve is opened. At this time the EGR control solenoids either maintain or alter the EGR flow depending on engine operating conditions. Source vacuum is manifold vacuum and is applied or vented depending on the computer commands. A cooler is frequently used to reduce exhaust gas temperatures, which enables the exhaust gas to flow better and in turn reduce the amount of detonation. Early EGR systems used an in-line cooler. Later systems use a cooler sandwiched between the EGR valve and carburetor spacer.

This system is often controlled by two solenoids (Figure 8-37). The solenoids respond to voltage signals from the on-board computer. An EGR vent solenoid, normally an open vent solenoid valve, closes when it is energized. An EGR control solenoid, normally a closed solenoid valve, opens when it is energized.

Voltage signals from the on-board computer can trigger the solenoids to increase EGR flow by applying vacuum to the EGR valve, maintain EGR flow by trapping vacuum in the system, or

The EGR Vent solenoid is referred to as the EGRV.

The EGR Control solenoid is referred to as the EGRC.

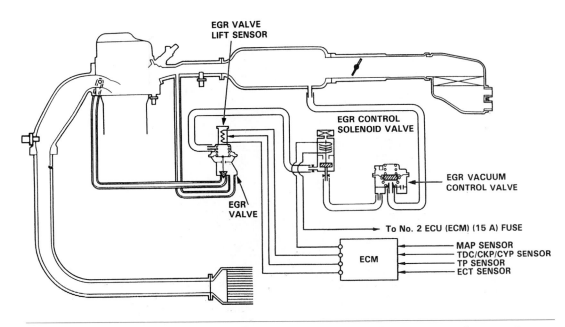

Figure 8-36 A twin solenoid EGR control circuit. (*Courtesy of American Honda Motor Company, Inc.*)

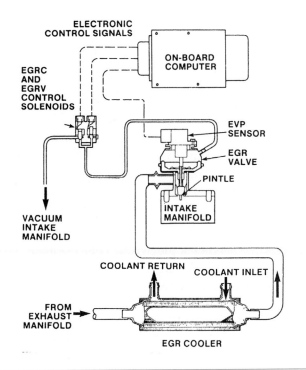

Figure 8-37 Typical EGR system used with an EEC. (*Reprinted with the permission of Ford Motor Company*)

decrease EGR flow by venting EGR vacuum. In actual operation, both solenoids constantly shift between the three operating conditions mentioned as engine operating conditions change.

EGR Vacuum Regulator (EVR). In many EGR systems the PCM operates a normally closed EGR vacuum regulator solenoid (Figure 8-38), which supplies vacuum to the EGR valve. If the EVR solenoid is not energized, the solenoid plunger tip is seated in the vacuum passage and shuts off vacuum to the EGR valve. When the EVR solenoid plunger is shutting off vacuum to the EGR

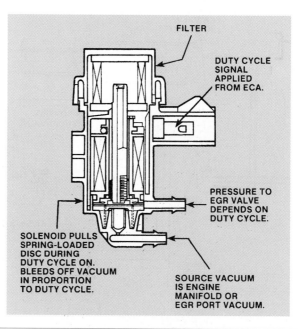

FILTER

DUTY CYCLE
SIGNAL
APPLIED
FROM ECA.

PRESSURE TO
EGR VALVE
DEPENDS ON
DUTY CYCLE.

SOLENOID PULLS
SPRING-LOADED
DISC DURING
DUTY CYCLE ON.
BLEEDS OFF VACUUM
IN PROPORTION
TO DUTY CYCLE.

SOURCE VACUUM
IS ENGINE
MANIFOLD OR
EGR PORT VACUUM.

Figure 8-38 Electronic vacuum regulator (EVR) operation. (*Reprinted with the permission of Ford Motor Company*)

valve, any vacuum in the EGR valve and hose is vented through the EVR solenoid to prevent vacuum from being locked in the system, which could hold the EGR valve open.

When the PCM inputs indicate the EGR valve should be open, the PCM provides a ground for the EVR solenoid winding. This action moves the solenoid plunger and opens the vacuum passage through the solenoid to the EGR valve. In some systems the PCM pulses the EVR on and off to supply the precise vacuum and EGR valve opening required by the engine.

The PCM uses inputs such as engine temperature, throttle position, and vehicle speed to operate the EGR valve. The PCM will not energize the EVR solenoid if the engine coolant temperature is below a preset value. The EGR valve is opened by the PCM when the vehicle is operating at normal temperature in the cruising speed range. When the vehicle is operating at low speed or near wide-open throttle, the PCM does not open the EGR valve. If the EGR valve is open while the engine is idling or operating at low rpm, engine operation is erratic.

EGR System with Pressure Feedback Electronic (PFE) Sensor. Some EGR systems have a pressure feedback electronic sensor. These systems have an orifice located in the exhaust passage below the EGR valve. A small pipe connected from this orifice chamber supplies exhaust pressure to the PFE sensor (Figure 8-39).

The PFE sensor changes the exhaust pressure signal to a voltage signal that is sent to the PCM. Three wires are connected from the PFE sensor to the PCM (Figure 8-40). These wires include ground, 5 V reference, and signal wires. The exhaust pressure in the orifice chamber is proportional to the EGR valve flow. The PFE signal informs the PCM regarding the amount of EGR flow and the PCM compares this signal to the EGR flow requested by the input signals. If there is some difference between the actual EGR flow indicated by the PFE signal and the requested EGR flow, the PCM makes the necessary correction to the EVR output signal.

In some EGR systems, two pipes supply exhaust pressure from above and below the orifice under the EGR valve to the differential PFE (DPFE) sensor (Figure 8-41).

EGR System with Pressure Transducer (EPT). Some EGR systems have a pressure transducer connected in the vacuum hose to the EVR solenoid. A small pressure pipe is connected from the exhaust passage under the EGR valve to the EGR pressure transducer. The exhaust pressure is supplied to a bleed valve diaphragm in the EPT and ported vacuum from a port above the throttle is connected through a hose to the upper side of this diaphragm.

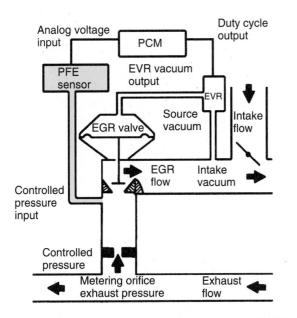

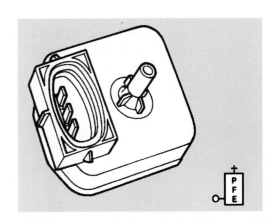

Figure 8-40 PFE sensor. (*Reprinted with the permission of Ford Motor Company*).

Figure 8-39 PFE sensor and related circuit. (*Reprinted with the permission of Ford Motor Company*)

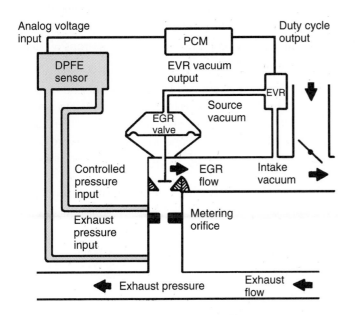

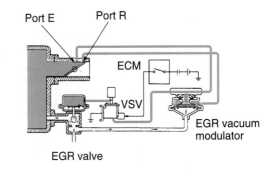

Figure 8-42 EGR system with exhaust pressure transducer.

Figure 8-41 Differential PFE with dual exhaust pressure pipes connected above and below the EGR orifice to the DPFE. (*Reprinted with the permission of Ford Motor Company*)

When the engine is operating at low rpm, the exhaust pressure is reduced. Under this condition, the exhaust pressure does not move the bleed valve in the EPT and the vacuum to the EVR is bled off through the open bleed valve in the modulator. At a preset engine rpm the combination of exhaust pressure below the EPT diaphragm and vacuum above the diaphragm moves the diaphragm upward and closes the bleed valve. When this action occurs, vacuum is supplied through the EPT to the EVR solenoid (Figure 8-42). The PCM operates the EVR solenoid as explained previously.

Intake Heat Control Systems

Hydrocarbon and carbon monoxide exhaust emissions are highest when the engine is cold. The introduction of warm combustion air improves the vaporization of the fuel in the carburetor, fuel injector body, or intake manifold. The three systems used on various gasoline engines to heat the inlet air and the air-fuel mixture are heated air inlet, manifold heat control valves, and early fuel evaporation (**EFE**) heaters.

Heat Air Inlet

An air cleaner with a heated air inlet is often called a **thermostatic air cleaner**.

Shop Manual
Chapter 8, page 348

A heated air inlet control is used on gasoline engines with carburetion or central fuel injection. It is not used with turbocharging or ported fuel injection. This system controls the temperature of the air on its way to the carburetor or fuel injection body. By warming the air, it reduces HC and CO emissions by improved fuel vaporization and faster warmup.

The principal components (Figure 8-43) and functions of a conventional air cleaner system follow.

❏ Heated air inlet duct. Directs air that has been warmed by the heat stove (shroud) on the manifold to the snorkel of the air cleaner.

❏ Air inlet door vacuum motor. Controls a flapper door inside the snorkel to admit manifold heated air, fresh air, or a mixture of both into the air cleaner.

❏ Air cleaner bimetal sensor. Regulates vacuum to the vacuum motor to determine the position of the air door. It is sensitive to air cleaner temperature.

❏ Cold weather modulator. Traps vacuum to the motor if manifold vacuum drops off due to the throttle opening while the air cleaner is cold.

The air cleaner bimetal sensor (Figure 8-44), which is installed in the air cleaner body or air horn, senses the air cleaner temperature. The sensing element is a bimetal spring linked to a sensing valve. Depending on the calibration, the sensor can be set to operate at 50°, 75°, 90°, or 120°F. It is what controls the operating modes shown in Figure 8-45. The sensor is calibrated to provide a specific output vacuum to the air door motor as it warms to its temperature setting. The calibration is based on 16 inches of source vacuum.

Some vacuum control systems use a retard **delay valve** instead of a cold weather modulator. The difference is that the retard delay valve traps the vacuum for a few seconds when the throttle

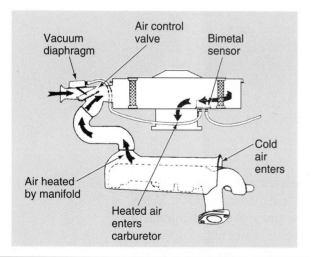

Figure 8-43 Typical heated air inlet system with conventional air cleaner. (*Courtesy of Chrysler Corporation*)

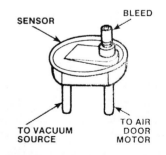

Figure 8-44 Air cleaner bimetal sensor.

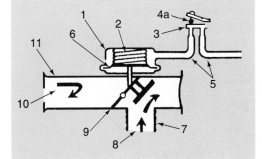
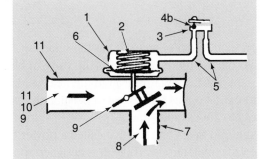
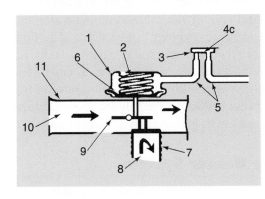

1	Vacuum diaphragm motor
2	Diaphragm spring
3	Temperature sensor
4a	Air bleed valve–closed
4b	Air bleed valve–partially open
4c	Air bleed valve–open
5	Vacuum hoses
6	Diaphragm
7	Heat stove
8	Hot air (exhaust manifold)
9	Damper door
10	Outside inlet air
11	Snorkel

A - Hot air delivery mode
B - Regulating mode
C - Outside air delivery mode

Figure 8-45 Air cleaner bimetal sensor at (A) cold start-up; (B) modulating partial warm-up; (C) hot engine/hot ambient air. (*Courtesy of Chevrolet Motor Division—GMC*)

opens. Its function is to prevent a change in the air door position if vacuum drops off because the throttle opens.

With a remote air cleaner the functions are the same. However, the bimetal sensor is in the air horn assembly and the air inlet vacuum motor and door are in the air cleaner assembly instead of the inlet snorkel.

Manifold Heat Control Valves

The exhaust manifold heat control valve routes exhaust gases to warm the intake manifold when the engine is cold. This heats the air-fuel mixture in the intake manifold and improves ventilation. The result is reduced HC and CO emissions.

The two general types of valves are vacuum-operated and thermostat-operated. Some V8 engines use a vacuum-operated valve, which is bolted between the left exhaust manifold and exhaust pipe. The vacuum diaphragm connects to the manifold vacuum through a ported vacuum switch (PVS). On electronically controlled engines, the system also includes an electric solenoid-operated vacuum valve.

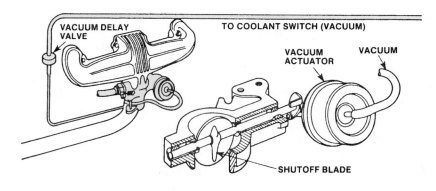

Figure 8-46 Power heat control valve.

A thermostat-operated valve is used mostly on six-cylinder engines. It has the same function as the vacuum-operated valve. When the engine is cold the thermostat closes the valve to block exhaust gas flow from the manifold, which forces the gas to flow up through the heat riser and then to the exhaust pipe. On a warm engine the thermostat opens the valve to a position that seals off the heat riser passage. Exhaust gases flow directly to the exhaust pipe.

Some V8 engines use a more complicated manifold heat control valve called a power heat control valve (Figure 8-46). It works similarly to the vacuum-controlled valve. It is designed specifically to work with a mini-catalyst and to preheat the air-fuel mixture for improved cold engine driveability. A vacuum actuator keeps the power heat control valve closed during warm-up. All right-side exhaust gas travels up through the intake manifold crossover to the left side of the engine. Then all exhaust gas from the engine passes through a mini-converter just down from the left manifold. This converter warms up rapidly because it is small and close to the engine. Its rapid warmup reduces exhaust emissions. As the engine and main converter warm up, a coolant-controlled engine vacuum switch closes. This cuts vacuum to the actuator and allows the valve to open. Exhaust gas flows through both manifolds into the exhaust system and main converter.

Mechanically Controlled Mixture Heater System

The mechanically controlled mixture heater system contains a mixture heater relay, water temperature switch, and a mixture heater (Figure 8-47). The water temperature switch is mounted in the

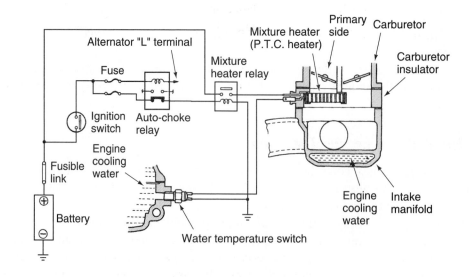

Figure 8-47 Mechanically controlled mixture heater system. (*Courtesy of Nissan Motors*)

cooling system, and the mixture heater is positioned between the carburetor and the intake manifold on the primary side of the carburetor.

When the ignition switch is turned on, current flows from the ignition switch through the auto-choke heater relay winding, an alternator terminal, and voltage regulator to ground. Under this condition, the auto-choke heater relay contacts open and prevent current flow through these relay contacts and the mixture heater relay winding. This action prevents current flow through the mixture heater and choke heater with the ignition switch on.

When the engine starts, equal voltage is supplied to both sides of the auto-choke heater relay winding. This action stops the current flow through this winding and allows the relay contacts to close, which supplies voltage to the mixture heater relay winding and the choke heater.

Current now flows through the mixture heater relay winding to ground, and these relay contacts close, supplying voltage to the mixture heater. When the coolant temperature is below a specified temperature, the water temperature switch is closed and current flows through the mixture heater and the temperature switch contacts to ground. Current flow through the heater warms the air-fuel mixture to prevent fuel condensation on the cold intake manifold and improve engine performance and emissions of HC and CO during engine warmup.

When the coolant temperature is above the specified temperature, the temperature switch opens and current flow through the mixture heater is stopped. This action shuts off the current flow through the mixture heater.

Computer-Controlled Mixture Heater System

In the computer-controlled mixture heater system, the PCM operates the mixture heater relay and the engine coolant temperature sensor sends a signal to the PCM in relation to coolant temperature. At a preset temperature, the PCM grounds the mixture heater relay winding and these relay contacts close. Voltage is then supplied through these relay contacts to the mixture heater. Above a specific coolant temperature, the PCM opens the ground circuit from the mixture heater winding to ground. Under this condition, the relay contacts open and shut off the current flow to the mixture heater.

Computer-Controlled Heat Riser Valve

CAUTION: The heat riser valve may be extremely hot if the engine has been running. Wear protective gloves to avoid burns when servicing this component.

Some V8 or V6 engines with computer-controlled carburetors or TBI have a heat riser valve positioned between the right exhaust manifold flange and the exhaust pipe. A linkage is connected from the heat riser valve to the vacuum diaphragm (Figure 8-48). In some systems, vacuum is sup-

Shop Manual
Chapter 8, page 350

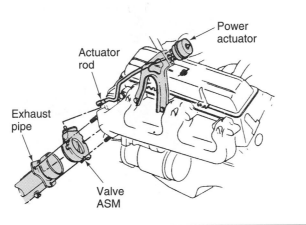

Figure 8-48 Computer-controlled heat riser. (*Courtesy of Pontiac Motor Division—GMC*)

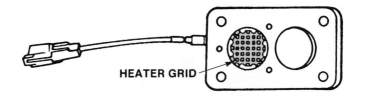

Figure 8-49 EFE heater resistance grid.

plied through a computer-controlled solenoid to this vacuum diaphragm. When the engine coolant is below a preset temperature, the PCM operates the solenoid and supplies vacuum to the heat riser diaphragm. Under this condition, the diaphragm moves the heat riser to the closed position, which forces more exhaust gas through the crossover passage in the intake manifold. This additional intake manifold heating improves fuel vaporization in the intake manifold and provides better engine performance.

At a preset temperature, the PCM closes the heat riser solenoid and shuts off the vacuum to the heat riser diaphragm. When this action occurs, the diaphragm spring moves the heat riser valve to the open position. In some systems vacuum is supplied to the heat riser valve diaphragm through a thermal vacuum valve, which is operated by coolant temperature.

Early Fuel Evaporation Control

Shop Manual
Chapter 8, page 348

The EFE heater contains a resistance grid that heats the mixture from the primary venturi of the carburetor (Figure 8-49). Its purpose is the same as a manifold heat control valve: to improve vaporization in a cold engine. The heater operates for about the first two minutes, permitting leaner choke calibrations for improved emissions without cold driveability problems.

The basic EFE system is similar from one engine to the next. In addition to the grid heater, EFE has two other important components.

Coolant Temperature Switch. The EFE temperature switch or solenoid mounts to the engine, usually on the bottom of the intake manifold (Figure 8-50). The switch is closed and provides a ground for the circuit when its temperature is below a specified temperature, generally between 130°–150°F. The switch opens as the engine coolant temperature goes above the specified temperature.

EFE Heater Relay. The temperature switch controls the EFE relay or valve. It powers the EFE heater when the temperature switch is cold and closed. After the engine has warmed up and the EFE is no longer needed, the relay de-energizes and the grid heater turns off. The EFE heater relay usually mounts on the body of the vehicle.

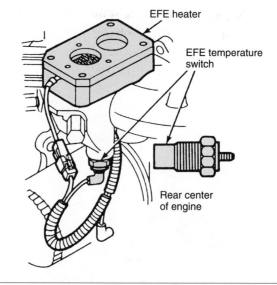

Figure 8-50 The EFE temperature switch is mounted in the engine.

In older engines equipped with a carburetor or TBI unit, the fuel is delivered above the throttles and the intake manifold is filled with a mixture of air and gasoline vapor. In these applications, some intake manifold heating is required to prevent fuel condensation, especially when the intake manifold is cool or cold. Therefore, these engines have intake manifold heat control devices such as heat riser valves or mixture heaters to provide proper fuel vaporization and engine performance.

Port Fuel Injection

In a modern port injected engine, the fuel is discharged from the injectors into the intake ports near the intake valves. Therefore, the intake ports are filled with fuel vapor, and the rest of the intake manifold passages are filled with air. Since the intake manifold passages are filled with air, the need for intake manifold heating is greatly reduced. Port injected engines are not equipped with intake manifold heating devices such as heat riser valves or mixture heaters. When designing intake manifolds for these engines, engineers can place more emphasis on increased airflow rather than on manifold heating. Many port injected engines have longer, curved intake manifold air passages that improve airflow and increase horsepower.

Post-Combustion Systems

Post-combustion emission control devices clean up the exhaust after the fuel has been burned but before the gases exit the vehicle's tailpipe. An excellent example of this is the catalytic converter. A converter is one of the most effective emission control devices on a vehicle for reducing HC, CO, and NO_X.

Another post-combustion system is the secondary air or air injection system. This system forces fresh air into the exhaust stream to reduce HC and CO emissions.

Catalytic Converters

Catalytic converters are the most effective devices for controlling exhaust emissions. Until 1975 car makers had done somewhat of an effective job of controlling emissions by the use of other systems—auxiliary air injection systems, exhaust gas recirculation systems, and positive crankcase ventilation. But controlling emissions with these systems alone also meant lean mixtures and exotic ignition timing that often severely penalized power and fuel economy. When catalytic converters were introduced, much of the emission control could be taken out of the engine and moved into the exhaust system. This change allowed manufacturers to retune the engine for better performance and improved fuel economy.

Many changes will be necessary in vehicles, engines, and emission systems to meet the emission standards of the 1990s. Since approximately 90 percent of HC emissions occur before the catalytic converter is hot enough to provide for proper HC oxidation, engineers are designing catalytic converters that are heated from the vehicle's electrical system. Since these heaters have high current flow requirements, some charging system modifications may be necessary.

A catalytic converter contains a ceramic element coated with a catalyst. A catalyst is something that causes a chemical reaction without being part of the reaction. A catalytic converter causes a chemical change to take place in the passing exhaust gases. Most of the harmful gases are changed to harmless gases.

Catalytic converters are often referred to as Cats.

Shop Manual
Chapter 8, page 352

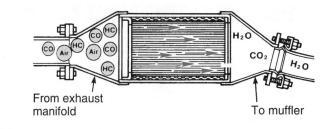

From exhaust manifold

To muffler

Figure 8-51 An oxidizing catalytic converter.

Three different materials are used as the catalyst in automotive converters: platinum, palladium, and rhodium. Platinum and palladium are the oxidizing elements of a converter (Figure 8-51). When HC and CO are exposed to heated surfaces covered with platinum and palladium, a chemical reaction takes place. The HC and CO are combined with oxygen to become H_2O and CO_2. Rhodium is a reducing catalyst. When NO_X is exposed to hot rhodium, oxygen is removed and NO_X becomes just N. The removal of oxygen is called reduction, which is why rhodium is a reducing catalyst.

Catalytic converters that contain all three catalysts and reduce HC, CO, and NO_X are called three-way converters (Figure 8-52). Catalytic converters that affect only HC and CO are called oxidizing converters. Three-way converters have the oxidizing catalysts in part of the container and the reducing catalyst in the other (Figure 8-53). Fresh air is injected by the secondary air system between the two catalysts. This air helps the oxidizing catalyst work by making extra oxygen available. The air from the secondary air system is not always forced into the converter, rather it is controlled by the secondary air system. Fresh air added to the exhaust at the wrong time could produce NO_X, something the catalytic converter is trying to destroy.

Air Injection Systems

One of the earliest methods used to reduce the amount of hydrocarbons and carbon monoxide in the exhaust was to force fresh air into the exhaust system after combustion. This additional fresh air causes further oxidation and burning of the unburned hydrocarbons and carbon monoxide. The

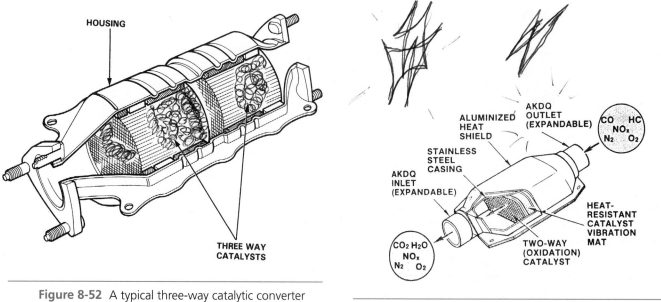

Figure 8-52 A typical three-way catalytic converter (TWC). (*Courtesy of American Honda Motor Company, Inc.*)

Figure 8-53 The action of a three-way catalytic converter.

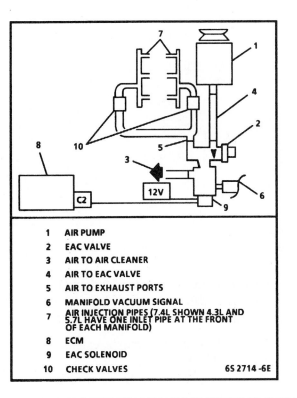

1	AIR PUMP
2	EAC VALVE
3	AIR TO AIR CLEANER
4	AIR TO EAC VALVE
5	AIR TO EXHAUST PORTS
6	MANIFOLD VACUUM SIGNAL
7	AIR INJECTION PIPES (7.4L SHOWN 4.3L AND 5.7L HAVE ONE INLET PIPE AT THE FRONT OF EACH MANIFOLD)
8	ECM
9	EAC SOLENOID
10	CHECK VALVES

6S 2714 -6E

Figure 8-54 A typical pump-type air injection system. (*Courtesy of Chevrolet Motor Division—GMC*)

process is much like blowing on a dwindling fire. Oxygen in the air combines with the HC and CO to continue the burning that reduces the HC and CO concentrations. This allows them to oxidize and produce harmless water vapor and carbon dioxide.

Pump Type

A typical system with an air pump is shown in Figure 8-54. On some engines, air from the air injection system is used to pressure purge the charcoal canister, in addition to reacting with the exhaust gases. The typical pump-type air injection system includes the following.

Air Pump. The air pump produces pressurized air that is sent to the exhaust manifold and to the catalytic converter. The air pump is driven by a belt from the crankshaft.

Air Control Valve (or Air-Switching Valve). This vacuum-operated valve is used to route the air from the pump either to the exhaust manifold or to the catalytic converter. During engine warm-up, the valve directs the air into the exhaust manifold. Once the engine is warm, the extra air in the manifold would affect EGR operation, so the air control valve directs the air to the converter where it aids the converter in oxidizing emissions.

Thermal Vacuum Switch. This switch controls the vacuum to the air control valve. When the coolant is cold, it signals the valve to direct air to the exhaust manifold. Then when the engine warms to normal operating temperature, the thermal vacuum switch signals the air control valve to reroute the air to the converter.

Air Bypass Valve (or Diverter Valve). This device is located between the air pump and the air control valve, or sometimes it is combined with the air control valve to make one component. The air bypass valve diverts or detours air during deceleration. Excess air in an exhaust rich with fuel can produce a backfire or explosion in a muffler. A vacuum signal operates the bypass valve during deceleration and compressed air is diverted to the atmosphere. The diverter valve prevents

Shop Manual
Chapter 8, page 353

Prior to the SAE J1930 terminology, some secondary air injection reactor pumps were called thermactor pumps.

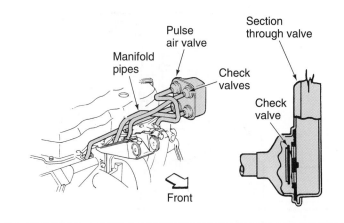

Figure 8-55 Pulsed secondary air injection system. (*Courtesy of Pontiac Motor Division—GMC*)

backfiring in the exhaust system during sudden deceleration. A very rich mixture is present the moment the throttle closes. The HCs in the exhaust can burn with the air from the air pump, causing a backfire. To prevent this, the air from the air pump is quickly diverted to the atmosphere.

One-Way Check Valves. These valves allow air into the exhaust but prevent exhaust from entering the pump in the event the drive belt breaks. Their location in the system is behind the air control valve and before the exhaust manifold and catalytic converter.

Hoses and Nozzles. These are necessary to distribute and inject the air.

Shop Manual
Chapter 8, page 354

Pulse Type (Nonpump Type)

This system uses the natural exhaust pressure pulses to pull air from the air cleaner into the exhaust manifolds and/or the catalytic converter. A manifold pipe is installed in the exhaust manifold for each engine cylinder. The inner end of these pipes is positioned near the exhaust port. The outer ends of the manifold pipes are connected to a metal container and a one-way check valve is mounted between the outer end of each pipe and the metal container (Figure 8-55).

The one-way check valves allow airflow from the metal container through the manifold pipes, but these valves prevent exhaust flow from the pipes into the container. A clean air hose is connected from the metal container to the air cleaner. At lower engine speeds, each negative pressure pulse in the exhaust manifold moves air from the air cleaner through the metal container, check valve, and manifold pipe into an exhaust port. High-pressure pulses in the exhaust manifold close the one-way check valves and prevent exhaust from entering the system. When the air is injected into the exhaust ports by the pulsed secondary air injection system, most of the unburned hydrocarbons coming out of the exhaust ports are ignited and burned in the exhaust manifold. Since the duration of low-pressure pulses in the exhaust ports decreases with engine speed, this system is more effective in reducing HC emissions at lower engine speeds.

If the system is used with a three-way catalytic converter, an air check valve and silencer (Figure 8-56) connects in-line between the air cleaner and the catalytic converter. When pressure in the exhaust system is greater than the pressure in the air cleaner, the reed valve (Figure 8-57) inside the air check valve closes. When the pressure in the exhaust system is less than the pressure in the air cleaner, the reed valve opens and draws air into the catalytic converter. The incoming oxygen reduces the hydrocarbons and carbon monoxide content of the exhaust gases by continuing the combustion of unburned gases in the same manner as the conventional system (with an air pump).

Electronic Secondary Air System

The typical electronic secondary air system, like the conventional air injection system, consists of an air pump to a secondary air bypass valve that directs the air either to the atmosphere or to the

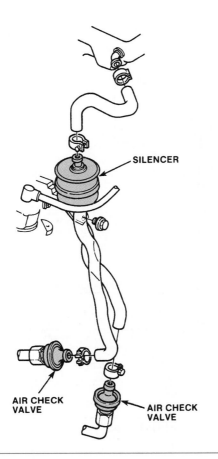

Figure 8-56 Typical pulse air system's check valves.

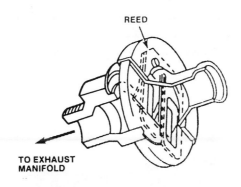

Figure 8-57 A reed valve opens and closes the air-check valve.

catalytic converter. The major difference between the two systems is the on-board computer and diverter valve.

The air pump is driven by a belt on the front of the engine and supplies air to the system. Some cars, such as the Corvette, have an air pump driven by an electric motor (Figure 8-58). Intake air passes through a centrifugal filter fan at the front of the pump where foreign materials are separated from the air by centrifugal force. In a commonly used system, air flows from the pump to a secondary **air bypass (AIRB)** valve which directs the air either to the atmosphere or to the secondary **air diverter (AIRD)** valve (Figure 8-59). The AIRD valve directs the air either to the exhaust manifold or to the catalytic converter. Therefore, secondary airflow can be directed to three points:

1. Vented (or bypassed) to the atmosphere via the air filter.
2. Upstream to the exhaust manifold.
3. Downstream to the catalytic converter.

Prior to the SAE J1930 terminology, some AIRB and AIRD valves were called the SAB and SAD valves.

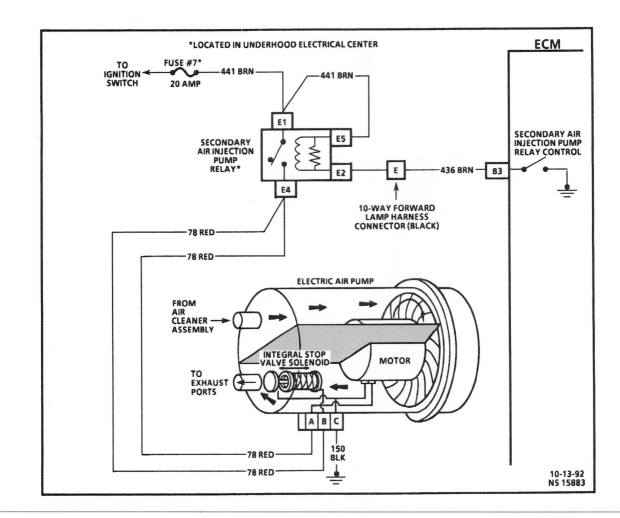

Figure 8-58 An electric air pump circuit. (*Courtesy of Chevrolet Motor Division—GMC*)

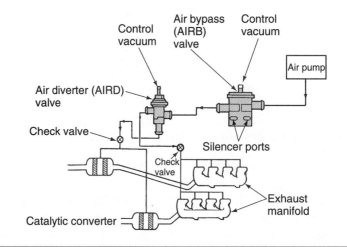

Figure 8-59 Secondary air injection system with AIRB and AIRD valves. (*Reprinted with the permission of Ford Motor Company*)

Both the AIRB and AIRD valves have solenoids that are controlled by the on-board computer. When either solenoid is energized by the computer, vacuum is applied to the AIRB valve (Figure 8-60) and secondary air is vented to the atmosphere. When no vacuum is applied to the AIRD valve, secondary air (if present) is directed to the catalytic converter (Figure 8-61).

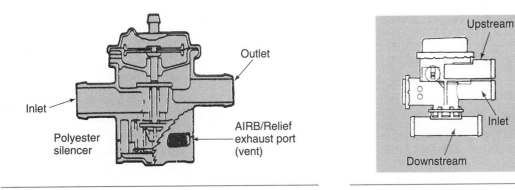

Figure 8-60 AIRB valve. (*Reprinted with the permission of Ford Motor Company*)

Figure 8-61 AIRD valve designs. (*Reprinted with the permission of Ford Motor Company*)

There are two check valves in the secondary air system. Secondary air must flow through a check valve before it reaches either the exhaust manifold or the catalytic converter. These check valves prevent the backflow of exhaust gases into the pump in the event of an exhaust backfire or if the pump drive belt fails.

Bypass Mode

In the bypass mode, vacuum is not applied either to the AIRB or the AIRD valve and secondary air is vented to the atmosphere. Secondary air may be vented or bypassed due to a fuel-rich condition, which the on-board computer recognizes as a problem in the system or during deceleration. Secondary air is also typically bypassed during cold engine cranking and cold idle conditions.

When engine coolant temperature is below 55°F at startup, secondary air is automatically bypassed. The system maintains a bypass mode of operation until the coolant temperature reaches 170°F. The computer has an internal electronic timer that keeps track of the length of time since the engine was started. Secondary air is also maintained in the bypass mode until a preset length of time has elapsed.

Upstream Mode

During the downstream mode, secondary air is routed through the AIRB valve and the AIRD valve to the exhaust manifold. The upstream mode is actuated when the computer senses a warm crank or start-up condition. The secondary airflow remains upstream for one to three minutes after start-up to help control emissions.

The air-fuel mixture at startup is typically very rich. This rich mixture results in unburned HC and CO in the exhaust after combustion. By switching to the upstream mode, the hot HC and CO mix with the incoming secondary air and are burned up.

This reburning of HC and CO compounds causes the exhaust gases to get hotter, which in turn heats up the oxygen sensor. Switching to the upstream mode allows the electronic engine control system to switch to the closed loop operation sooner because the oxygen sensor is ready to function sooner.

The warm oxygen sensor sends exhaust gas oxygen information to the computer. When the electronic engine control system is in the closed loop operation, the computer uses this information to adjust the air-fuel mixture.

It should be noted that the upstream mode of operation increases the oxygen level in the exhaust gases. The voltage signal from the oxygen level in the exhaust gases. The voltage signal from the oxygen sensor to the computer maintains a continuous low level. The computer interprets this signal as a continuous lean condition i.e., too much oxygen in the air-fuel mixture. It can readily be seen, then, that the upstream mode results in inaccurate exhaust gas oxygen measurements.

To solve this dilemma, the computer automatically switches to the open loop fuel control whenever the upstream mode is activated. It ignores the oxygen sensor input.

Downstream Mode

During the downstream mode, secondary air is routed through the AIRB valve and the AIRD valve to the catalytic converter. The secondary air system operates in the downstream mode during a majority of engine conditions.

The catalytic converter is most efficient at reducing NO_X when the air-fuel mixture is near stoichiometric. This plays a big part in producing that optimum air-fuel mixture in the closed loop fuel control mode. When secondary air is diverted downstream, the oxygen sensor can provide accurate information about the level of oxygen in the exhaust gases to the computer. Therefore, the electronic engine control system can operate in the closed loop fuel control mode only when the secondary air system is diverted downstream.

After the engine has warmed up sufficiently, the air-fuel mixture tends to run lean leaving fewer excess hydrocarbons remaining after combustion. Therefore, it is not necessary to run the secondary air system in the upstream mode of operation. The computer automatically switches the system to the downstream mode to allow the secondary air to mix with the exhaust gases inside the catalytic converter. The fresh secondary air diverted downstream allows the converter to reduce the NO_X emissions, which are also reduced by the closed loop mode.

Summary

Terms to Know

Air bypass (AIRB)

Air diverter (AIRD)

Air injector reactor (AIR)

Blowby

California Air Research Board (ARB)

Canister purge

Carbon monoxide (CO)

Charcoal carbon canister

Delay valve

Diverter valve

EFE

Evaporated emission system (EVAP)

Evaporative control

Hydrocarbons (HC)

Inspection/Maintenance (I/M)

Low emission vehicles (LEV)

Negative backpressure

Oxides of nitrogen (NO$_X$)

❏ Unburned hydrocarbons, carbon monoxide, and oxides of nitrogen are three types of emissions controlled in gasoline engines.

❏ HC emissions are unburned gasoline released by the engine because of incomplete combustion.

❏ CO emissions are a byproduct of combustion.

❏ CO emissions are caused by a rich air-fuel ratio.

❏ Oxides of nitrogen (NO_X) are formed when combustion temperatures reach more than 2,500°F (1,371°C).

❏ Precombustion control systems prevent emissions from being created in the engine, either during or before the combustion cycle. Post-combustion control systems clean up exhaust gases after the fuel has been burned. The evaporative control system traps fuel vapors that would normally escape from the fuel and carburetor into the air.

❏ The PCV system removes blowby gases from the crankcase and recirculates them to the engine intake. The PCV system benefits the vehicle's driveability by eliminating harmful crankcase gases, reducing air pollution, and promoting fuel economy.

❏ With the engine running at idle speed, the high intake manifold vacuum moves the PCV valve toward the closed position.

❏ During part-throttle operation, the intake manifold vacuum decreases and the PCV valve spring moves the valve toward the open position.

❏ As the throttle approaches the wide-open position, intake manifold vacuum decreases and the spring moves the PCV valve further toward the open position.

❏ When the engine backfires into the intake manifold, the PCV valve seats and prevents the backfire from entering the engine.

❏ If the engine has excessive blowby or the PCV valve is restricted, crankcase pressure forces crankcase gases through the clean air hose into the air cleaner.

❏ The thermal vacuum valve (TVV) connected in the distributor vacuum advance hose vents the vacuum supplied to the vacuum advance at certain temperatures during engine warmup.

❑ A vacuum delay valve delays the vacuum supplied to the distributor advance when the throttle is opened.

❑ A detonation sensor changes a vibration caused by engine detonation to an analog voltage signal. The electronic spark control module changes the analog detonation sensor signal to a digital signal and sends this signal to the PCM.

❑ An evaporative (EVAP) emission system stores vapors from the fuel tank in a charcoal canister until certain engine operating conditions are present. When the proper conditions are present, fuel vapors are purged from the charcoal canister into the intake manifold.

❑ A port EGR valve is opened when vacuum is supplied to the chamber above the diaphragm.

❑ A positive backpressure EGR valve has a normally open bleed valve in the center of the diaphragm. This bleed valve is closed by exhaust pressure at a specific throttle opening.

❑ A negative backpressure EGR valve contains a normally closed bleed valve in the center of the diaphragm. This bleed valve is opened by negative pressure pulses in the exhaust at low engine speed.

❑ A digital EGR valve has up to three electric solenoids operated by the PCM.

❑ A linear EGR valve contains an electric solenoid that is operated by the PCM with a pulse width modulation (PWM) signal.

❑ Some EGR valves contain an exhaust gas temperature sensor signal, which sends a voltage signal to the PCM in relation to exhaust temperature.

❑ A pressure feedback electronic (PFE) sensor sends a voltage signal to the PCM in relation to the exhaust pressure under the EGR valve.

❑ The pulsed secondary air injection system uses the negative pressure pulses in the exhaust system at low speeds to move air into the exhaust ports.

❑ Many secondary air injection systems pump air into the exhaust ports during engine warm-up, and deliver air to the catalytic converters with the engine at normal operating temperature.

Terms to Know (continued)

Periodic motor vehicle inspection (PMVI)

Positive backpressure

Post-combustion control

Precombustion control

Ported vacuum switch (PVS)

Temperature vacuum switch (TVS)

Thermal vacuum valve (TVV)

Thermostatic air cleaner

Transducer

Transitional low emission vehicles (TLEV)

Ultra low emission vehicles (ULEV)

Vacuum amplifier

Zero emission vehicles (ZEV)

Review Questions

Short Answer Essays

1. Explain why a small PCV valve opening is adequate at idle speed.

2. Describe the purpose of the ESC module.

3. Explain how a port EGR valve is opened.

4. Describe the operation of a digital EGR valve.

5. Explain why a secondary air injection system pumps air into the exhaust ports during engine warmup.

6. Describe the causes of high HC emissions.

7. Describe carbon monoxide (CO) emissions in relation to air-fuel ratio.

8. Name the three types of emissions being controlled in gasoline engines.

9. Name the three basic types of emission control systems used in modern vehicles.

10. List three ways the PCV system benefits the vehicle's driveability.

Fill-in-the-Blanks

1. The PCV system prevents _____ _____ from escaping to the atmosphere.

2. On some systems, the detonation sensor signal is sent through the _____ _____ _____ to the PCM.

3. In a negative backpressure EGR valve, if the exhaust pressure passage in the stem is plugged, the bleed valve remains _____.

4. In a pulsed secondary air injection system, the one-way check valves prevent _____ from entering the metal container and air pipe to the air cleaner.

5. In a secondary air injection system, the airflow from the pump is directed into the _____ _____ with the engine at normal operating temperature.

6. HC emissions may come from the tailpipe or _____ sources.

7. A lean air-fuel ratio causes HC emissions to _____.

8. CO emissions are a good indicator of a _____ air-fuel ratio.

9. _____ emissions are formed by high combustion temperatures.

10. NO_x, HC, and CO are changed into harmless gases by the _____ in the catalytic converter.

ASE Style Review Questions

1. While discussing PCV valve operation:
   Technician A says the PCV valve opening is decreased at part-throttle operation compared to idle operation.
   Technician B says the PCV valve opening is decreased at wide-open throttle compared to part throttle.
   Who is correct?
   A. A only
   B. B only
   C. Both A and B
   D. Neither A nor B

2. While discussing detonation sensors used with electronic spark control (ESC) modules:
   Technician A says the detonation sensor produces an analog signal.
   Technician B says the detonation sensor signal is sent directly to the PCM.
   Who is correct?
   A. A only
   B. B only
   C. Both A and B
   D. Neither A nor B

3. While discussing computer-controlled heat riser valves:

   Technician A says vacuum is supplied through a computer-operated solenoid to the heat riser diaphragm.

   Technician B says at a specific temperature during engine warmup, the heat riser valve remains in the half-open position.

   Who is correct?

   A. A only   **C.** Both A and B

   B. B only   **D.** Neither A nor B

4. While discussing EGR systems with a pressure feedback electronic (PFE) sensor:

   Technician A says the PFE sensor sends a signal to the PCM in relation to intake manifold pressure.

   Technician B says the PCM corrects the EGR flow if the actual flow does not match the requested flow.

   Who is correct?

   A. A only   **C.** Both A and B

   B. B only   **D.** Neither A nor B

5. While discussing exhaust pressure transducers:

   Technician A says the exhaust pressure transducer is connected in the vacuum hose between the intake manifold and the EGR solenoid.

   Technician B says the exhaust pressure transducer bleeds off vacuum to the EGR valve when the engine is operating at low rpm.

   Who is correct?

   A. A only   **C.** Both A and B

   B. B only   **D.** Neither A nor B

6. While discussing evaporative (EVAP) systems:

   Technician A says the coolant temperature has to be above a preset value before the PCM will operate the canister purge solenoid.

   Technician B says the vehicle speed has to be above a preset value before the PCM will operate the canister purge solenoid.

   Who is correct?

   A. A only   **C.** Both A and B

   B. B only   **D.** Neither A nor B

7. While discussing tailpipe emissions:

   Technician A says CO emissions increase as the air-fuel ratio becomes richer.

   Technician B says CO emissions increase as the air-fuel ratio becomes leaner.

   Who is correct?

   A. A only   **C.** Both A and B

   B. B only   **D.** Neither A nor B

8. While discussing tailpipe emissions and cylinder misfiring:

   Technician A says cylinder misfiring causes a significant increase in HC emissions.

   Technician B says cylinder misfiring results in a large increase in CO emissions.

   Who is correct?

   A. A only   **C.** Both A and B

   B. B only   **D.** Neither A nor B

9. *Technician A* says that the EGR vent solenoid is normally open.

   Technician B says that the EGR control solenoid is normally open.

   Who is correct?

   A. A only   **C.** Both A and B

   B. B only   **D.** Neither A nor B

10. *Technician A* says that the AIRB valve directs secondary air either to the exhaust manifold or to the catalytic converter.

    Technician B says that secondary air may be vented during deceleration.

    Who is correct?

    A. A only   **C.** Both A and B

    B. B only   **D.** Neither A nor B

Computers and Input Sensors

Upon completion and review of this chapter, you should be able to:

❏ Explain the difference between analog and digital voltage signals.

❏ Explain binary coding as it relates to computer input signals.

❏ Describe input signal amplification in a computer.

❏ Explain why an analog/digital (A/D) converter is necessary in a computer.

❏ Describe briefly the design of a microprocessor chip and state the basic purpose of this chip.

❏ Explain briefly how the microprocessor stores and retrieves information.

❏ Describe the purpose of a random access memory (RAM).

❏ Explain the terms volatile and nonvolatile memory.

❏ Describe the purpose of the read only memory (ROM).

❏ Explain the purpose of a programmable read only memory (PROM).

❏ Describe the purpose of a keep alive memory (KAM).

❏ Define the term adaptive strategy.

❏ Explain how the computer output drivers operate most output actuators.

❏ Describe the design and operation of the oxygen sensor.

❏ Explain the importance of the engine coolant temperature sensor signal in relation to the output control functions of the computer.

❏ Describe the operation of the air charge temperature sensor.

❏ Explain the wiring connections located on the throttle position sensor.

❏ Describe two types of voltage signals produced by manifold absolute pressure sensors.

❏ Explain the operation of three different types of mass air flow sensors.

❏ Describe the operation of a knock sensor.

❏ Explain the purpose of the exhaust gas recirculation valve position sensor.

❏ Identify the output control functions affected by the vehicle speed sensor signal.

Introduction

Students must have an understanding of basic computer operation prior to a study of complete computer systems. When students understand this basic information, sophisticated computer systems are much easier to comprehend. Computers manufactured by various electronics companies or vehicle manufacturers operate in a similar way. This chapter provides a generic explanation of computer operation and a discussion of computer memories that are used by certain vehicle manufacturers. The actual computers used by various car manufacturers are explained with the correct terminology. The discussion about computers is followed by information on computer input sensors.

Computer Operation

A computer is an electronic device that stores and processes data. It is also capable of operating other computers. The operation of a computer is divided into four basic functions:

1. Input: A voltage signal sent from an input device. The device can be a sensor or a button activated by the driver, technician, or mechanical part.

2. Processing: The computer uses the input information and compares it to programmed instructions. This information is processed by logic circuits in the computer.

3. Storage: The program instructions are stored in the computer's memory. Some of the input signals are also stored in memory for processing later.

4. Output: After the computer has processed the sensor input and checked its programmed instructions, it will put out control commands to various output devices. These output devices may be instrument panel displays or output actuators. The output of one computer may also be an input to another computer.

Computers have taken over many of the tasks in cars and trucks that were formerly performed by vacuum, mechanical, or electromechanical devices. When properly programmed, they can carry out explicit instructions with blinding speed and almost flawless consistency.

A typical electronic control system is made up of sensors, actuators, and related wiring that is tied into a central processor called a **microprocessor** or microcomputer (a smaller version of a computer).

The central processing unit (CPU) is the brain of a computer. The CPU is constructed of thousands of transistors that are placed on a small chip. The CPU brings information into and out of the computer's memory. The input information is processed in the CPU and checked against the program in the computer's memory. The CPU also checks the memory for any other information regarding programmed parameters. The information obtained by the CPU can be altered according to the instructions of the program. The CPU may be ordered to make logic decisions on the information received. Once these decisions, or calculations, are made, the CPU sends out commands to make the required corrections or adjustments to the system being controlled (Figure 9-1).

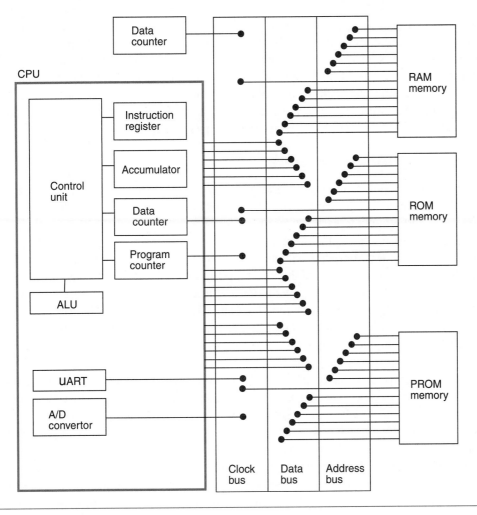

Figure 9-1 Major components of a computer and its CPU.

Sensors

The CPU receives inputs that it checks with programmed values. Depending on the input, the computer will control the actuator(s) until the programmed results are obtained. The inputs can come from other computers, the driver, the technician, or through a variety of sensors.

Driver input signals are usually provided by momentarily applying a ground through a switch. The computer receives this signal and performs the desired functions. For example, if the driver wishes to reset the trip odometer on a digital instrument panel, a reset button is depressed. This switch provides a momentary ground that the computer receives as an input and sets the trip odometer to zero.

Switches can be used as an input for any operation that only requires a yes-no, or on-off, condition. Other inputs include those supplied by means of a sensor and those signals returned to the computer in the form of **feedback**. Feedback means that data concerning the effects of the computer's commands are fed back to the computer as an input signal.

If the computer sends a command signal to actuate an output device, a feedback signal may be sent back from the actuator to inform the computer that the task was performed. The feedback signal will confirm both the position of the output device and the operation of the actuator. Another form of feedback is for the computer to monitor voltage as a switch, relay, or other actuator is activated. Changing positions of an actuator should result in predictable changes in the computer's voltage sensing circuit. The computer may set a diagnostic code if it does not receive the correct feedback signal.

All sensors perform the same basic function. They detect a mechanical condition (movement or position), chemical state, or temperature condition and change it into an electrical signal that can be used by the computer to make decisions. The CPU makes decisions based on information it receives from sensors. Each sensor used in a particular system has a specific job to do (for example, monitor throttle position, vehicle speed, manifold pressure). Together these sensors provide enough information to help the computer form a complete picture of vehicle operation. Even though there are a variety of different sensor designs, they all fall under one of two operating categories: reference voltage sensors or voltage generating sensors.

Reference Voltage Sensors

Reference voltage (Vref) sensors provide input to the computer by modifying or controlling a constant, predetermined voltage signal. This signal, which can have a reference value from 5 V to 9 V is generated and sent out to each sensor by a reference voltage regulator located inside the CPU. Because the computer knows that a certain voltage value has been sent out, it can indirectly interpret things like motion, temperature, and component position based on what comes back. For example, consider the operation of the throttle position sensor (TPS). During acceleration (from idle to wide-open throttle), the computer monitors throttle plate movement based on the changing reference voltage signal returned by the TPS. (The TPS is a type of variable resistor known as a rotary potentiometer that changes circuit resistance based on throttle shaft rotation.) As TPS resistance varies, the computer is programmed to respond in a specific manner (for example, increase fuel delivery or alter spark timing) to each corresponding voltage change.

Most sensors presently in use are variable resistors or potentiometers (Figure 9-2). They modify a voltage to or from the computer, indicating a constantly changing status that can be calculated, compensated for, and modified. That is, most sensors simply control a voltage signal from the computer. When varying internal resistance of the sensor allows more or less voltage to ground, the computer senses a voltage change on a monitored signal line. The monitored signal line may be the output signal from the computer to the sensor (one and two-wire sensors), or the computer may use a separate return line from the sensor to monitor voltage changes (three-wire sensors).

Another commonly used variable resistor is a thermistor. A thermistor is a solid-state variable resistor made from a semiconductor material that changes resistance in relation to temperature

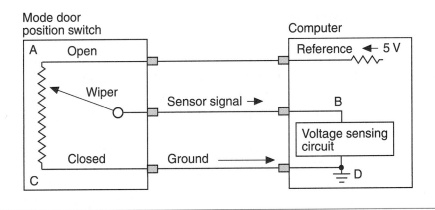

Figure 9-2 A potentiometer sensor circuit measures the amount of voltage drop to determine the position of something, such as the A/C door shown in the illustration.

changes. Thermistors are used to monitor engine coolant, intake air, and ambient temperatures. By monitoring the thermistor's resistance value, the CPU is capable of observing small changes in temperature. The CPU sends a reference voltage to the thermistor, normally 5 volts, through a fixed resistor. As the current flows through the thermistor to ground, voltage is dropped by the thermistor. A voltage sensing circuit compares the voltage sent out by the CPU to the voltage returned to the CPU and determines the voltage drop. Using its programmed values, the computer is able to translate the voltage drop into a temperature reading.

There are two basic types of thermistors: NTC and PTC. A **negative temperature coefficient** (**NTC**) thermistor reduces its resistance as temperature increases. This type is the most commonly used thermistor. A **positive temperature coefficient** (**PTC**) thermistor increases its resistance with an increase in temperature. In order to diagnose thermistors accurately, you must be able to identify the type by looking at a wiring diagram.

Wheatstone bridges (Figure 9-3) are also used as variable resistance sensors. These are typically constructed of four resistors, connected in series-parallel between an input terminal and a ground terminal. Three of the resistors are kept at the same value. The fourth resistor is a sensing resistor. When all four of the resistors have the same value, the bridge is balanced and the voltage sensor will have a value of 0 volt. If the sensing resistor changes value, a change will occur in the circuit's balance. The sensing circuit will receive a voltage reading that is proportional to the amount of resistance change. If the Wheatstone bridge is used to measure temperature, temperature changes will be indicated as a change in voltage by the sensing circuit. Wheatstone bridges are also used to measure pressure (**piezoresistive**) and mechanical strain.

In addition to variable resistors, another commonly used reference voltage sensor is a switch. By opening and closing a circuit, switches provide the necessary voltage information to the computer so that vehicles can maintain the proper performance and driveability.

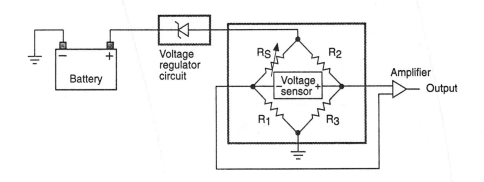

Figure 9-3 Wheatstone bridge.

Voltage Generating Sensors

While most sensors are variable resistance/reference voltage, there is another category of sensors—the voltage generating devices. These sensors include components like the magnetic pulse generators, Hall effect switches, oxygen sensors (zirconium dioxide), and knock sensors (piezoelectric), which are capable of producing their own input voltage signal. This varying voltage signal, when received by the computer, enables the computer to monitor and adjust for changes in the computerized engine control system.

Magnetic pulse generators use the principle of magnetic induction to produce a voltage signal. They are also called permanent magnet (PM) generators. These sensors are commonly used to send data to the computer about the speed of the monitored component. This data provides information about vehicle speed, shaft speed, and wheel speed. The signals from speed sensors are used for instrumentation, cruise control systems, anti-lock brake systems, ignition systems, speed sensitive steering systems, and automatic ride control systems. A magnetic pulse generator is also used to inform the computer about the position of a monitored device. This is common in engine controls where the CPU needs to know the position of the crankshaft in relation to rotational degrees.

The major components of a pulse generator are a timing disc and a pick-up coil. The timing disc is attached to a rotating shaft or cable. The number of teeth on the timing disc is determined by the manufacturer and the application. If only the number of revolutions is required, the timing disc may have only one tooth. Whereas, if it is important to track quarter revolutions, the timing disc needs at least four teeth. The teeth will cause a voltage generation that is constant per revolution of the shaft. For example, a vehicle speed sensor may be designed to deliver 4,000 pulses per mile. The number of pulses per mile remains constant regardless of speed. The computer calculates how fast the vehicle is going based on the frequency of the signal. The timing disc is also known as an armature, reluctor, trigger wheel, pulse wheel, or timing core.

The pick-up coil is also known as a stator, sensor, or pole piece. It remains stationary while the timing disc rotates in front of it. The changes of magnetic lines of force generate a small voltage signal in the coil. A pick-up coil consists of a permanent magnet with fine wire wound around it.

An air gap is maintained between the timing disc and the pick-up coil. As the timing disc rotates in front of the pick-up coil, the generator sends a pulse signal (Figure 9-4). As a tooth on the timing disc aligns with the core of the pick-up coil, it repels the magnetic field. The magnetic field is forced to flow through the coil and pick-up core (Figure 9-5). When the tooth passes the core, the magnetic field is able to expand (Figure 9-6). This action is repeated every time a tooth passes the core. The moving lines of magnetic force cut across the coil windings and induce a voltage signal.

When a tooth approaches the core, a positive current is produced as the magnetic field begins to concentrate around the coil. When the tooth and core align, there is no more expansion or contraction of the magnetic field and the voltage drops to zero. When the tooth passes the core, the magnetic field expands and a negative current is produced (Figure 9-7). The resulting pulse signal is sent to the CPU.

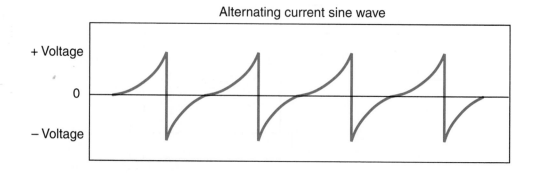

Figure 9-4 Pulse signal sine wave.

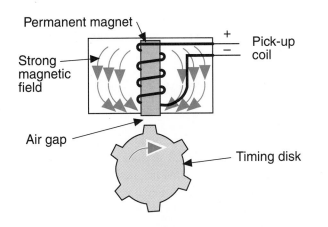

Figure 9-5 A strong magnetic field is produced in the pick-up coil as the teeth align with the core.

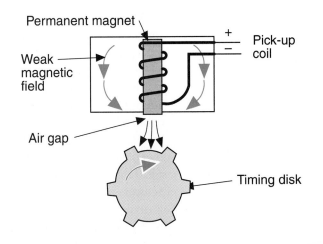

Figure 9-6 The magnetic field expands and weakens as the teeth pass the core.

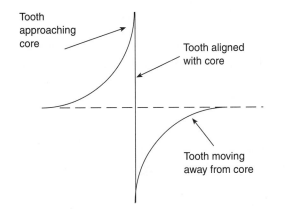

Figure 9-7 Waveform produced by a magnetic pulse generator.

Hall Effect Sensors

The Hall effect switch performs the same function as a magnetic pulse generator. It operates on the principle that if a current is allowed to flow through thin conducting material that is exposed to a magnetic field, another voltage is produced (Figure 9-8).

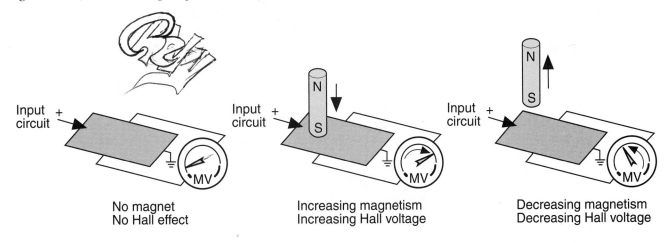

Figure 9-8 Hall effect principles of voltage induction.

A Hall effect switch contains a permanent magnet and a thin semiconductor layer made of Gallium arsenate crystal (Hall layer) and a shutter wheel. The Hall layer has a negative and a positive terminal connected to it. Two additional terminals located on either side of the Hall layer are used for the output circuit.

The permanent magnet is located directly across from the Hall layer so that its lines of flux will bisect at right angles to the current flow. The permanent magnet is mounted so that a small air gap is between it and the Hall layer.

A steady current is applied to the crystal of the Hall layer. This produces a signal voltage that is perpendicular to the direction of current flow and magnetic flux. The signal voltage produced is a result of the effect the magnetic field has on the electrons. When the magnetic field bisects the supply current flow, the electrons are deflected toward the Hall layer negative terminal (Figure 9-9). This results in a weak voltage potential being produced in the Hall switch.

The shutter wheel consists of a series of alternating windows and vanes. It creates a magnetic shunt that changes the strength of the magnetic field from the permanent magnet. The shutter wheel is attached to a rotational component. As the wheel rotates, the vanes pass through the air gap. When a shutter vane enters the gap, it intercepts the magnetic field and shields the Hall layer from its lines of force. The electrons in the supply current are no longer disrupted and return to a normal state. This results in low voltage potential in the signal circuit of the Hall switch.

The signal voltage leaves the Hall layer as a weak analog signal. To be used by the CPU, the signal must be conditioned. It is first amplified because it is too weak to produce a desirable result. The signal is also inverted so that a low input signal is converted into a high output signal. It is then sent through a Schmitt trigger, which is a type of A/D converter, where it is digitized and conditioned into a clean square wave signal. The signal is finally sent to a switching transistor. The computer senses the turning on and off of the switching transistor to determine the frequency of the signals and calculates speed.

Regardless of the type of sensors used in electronic control systems, the computer is incapable of functioning properly without input signal voltage from sensors.

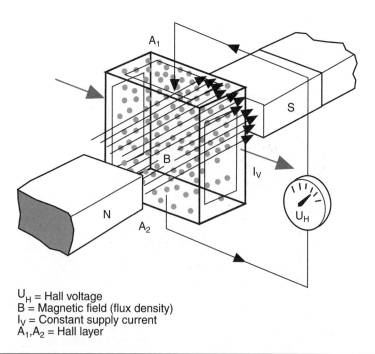

U_H = Hall voltage
B = Magnetic field (flux density)
I_V = Constant supply current
A_1, A_2 = Hall layer

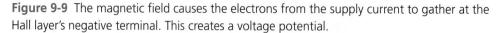

Figure 9-9 The magnetic field causes the electrons from the supply current to gather at the Hall layer's negative terminal. This creates a voltage potential.

Voltage Signals

Voltage does not flow through a conductor, current flows while voltage is the pressure that pushes the current. However, voltage can be used as a signal, for example, difference in voltage levels, frequency of change, or switching from positive to negative values can be used as a signal.

A computer is capable of reading voltage signals. The programs used by the CPU are "burned" into IC chips using a series of numbers. These numbers represent various combinations of voltages that the computer can understand. The voltage signals to the computer can be either analog or digital.

Analog Voltage Signals

An **analog voltage signal** is continuously variable within a certain range. When a rheostat is used to control a 5-V bulb, the rheostat voltage may be anywhere between 0 V and 5 V. If the rheostat voltage is low, a small amount of current flows through the bulb to produce a dim light from the bulb. If the rheostat voltage is 5 V, higher current flow produces increased light brilliance. As the rheostat voltage is decreased, the light becomes dimmer. This is an example of an analog voltage (Figure 9-10). Most of the sensors in an automotive computer system produce analog voltages.

An analog voltage signal is continuously variable in a specific range.

Digital Voltage Signal

If an ordinary on/off switch is connected to a 5-V bulb and the switch is off, 0 V is available at the bulb. When the switch is turned on, a 5-V signal is sent to the bulb, and the bulb is illuminated to full brilliance. If the switch is turned off, the voltage at the bulb returns to 0 V, and the bulb goes out. The voltage signal supplied through the switch is either 0 V or 5 V, or we could say the voltage signal is either high or low. This type of voltage signal is referred to as a digital signal. If the switch is turned on and off rapidly, a square-wave digital voltage signal is applied from the switch to the bulb (Figure 9-11).

A **digital voltage signal** is either high or low.

In an automotive computer, the microprocessor contains a large number of miniature switches that are capable of producing many digital voltage signals per second. These digital voltage signals are used to control various relays and components in the system. The microprocessor can vary the length of time that the digital signal is high or low for precise control (Figure 9-12).

A digital signal may be called a square-wave signal.

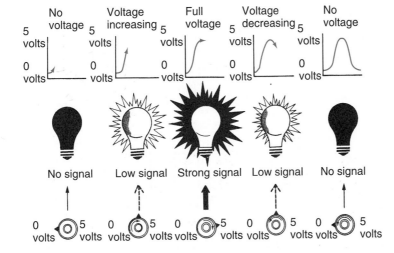

Figure 9-10 Analog voltage signal. *(Reprinted with the permission of Ford Motor Company)*

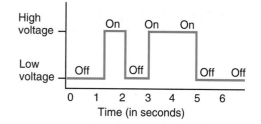

Figure 9-12 Digital voltage signal with variable on-time. *(Reprinted with the permission of Ford Motor Company)*

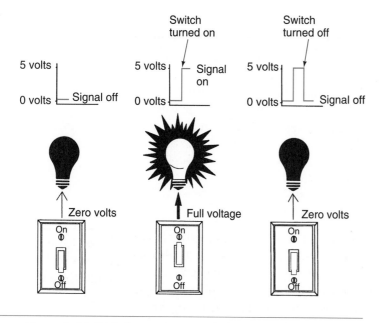

Figure 9-11 Digital square-wave voltage signal. *(Reprinted with the permission of Ford Motor Company)*

Binary Code

We have explained that a digital signal is either high or low. A numeric value may be assigned to digital signals. For example, a low digital signal may be given a value of 0, and a high digital signal may be given a value of 1. This assignment of numeric values to digital signals is called binary coding. The word *binary* means two values, and in the binary coding system, the two values are 0 and 1 (Figure 9-13). In an automotive computer, information is transmitted in binary code form. Conditions, numbers, and letters can be represented by a series of zeros and ones.

Many of the input sensors operate in the 0-V to 5-V range. The throttle position sensor (TPS) may produce these voltages:

> Closed throttle — 0 to 2 volts
>
> Part open throttle — 2 to 4 volts
>
> Wide open throttle — 4 to 5 volts

A numeric value may be assigned by the computer to each of these voltages:

> 0 to 2 volts — 1
>
> 2 to 4 volts — 2
>
> 4 to 5 volts — 3

A **binary code** is the application of numeric values to digital signals.

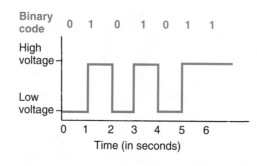

Figure 9-13 Applying numbers to digital voltage signals in a binary code. *(Reprinted with the permission of Ford Motor Company)*

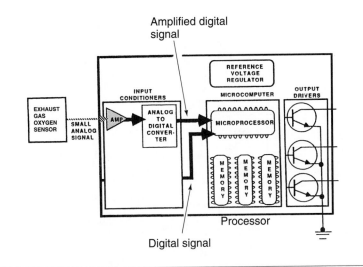

Figure 9-14 Amplification circuit in the computer input conditioning chip. *(Reprinted with the permission of Ford Motor Company)*

Input Conditioning

Amplification

Some input sensors such as the oxygen (O_2) sensor produce a very low voltage signal of less than 1 V. This signal also has an extremely low current flow. Therefore, this type of signal must be amplified, or increased, before it is sent to the microprocessor. This amplification is accomplished by the amplification circuit in the input conditioning chip inside the computer (Figure 9-14).

Input signal amplification means increasing these signals so they are useful to the computer.

Analog to Digital (A/D) Conversion

Since the input sensors produce analog signals and the microprocessor operates on digital signals, something must change the sensor analog signals to digital signals. This job is done by the A/D converter in the computer input conditioning chip (Figure 9-15).

The A/D converter continually scans the analog input signals at regular intervals. If the A/D converter scans the TPS signal and finds this signal at 5 V, the A/D converter assigns a numeric

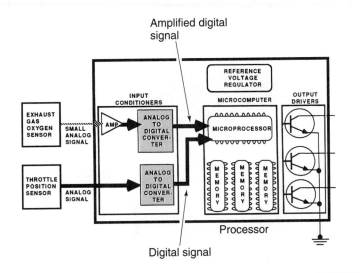

Figure 9-15 Analog/digital converter in the input conditioning chip. *(Reprinted with the permission of Ford Motor Company)*

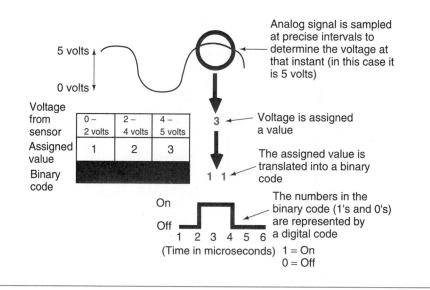

Analog signal is sampled at precise intervals to determine the voltage at that instant (in this case it is 5 volts)

5 volts

0 volts

Voltage from sensor	0 – 2 volts	2 – 4 volts	4 – 5 volts
Assigned value	1	2	3
Binary code			

3 ← Voltage is assigned a value

1 1 ← The assigned value is translated into a binary code

On
Off
1 2 3 4 5 6
(Time in microseconds)

The numbers in the binary code (1's and 0's) are represented by a digital code

1 = On
0 = Off

Figure 9-16 The A/D converter in the computer assigns a numeric value to input voltages and changes the numeric value to a binary code. *(Reprinted with the permission of Ford Motor Company)*

value of 3 to this specific voltage. The A/D converter then changes this numeric value to a binary code of 1 1 (Figure 9-16).

Therefore, we can understand that the A/D converter continually scans the input sensor signals and assigns numeric values to these voltages. The A/D converter then translates the numeric value to a binary code. In some automotive computers, the input conditioning chip is combined with the microprocessor.

Microprocessors

Design

The microprocessor chip is the calculating and decision-making chip in a computer.

The microprocessor is the calculating and decision-making chip in a computer. Thousands of miniature transistors and diodes are contained in the microprocessor. These transistors act as electronic switches that are either on or off. The components in the microprocessor are etched on an integrated circuit (IC) that is small enough to fit on a fingertip (Figure 9-17). The silicon chip containing

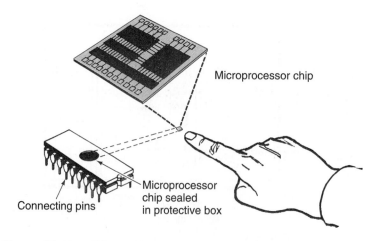

Microprocessor chip

Connecting pins

Microprocessor chip sealed in protective box

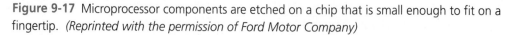

Figure 9-17 Microprocessor components are etched on a chip that is small enough to fit on a fingertip. *(Reprinted with the permission of Ford Motor Company)*

the IC is mounted in a flat, rectangular, protective box. Metal connecting pins extend from each side of the microprocessor container. These pins connect the microprocessor to the circuit board in the computer.

The microprocessor is supported by various memory chips that store information and help the microprocessor in making decisions. These memory chips are similar in appearance to the microprocessor chip. We will explain the function of the memory chips later in this chapter.

Program

A program is a group of instructions that is followed by the microprocessor. The program guides the microprocessor in decision making. For example, the program may inform the microprocessor when sensor information should be retrieved, and then tell the microprocessor how to process this information. Finally, the program will guide the microprocessor regarding the activation of output control devices such as relays and solenoids. The various memories contain the programs and other vehicle data that the microprocessor refers to as it performs calculations. As the microprocessor performs calculations and makes decisions, it works with the memories in the following ways:

1. The microprocessor can read information from the memories.

2. The microprocessor can write new information into the memories.

Information Storage

The memories contain many different locations. These locations may be compared to file folders in a filing cabinet, and each location contains one piece of information. An address is assigned to each memory location. This address may be compared to the lettering or numbering arrangement on file folders. Each address is written in a binary code, and these codes are numbered sequentially beginning with 0.

While the engine is running, the computer receives a large quantity of information from a number of sensors. The computer may not be able to process all this information immediately. In some instances, the computer may receive sensor inputs that the computer requires to make a number of decisions. In these cases, the microprocessor writes information into memory by specifying a memory address and sending information to this address (Figure 9-18).

Information Retrieval

When stored information is required, the microprocessor specifies the stored information address and requests the information. When stored information is requested from a specific address, the

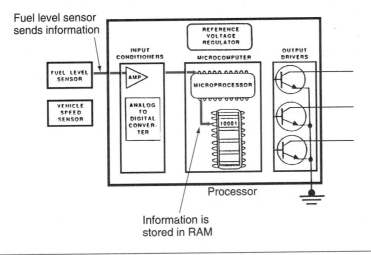

Figure 9-18 Information storage. *(Reprinted with the permission of Ford Motor Company)*

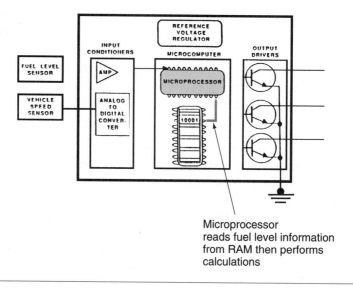

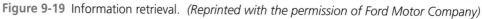

Microprocessor
reads fuel level information
from RAM then performs
calculations

Figure 9-19 Information retrieval. *(Reprinted with the permission of Ford Motor Company)*

memory sends a copy of this information to the microprocessor (Figure 9-19). However, the original stored information is still retained in the memory address.

The memories store information regarding the ideal air-fuel ratios for various operating conditions. The sensors inform the computer about the engine and vehicle operating conditions. The microprocessor reads the ideal air-fuel ratio information from memory and compares this information with the sensor inputs. After this comparison, the microprocessor makes the necessary decision and operates the injectors to provide the exact air-fuel ratio required by the engine.

Types of Computer Memories

A computer's memory holds the programs and other data, such as vehicle calibrations, that the microprocessor refers to in performing calculations. To the CPU, the program is a set of instructions or procedures that it must follow. Included in the program is information that tells the microprocessor when to retrieve input (based on temperature, time, etc.), how to process the input, and what to do with it once it has been processed. The microprocessor works with memory in two ways: it can read information from memory or change information in memory by writing in or storing new information.

Random Access Memory

Information that requires temporary storage is sent from the microprocessor to the **random access memory (RAM)**. The information stored in the RAM is subject to change. Since the sensor input information changes frequently in relation to various operating conditions, this information is stored in the RAM (Figure 9-20). The microprocessor may write the results of calculations and other changeable data into the RAM. The microprocessor can write information into the RAM, read information from the RAM, and erase RAM information.

The term *random access* indicates that the microprocessor can retrieve information from any RAM address in any order. If the RAM has a volatile memory, each time the ignition switch is turned off, the information stored in the RAM is erased. RAMs may also be designed with a nonvolatile memory. This type of RAM retains information when the ignition switch is turned off. If the RAM has a volatile memory, new information will be written into the RAM when the engine is restarted.

The microprocessor may write information into a random access memory (RAM) chip and read information from this chip.

A RAM with a volatile memory erases stored information when the ignition switch is turned off.

A RAM with a nonvolatile memory retains stored information when the ignition switch is turned off.

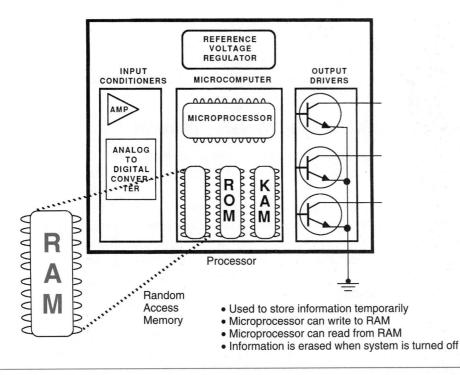

Figure 9-20 Random access memory (RAM). *(Reprinted with the permission of Ford Motor Company)*

Read Only Memory

The microprocessor can read information from the **read only memory** (**ROM**), but information cannot be written into the ROM by the microprocessor, and the microprocessor cannot erase ROM information (Figure 9-21). During the chip manufacturing process, information is programmed into the ROM. This information is not erased even if the battery terminals are disconnected.

The ROM contains look-up tables that contain information about how a vehicle should perform. For example, the look-up table would contain the ideal manifold vacuum under various

The computer can only read information from a read only memory (ROM).

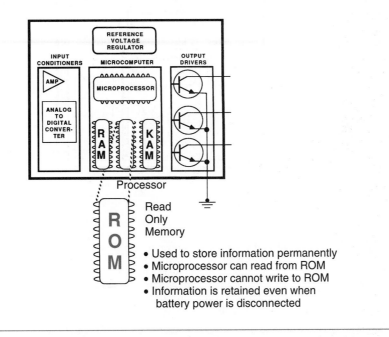

Figure 9-21 Read only memory (ROM). *(Reprinted with the permission of Ford Motor Company)*

engine operating conditions. The microprocessor compares the sensor inputs to the ideal vacuum in this table. If the sensor inputs indicate that the actual manifold vacuum is different from the ideal vacuum in look-up tables, the microprocessor will take some appropriate action.

The ROM also contains calibration tables regarding specific engine, transaxle or transmission, and differential specifications.

Programmable Read Only Memory (PROM)

A **programmable read only memory** (**PROM**) is usually a removable chip containing a vehicle specific program.

Many General Motors computers have a removable programmable read only memory (PROM), which is serviced separately from the computer. The PROM contains specific programs such as the spark advance program, which is designed for the specific requirements of each vehicle. For example, this spark advance program varies with different transmissions or rear axle ratios.

Some computers are equipped with an **electronically erasable programmable read only memory** (**EEPROM**). This type of memory chip may be reprogrammed easily by the manufacturer, and these chips usually are not serviced separately from the computer.

Keep Alive Memory

The microprocessor can write information into and read information from a **keep alive memory** (**KAM**).

The keep alive memory (KAM) has characteristics similar to the RAM. For example, the microprocessor can read and write information to and from the KAM and it can erase KAM information (Figure 9-22). However, the KAM retains information when the ignition switch is turned off. When the battery power is disconnected from the computer, the KAM memory is erased. The KAM is used for adaptive strategies that are explained under the next heading.

The KAM retains information when the ignition switch is turned off, but it erases information when battery power is disconnected from the computer.

Adaptive Strategy

Operation

If a computer has adaptive strategy capabilities, the computer can actually learn from past experience. For example, the normal voltage input range from a throttle position sensor (TPS) may be 0.6 V to 4.5 V. If a worn TPS sends a 0.3-V signal to the computer, the microprocessor interprets

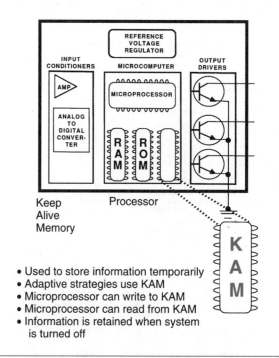

Figure 9-22 Keep alive memory (KAM). *(Reprinted with the permission of Ford Motor Company)*

this signal as an indication of component wear. The microprocessor stores this altered calibration in the KAM. The microprocessor now refers to this new calibration during calculations; thus, normal engine performance is maintained. If a sensor output is erratic or considerably out of range, the computer may ignore this input. When a computer has adaptive strategy, a short learning period is necessary:

1. after the battery has been disconnected.

2. when a computer system component has been replaced or disconnected.

3. on a new vehicle.

During this learning period, the engine may surge, idle fast, or have a loss of power. The average learning period lasts for five miles of driving.

Most adaptive strategies have two parts: Short Term Fuel Trim and Long Term Fuel Trim. Short term strategies are those immediately enacted by the computer to overcome a change in operation. These changes are temporary. Long term strategies are based on the feedback about the short term strategies. These changes are more permanent. To understand how a computer adapts itself to certain conditions, you must understand the differences between Short Term Fuel Trim (Adaptive) memory and Long Term Fuel Trim (Adaptive) memory.

Short term changes are not saved in the computer's memory. All changes to the fuel system happen immediately and occur in direct response to the O_2 sensor and/or other sensors. These changes are also designed to keep the O_2 sensor switching at its middle point.

Long term changes are saved in the computer's memory. These stored values are used the next time the engine is operated in a similar situation and under similar conditions. Long term changes are triggered to keep all short term strategies within particular parameters. These strategies are not based on the activity of the O_2 sensor, rather they are based on the results of the O_2 sensor.

Actuators

Once the computer's programming instructs that a correction or adjustment must be made in the controlled system, an output signal is sent to control devices called actuators. These actuators, which are solenoids, switchers, relays, or motors, physically act or carry out the command sent by the computer.

Actuators are electromechanical devices that convert an electrical current into mechanical action. This mechanical action can then be used to open and close valves, control vacuum to other components, or open and close switches. When the CPU receives an input signal indicating a change in one or more of the operating conditions, the CPU determines the best strategy for handling the conditions. The CPU then controls a set of actuators to achieve a desired effect or strategy goal. In order for the computer to control an actuator, it must rely on a component called an output driver.

The circuit driver usually applies the ground circuit of the actuator (Figure 9-23). The ground can be applied steadily if the actuator must be activated for a selected amount of time. Or the ground can be pulsed to activate the actuator in pulses.

Output drivers operate by the digital commands issued by the CPU. Basically, the output driver is nothing more than an electronic on/off switch used to control a specific actuator.

To illustrate this relationship, let us suppose the computer wants to turn on the engine's cooling fan. Once it makes a decision, it sends a signal to the output driver that controls the cooling fan relay (actuator). In supplying the relay's ground, the output driver completes the power circuit between the battery and cooling fan motor and the fan operates. When the fan has run long enough, the computer signals the output driver to open the relay's control circuit (by removing its ground), thus opening the power circuit to the fan.

For actuators that cannot be controlled by a digital signal, the CPU must turn its digitally coded instructions back into an analog signal. This conversion is completed by the A/D converter.

Adaptive strategy allows the computer to adapt to minor defects in the inputs and outputs.

Shop Manual
Chapter 9, page 433

Output drivers are often called quad drivers.

Output drivers are transistors or groups of transistors that control the output actuators.

Output actuators are usually solenoids and relays that are controlled by the computer.

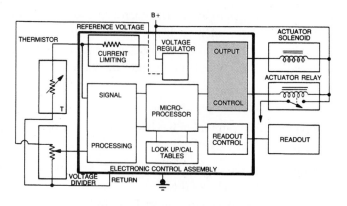

Figure 9-23 Output drivers in the computer usually supply a ground for the actuator solenoids and relays. *(Reprinted with the permission of Ford Motor Company)*

Displays can be controlled directly by the CPU. They do not require digital-to-analog conversion or output drivers because they contain circuitry that decodes the microprocessor's digital signal. The decoded information is then used to indicate such things as vehicle speed, engine rpm, fuel level, or scan tool values. Common types of electronic readout devices used as displays include light-emitting diodes (LED), liquid crystal display (LCD), and vacuum fluorescent display (VFD).

Some systems require the actuator to either be turned on and off very rapidly or for a set amount of cycles per second. It is duty cycled if it is turned on and off a set amount of cycles per second. **Duty cycle** is the percentage of on-time to total cycle time. Most duty cycled actuators cycle ten times per second. To complete a cycle, it must go from off to on to off again. If the cycle rate is ten times per second, one actuator cycle is completed in 1/10th of a second. If the actuator is turned on for 30 percent of each tenth of a second and off for 70 percent, it is referred to as a 30 percent duty cycle (Figure 9-24).

If the actuator is cycled on and off very rapidly, the pulse width can be varied to provide the desired results. **Pulse width** is the length of time in milliseconds that an actuator is energized. For example, the computer program will select an illumination level of a digital instrument panel based on the intensity of the ambient light in the vehicle. The illumination level is achieved through pulse width modulation of the lights. If the lights need to be bright, the pulse width is increased, which increases the length of on-time. As light intensity needs to be reduced, the pulse width is decreased (Figure 9-25).

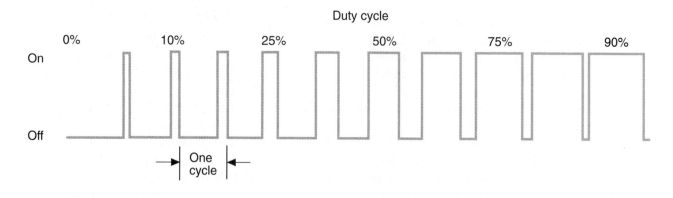

Figure 9-24 Duty cycle is the percentage of on-time per cycle. Duty cycle can be changed, however total cycle time remains constant.

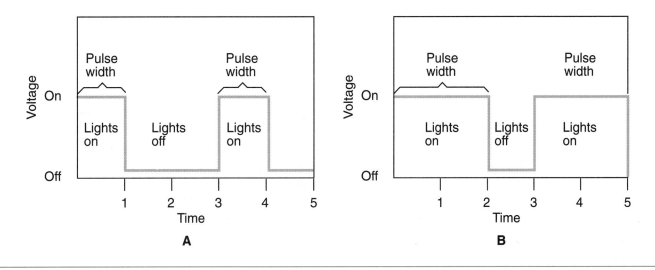

Figure 9-25 Pulse width is the duration of on-time: (A) pulse width modulation to achieve dimmer lights, (B) pulse width modulation to achieve brighter lights.

Input Sensors

Oxygen Sensor Design

The oxygen (O_2) sensor is threaded into the exhaust manifold or into the exhaust pipe near the engine. Some manufacturers refer to this sensor as an exhaust gas oxygen (EGO) sensor or a heated exhaust gas oxygen (HEGO) sensor. An oxygen-sensing element in the center of the O_2 sensor is surrounded by a steel shell. The oxygen-sensing element is made from zirconia in many O_2 sensors. The steel shell has a hex-shaped area on which a socket may be installed for removal and installation purposes. Threads on the lower end of the steel shell match the threads in the exhaust pipe or manifold opening where the sensor is installed (Figure 9-26).

A steel cover or a neoprene boot is installed over the top of the sensor. On many O_2 sensors, the steel cover is loosely installed on the sensor, which allows a constant supply of oxygen from the atmosphere inside the oxygen-sensing element. If a neoprene boot is installed on the sensor, the boot has grooves cut on the inner surface to allow air inside the sensing element. On some later model O_2 sensors, the top of the sensor is tightly sealed, and oxygen enters the sensor through the signal wire.

The **oxygen (O_2) sensor** sends an analog voltage signal to the computer in relation to air-fuel ratio and the amount of oxygen in the exhaust stream. A rich air-fuel ratio causes high oxygen sensor voltage.

Shop Manual
Chapter 9, page 415

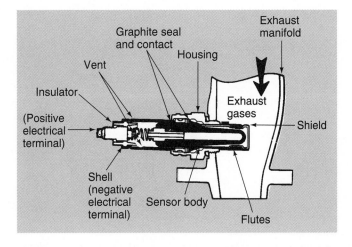

Figure 9-26 Oxygen sensor design. *(Reprinted with the permission of Ford Motor Company)*

A shield covers the bottom end of the sensor that protrudes into the exhaust manifold or pipe. Flutes in this shield help to swirl the exhaust gas continually around the sensor element when the engine is running.

Some O_2 sensors have a single wire that is connected from the oxygen-sensing element to the computer, and this wire acts as a signal wire. If an O_2 sensor has two wires, the second wire is a ground wire, which is connected back to the computer. Many O_2 sensors have three wires, and the third wire is connected to an electric heating element in the sensor. Voltage is supplied from the ignition switch to this heater when the ignition switch is on. Since the O_2 sensor does not produce a satisfactory signal until it reaches about 600°F (315°C), the internal heater provides faster sensor warm-up time and helps to keep the sensor hot during prolonged idle operation. The internal O_2 sensor heater maintains higher sensor temperatures, which helps to burn deposits off the sensor. When the O_2 sensor has an internal heater, the sensor may be placed further away from the engine in the exhaust stream, thus giving engineers more flexibility in sensor location. Some O_2 sensors have four wires—a signal wire, heater wire, and two ground wires. In these four-wire sensors, the heater and the sensing element have individual ground wires. A replacement O_2 sensor must have the same number of wires as the original sensor.

Oxygen Sensor Operation

A lean air-fuel ratio provides excess quantities of oxygen in the exhaust stream, because the mixture entering the cylinders has an excessive amount of air in relation to the amount of fuel. Therefore, air containing oxygen is left over after the combustion process. When the exhaust stream has high oxygen content, oxygen from the atmosphere is also present inside the O_2 sensor element. When oxygen is present on both sides of the sensor element, the sensor produces a low voltage.

A rich air-fuel ratio contains excessive fuel in relation to the amount of air entering the cylinders. A rich air-fuel mixture produces very little oxygen in the exhaust stream because the oxygen in the air is all mixed with fuel, and excess fuel is left over after combustion is completed. When the exhaust stream with very low oxygen content strikes the O_2 sensor, there is high oxygen content from the atmosphere inside the sensor element. With different oxygen levels inside and outside of the sensor element, the sensor produces up to 1 V. As the air-fuel ratio cycles from lean to rich, the O_2 sensor voltage changes in a few milliseconds.

In a gasoline fuel system, the stoichiometric, or ideal, air-fuel mixture is 14.7:1. This indicates that for every 14.7 pounds of air entering the air intake, the carburetor or fuel injectors supply 1 pound of fuel. At the stoichiometric air-fuel ratio, combustion is most efficient and nearly all the oxygen in the air is mixed with fuel and burned in the combustion chambers. Computer-controlled carburetor or fuel injection systems always maintain the air-fuel ratio at stoichiometric under most operating conditions. As the air-fuel ratio cycles slightly rich and lean from the stoichiometric ratio, the O_2 sensor voltage cycles from high to low (Figure 9-27).

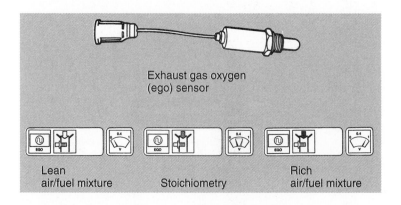

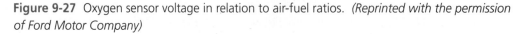

Figure 9-27 Oxygen sensor voltage in relation to air-fuel ratios. *(Reprinted with the permission of Ford Motor Company)*

WARNING: The use of some RTV sealants will contaminate the O_2 sensor.

 WARNING: The use of leaded gasoline or coolant leaks into the combustion chamber will contaminate the O_2 sensor.

If leaded regular fuel is used in an engine with an O_2 sensor, the sensor becomes lead-coated in a short time. Under this condition, the sensor signal is no longer satisfactory, and sensor replacement is likely required. Certain types of room temperature vulcanizing (RTV) sealant or antifreeze entering the combustion chambers will contaminate the O_2 sensor. Always use the RTV sealant recommended by the vehicle manufacturer.

WARNING: The O_2 sensor will be damaged if it is tested with an analog voltmeter. A digital meter must be used to test this sensor.

An analog voltmeter must never be used to check the O_2 sensor voltage, because this type of meter draws too much current and damages the sensor. Always use a digital voltmeter to check the O_2 sensor. The sensor threads should be coated with an antiseize compound prior to sensor installation, or sensor removal may be difficult the next time it is necessary.

Titania-Type O_2 Sensor

Some vehicles are now equipped with a titania-type O_2 sensor. This sensor design is similar to the zirconia-type sensor, but the sensing element is made from titania rather than zirconia (Figure 9-28).

The titania-type sensor modifies voltage, whereas the zirconia-type sensor generates voltage. The computer supplies battery voltage to the titania-type sensor, and this voltage is lowered by a resistor in the circuit. The resistance of the titania varies as the air-fuel ratio cycles from rich to lean. When the air-fuel ratio is rich, the titania resistance is low. This provides a higher voltage signal to the computer. If the air-fuel ratio is lean, the titania resistance is high, and a lower voltage signal is sent to the computer (Figure 9-29).

The titania-type O_2 sensor provides a satisfactory signal almost immediately after a cold engine is started. This action provides improved air-fuel ratio control during engine warmup.

Engine Coolant Temperature Sensor

The engine coolant temperature (ECT) sensor is often threaded into the intake manifold, and the lower end of the sensor is immersed in engine coolant. A typical ECT sensor contains a thermistor that provides high resistance when cold and much lower resistance at higher temperatures. A typical ECT sensor may have 269,000 Ω at -40°F (-40°C) and 1,200 Ω at 248°F (120°C). The ECT sen-

The **engine coolant temperature (ECT) sensor** sends an analog voltage signal to the computer in relation to coolant temperature. This signal is high at low coolant temperature and low at high coolant temperature.

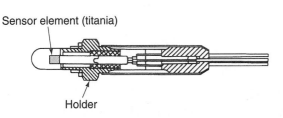

Sensor element (titania)

Holder

Figure 9-28 Titania-type O_2 sensor. *(Reprinted with the permission of Ford Motor Company)*

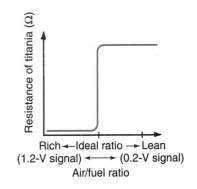

Figure 9-29 Titania-type O_2 sensor resistance and voltage signal. *(Reprinted with the permission of Ford Motor Company)*

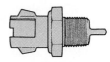

Engine coolant temperature
(ECT) sensor

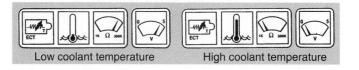

Low coolant temperature   High coolant temperature

Figure 9-30 Engine coolant temperature sensor resistance and voltage drop. *(Reprinted with the permission of Ford Motor Company)*

Shop Manual
Chapter 9, page 410

sor has two wires connected between the sensor and the computer. One of these wires is a voltage signal wire and the other wire provides a ground. The computer senses the voltage drop across the ECT sensor. This voltage changes in relation to the coolant temperature and sensor resistance. For example, at low coolant temperature and high sensor resistance, the voltage drop across the sensor might be 4.5 V, whereas at high coolant temperature and low sensor resistance, this voltage drop would be approximately 0.3 V (Figure 9-30).

The computer must know the coolant temperature to complete many of the necessary decisions regarding output control functions. For example, the computer must provide a richer air-fuel ratio when the engine is cold, and a leaner stoichiometric air-fuel ratio once the engine is at normal operating temperature. Therefore, the computer must know engine coolant temperature from the ECT signal to provide the correct air-fuel ratio.

A cold engine requires more spark advance for improved performance, whereas a hot engine needs less spark advance to prevent detonation. The computer must know engine coolant temperature and other inputs to provide the correct spark advance.

Shop Manual
Chapter 9, page 411

Some emission devices, such as the exhaust gas recirculation (EGR) valve, are not required on a cold engine, but this valve must be operational on a hot engine under certain operating conditions. The computer uses the ECT signal and other inputs to control the EGR valve.

Air Charge Temperature Sensor

The air charge temperature sensor signal is very similar to the engine coolant temperature sensor signal.

Some **air charge temperature (ACT) sensors** are threaded into the intake manifold, and the lower end of the sensor protrudes into one of the air passages in the intake manifold. On some applications, the ACT sensor is mounted in the air cleaner and senses air intake temperature at this location. The air charge temperature sensor contains a thermistor that has ohm and voltage drop readings that are similar to the ECT sensor (Figure 9-31).

The ACT sensor may be referred to as a charge temperature sensor or a manifold air temperature (MAT) sensor.

A voltage signal wire and a ground wire are connected between the air charge temperature sensor and the computer. Cold intake air is denser; therefore, a richer air-fuel ratio is required. When the ACT sensor signal indicates colder intake air temperature, the computer provides a richer air-fuel ratio.

Throttle Position Sensor

The throttle position sensor analog voltage signal increases in relation to throttle opening.

A **rotary-type throttle position sensor (TPS)** contains a potentiometer with a pointer that is rotated by the throttle shaft. Therefore, the TPS is mounted on the end of the throttle shaft at the carburetor on computer-controlled carburetor systems. On computer-controlled fuel injection systems, this sensor is positioned on the end of the throttle shaft in the throttle body. On some General Motors computer-controlled carburetor systems, the TPS is located inside the carburetor, and then it

Air charge temperature
(ACT) sensor

Low air temperature   High air temperature

Figure 9-31 Air charge temperature (ACT) sensor resistance and voltage drop. *(Reprinted with the permission of Ford Motor Company)*

is operated by the accelerator pump linkage. This linear type of TPS has a pointer moving up and down on a variable resistor.

Three wires are connected from the TPS to the computer. When the ignition switch is on, the computer supplies a constant 5-V reference voltage to the sensor on one of these wires. One of the other wires is a TPS signal wire from the sensor to the computer, and the third wire is a ground wire between these two components. Sensor ground wires are usually black, or black with a colored tracer.

A typical TPS has 1,000 Ω with the throttle in the idle position, and 4,000 Ω at wide-open throttle. The voltage signal from a typical TPS is 0.5 V to 1 V at idle and 4.5 V at wide-open throttle. This signal informs the computer regarding the exact throttle position (Figure 9-32).

The computer also knows how fast the throttle is opened from the TPS signal. When the engine is accelerated suddenly, a richer air-fuel ratio is required with the additional airflow into the engine. If the computer receives a TPS signal indicating sudden acceleration, the computer supplies the necessary richer air-fuel ratio. The computer also uses the TPS signal to control other outputs.

Elongated mounting holes on some TPSs provide a sensor adjustment. When a TPS adjustment is necessary, the mounting bolts are loosened and the sensor is rotated until the specified signal voltage is available with the engine idling. On many TPSs there is no provision for sensor adjustment.

A rotary-type throttle position sensor (TPS) has a potentiometer with a pointer that is rotated around the potentiometer.

A **linear-type TPS** has a pointer that moves vertically or horizontally on a variable resistor.

Shop Manual
Chapter 9, page 411

Manifold Absolute Pressure Sensor

The **manifold absolute pressure (MAP) sensor** is usually mounted in the engine compartment. A hose is connected from the intake manifold to a vacuum inlet on the MAP sensor, and three wires

Some manifold absolute pressure sensors produce a digital voltage with a continually varying frequency. This frequency increases as throttle opening and engine load increase.

Throttle position (TP) sensor

Closed throttle   Wide open throttle

Figure 9-32 Throttle position sensor (TPS) resistance and voltage signal. *(Reprinted with the permission of Ford Motor Company)*

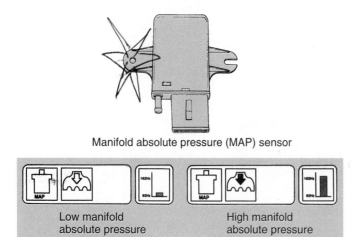

Manifold absolute pressure (MAP) sensor

| Low manifold absolute pressure | High manifold absolute pressure |

Figure 9-33 Manifold absolute pressure (MAP) sensor and its voltage signal with varying frequency in response to manifold pressure. *(Reprinted with the permission of Ford Motor Company)*

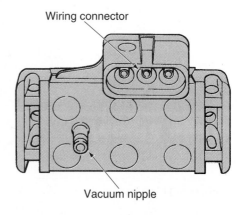

Wiring connector

Vacuum nipple

Figure 9-34 Manifold absolute pressure sensor with a silicon diaphragm. *(Courtesy of Chrysler Corporation)*

Other manifold absolute pressure sensors produce an analog voltage signal in relation to intake manifold vacuum. This signal is low when the intake manifold vacuum is high, and the signal increases as the vacuum decreases in relation to engine load.

Shop Manual
Chapter 9, page 423

are connected from the sensor to the computer. The computer supplies a constant 5-V reference voltage to the sensor on one of these wires, and the other wires are a signal wire and a ground wire. These wiring connections are the same for most MAP sensors.

Ford MAP sensors contain a pressure-sensitive disk capacitor. This type of MAP sensor changes manifold pressure to a digital voltage signal of varying frequency. The MAP sensor actually senses the difference between atmospheric pressure and manifold vacuum. When the engine is idling, high intake manifold vacuum is about 18 in. Hg. Under this condition, the MAP sensor signal is approximately 109 hertz. Ford refers to this condition as low MAP, because there is a greater difference between atmospheric pressure and manifold vacuum (Figure 9-33).

If the engine is operating at, or near, wide-open throttle, the manifold vacuum may be 2 in. Hg, and the MAP sensor hertz approximates 153. This condition may be called high MAP, because the manifold vacuum is closer to atmospheric pressure.

WARNING: Never connect any type of voltmeter directly to the voltage signal wire on a Ford MAP sensor. This action causes MAP sensor damage!

Never connect any type of voltmeter to the signal wire on a Ford MAP sensor with the ignition switch on, or the MAP sensor will be damaged. Ford MAP sensor testers read the sensor hertz directly, and other Ford MAP sensor testers convert the hertz to a voltage signal.

The computer uses the MAP sensor signal to determine the engine load. When the MAP sensor signal indicates wide-open throttle, heavy load conditions, the computer provides a richer air-fuel ratio. The computer supplies a leaner air-fuel ratio if the MAP sensor signal indicates light load, moderate cruising speed conditions.

When the ignition switch is turned on before the engine is started, many MAP sensors act as a barometric pressure sensor. Under this condition, the MAP sensor signal informs the computer regarding atmospheric pressure that varies in relation to altitude and atmospheric conditions such as the humidity. Some applications have a separate barometric pressure sensor.

Other car manufacturers use MAP sensors that contain a silicon diaphragm. These sensors have the same three wires as the Ford MAP sensors (Figure 9-34).

When the manifold vacuum stresses the silicon diaphragm, a voltage signal is sent from the sensor to the computer in relation to the amount of vacuum. On a typical MAP sensor, the signal voltage changes from 1 V to 1.5 V at idle to 4.5 V at wide-open throttle. The signal voltage may be checked with a digital voltmeter on MAP sensors with a silicon diaphragm.

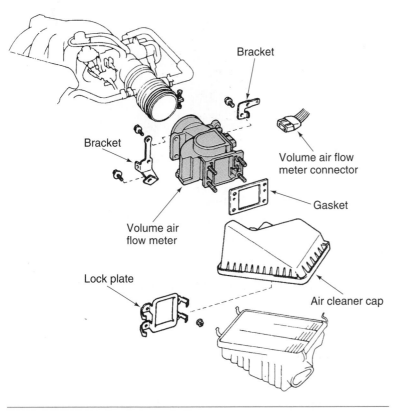

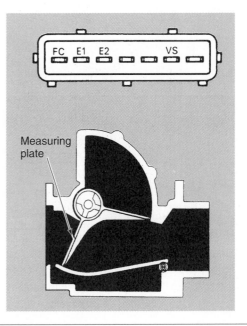

Figure 9-36 The measuring plate movement in a MAF sensor is proportional to intake air flow.

Figure 9-35 A vane-type mass air flow (MAF) sensor which may be referred to as a volume air flow meter.

Mass Air Flow Sensors

Vane-Type Mass Air Flow Sensor. Domestic and off-shore vehicle manufacturers commonly use mass air flow (MAF) sensors in electronic fuel injection (EFI) systems, but computer-controlled carburetor systems usually have a MAP sensor. MAF sensors may be classified as vane-type, heated grid-type, or hot wire-type. The MAF sensor is often mounted in the hose between the air cleaner and the throttle body so that all the intake air must flow through the sensor. Some vane-type MAF sensors are referred to as volume air flow meters (Figure 9-35).

In a vane-type MAF sensor, a pivoted air measuring plate is lightly spring-loaded in the closed position. As the intake air flows through the sensor, the air measuring plate moves toward the open position. This plate movement is proportional to intake airflow (Figure 9-36).

A movable pointer is attached to the measuring plate shaft, and this pointer contacts a resistor. The movable pointer and resistor form a potentiometer, which sends a voltage signal to the power train control module (PCM) in relation to measuring plate movement and intake airflow. A thermistor in the air flow meter sends a signal to the PCM in relation to air intake temperature (Figure 9-37).

The **mass air flow (MAF) sensor** sends a voltage signal to the computer in relation to the total volume of air entering the engine.

Shop Manual
Chapter 9, page 428

Heated Resistor-Type MAF Sensor. A heated resistor is mounted in the center of the air passage in some MAF sensors, and an electronic module is mounted on the side of the sensor (Figure 9-38). When the ignition switch is turned on, voltage is supplied to the module, and the module sends enough current through the resistor to maintain a specific resistor temperature.

If a cold engine is accelerated suddenly, the rush of cold air tries to cool the resistor. Under this condition, the module supplies more current to maintain the resistor temperature. The module sends the increasing current signal to the PCM, and this signal is proportional to the airflow entering the engine. When the PCM receives this increasing current flow signal from the MAF module, the PCM supplies a richer air-fuel ratio to go with the additional cold airflow entering the engine.

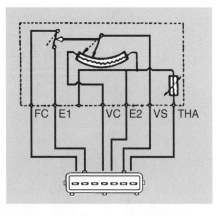

Figure 9-37 Potentiometer and thermistor in a MAF sensor.

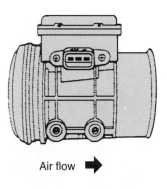

Air flow ➡

Figure 9-38 Heated resistor-type MAF sensor and module. *(Reprinted with the permission of Ford Motor Company)*

The MAF module reacts in a few milliseconds to maintain the resistor temperature. Some MAF sensors have an electric grid in place of the resistor.

Hot Wire-Type MAF Sensor. In a hot wire-type MAF sensor, a hot wire is positioned in the airstream through the sensor, and an ambient temperature sensor wire is located beside the hot wire. This ambient temperature sensor wire senses intake air temperature, and this wire may be referred to as a cold wire (Figure 9-39).

When the ignition switch is turned on, the module in the MAF sensor sends enough current through the hot wire to maintain the temperature of this wire 392°F (200°C) above the ambient temperature sensed by the cold wire. If the engine is accelerated suddenly, the rush of cold air tries to cool the hot wire. The module immediately sends more current through the wire to maintain the wire temperature at 392°F (200°C) above the cold wire temperature. The module sends this increasing current signal to the PCM. This signal is directly proportional to the intake airflow. When this signal is received, the PCM supplies more fuel to go with the increased intake airflow, and this action maintains the exact air-fuel ratio required by the engine. Some MAF sensors have a burn-off relay and related circuit. When the ignition switch is turned off after the engine has been running for a specific length of time, the computer closes the burn-off relay. This relay activates a burn-off circuit in the MAF, which heats the hot wire to a very high temperature for a short time period. This action burns contaminants off the hot wire.

Knock Sensor

The **knock sensor** changes a vibration to an analog voltage signal and sends this signal to the computer.

The knock sensor is threaded into the engine block, intake manifold, or cylinder head. In some applications, the knock sensor is called a detonation sensor. Many engines now contain two knock

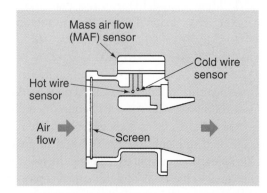

Figure 9-39 Hot wire-type MAF sensor. *(Reprinted with the permission of Ford Motor Company)*

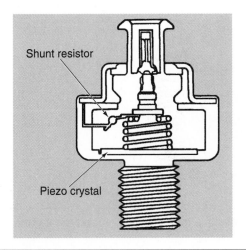

Figure 9-40 Knock sensor with a piezoelectric sensing device and parallel resistor. *(Courtesy of Oldsmobile Division—GMC)*

Knock sensor (KS)

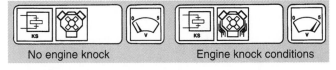

Figure 9-41 Knock sensor and related voltage signal. *(Reprinted with the permission of Ford Motor Company)*

sensors for improved detonation control. The knock sensor contains a piezoelectric sensing element that changes a vibration to a voltage signal (Figure 9-40). An internal resistor is connected in parallel with the piezoelectric sensing element.

When the engine detonates, a vibration is present in the engine block and cylinder head castings. The knock sensor changes this vibration to a voltage signal and sends the signal to the computer. When this signal is received, the computer reduces spark advance to eliminate the detonation. A typical knock sensor signal would be 300 millivolts (mV) to 500 mV depending on the severity of detonation (Figure 9-41). On some General Motors cars and light duty trucks, the knock sensor signal is sent to an electronic spark control (ESC) module, which changes the analog sensor signal to a digital signal.

Exhaust Gas Recirculation Valve Position Sensor

CAUTION: The EGR valve and EVP sensor may be very hot if the engine has been running. Wear protective gloves to avoid burns if it is necessary to remove and handle these components after the engine has been running.

The exhaust gas recirculation valve position (EVP) sensor is mounted on top of the EGR valve. A linear potentiometer is mounted in the EVP sensor, and a stem extends from this potentiometer to the top of the EGR valve. When the EGR valve opens, the potentiometer stem moves upward, and a higher voltage signal is sent to the computer (Figure 9-42).

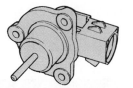

EGR valve position (EVP) sensor

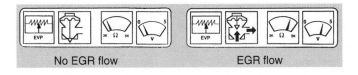

No EGR flow   EGR flow

Figure 9-42 EVP sensor and related voltage signal. *(Reprinted with the permission of Ford Motor Company)*

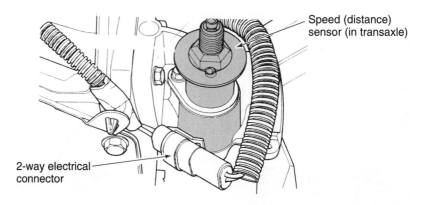

Speed (distance)
sensor (in transaxle)

2-way electrical
connector

Figure 9-43 Vehicle speed sensor (VSS) mounted in a transaxle case. *(Courtesy of Chrysler Corporation)*

The EVP sensor acts as a feedback signal to inform the computer regarding the EGR valve position. Resistance in the EVP linear potentiometer changes from 3,000 Ω with the EGR valve closed to 5,000 Ω when the EGR valve is open. The voltage signal from the EVP sensor changes from 0.8 V with the EGR valve closed to 4.5 V with the EGR valve open. A 5-V reference wire, a ground wire, and a signal wire are connected from the EVP sensor to the computer. The EVP sensor is used on some Ford products; however, other manufacturers have similar sensors.

Vehicle Speed Sensor

The vehicle speed sensor (VSS) is connected in the speedometer cable or mounted in the transaxle opening where the speedometer cable was located previously (Figure 9-43). In the latter case, the speedometer cable is connected to the VSS. The speedometer cable is not required if the vehicle has an electronic speedometer.

In some VSSs the speedometer drive rotates a magnet inside a coil of wire. This type of sensor produces an ac voltage signal proportional to vehicle speed. In some other VSSs, a magnet with eight poles rotates past a set of reed points, and these rotating poles open the points eight times per sensor revolution. Each time the points open, a signal is sent to the computer. The computer uses the VSS signal for converter clutch lockup and cruise control operation. In some applications, the VSS is located in the speedometer head.

Neutral Drive Switch

The neutral drive switch (NDS) is operated by the transmission linkage. If the transmission is in park or neutral, the NDS is closed. When the transmission is placed in drive or reverse, the NDS opens. A closed NDS sends a voltage signal below 1 V to the computer, whereas an open NDS provides a signal above 5 V to the computer (Figure 9-44).

The NDS switch signal informs the computer regarding gear shift selector position, and the computer uses this signal to control idle speed. The NDS may be referred to as a neutral-park switch.

The **exhaust gas recirculation valve position (EVP) sensor** sends an analog voltage signal to the computer in relation to EGR valve opening.

The **vehicle speed sensor (VSS)** sends a voltage signal to the computer in relation to vehicle speed.

The **neutral drive switch (NDS)** sends a voltage signal to the computer in relation to gear selector position.

A BIT OF HISTORY

Computer technology is evolving rapidly. In 1980, the average computer could handle about 150,000 instructions per second. At the present time, computers can handle over 1 million instructions per second, and by the year 2000 it is predicted that computers will be able to process 7 million instructions per second.

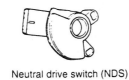

Neutral drive switch (NDS)

Transmission in park or neutral – switch closes, low voltage signal

Transmission in drive or reverse – switch opens, high voltage signal

Figure 9-44 Neutral-drive switch and related voltage signal. *(Reprinted with the permission of Ford Motor Company)*

Summary

❏ Reference voltage (Vref) sensors provide input to the computer by modifying or controlling a constant, predetermined voltage signal.

❏ Most sensors presently in use are variable resistors or potentiometers. They modify a voltage to or from the computer, indicating a constantly changing status that can be calculated, compensated for, and modified.

❏ A thermistor is a solid-state variable resistor made from a semiconductor material that changes resistance in relation to temperature changes.

❏ There are two basic types of thermistors: NTC and PTC. A negative temperature coefficient (NTC) thermistor reduces its resistance as temperature increases. This type is the most commonly used thermistor. A positive temperature coefficient (PTC) thermistor increases its resistance with an increase in temperature.

❏ Voltage generating sensors include components like the magnetic pulse generators, Hall effect switches, oxygen sensors (zirconium dioxide), and knock sensors (piezoelectric), which are capable of producing their own input voltage signal.

❏ An analog voltage signal is continually variable within a certain range.

❏ A digital voltage signal is either high or low.

❏ The process of assigning numeric values to digital voltage signals is called binary coding.

❏ Some input sensor signals must be amplified by the amplification circuit in the computer input conditioning chip.

❏ Since the input sensors produce analog voltage signals and the computer operates on digital signals, an analog/digital (A/D) signal conversion must be completed by the A/D converter in the input conditioning chip.

❏ The microprocessor chip is the calculating and decision-making chip in a computer.

❏ The microprocessor can write information into a random access memory (RAM) or read information from this memory.

❏ A RAM may have a volatile or nonvolatile memory.

❏ Some of the computer program is permanently programmed into the read only memory (ROM), and the microprocessor can only read information from this memory.

❏ Some computers have a removable programmable read only memory (PROM), which contains specific programs such as the spark advance program.

Terms to Know

Adaptive strategy

Air charge temperature (ACT) sensor

Analog voltage signal

Analog/digital conversion

Binary code

Digital voltage signal

Duty cycle

Electronically erasable programmable read only memory (EEPROM)

Engine coolant temperature (ECT) sensor

Exhaust gas recirculation valve position (EVP) sensor

Feedback

Information retrieval

Information storage

Input signal amplification

Keep alive memory (KAM)

Knock sensor

Linear TPS

Manifold absolute pressure (MAP) sensor

Mass air flow (MAF) sensor

❏ A keep alive memory (KAM) is erased when battery voltage is removed from the computer.

❏ Adaptive strategy allows a computer to adapt to minor defects in the computer inputs and outputs.

❏ The computer output drivers provide a ground to operate most output actuators.

❏ Once the computer's programming instructs that a correction or adjustment must be made in the controlled system, an output signal is sent to control devices called actuators. These actuators, which are solenoids, switchers, relays, or motors, physically act or carry out the command sent by the computer.

❏ In order for the computer to control an actuator, it must rely on a component called an output driver. These drivers usually apply the ground circuit of the actuator.

❏ Some systems require the actuator to either be turned on and off very rapidly or for a set amount of cycles per second. It is duty cycled if it is turned on and off a set amount of cycles per second. Duty cycle is the percentage of on-time to total cycle time.

❏ If the actuator is cycled on and off very rapidly, the pulse width can be varied to provide the desired results. Pulse width is the length of time in milliseconds that an actuator is energized.

❏ Oxygen (O₂) sensors may be made from zirconia or titania.

❏ The voltage signal from the O₂ sensor increases as the air-fuel ratio becomes richer, and the computer adjusts the air-fuel ratio in response to this signal.

❏ O₂ sensors may be contaminated with leaded gasoline, some RTV sealants, or antifreeze entering the combustion chambers.

❏ O₂ sensor damage occurs if the sensor is tested with an analog voltmeter.

❏ The engine coolant temperature (ECT) sensor resistance increases as the coolant temperature decreases.

❏ The computer senses the voltage drop across the (ECT) sensor.

❏ The air charge temperature (ACT) sensor has ohm and voltage drop values that are similar to those of the engine coolant temperature sensor.

❏ The throttle position sensor (TPS) sends a voltage signal to the computer in relation to the amount and speed of throttle opening.

❏ The manifold absolute pressure (MAP) sensor sends a voltage signal to the computer in relation to manifold vacuum.

❏ Some MAP sensors provide an ordinary analog voltage signal to the computer, whereas other MAP sensors produce a digital voltage signal of varying frequency.

❏ Mass air flow (MAF) sensors send a signal to the computer in relation to the total volume of air entering the engine.

❏ MAF sensors may be vane-type, heated resistor-type, or hot wire-type.

❏ The knock sensor changes the vibration caused by engine detonation to a voltage signal.

❏ The exhaust gas recirculation valve position (EVP) sensor informs the computer regarding the EGR valve position.

❏ The vehicle speed sensor (VSS) sends a voltage signal to the computer in relation to vehicle speed.

❏ The neutral drive switch informs the computer regarding gear selector position.

Review Questions

Short Answer Essays

1. Compare a digital voltage signal with an analog signal.

2. Explain the purpose of a read only memory (ROM) chip in an automotive computer.

3. Describe the differences between a programmable read only memory (PROM) and an electronically erasable programmable read only memory (EEPROM).

4. Explain adaptive strategy in an automotive computer, including the conditions required for a learning period.

5. Describe how the computer output drivers operate most actuators.

6. Explain the conditions that cause an oxygen (O_2) sensor to produce high and low voltage.

7. Explain the advantage of the titania-type O_2 sensor compared to the zirconia-type O_2 sensor.

8. What is meant by the term pulse width?

9. Explain the operation of a hot wire-type mass air flow (MAF) sensor.

10. Describe the voltage signals from the neutral drive switch to the computer with this switch open and closed.

Fill-in-the-Blanks

1. Most computer input sensors produce _____ voltage signals.

2. A digital voltage signal may be called a _____ _____ signal.

3. The term binary means _____ values.

4. A random access memory may have a _____ or _____ memory.

5. _____ means data concerning the effects of the computer's commands are fed back to the computer as an input.

6. When the air-fuel ratio is lean, the oxygen (O_2) sensor voltage is _____.

7. If the coolant temperature increases, the voltage drop across the engine coolant temperature (ECT) sensor _____.

8. The three wires connected to a manifold absolute pressure (MAP) sensor are the ground wire, signal wire, and _____ _____ wire.

9. In a vane-type mass air flow (MAF) sensor, the pointer on the potentiometer is connected to the _____ _____.

10. In some MAF sensors, the burn-off circuit is activated when the ignition switch is turned _____.

ASE Style Review Questions

1. While discussing voltage signals:
   Technician A says an analog voltage signal is continuously variable.
   Technician B says most input sensors produce analog voltage signals.
   Who is correct?
   - **A.** A only
   - **B.** B only
   - **C.** Both A and B
   - **D.** Neither A nor B

2. While discussing voltage signals:
   Technician A says a digital voltage signal is either high or low.
   Technician B says a digital signal must be converted to an analog signal before the computer can use it.
   Who is correct?
   - **A.** A only
   - **B.** B only
   - **C.** Both A and B
   - **D.** Neither A nor B

3. While discussing the assignment of numeric values to digital signals:
   Technician A says the assignment of numeric values to digital voltage signals is called binary coding.
   Technician B says the word binary means three values.
   Who is correct?
   - **A.** A only
   - **B.** B only
   - **C.** Both A and B
   - **D.** Neither A nor B

4. While discussing input signal amplification:
   Technician A says all the sensor signals must be amplified.
   Technician B says the random access memory (RAM) chip provides input signal amplification.
   Who is correct?
   - **A.** A only
   - **B.** B only
   - **C.** Both A and B
   - **D.** Neither A nor B

5. While discussing the conversion of analog signals to digital signals:
   Technician A says the analog input signals are changed to digital signals in the read only memory (ROM).
   Technician B says this analog to digital (A/D) conversion is done by the keep alive memory (KAM).
   Who is correct?
   - **A.** A only
   - **B.** B only
   - **C.** Both A and B
   - **D.** Neither A nor B

6. While discussing computer memory chips:
   Technician A says the microprocessor can write information into the RAM.
   Technician B says the microprocessor can read information from the RAM.
   Who is correct?
   - **A.** A only
   - **B.** B only
   - **C.** Both A and B
   - **D.** Neither A nor B

7. While discussing computer memory chips:
   Technician A says the microprocessor can erase ROM information.
   Technician B says information in the ROM is erased when the ignition switch is turned off.
   Who is correct?
   - **A.** A only
   - **B.** B only
   - **C.** Both A and B
   - **D.** Neither A nor B

8. While discussing input sensors:
   Technician A says the engine coolant temperature sensor resistance increases when the coolant temperature becomes higher.
   Technician B says the engine coolant temperature sensor voltage drop decreases as the coolant temperature increases.
   Who is correct?
   - **A.** A only
   - **B.** B only
   - **C.** Both A and B
   - **D.** Neither A nor B

9. While discussing input sensors:
   Technician A says the throttle position sensor voltage signal increases in relation to throttle opening.
   Technician B says the throttle position sensor voltage signal informs the computer how fast the throttle was opened.
   Who is correct?
   - **A.** A only
   - **B.** B only
   - **C.** Both A and B
   - **D.** Neither A nor B

10. While discussing hot wire-type mass air flow sensors:
    Technician A says the sensor module varies the temperature of the hot wire.
    Technician B says the sensor module maintains the hot wire at a specific temperature.
    Who is correct?
    - **A.** A only
    - **B.** B only
    - **C.** Both A and B
    - **D.** Neither A nor B

Computer-Controlled Carburetors

Upon completion and review of this chapter, you should be able to:

❏ Explain the reason for installing computer-controlled carburetor systems on vehicles made in the late 1970s and 1980s.

❏ Describe the operation of a nonadjustable oxygen feedback solenoid.

❏ Explain the term duty cycle as it relates to an oxygen feedback solenoid.

❏ Describe how a nonadjustable oxygen feedback solenoid controls idle air-fuel ratio.

❏ Explain how the computer operates a stepper motor to control air-fuel ratio.

❏ Describe the two adjustments on the adjustable mixture control solenoid.

❏ Define open loop and closed loop in relation to computer-controlled carburetor systems.

❏ Explain the input sensor signals required to enter closed loop.

❏ Describe the operation of the idle speed control motor on a computer-controlled carburetor system.

❏ Explain the operation of a vacuum-operated throttle kicker and solenoid.

❏ Describe the operation of a temperature-compensated pump (TCP) in a computer-controlled carburetor.

❏ Explain the operation of a variable voltage choke (VVC) relay in a computer-controlled carburetor.

❏ Explain the operation of a vacuum-operated decel valve.

❏ Describe the operation of a throttle kicker and idle stop solenoid.

❏ Explain the operation of an idle air control motor, including the idle contact switch.

❏ Describe the inputs used by the computer to control the idle air control and idle air.

Introduction

The carburetor must supply the proper air-fuel ratio under all engine operating conditions. This job sounds relatively simple, but when one considers the tremendous variation in engine operating conditions, the job becomes extremely complex. For example, the engine may be idling and a second later it may be operating at wide-open throttle. The carburetor must supply the proper air-fuel ratio for both these operating conditions. During severely cold starting conditions, the carburetor must supply a very rich air-fuel ratio, and a few minutes after the cold start the carburetor must supply the proper air-fuel ratio when the engine is at normal operating temperature. The carburetor also plays a significant role in meeting emission standards and supplying a reasonable amount of fuel economy. The last generation of carburetors is the computer-controlled carburetor.

In the late 1970s and early 1980s, automotive engineers were faced with the problems of increasingly stringent emission standards and **corporate average fuel economy (CAFE)** requirements imposed by the United States Congress. These CAFE standards demanded the following fuel mileage for passenger cars:

Computer-controlled carburetors are called feedback carburetors.

Model Year	MPG	Model Year	MPG
1978	18	1984	27
1979	19	1985	27.5
1980	20	1986	26
1981	22	1987	27.5
1982	24	1988	26
1983	26	1989 to present	27.5

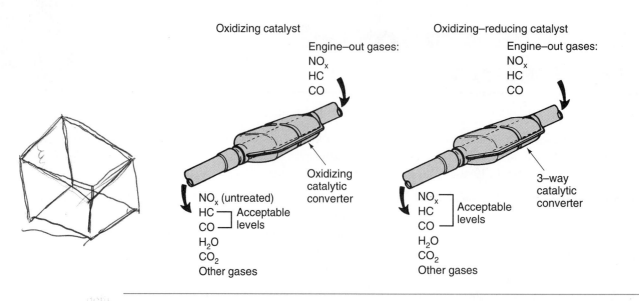

Figure 10-1 Oxidizing and three-way catalytic converters. *(Courtesy of Chrysler Corporation)*

During this time period, engineers were mainly concerned with the reduction of carbon monoxide (CO), hydrocarbons (HC), and oxides of nitrogen (NO_x) emissions. When three-way catalytic converters were installed, engineers discovered that air-fuel ratios must be maintained very close to the stoichiometric ratio of 14.7:1 in order for the converters to be effective in lowering emission levels. If the air-fuel ratio is richer than stoichiometric, HC and CO levels are high and the converter cannot lower these emission levels to the desired limit.

With a rich air-fuel ratio, the combustion temperature is lowered, and since NO_x emissions are caused by high combustion chamber temperatures, the converter is very effective in controlling NO_x emissions under this condition.

When the air-fuel ratio is leaner than the stoichiometric ratio, the levels of CO and HC are low and the converter is very effective in controlling these pollutants. However, a lean air-fuel ratio burns hotter in the combustion chambers and NO_x emissions become very high. Under this condition, the converter is ineffective in controlling NO_x emissions. Therefore, the air-fuel ratio must be controlled at, or very close to, the stoichiometric ratio for the three-way converter to reduce all three emission levels to the desired limit (Figure 10-1).

Exhaust Gas Oxygen Sensor

It was discovered that conventional carburetors did not provide accurate air-fuel ratio control under all conditions; therefore, the three-way converter was not effective in reducing emission levels on engines with these carburetors. Computer-controlled carburetors were designed to maintain the air-fuel ratio at, or near, the stoichiometric ratio under most operating conditions, which allows the catalytic converter to provide effective control of exhaust emissions. Monitoring the air-fuel ratio is the job of the exhaust gas oxygen sensor (Figure 10-2).

An oxygen sensor senses the amount of oxygen present in the exhaust stream. A lean mixture produces a high level of oxygen in the exhaust. A rich mixture produces little oxygen in the exhaust. The oxygen sensor, placed in the exhaust before the catalytic converter, produces a voltage signal that varies with the amount of oxygen the sensor detects in the exhaust. If the oxygen level is high, the voltage output is low. If the oxygen level is low, the voltage output is high.

The electrical output of the oxygen sensor is monitored by an electronic control unit (PCM). This microprocessor is programmed to interpret the input signals from the sensor and in turn generate output signals to a mixture control device that meters more or less fuel into the air charge as it is needed to maintain the 14.7:1 ratio.

Manufacturers used different names to describe their computer-controlled carburetor systems. General Motors referred to theirs as computer command control (3C) systems. Chrysler Corporation refers to their computer-controlled carburetor systems as oxygen (O_2) feedback systems. Ford Motor Company refers to their computer-controlled carburetor systems as microprocessor control unit (MCU) systems, electronic engine control III (EEC III), or electronic engine control IV (EEC IV) systems.

Low O_2 sensor voltage means a lean condition, whereas a high O_2 sensor voltage means a rich condition.

Figure 10-2 Exhaust gas oxygen sensor.

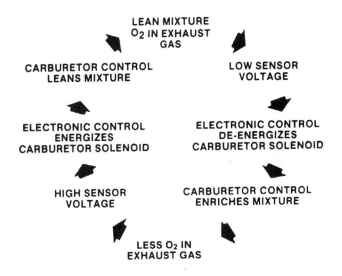

Figure 10-3 Closed loop operation.

Open Loop vs. Closed Loop

Whenever these components are working to control the air-fuel ratio, the carburetor is said to be operating in **closed loop**. Closed loop is illustrated in Figure 10-3. The oxygen sensor is constantly monitoring the oxygen in the exhaust and the control module is constantly making adjustments to the air-fuel mixture based on the fluctuations in the sensor's voltage output. However, there are certain conditions under which the control module ignores the signals from the oxygen sensor and does not regulate the ratio of fuel to air. During these times, the carburetor is functioning in a conventional manner and is said to be operating in **open loop**.

Shop Manual
Chapter 10, page 448

The carburetor operates in open loop until the oxygen sensor reaches a certain temperature (approximately 600°F). The carburetor also goes into open loop when a richer than normal air-fuel mixture is required, such as during warmup and heavy throttle application. Several other sensors are needed to alert the electronic control module of these conditions.

Open loop also occurs when the engine coolant is cold. Under this condition, the O_2 sensor is too cold to produce a satisfactory signal, and the computer program controls the air-fuel ratio without the O_2 sensor input. During open loop mode, the computer provides a rich air-fuel ratio. Since a computer-controlled carburetor also has a conventional choke, a richer air-fuel ratio is supplied.

As the engine approaches normal operating temperature, the computer enters the closed loop mode in which the computer uses the O_2 sensor signal to control the air-fuel ratio. Closed loop in many computer-controlled carburetor systems occurs when the engine coolant temperature (ECT) sensor signal informs the computer that the coolant temperature is approximately 175°F (79.4°C) and the O_2 sensor signal is valid.

Therefore, the ECT sensor signal is very important because it determines the open or closed loop status. For example, if the engine thermostat is defective and the coolant temperature never reaches 175°F (79.4°C), the computer-controlled carburetor system never enters closed loop. When this open loop condition occurs, the air-fuel ratio is continually rich and fuel economy is reduced. Some systems go back into open loop during prolonged periods of idle operation when the O_2 sensor cools down. Many systems revert to open loop at, or near, wide-open throttle to provide a richer air-fuel ratio.

Early feedback systems used a vacuum switch to control metering devices on the carburetor. Closed loop signals from the electronic control module are sent to a vacuum solenoid regulator (Figure 10-4), which in turn controls vacuum to a piston and diaphragm assembly in the carburetor. The vacuum diaphragm and a spring above the diaphragm work together to lift and lower a tapered fuel metering rod that moves in and out of an auxiliary fuel jet in the bottom of the fuel

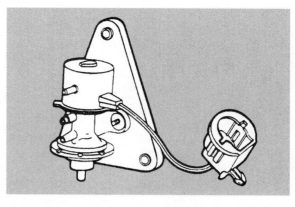

Figure 10-4 Remote-mounted fuel control solenoid.

Shop Manual
Chapter 10, page 448

bowl. The position of the metering rod in the jet controls the amount of fuel allowed to flow into the main fuel well.

The more advanced feedback systems use electrical solenoids on the carburetor to control the metering rods (Figure 10-5). These solenoids are generally referred to as duty-cycle solenoids or **mixture control (M/C) solenoids**. The solenoid is normally wired through the ignition switch and grounded through the electronic control module. The solenoid is energized when the electronic control module completes the ground. The control module is programmed to cycle (turn on and off) the solenoid ten times per second. Each cycle lasts 100 milliseconds. The amount of fuel metered into the main fuel well is determined by how many milliseconds the solenoid is on during each cycle. The solenoid can be on almost 100 percent of the cycle or it can be off nearly 100 percent of the time. The M/C solenoid can control a fuel metering rod, an air bleed, or both.

Nonadjustable Oxygen Feedback Solenoid Systems

Mixture control (MC) solenoid is another name for the O₂ feedback solenoid.

Computer-controlled carburetors have basically the same systems as conventional carburetors except for one important difference. In most computer-controlled carburetors, the conventional vacuum-operated power piston is replaced with an O_2 feedback solenoid. This solenoid contains a movable plunger surrounded by a winding. When the ignition switch is turned on, 12 V is supplied to the O_2 feedback solenoid winding, and the other end of this winding is grounded through the computer. The O_2 feedback solenoid plunger is spring-loaded in the upward position. Under this condition, additional fuel moves past the end of the solenoid plunger into the carburetor main system.

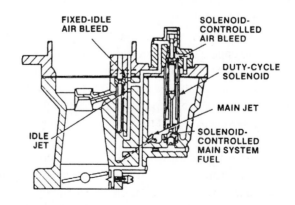

Figure 10-5 Electronic feedback carburetor.

When fuel flows through the main metering jet and the O_2 feedback solenoid plunger into the main system, a rich air-fuel ratio results, causing a high O_2 sensor voltage. When this O_2 sensor signal is received, the computer grounds the O_2 feedback solenoid winding and the magnetism of the winding moves the solenoid plunger downward. Under this condition, the solenoid plunger blocks the additional fuel passage into the main system and the air-fuel ratio becomes leaner.

The lean air-fuel ratio causes a low O_2 sensor voltage signal to be sent to the computer. The computer responds to this signal by opening the O_2 feedback solenoid ground circuit. This action allows the solenoid to move upward, and the air-fuel ratio returns to a rich condition.

Duty Cycle

On some computer-controlled carburetor systems, the computer grounds the O_2 feedback solenoid winding at a constant rate of ten times per second. This rate of O_2 feedback solenoid grounding is referred to as **duty cycle**. Although the duty cycle remains constant, the computer has the capability to change the length of time that the solenoid winding remains grounded. This is referred to as dwell time or on time. For example, if a manifold vacuum leak is making the air-fuel ratio lean, the computer receives a continuously low O_2 sensor voltage. When this continually lean signal is received, the computer provides a rich command to the O_2 feedback solenoid winding. This rich command reduces the O_2 feedback dwell time and leaves the solenoid plunger up longer to try to provide a stoichiometric air-fuel ratio. Under this condition, the computer may leave the solenoid winding off for 90 percent of the time and on for 10 percent of the time on each cycle (Figure 10-6).

The computer operates the O_2 feedback solenoid to maintain the air-fuel ratio at stoichiometric under most operating conditions. If the O_2 sensor sends a continually rich signal to the computer, the computer provides a lean command to the O_2 feedback solenoid. When this action occurs, the computer may keep the solenoid winding on for 90% of the time and off for 10% of the time on each cycle to bring the mixture to the stoichiometric ratio (Figure 10-7).

Duty cycle refers to the number of times per second that the computer grounds the O_2 feedback solenoid.

Idle System Air-Fuel Ratio Control

The computer and the O_2 feedback solenoid also control the air-fuel ratio when the engine is idling. An air passage extends from the top of the solenoid plunger into the idle system (Figure 10-8).

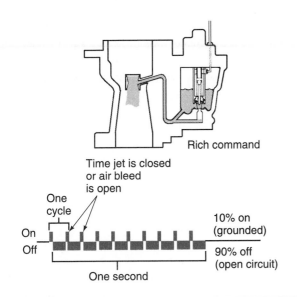

Figure 10-6 Lean oxygen sensor signal and the resultant rich command from the computer to the oxygen feedback solenoid. *(Courtesy of Chrysler Corporation)*

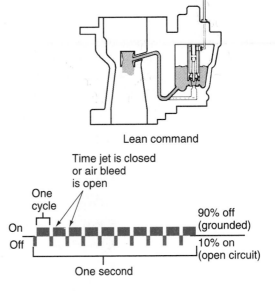

Figure 10-7 Rich oxygen sensor signal and the resultant lean command from the computer to the oxygen feedback solenoid. *(Courtesy of Chrysler Corporation)*

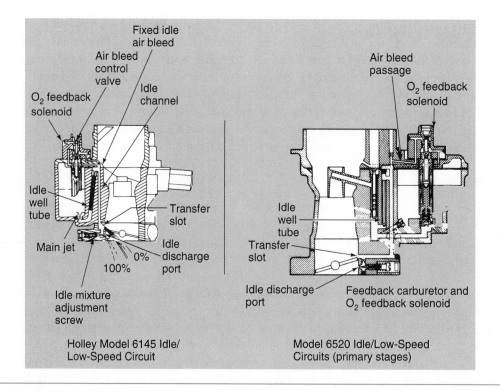

Figure 10-8 Air passage from oxygen feedback solenoid plunger to the idle system. *(Courtesy of Chrysler Corporation)*

When the solenoid plunger is moved upward by the spring, the top of the plunger blocks the air passage into the idle system, which provides a richer idle air-fuel ratio. If the computer grounds the solenoid winding and moves the plunger downward, the air passage is opened past the top of the plunger into the idle system. Under this condition, the idle air-fuel ratio is leaner. At idle speed, the computer operates the O_2 feedback solenoid plunger to maintain the air-fuel ratio at, or near, stoichiometric.

If a defect occurs in the O_2 feedback system so the computer no longer operates the O_2 feedback solenoid, the spring moves the plunger upward and the air-fuel mixture is continually rich. Under this condition, fuel economy and performance are reduced and emission levels increase, but the vehicle can be driven to a service center.

The inputs on the computer-controlled carburetor system with a nonadjustable mixture control solenoid are (Figure 10-9):

1. Vacuum sensor

2. Barometric (BARO) pressure sensor

3. Throttle position sensor (TPS)

4. Coolant sensor

5. Oxygen (O_2) sensor

6. Park-neutral (P/N) switch

7. Fourth-gear switch

8. Vehicle speed sensor (VSS)

The barometric (BARO) pressure sensor is a separate sensor in this system. The signal from this sensor informs the PCM regarding atmospheric pressure changes in relation to weather conditions and altitude. On many systems, the BARO sensor is combined with the manifold absolute pressure (MAP) sensor.

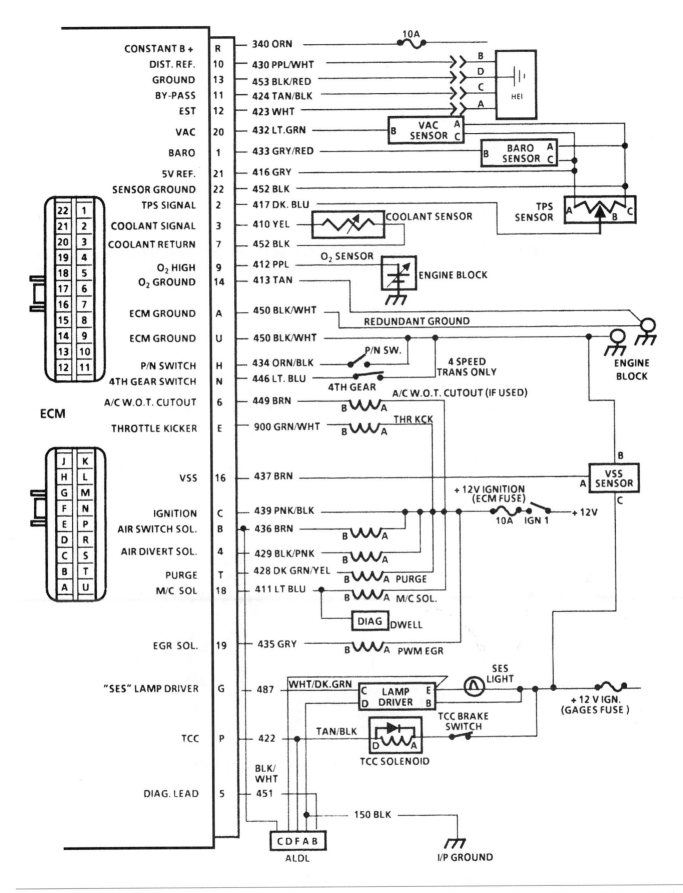

Figure 10-9 GM's 3C system with a nonadjustable mixture control solenoid. *(Courtesy of Chevrolet Motor Division—GMC)*

The outputs in the 3C system with a nonadjustable mixture control solenoid are:

1. A/C wide open throttle (WOT) cutout relay
2. Throttle kicker solenoid
3. Air switch solenoid
4. Air divert solenoid
5. Purge solenoid
6. Mixture control (MC) solenoid
7. EGR solenoid
8. Malfunction indicator lamp (MIL) driver module
9. TCC solenoid

Adjustable Mixture Control Solenoid Systems

Shop Manual
Chapter 10, page 456

Malfunction Indicator Lamp (MIL) is the standard term for Check Engine, Service Engine Soon, and other such warning lights.

The adjustable mixture control solenoid controls the air-fuel ratio in the idle and main systems in the same way as the nonadjustable O_2 feedback solenoid. In many carburetors, the O_2 feedback solenoid is not adjustable; however, in General Motors computer-controlled Dualjet and Quadrajet carburetors, adjustments on the mixture control (MC) solenoid are possible (Figure 10-10). These adjustments are set at the factory and field service adjustment should only be required if major carburetor service is necessary.

A diagnostic dwell connector is attached to wire from the mixture control solenoid to the PCM. This wire is used for diagnostic purposes. In the 3C system, the mixture control solenoid is controlled by the PCM with a duty cycle. This solenoid controls the amount of fuel supplied to the

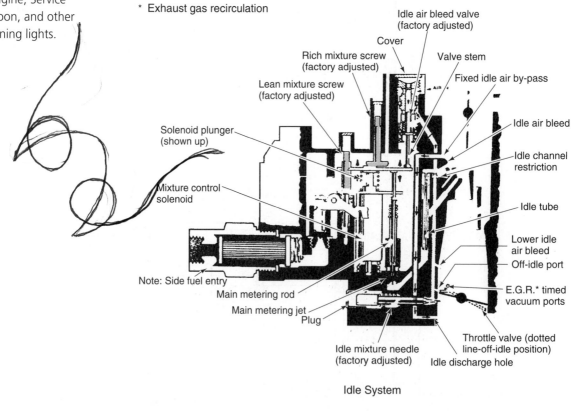

Idle System

Figure 10-10 Mixture control (MC) solenoid adjustments. *(Courtesy of Chevrolet Motor Division—GMC)*

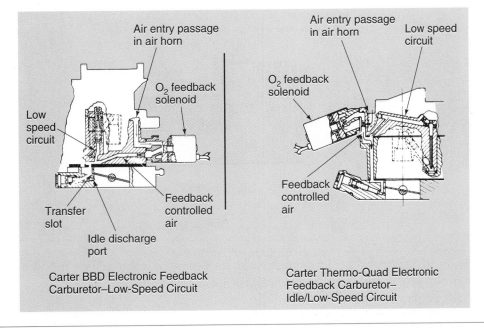

Figure 10-11 Air passage from the O_2 feedback solenoid into the carburetor's idle circuit. *(Courtesy of Chrysler Corporation)*

main system and the airflow to the idle system to provide the correct air-fuel ratio. The malfunction indicator lamp comes on if a defect occurs in the 3C system and a fault code is set in the PCM memory. The MIL lamp is operated by an external lamp driver module mounted under the dash and the PCM. On some other systems, the MIL lamp is operated directly by the PCM.

Air Control Solenoid Systems

Some computer-controlled carburetors have an O_2 feedback solenoid that allows air into the idle system (Figure 10-11) or main system (Figure 10-12) to control the air-fuel ratio.

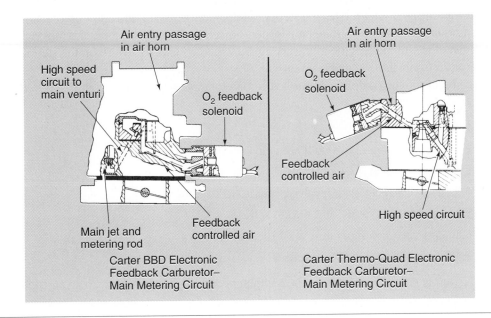

Figure 10-12 Air passage from the O_2 feedback solenoid into the carburetor's main circuit. *(Courtesy of Chrysler Corporation)*

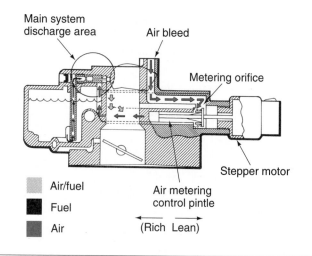

Figure 10-13 Variable venturi carburetor with stepper motor, which controls air flow into the main circuit. *(Reprinted with the permission of Ford Motor Company)*

The computer uses a duty-cycle principle to control this solenoid. When the O_2 sensor signal indicates a rich air-fuel ratio, the computer energizes the solenoid longer, which allows more air into the idle or main system to make the air-fuel ratio leaner. Conversely, if the O_2 sensor signal is low, indicating a lean air-fuel ratio, the computer reduces the solenoid on time and shuts off some of the airflow into the idle or main system to provide a richer air-fuel ratio.

Other computer-controlled carburetors have a **stepper motor** that controls the airflow past a tapered valve on the motor stem into the main system. This stepper motor is popular on some variable venturi carburetors (Figure 10-13).

The stepper motor has four field windings and the travel on the motor stem and valve depends on which field winding the computer grounds. As the computer grounds the various field windings, the motor stem and valve move horizontally in steps with a maximum travel of 0.400 in .(1.01 cm).

A lean O_2 sensor signal causes the computer to move the stepper motor stem and valve inward to close off some airflow to provide a richer air-fuel ratio. If the O_2 sensor indicates a rich air-fuel ratio, the computer moves the stepper motor stem outward and allows more airflow into the main system to make the air-fuel ratio leaner. Since a certain amount of fuel is moving through the main system with the engine idling on a Ford variable venturi carburetor, the stepper motor controls airflow only into the main system.

The inputs on the computer-controlled carburetor system with an air control solenoid follow (Figure 10-14):

Shop Manual
Chapter 10, page 463

1. Knock sensor

2. Air charge temperature (ACT) sensor

3. Throttle position sensor (TPS)

4. Exhaust gas recirculation valve position (EVP) sensor

5. Manifold absolute pressure (MAP) sensor

6. Engine coolant temperature (ECT) sensor

7. Exhaust gas oxygen (EGO) sensor

The output control functions on the computer-controlled carburetor with an air control solenoid are:

1. Temperature compensated pump (TCP) solenoid

2. Feedback carburetor solenoid (FBC)

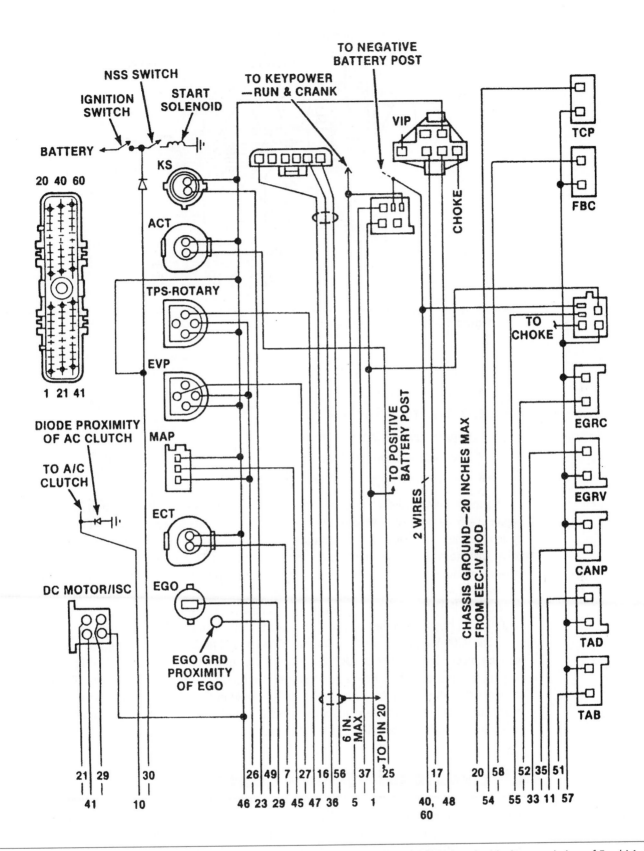

Figure 10-14 Ford EEC-IV system with air-type feedback control solenoid. *(Reprinted with the permission of Ford Motor Company)*

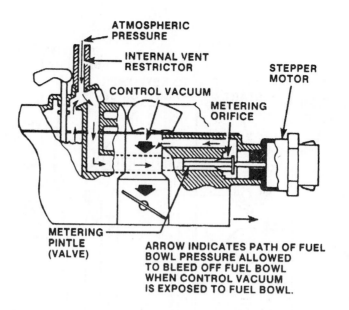

ATMOSPHERIC
PRESSURE

INTERNAL VENT
RESTRICTOR

CONTROL VACUUM

STEPPER
MOTOR

METERING
ORIFICE

METERING
PINTLE
(VALVE)

ARROW INDICATES PATH OF FUEL
BOWL PRESSURE ALLOWED
TO BLEED OFF FUEL BOWL
WHEN CONTROL VACUUM
IS EXPOSED TO FUEL BOWL.

Figure 10-15 Back suction feedback system.

3. Variable voltage choke (VVC) relay

4. EGR control (EGRC) solenoid

5. EGR vent (EGRV) solenoid

6. Canister purge (CANP) solenoid

7. Thermactor air divert (TAD) solenoid

8. Thermactor air bypass (TAB) solenoid

A less common method to control the air-fuel mixture is with a back suction system feedback. Figure 10-15 shows a variable venturi carburetor with an electric stepper motor rather than a duty-cycle solenoid. The back suction system consists of an electric stepper motor, a metering pintle valve, an internal vent restrictor, and a metering orifice. The stepper motor regulates the pintle movement in the metering orifice, thereby varying the area of the opening communicating control vacuum to the fuel bowl. The larger this area is, the leaner the air-fuel mixture. Some of the control vacuum is bled off through the internal vent restrictor. The internal vent restrictor also serves to vent the fuel bowl when the back suction control pintle is in the closed position.

The 7200 VV carburetor was also produced with a feedback stepper motor that controls the main air bleed. The stepper motor controls the pintle movement in the air metering orifice thereby varying the amount of air being metered into the main system discharge area. The greater the amount of air, the leaner the air-fuel mixture is. A hole in the upper body casting of the carburetor allows air from beneath the air cleaner to be channeled into the main system discharge area. The metered air lowers the metering signal at the main fuel metering jets.

Electronic Idle-Speed Control

Shop Manual
Chapter 10, page 450

In order to maintain federally mandated emission levels, it is necessary to control the idle speed. Most feedback systems operate in open loop when the engine is idling. To reduce emissions during idle, most feedback carburetors idle faster and leaner than non-feedback carburetors.

To adjust idle speed, many feedback carburetors have an idle speed control (ISC) motor that is controlled by an electronic control module. The ISC motor is a small, reversible, electric motor. It is part of an assembly that includes the motor, gear drive, and plunger (Figure 10-16). When the

Figure 10-16 An idle speed control motor. *(Reprinted with the permission of Ford Motor Company)*

motor turns in one direction, the gear drive extends the plunger. When the motor turns in the opposite direction, the gear drive retracts the plunger. The ISC motor is mounted so that the plunger can contact the throttle level. The PCM controls the ISC motor and can change the polarity applied to the motor's armature in order to control the direction it turns. When the idle tracking switch is open (throttle closed), the PCM commands the ISC motor to control idle speed. The ISC provides the correct throttle opening for cold or warm engine idle.

According to SAE J1930's list of acceptable terms, the ISC can also be referred to as an IAC (Idle Air Control) motor.

The electronic control module receives input from various switches and sensors to determine the best idle speed. Some of the possible inputs follow.

- ❏ Engine coolant temperature sensor
- ❏ Intake Air Temperature (IAT) sensor
- ❏ Manifold absolute pressure (MAP) sensor
- ❏ Barometric pressure (BARO) sensor
- ❏ Park/neutral or neutral gear switch
- ❏ Clutch engaged switch
- ❏ Power steering pressure switch
- ❏ A/C clutch compressor switch
- ❏ Idle tracking switch (ITS)

Based on the input signals from the system's sensors, the PCM increases the curb idle speed if the coolant is below a specific temperature, if a load (such as air conditioning, transmission, power steering) is placed on the engine, or when the vehicle is operated above a specific altitude.

During closed choke idle, the fast-idle cam holds the throttle blade open enough to lift the throttle linkage off the ISC plunger. This allows the ISC switch to open so that the PCM does not monitor idle speed. As the choke spring allows the fast-idle cam to fall away and the throttle returns to the warm idle position, the PCM notes the still low coolant temperature and commands a slightly higher idle speed.

As the engine warms up, the plunger is retracted by the electronic control module. If the A/C compressor is turned on, the PCM extends the plunger a certain distance to increase engine idle speed to compensate for the added load. When the throttle is opened and the lever leaves contact with the plunger, an idle tracking switch (ITS) in the end of the plunger signals the PCM. The electronic control module then fully extends the plunger where, upon contact with the lever (during deceleration), it acts as a dashpot, slowing the return of the throttle lever. When the engine is shut down, the plunger retracts to prevent the engine from dieseling. It then extends for the next engine startup.

In some systems if the engine starts to overheat, the PCM commands a higher idle speed to increase coolant flow. Also, if system voltage falls below a predetermined value, the PCM commands a higher idle speed in order to increase alternator speed and output.

Other output functions may be affected by the idle tracking switch signal. These output functions may vary depending on the computer programming. Typical output functions affected by the idle contact switch closed signal are:

Upstream air pump flow refers to airflow from the air pump to the exhaust manifold.

1. Air pump system is upstream.
2. Computer is in open loop.
3. Exhaust gas recirculation (EGR) system is disabled.
4. Torque converter clutch is disabled.
5. Spark advance capability is reduced.
6. Overdrive 3-4 shift is disabled.
7. PCM operates the ISC motor to maintain target idle speed.

Typical output functions affected by the idle contact switch open signal are:

Downstream air pump flow refers to airflow from the air pump to the catalytic converter.

1. Air pump system is downstream if the engine temperature is high enough.
2. Computer is in closed loop if the engine temperature is high enough.
3. EGR system is enabled.
4. Torque converter clutch is enabled.
5. Overdrive 3-4 shift is enabled.
6. Spark advance is completely enabled.
7. PCM does not operate ISC motor.

Vacuum-Operated Decel Valves

A vacuum-operated decel valve may be called a gulp valve.

Some engines with computer-controlled carburetors are equipped with vacuum-operated decel valves. A tapered valve in the lower end of the valve assembly is connected through a stem to a diaphragm in the upper area of the assembly. A vacuum signal outlet extends from the area under the diaphragm, and a hose connects this outlet to the intake manifold. A clean air intake fitting is located below the tapered valve, and an intake manifold fitting is positioned above this valve (Figure 10-17). The clean air fitting is connected to the air cleaner inside the element, and the intake manifold fitting is connected to the intake manifold.

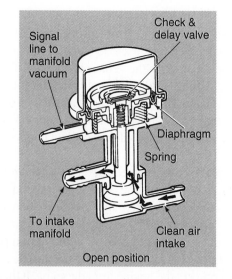

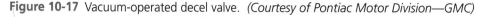

Figure 10-17 Vacuum-operated decel valve. *(Courtesy of Pontiac Motor Division—GMC)*

Idle speed devices

Figure 10-18 Throttle kicker and idle speed solenoid. *(Courtesy of Chevrolet Motor Division—GMC)*

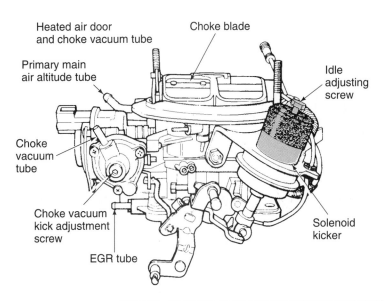

Heated air door and choke vacuum tube

Choke blade

Primary main air altitude tube

Idle adjusting screw

Choke vacuum tube

Choke vacuum kick adjustment screw

EGR tube

Solenoid kicker

Figure 10-19 Combined kicker and idle stop solenoid. *(Courtesy of Chrysler Corporation)*

When the engine is idling or operating at cruising speed, the tapered valve remains closed. During engine deceleration, high manifold vacuum supplied through the signal line pulls the diaphragm downward and opens the tapered valve. Once this valve opens, the intake manifold vacuum above the tapered valve pulls air through the clean air hose into the intake manifold. This action provides a leaner air-fuel ratio and reduced emissions during deceleration.

Throttle Kicker and Idle Stop Solenoid

Some computer-controlled carburetors have a vacuum-operated throttle kicker and an idle stop solenoid. The throttle kicker maintains engine idle speed when the engine accessory load is increased, such as during A/C compressor clutch operation. This kicker also maintains idle speed during warmup, after the fast idle cam has dropped away from the fast idle screw. Vacuum to the throttle kicker is controlled by an electric solenoid, which in turn is controlled by the powertrain control module (PCM).

One end of the idle stop solenoid winding is connected to a terminal on the solenoid housing and the other end of this winding is connected to ground on the solenoid housing. The idle stop solenoid terminal is connected to the ignition switch. When the ignition switch is turned on, current flows through the solenoid winding to ground. The magnetic field from this current flow pulls the solenoid plunger outward. Since the solenoid plunger contacts the throttle linkage, this plunger action moves the throttle linkage to provide the specified idle rpm (Figure 10-18). When the ignition switch is turned off, the idle stop solenoid plunger retracts and allows the throttle to move toward the closed position until the linkage contacts the throttle stop screw. This action prevents the engine from dieseling.

On some computer-controlled carburetors, the throttle kicker and idle stop solenoid are combined in one unit (Figure 10-19), but the operation of both components is basically the same.

Vacuum-Operated Throttle Kicker

When a vacuum-operated throttle kicker (TK) and solenoid are used to control idle speed, the TK stem pushes against the throttle linkage (Figure 10-20). Under these operating conditions, the PCM energizes the solenoid and vacuum is supplied from the intake manifold through the solenoid to the TK diaphragm: (1) During engine warmup, (2) when the air conditioning (A/C) is turned on, and (3) when the engine begins overheating.

Shop Manual
Chapter 10, page 450

Engine dieseling may be called afterrunning.

327

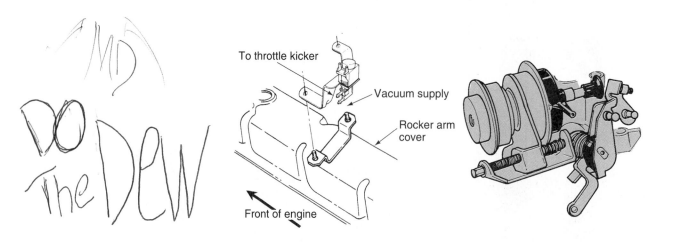

Figure 10-20 Throttle kicker diaphragm and solenoid. *(Reprinted with the permission of Ford Motor Company)*

When vacuum is supplied to the TK diaphragm, the stem is extended and the engine rpm is increased. During other engine operating conditions, the TK solenoid is de-energized and vacuum is shut off to the TK diaphragm. Under this condition, the TK diaphragm stem is retracted and the idle speed is controlled by the idle speed adjustment screw.

Temperature Compensated Pump

The temperature compensated pump (TCP) solenoid is operated by the PCM. When the engine coolant temperature is cold, the TCP solenoid is de-energized, and this solenoid shuts off vacuum to the TCP diaphragm. When this action is taken, the TCP diaphragm seats the ball check, and full accelerator pump output is available for improved cold acceleration. Once the engine coolant temperature is warmed up, the PCM energizes the TCP solenoid, which supplies vacuum to the TCP diaphragm (Figure 10-21). Under this condition, the diaphragm moves upward and the ball check is unseated, which allows some of the accelerator pump output to return to the float bowl.

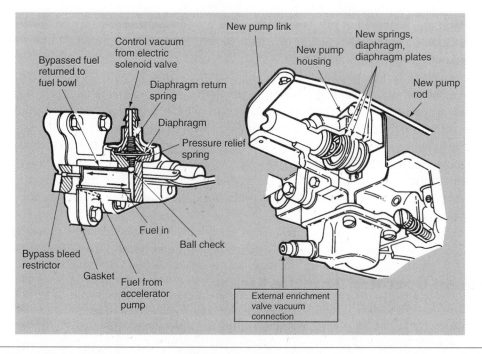

Figure 10-21 Temperature compensated pump (TCP) system. *(Reprinted with the permission of Ford Motor Company)*

Variable Voltage Choke Relay

The variable voltage choke (VVC) relay is operated by the PCM and connected to the electric choke cover terminal. If the engine coolant temperature and intake air temperature are cold, the PCM energizes the choke relay and supplies voltage to the choke heater in the choke cover every 2.5 seconds. This action reduces current to the choke heater and slows the choke opening. As the engine coolant and intake air temperature increase, the PCM increases the choke relay duty on time, which increases current to the choke heater and speeds up the choke opening. When the intake air temperature reaches 80°F (26.6°C), the PCM energizes the choke relay at 100 percent on time, and current is supplied to the choke heater continually to ensure that the choke continues to open.

Summary

❏ When emission standards required the installation of three-way catalytic converters, computer-controlled carburetors had to be installed to provide accurate air-fuel ratio control and adequate catalytic converter efficiency.

❏ Some mixture control solenoids regulate fuel flow into the carburetor main system and airflow into the idle system to control air-fuel ratio at all engine speeds.

❏ Other mixture control solenoids control the airflow into the idle and main systems to control the air-fuel ratio at all engine speeds.

❏ Some computer-controlled carburetors have a computer-operated stepper motor, which controls airflow into the main system to control the air-fuel ratio.

❏ When the O_2 sensor detects a rich air-fuel ratio, the sensor voltage is high. In response to this signal, the computer energizes the mixture control solenoid, which makes the air-fuel mixture leaner in the idle and main systems.

❏ If the O_2 sensor detects a lean air-fuel ratio, the sensor voltage is low. In response to this signal, the computer opens the mixture control solenoid ground circuit. This action allows the mixture control solenoid plunger to move upward and provide a richer air-fuel mixture in the idle and main systems.

❏ If a defect occurs in the mixture control solenoid electrical circuit, the spring moves the plunger upward to provide a rich air-fuel ratio.

❏ The computer operates the mixture control solenoid with a duty cycle. The computer can vary the on time of the solenoid to maintain the stoichiometric air-fuel ratio.

❏ Some mixture control solenoids are adjustable.

❏ Open loop occurs when the engine coolant is cold or cool.

❏ During open loop operation, the computer ignores the O_2 sensor signal, and the computer program provides a richer air-fuel ratio.

❏ Closed loop occurs when the coolant temperature sensor indicates approximately 175°F (79°C) and the O_2 sensor signal is valid.

❏ In closed loop, the computer controls the air-fuel ratio in response to the O_2 sensor signal.

❏ On a computer-controlled carburetor, idle speed may be controlled with an idle speed control motor or a throttle kicker diaphragm assembly.

❏ A temperature compensated accelerator pump system allows the accelerator pump to deliver full pump output when the engine is cold and a reduced output when the engine is at normal operating temperature.

❏ In a variable voltage choke system, the computer increases the length of time that the choke relay is on and supplies an increasing current to the choke heater as the engine warms up.

Terms to Know

Closed loop

Corporate average fuel economy (CAFE)

Duty cycle

Feedback

Mixture control (M/C) solenoid

Malfunction indicator lamp (MIL)

Open loop

Stepper motor

❏ Some computer-controlled carburetor systems have a separate barometric (BARO) pressure sensor. On other systems, this sensor is combined with the MAP sensor.

❏ A vacuum-operated decel valve allows additional airflow into the intake manifold during deceleration to provide a leaner air-fuel ratio and reduce emissions.

❏ The vacuum-operated throttle kicker increases idle speed under certain engine operating conditions such as during warmup or when the A/C is on.

❏ The idle stop solenoid holds the throttle in the idle position while the engine is idling and prevents dieseling by allowing the throttle to move toward the closed position when the ignition switch is turned off.

❏ The IAC motor is a reversible DC motor operated by the PCM. This motor stem contacts the throttle linkage to control idle speed.

❏ The PCM uses input sensor signals such as the engine coolant temperature sensor signal to control the IAC motor.

❏ The PCM also uses switch inputs such as the A/C, park/neutral, brake, and power steering pressure switches to control idle speed.

❏ A switch in the IAC motor stem signals the PCM when the throttle linkage is contacting the stem, and the PCM only operates the IAC motor under this condition.

Review Questions

Short Answer Essays

1. Explain the operation and purpose of a vacuum-operated decel valve.

2. Explain how the vacuum is supplied to a vacuum-operated throttle kicker.

3. Explain the purpose of an idle stop solenoid.

4. Describe how an idle air control motor controls idle speed.

5. Explain why an idle air control motor only operates when the throttle linkage is contacting the motor stem.

6. List some of the input sensors and switches used by the PCM to control the idle air control motor.

7. Describe the conditions required for the computer to increase the on time of the mixture control solenoid.

8. Define open loop, including the engine operating conditions during this mode.

9. Explain the engine operating conditions required to enter closed loop.

10. Describe the engine operating conditions required for the throttle kicker to increase the idle rpm.

Fill-in-the-Blanks

1. Monitoring the air-fuel ratio of a running engine is the job of a(n) _____ sensor.

2. The on time of a feedback solenoid can also be called its _____ time.

3. When an engine is cold, its control computer is probably in _____ _____.

4. The idle stop solenoid winding is energized when the _____ _____ is turned on.

5. The PCM uses _____ and _____ inputs to control the IAC motor.

6. The idle contact switch in the IAC motor is closed when the throttle linkage is _____ the _____.

7. One end of the mixture control solenoid winding is connected to the computer and the other end of this winding is connected to the _____ _____.

8. When a stepper motor is used in a computer-controlled carburetor, the plunger and valve on this motor move _____.

9. On a computer-controlled carburetor system immediately after a cold engine is started, the fast idle speed is controlled by the _____ _____ _____.

10. Lean air-fuel ratios increase the possibility of _____ being formed during combustion.

ASE Style Review Questions

1. While discussing throttle kickers:
   Technician A says ported vacuum above the throttle is supplied to the throttle kicker.
   Technician B says vacuum is continually supplied to the throttle kicker diaphragm while the engine is running.
   Who is correct?
   A. A only
   B. B only
   C. Both A and B
   D. Neither A nor B

2. While discussing idle stop solenoids:
   Technician A says the solenoid increases throttle opening on deceleration.
   Technician B says the solenoid maintains the throttle in the specified idle speed position.
   Who is correct?
   A. A only
   B. B only
   C. Both A and B
   D. Neither A nor B

3. While discussing IAC motors:
   Technician A says the IAC motor is a reversible DC motor.
   Technician B says the IAC motor controls throttle opening during deceleration to reduce emissions.
   Who is correct?
   A. A only
   B. B only
   C. Both A and B
   D. Neither A nor B

4. While discussing IAC motors:
   Technician A says on some systems if the IAC motor is maintaining the idle speed within 35 rpm of the target idle rpm programmed in the PCM, the PCM uses ignition timing variation to control slight variations in idle speed.
   Technician B says the PCM uses input information from input switches and sensors to control the IAC motor.
   Who is correct?
   A. A only
   B. B only
   C. Both A and B
   D. Neither A nor B

5. While discussing the idle contact switch:
   Technician A says the idle contact switch is mounted on the side of the throttle body.
   Technician B says the idle contact switch is closed with the engine operating at normal cruising speed.
   Who is correct?
   A. A only
   B. B only
   C. Both A and B
   D. Neither A nor B

6. While discussing emission levels and catalytic converters:

Technician A says a three-way catalytic converter controls levels of CO, HC, and NO_x regardless of the air-fuel ratio.

Technician B says the air-fuel ratio must be controlled near the stoichiometric ratio of 14.7:1 to allow a three-way catalytic converter to control CO, HC, and NOx emission levels effectively.

Who is correct?

A. A only   **C.** Both A and B

B. B only   **D.** Neither A nor B

7. While discussing computer-controlled carburetor operation:

Technician A says if the O_2 sensor voltage is low, the computer increases the on time of the mixture control solenoid winding.

Technician B says if the O_2 sensor voltage is low, the air-fuel ratio is rich.

Who is correct?

A. A only   **C.** Both A and B

B. B only   **D.** Neither A nor B

8. While discussing computer-controlled carburetor operation:

Technician A says if a defect occurs in the mixture control solenoid winding, the air-fuel mixture is lean.

Technician B says if a defect occurs in the computer so the computer cannot ground the mixture control solenoid winding, the air-fuel ratio is lean.

Who is correct?

A. A only   **C.** Both A and B

B. B only   **D.** Neither A nor B

9. While discussing the duty cycle provided by the computer for mixture control solenoid operation:

Technician A says some computers operate the mixture control solenoid with a duty cycle of 10 times per second.

Technician B says the computer is capable of varying the on time of the mixture control solenoid.

Who is correct?

A. A only   **C.** Both A and B

B. B only   **D.** Neither A nor B

10. While discussing open and closed loop:

Technician A says the computer may enter open loop with the engine at normal operating temperature and the throttle in the wide-open position.

Technician B says the computer may enter open loop during extended engine idle operation with the engine at normal operating temperature.

Who is correct?

A. A only   **C.** Both A and B

B. B only   **D.** Neither A nor B

Electronic Fuel Injection

CHAPTER 11

Upon completion and review of this chapter, you should be able to:

❏ Explain the principles of operation of a fuel injection system.

❏ Explain the design and function of major electronic fuel injection (EFI) components.

❏ Describe how the computer supplies the correct air-fuel ratio on a throttle body injection (TBI) system.

❏ Explain how the clear flood mode operates on a TBI system.

❏ Describe the effect of high fuel pressure in a TBI system.

❏ Explain the purpose of the throttle body temperature sensor.

❏ Explain how the computer provides air-fuel ratio enrichment while starting a cold engine equipped with a TBI system.

❏ Describe the difference between a sequential fuel injection (SFI) system and a multiport fuel injection (MFI) system.

❏ Explain why manifold vacuum is connected to the pressure regulator in an MFI system.

❏ Describe the operation of the pressure regulator in a returnless EFI system.

❏ Explain the operation of a cold start injector.

❏ Describe the operation of the central injector and poppet nozzles in a central port injection (CMFI) system.

❏ Explain the purpose of the intake manifold tuning valve (IMTV)

❏ List the components located in a constant control relay module.

❏ Explain the purpose of a dropping resistor assembly connected in series with the fuel injectors.

❏ Describe the inputs used by the computer to control the idle air control and idle air control bypass air motors.

❏ Explain how an idle air control bypass air motor controls idle speed.

❏ Explain the operation of the fast idle thermo valve.

❏ Describe the operation of the starting air valve.

Introduction

Fuel injection involves spraying or injecting fuel directly into the engine's intake manifold (Figure 11-1). Fuel injection, especially when it is electronically controlled, has several major advantages over carbureted systems. These include improved driveability under all conditions, improved fuel control and economy, decreased exhaust emissions, and an increase in engine efficiency and power.

During the 1980s, automotive engineers changed many engines from carburetors or computer-controlled carburetors to electronic fuel injection systems. This action was taken to improve fuel economy, performance, and emissions levels. Automotive manufacturers had to meet increasingly stringent corporate average fueled economy (CAFE) regulations and comply with emission standards at the same time.

Many of the early electronic fuel injection (EFI) systems were throttle body injection systems in which the fuel is injected above the throttles. The throttle body injection (TBI) systems have been gradually changed to port fuel injection systems with the injectors located in the intake ports. On carbureted and throttle body injected engines, some intake manifold heating was required to prevent fuel condensation on the intake manifold passages. When the injectors are positioned in the intake ports, intake manifold heating is not required. This provided engineers with increased intake manifold design flexibility. Intake manifolds could now be designed with longer, curved air passages that increased airflow and improved torque and horsepower.

Since intake manifolds no longer require heating, they can now be made from plastic materials such as glass-fiber reinforced nylon resin. This material is considerably lighter than cast iron or even aluminum. Saving weight means an improvement in fuel economy. The plastic-type intake

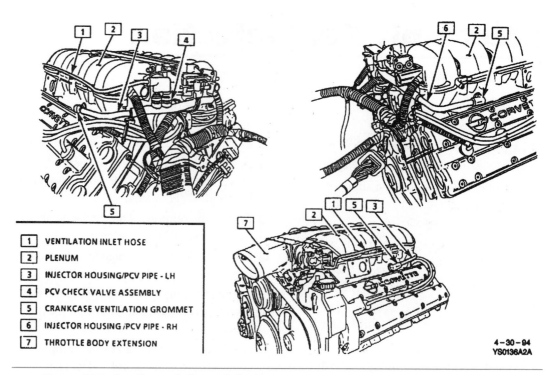

1	VENTILATION INLET HOSE
2	PLENUM
3	INJECTOR HOUSING/PCV PIPE - LH
4	PCV CHECK VALVE ASSEMBLY
5	CRANKCASE VENTILATION GROMMET
6	INJECTOR HOUSING /PCV PIPE - RH
7	THROTTLE BODY EXTENSION

4-30-94
YS0136A2A

Figure 11-1 An advanced electronic fuel injection system used in a Corvette. *(Courtesy of Chevrolet Motor Division—GMC)*

manifold does not transfer heat to the air and fuel vapor in the intake passages, which improves economy and hot start performance.

The objective of this chapter is to present the common components found in most EFI systems and the principles of various types of EFI. Some typical EFI systems are included, but not all the EFI systems used on domestic and imported vehicles are covered. If you are familiar with EFI components and principles, you will be able to understand the many different EFI systems when you encounter them.

The stoichiometric air-fuel ratio is 14.7:1. This is the mixture at which combustion can be the most complete.

Electronic fuel injection has proven to be the most precise, reliable, and cost effective method of delivering fuel to the combustion chambers of today's vehicles. EFI systems must provide the correct air-fuel ratio for all engine loads, speeds, and temperature conditions. To accomplish this, an EFI system uses a fuel delivery system, air induction system, input sensors, control computer, fuel injectors, and some sort of idle speed control.

Idle speed control is a computer output function on electronic fuel injection systems. It is very important that technicians understand how these systems operate. Since improper idle rpm is often the result of faulty input sensor or switch signals, it is even more important that technicians understand the input sensor signals used by the computer control idle air control motor.

A BIT OF HISTORY

If the idle or fast idle speed was not within specifications on a carbureted engine, the problem was usually corrected by turning a slow idle or fast idle screw with a screwdriver. However, the computer has changed this procedure! If the idle speed is not within specifications on an engine with electronic fuel injection, the technician needs to determine if the problem is in the idle air control motor, input sensors and switches, or the powertrain control module (PCM).

Input sensors

The ability of the fuel injection system to control the air-fuel ratio depends on its ability to properly time the injector pulses with the compression stroke of each cylinder and its ability to vary the injector "on" time, according to changing engine demands. Both tasks require the use of electronic sensors that monitor the operating conditions of the engine. Many of the same sensors used with computer-controlled carburetors are used with EFI systems.

Air Flow Sensors

In order to control the proportion of fuel to air in the air-fuel charge, the fuel system must be able to measure the amount of air entering the engine. Several sensors have been developed to do just that.

Shop Manual
Chapter 11, page 000

Volume Air flow Sensor. The air flow sensor shown in Figure 11-2 measures airflow, or air volume. The sensor consists of a spring-loaded flap, potentiometer, damping chamber, backfire protection valve, and idle bypass channel. As air is drawn into the engine, the flap is deflected against the spring. A potentiometer attached to the flap shaft, monitors the flap movement and produces a corresponding voltage signal (Figure 11-3). The strength of the signal increases as the flap opens. The signal increases as the flap opens. The signal voltage is relayed to the electronic control module.

Damping Chamber. The curved shape of the air flow sensor is the damping chamber. The damping flap in this chamber is on the same shaft as the air flow sensing flap and is also about the same area. As a result, the damping flap smoothens out any possible pulsations caused by opening and closing of the intake valves. Airflow measurement can be a steady signal, closely related to airflow as controlled by the movement of the flap.

Backfire Protection. The air flow sensor flap provides for backfire protection with a spring-loaded valve. If the intake manifold pressure suddenly rises because of a backfire, this valve releases the pressure and prevents damage to the system.

Idle Bypass. The air flow sensor assembly includes an extra air passage for idle, bypassing the air flow sensor plate. When the throttle is closed at idle, the opening and closing of intake valves can cause pulsations in the intake manifold. Without the idle bypass, such pulsations could cause the flap to shudder resulting in an uneven air-fuel mixture. The idle bypass smoothens the flow of the idle intake air, ensuring regular signals to the electronic control.

Another design of air flow sensor, called a Karman Vortex sensor, works on a different operating principle. Air entering the air flow sensor assembly passes through vanes arranged around the

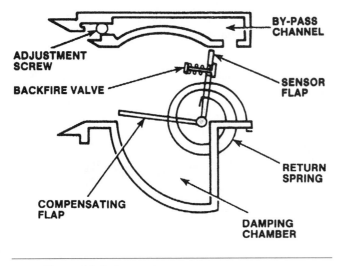

Figure 11-2 A vane-type sensor.

Figure 11-3 In a typical air flow sensor, the strength of the voltage signal produced by the potentiometer varies in response to the amount the sensor flap moves.

inside of a tube. As the air flows through the vanes, it begins to swirl. The outer part of the swirling air exerts high pressure against the outside of the housing. There is a low-pressure area in the center.

The low-pressure area moves in a circular motion as the air swirls through the intake tube. Two pressure-sensing tubes near the end of the tube sense the low-pressure area as it moves around. An electronic sensor counts how many times the low-pressure area is sensed.

The faster the airflow, the more times the low-pressure area is sensed. This is translated into a signal that indicates to the combustion control computer how much air is flowing into the intake manifold.

Air Temperature Sensor

The denser air gets, the more the air molecules are packed into a volume of air and the more that volume of air weighs.

Cold air is denser than warm air. Cold, dense air can burn more fuel than the same volume of warm air because it contains more oxygen. This is why air flow sensors that only measure air volume must have their readings adjusted to account for differences in air temperature.

Most systems do this by using an air temperature sensor mounted in the throttle body of the induction system. The air sensor measures air temperature and sends an electronic signal to the control computer. The computer uses this input along with the air volume input in determining the amount of oxygen entering the engine.

In some early EFI systems, the incoming air is heated to a set temperature. In these systems an air temperature sensor is used to ensure this predetermined operating temperature is maintained.

Mass Air Flow Sensor

A **mass air flow sensor** (Figure 11-4) does the job of a volume airflow sensor and an air temperature sensor; it measures air mass. The mass of a given amount of air is calculated by multiplying its volume by its density. As explained previously, the denser the air, the more oxygen it contains. Monitoring the oxygen in a given volume of air is important, since oxygen is a prime catalyst in the combustion process. From a measurement of mass, the electronic control unit adjusts the fuel delivery for the oxygen content in a given volume of air. The accuracy of air-fuel ratios is greatly enhanced when matching fuel to air mass instead of fuel to air volume.

The mass air flow sensor converts air flowing past a heated sensing element into an electronic signal. The strength of this signal is determined by the energy needed to keep the element at a constant temperature above the incoming ambient air temperature. As the volume and density (mass) of airflow across the heated element changes, the temperature of the element is affected and the current flow to the element is adjusted to maintain the desired temperature of the heating element. The varying current flow parallels the particular characteristics of the incoming air (hot, dry, cold, humid, high/low pressure). The electronic control unit monitors the changes in current to determine air mass and to calculate precise fuel requirements.

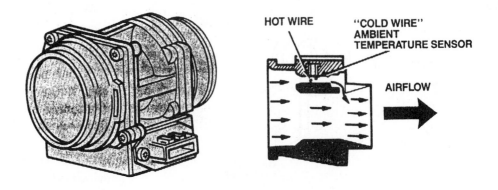

Figure 11-4 An MAF sensor. *(Reprinted with the permission of Ford Motor Company)*

There are two basic types of mass air flow sensors: hot wire and hot film. In the first type, a very thin wire (about 0.2 mm thick) is used as the heated element. The element temperature is set at 100 to 200°C above the incoming air temperature. Each time the ignition switch is turned to the off position the wire is heated to approximately 1,000°C for 1 second to burn off any accumulated dust and contaminants.

The second type uses a nickel foil sensor, which is kept 75°C above ambient air temperatures. It does not require a burnoff period, thus, it is potentially longer lasting than the hot wire type.

Manifold Absolute Pressure Sensor

Some EFI systems do not use airflow or air mass to determine the base pulse of the injector(s). Instead, the base pulse is calculated on manifold absolute pressure (MAP).

The MAP sensor (Figure 11-5) measures changes in the intake manifold pressure that result from changes in engine load and speed. The pressure measured by the MAP sensor is the difference between barometric pressure and manifold pressure. At closed throttle, the engine produces a low MAP value. A wide-open throttle produces a high value. This high value is produced when the pressure inside the manifold is the same as pressure outside the manifold, and 100% of the outside air is being measured. This MAP output is the opposite of what is measured on a vacuum gauge. The use of this sensor also allows the control computer to adjust automatically for different altitudes.

The control computer sends a voltage reference signal to the MAP sensor. As the MAP changes, the electrical resistance of the sensor also changes. The control computer can determine the manifold pressure by monitoring the sensor output voltage. A high pressure, low vacuum requires more fuel. A low pressure, high vacuum requires less fuel. Like an air flow sensor, a MAP sensor relies on an air temperature sensor to adjust its base pulse signal to match incoming air density.

Many EFI systems with MAF sensors do not have MAP sensors. However, there are a few engines with both of these sensors. In these cases, the MAP is used mainly as a backup if the MAF fails. When the EFI system has a MAF, the computer calculates the intake airflow from the MAF and rpm inputs.

Other EFI Systems Sensors

In addition to airflow, air mass, or manifold absolute pressure readings, the control computer relies on input from a number of other system sensors. This input further adjusts the injector pulse width

Shop Manual
Chapter 11, page 474

Barometric pressure is the air around us.

Manifold pressure is a negative pressure or a vacuum.

When a MAP has a low voltage signal, vacuum is high. When vacuum is low, the voltage signal is high.

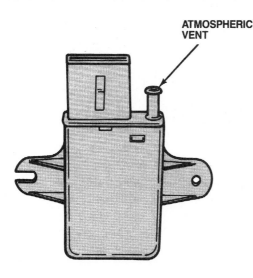

ATMOSPHERIC VENT

Figure 11-5 A typical Ford MAP sensor. *(Reprinted with the permission of Ford Motor Company)*

to match engine operating conditions. Operating conditions are communicated to the control computer by the following types of sensors.

Coolant Temperature. The coolant temperature sensor signals the PCM when the engine needs cold enrichment, as it does during warmup. This adds to the base pulse, but decreases to zero as the engine warms up.

Throttle Position. The switches on the throttle shaft signal the PCM for idle enrichment when the throttle is closed. These same throttle switches signal the PCM when the throttle is near the wide-open position to provide full load enrichment.

Engine Speed. The ignition system sends a tachometer signal reference pulse corresponding to engine speed to the electronic control unit. This signal advises the electronic control unit to adjust the pulse width of the injectors for engine speed. This also times the start of the injection according to the intake stroke cycle.

Cranking Enrichment. The starter circuit sends a signal for fuel enrichment during cranking operations even when the engine is warm. This is independent of any cold-start fuel enrichment demands.

Altitude Compensation. As the car operates at higher altitudes, the thinner air needs less fuel. Altitude compensation in a fuel injection system is accomplished by installing a sensor to monitor barometric pressure. Signals from the barometric pressure sensor are sent to the PCM to reduce the injector pulse width or reduce the amount of fuel injected.

Coasting Shutoff. Coasting shutoff can be found on a number of control systems. It can improve fuel economy as well as reduce emissions of hydrocarbons and carbon monoxide. Fuel shutoff is controlled in different ways depending on the type of transmission (manual or automatic). The PCM makes a coasting shutoff decision based on a closed throttle as indicated by the throttle position or idle switch or on engine speed as indicated by the signal from the ignition coil. When the PCM detects that power is not needed to maintain vehicle speed, the injectors are turned off until the need for power exists again.

Additional Input Information Sensors. Additional sensors are also used to provide the following information on engine conditions. Only the most common sensors that are used by all manufacturers are listed.

- ❏ Detonation
- ❏ Crankshaft position
- ❏ Camshaft position
- ❏ Air Charge Temperature
- ❏ Air conditioner operation
- ❏ Gearshift lever position
- ❏ Battery voltage
- ❏ Vehicle speed
- ❏ Oxygen in exhaust gases
- ❏ EGR valve position

Figure 11-6 shows the typical inputs and ouputs of the powertrain control module responsible for controlling the EFI system.

Basic EFI

In an **electronic fuel injection (EFI)** system, the computer must know the amount of air entering the engine so it can supply the stoichiometric air-fuel ration. In EFI systems with a MAP sensor, the

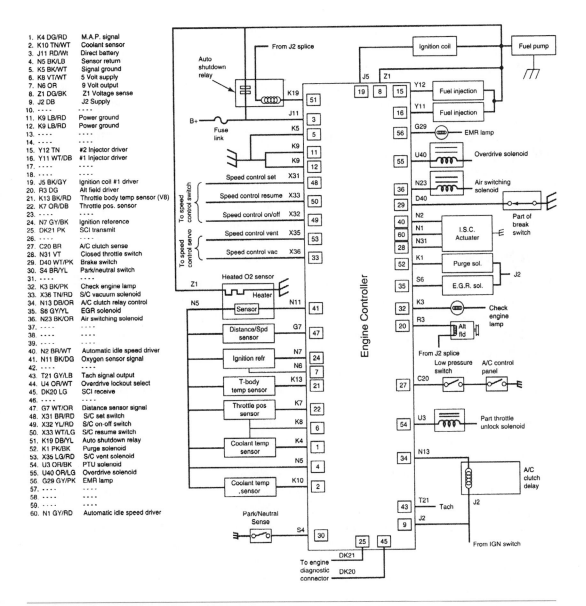

1. K4 DG/RD — M.A.P. signal
2. K10 TN/WT — Coolant sensor
3. J11 RD/Wt — Direct battery
4. N5 BK/LB — Sensor return
5. K5 BK/WT — Signal ground
6. K8 VT/WT — 5 Volt supply
7. N6 OR — 9 Volt output
8. Z1 DG/BK — Z1 Voltage sense
9. J2 DB — J2 Supply
10. ---- — ----
11. K9 LB/RD — Power ground
12. K9 LB/RD — Power ground
13. ---- — ----
14. ---- — ----
15. Y12 TN — #2 Injector driver
16. Y11 WT/DB — #1 Injector driver
17. ---- — ----
18. ---- — ----
19. J5 BK/GY — Ignition coil #1 driver
20. R3 DG — Alt field driver
21. K13 BK/RD — Throttle body temp sensor (V8)
22. K7 OR/DB — Throttle pos. sensor
23. ---- — ----
24. N7 GY/BK — Ignition reference
25. DK21 PK — SCI transmit
26. ---- — ----
27. C20 BR — A/C clutch sense
28. N31 VT — Closed throttle switch
29. D40 WT/PK — Brake switch
30. S4 BR/YL — Park/neutral switch
31. ---- — ----
32. K3 BK/PK — Check engine lamp
33. X36 TN/RD — S/C vacuum solenoid
34. N13 DB/OR — A/C clutch relay control
35. S6 GY/YL — EGR solenoid
36. N23 BK/OR — Air switching solenoid
37. ---- — ----
38. ---- — ----
39. ---- — ----
40. N2 BR/WT — Automatic idle speed driver
41. N11 BK/DG — Oxygen sensor signal
42. ---- — ----
43. T21 GY/LB — Tach signal output
44. U4 OR/WT — Overdrive lockout select
45. DK20 LG — SCI receive
46. ---- — ----
47. G7 WT/OR — Distance sensor signal
48. X31 BR/RD — S/C set switch
49. X32 YL/RD — S/C on-off switch
50. X33 WT/LG — S/C resume switch
51. K19 DB/YL — Auto shutdown relay
52. K1 PK/BK — Purge solenoid
53. X35 LG/RD — S/C vent solenoid
54. U3 OR/BK — PTU solenoid
55. U40 OR/LG — Overdrive solenoid
56. G29 GY/PK — EMR lamp
57. ---- — ----
58. ---- — ----
59. ---- — ----
60. N1 GY/RD — Automatic idle speed driver

Figure 11-6 Input sensors from all engine systems supply the control computer (PCM) with the data needed to calculate the correct injector pulse for driving conditions and driver demands. *(Courtesy of General Motors Corporation, Service Technology Group)*

computer program is designed to calculate the amount of air entering the engine from the MAP and rpm input signals. The distributor pickup supplies an rpm signal to the computer. This type of EFI system is referred to as a **speed density** system because the computer calculates the air intake flow from the engine rpm, or speed, input, and the density of intake manifold vacuum input. Therefore, the computer must have accurate signals from these inputs to maintain the stoichiometric air-fuel ration. The other inputs are used by the computer to "fine tune" the air-fuel ratio. For example, if the TPS input indicates sudden acceleration, the computer momentarily supplies a richer air-fuel ration.

A speed density EFI system is one in which the computer calculates the air entering the engine from the MAF or MAP and engine speed inputs.

Powertrain Control Module

The heart of the fuel injection system is the computer or powertrain control module (PCM). The PCM is a small computer that is usually mounted within the passenger compartment to keep it away from the heat and vibration of the engine. The PCM includes solid state devices, including integrated circuits and a microprocessor.

The PCM receives and processes signals from all the system sensors, then transmits programmed electrical pulses to the fueled injectors. Both incoming and outgoing signals are sent through a wiring harness and a multiple-pin connector.

Electronic feedback in the PCM means the unit is self-regulating and is controlling the injectors on the basis of operating performance or parameters rather than on preprogrammed instructions. A PCM with a feedback loop, for example, reads signal from the oxygen sensor, varies the pulse width of the injectors, and again reads the signals from the oxygen sensor. This is repeated until the injectors are pulsed for just the amount of time needed to get the proper amount of oxygen into the exhaust stream. While this interaction is occurring, the system is operating in closed loop. When conditions such as starting or wide-open throttle demand that the signals from the oxygen sensor be ignored, the system operates in open loop. During open loop, injector pulse length is controlled by set parameters contained in the PCM's memory.

Fuel Injectors

Shop Manual
Chapter 11, page 484

Fuel injectors are electromechanical devices that meter and atomize fuel so it can be sprayed into the intake manifold. Fuel injectors resemble a spark plug in size and shape. O-rings are used to seal the injector at the intake manifold, throttle body, and/or fuel rail mounting positions. These O-rings provide thermal insulation to prevent the formation of vapor bubbles and promote good hot start characteristics. They also dampen potentially damaging vibration.

When the injector is electrically energized, a solenoid-operated valve opens, and a fine mist of fuel sprays from the injector tip. Two different valve designs are commonly used.

The first consists of a valve body and a nozzle or needle valve that has a special ground **pintle** (Figure 11-7). A movable armature is attached to the nozzle valve, which is pressed against the nozzle body sealing seat by a helical spring. The solenoid winding is located at the back of the valve body.

When the solenoid winding is energized, it creates a magnetic field that draws the armature back and pulls the nozzle valve from its seat. When the solenoid is de-energized, the magnetic field collapses and the helical spring forces the nozzle valve back on its seat.

The second popular valve design uses a ball valve and valve seat. In this case, the magnetic field created by the solenoid coil pulls a plunger upward lifting the ball valve from its seat. Once again, a spring is used to return the valve to its seated or closed position.

Fuel injectors can be either top fuel feeding or bottom fuel feeding (Figure 11-8). Top feed injectors are primarily used in port injection systems that operate using high fuel system pressures.

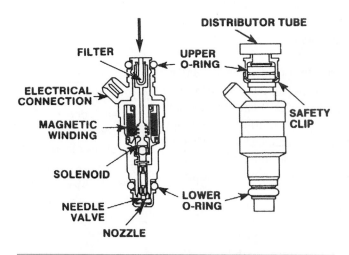

Figure 11-7 This fuel injector design is equipped with a nozzle or needle valve having a special ground pintle for precise fuel delivery control.

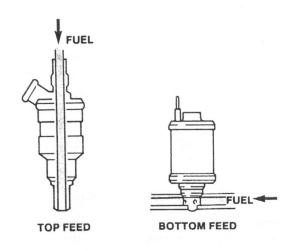

Figure 11-8 Examples of bottom and top feed injectors.

Bottom feed injectors are used in throttle body systems. Bottom feed injectors are able to use fuel pressures as low as 10 psi.

There have been some problems with deposits on injector tips. Since small quantities of gum are present in gasoline, injector deposits usually occur when this gum bakes onto the injector tips after a hot engine is shut off. Most oil companies have added a detergent to their gasoline to help prevent injector tip deposits. Car manufacturers and auto parts stores sell detergents to place in the fuel tank to clean injector tips.

Some manufacturers and auto parts suppliers have designed deposit-resistant injectors. These injectors have several different pintle tip and orifice designs to help prevent deposits. On one type of deposit-resistant injector, the pintle seat opens outward away from the injector body and more clearance is provided between the pintle and the body. Another type of deposit-resistant injector has four orifices in a metering plate rather than a single orifice. Some deposit-resistant injectors may be recognized by the color of the injector body. For example, regular injectors supplied by Ford Motor Company are painted black, whereas their deposit-resistant injectors have tan or yellow bodies.

Shop Manual
Chapter 11, page 500

Some engines equipped with port injection have an additional injector, the cold start injector. Unlike the individual injectors at the intake ports, the cold start injector is not operated directly by the PCM. Rather it is operated by a **thermo time switch** which senses coolant temperature (Figure 11-9).

When the engine is cranked, voltage is supplied from the starter solenoid to one terminal on the cold start injector. If the coolant temperature is below a certain amount, the thermo time switch completes the ground for the cold start injector. This energizes the injector while the engine is cranking. The fuel from the cold start injector sprays into the intake manifold and is delivered to the cylinders. Since the injectors at each intake port will also be providing fuel, the fuel from the cold start injector allows for a richer mixture.

A bimetal switch in the thermo time switch is heated as current flows through the injector coil. The bimetal switch opens the circuit through the thermo time switch after the engine has reached a certain temperature.

Each fuel injector is equipped with a two-wire connector. The connector is often equipped with a spring clip that must be unlocked before the connector can be removed from the injector.

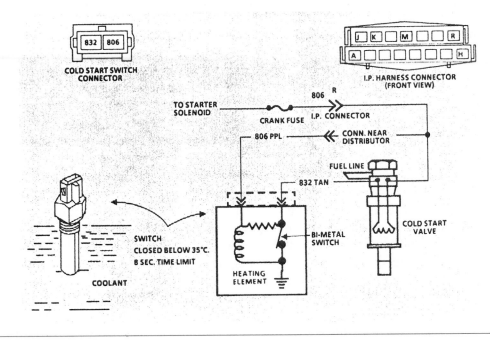

Figure 11-9 Cold start injector and thermo time switch. *(Courtesy of Chevrolet Motor Division—GMC)*

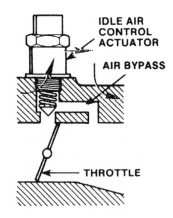

Figure 11-10 An IAC fitted in the casting of a TBI assembly.

One wire of the connector supplies voltage to the injector. This voltage supply wire may connect directly to the fuse panel or to the PCM, which in turn connects to the fuse panel. In some systems, a resistor at the fuse panel or PCM is used to reduce the 12-volt battery supply voltage to 3 volts or less. Most other injectors are fed battery voltage.

The second wire of the connector is a ground wire. This ground wire is routed to the PCM. The PCM energizes the injector by grounding its electrical circuit. The pulse width of the injector equals the length of time the injector circuit is grounded. Typical pulse widths range from 1 millisecond to 10 milliseconds at full load. Port fuel injection systems having four, six, or eight injectors use a special wiring harness to simplify and organize injector wiring.

Pulse width is the amount of time an injector is open and injecting fuel.

Idle Speed Control

In throttle body and port EFI systems, engine idle speed is controlled by bypassing a certain amount of airflow past the throttle valve in the throttle body housing. Two types of air bypass systems are used: auxiliary air valves and **idle air control (IAC)** valves. IAC valve systems are more common.

The IAC system consists of an electrically controlled stepper motor or actuator that positions the IAC valve in the air bypass channel around the throttle valve. The air bypass channel is part of the throttle body casting (Figure 11-10). The PCM calculates the amount of air needed for smooth idling based on input data such as coolant temperature, engine load, engine speed, and battery voltage. It then signals the actuator to extend or retract the idle air control valve in the air bypass channel.

If the engine speed is lower than desired, the PCM activates the motor to retract the IAC valve. This open the channel and diverts more air around the throttle valve. If engine speed is higher than desired, the valve is extended and the bypass channel is made smaller. Air supply to the engine is reduced and engine speed falls.

During the cold starts idle speed can be as high as 2,100 rpm to quickly raise the temperature of the catalytic converter for proper control of exhaust emissions. Idle speed that is attained after a cold start is controlled by the PCM. The PCM maintains idle speed for approximately 40 to 50 seconds even if the driver attempts to alter it by kicking the accelerator. After this preprogrammed time interval, depressing the accelerator pedal rotates the throttle position sensor (TPS) and signals the PCM to reduce idle speed.

Some engines are equipped with an auxiliary air valve to aid in the control of engine idle speed. The major difference between an IAC valve and an auxiliary air valve is that the auxiliary air valve is not controlled by the PCM. But like the IAC system, the auxiliary air valve provides additional air during cold engine starts and warmup.

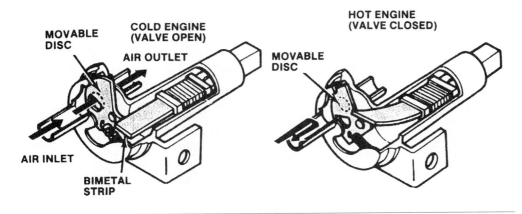

MOVABLE DISC   COLD ENGINE (VALVE OPEN)   HOT ENGINE (VALVE CLOSED)

AIR OUTLET   MOVABLE DISC

AIR INLET   BIMETAL STRIP

Figure 11-11 Operation of an IAC system. *(Courtesy of Robert Bosch Corporation)*

The auxiliary air valve consists of an air bypass channel or hose around the throttle valve, a movable plate or disc, and a heat sensitive bimetal strip. Figure 11-11 shows how an auxiliary air valve on a port injection system operates. When the plate opens the channel, extra air bypasses the throttle. Opening is controlled by the bimetal strip. As the bimetal heats up, it bends to rotate the movable plate, gradually blocking the opening. When the device is closed, there is no auxiliary airflow.

The bimetal strip is warmed by an electric heating element powered from the run circuit of the ignition switch. This bimetal element is not a switch, but a strip that moves the movable plate directly. The auxiliary air device is independent of the cold start injector. It is not controlled by the PCM but is continuously powered when the ignition key is set to the run position.

When the engine is cold, the passage opens for extra air when the engine starts. When the engine is running and still cold, the passage is open but the heater begins operating to close it gradually. If the engine is warm at startup, the passage is closed and normal air is delivered for idle.

Throttle Body Fuel Injection

Shop Manual
Chapter 11, page 480

For some auto manufacturers, TBI served as a stepping stone from carburetors to more advanced port fuel injection systems. TBI units were used on many engines during the 1980s and are still used on some engines. The throttle body unit is similar in size and shape to a carburetor, and like a carburetor, mounted on the intake manifold (Figure 11-12). The injector(s) spray fuel down into a throttle body chamber leading to the intake manifold. The intake manifold feeds the air-fuel mixture to all cylinders.

Throttle body injection (TBI) systems may be referred to as single -point or central point fuel injection systems.

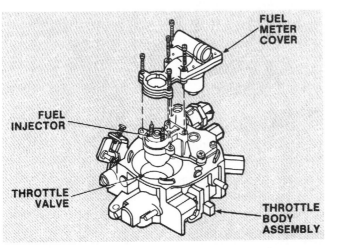

FUEL METER COVER

FUEL INJECTOR

THROTTLE VALVE

THROTTLE BODY ASSEMBLY

Figure 11-12 Parts of a throttle body injection unit.

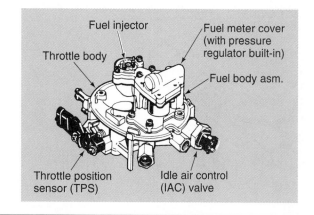

Figure 11-13 Single throttle body assembly used on a four-cylinder engine. *(Courtesy of Chevrolet Motor Division—GMC)*

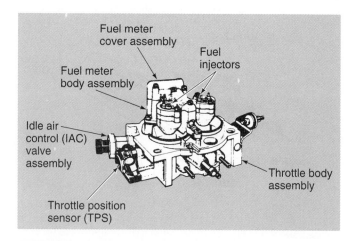

Figure 11-14 Dual throttle body assembly used on an eight-cylinder engine. *(Courtesy of Chevrolet Motor Division—GMC)*

TBI Advantages

Throttle body systems provide improved fuel metering when compared to carburetors. They are also less expensive and simpler to service. TBI units also have some advantages over port injection. They are less expensive to manufacture, simpler to diagnose and service, and don't have injector balance problems to the extent that port injection systems do when the injectors begin to clog.

However, throttle body units are not as efficient as port systems. The disadvantages are primarily manifold related. Like a carburetor system, fuel is still not distributed equally to all cylinders and a cold manifold may cause fuel to condense and puddle in the manifold. Like a carburetor, throttle body injection systems must be mounted above the combustion chamber level, which eliminates the possibility of tuning the manifold design for more efficient operation.

The throttle body assembly is mounted on top of the intake manifold where the carburetor was mounted on carbureted engines. Four-cylinders engines have a single throttle body assembly with one throttle (Figure 11-13), whereas V6 and V8 engines are equipped with dual throttle bodies with two throttles on a common throttle shaft (Figure 11-14).

The throttle body assembly contains a pressure regulator, injector or injectors, TPS, idle speed control motor, and throttle shaft and linkage assembly. A fuel filter is located in the fuel line under the vehicle or in the engine compartment. When the engine is cranking or running, fuel is supplied from the fuel pump through the lines and filter to the throttle body assembly. A fuel return line connected from the throttle body to the fuel tank returns excess fuel to the fuel tank (Figure 11-15).

The throttle body casting has ports that can be located above, below, or at the throttle valve depending on the manufacturer's design. These ports generate vacuum signals for the manifold absolute pressure sensor and for devices in the emission control system, such as the EGR valve and the canister purge system.

The **fuel pressure regulator** used on the throttle body assembly is similar to a diaphragm-operated relief valve (Figure 11-16). Fuel pressure is on one side of the diaphragm and atmospheric pressure is on the other side. The regulator is designed to provide a constant pressure on the fuel injector throughout the range of engine loads and speeds. If regulator pressure is too high, a strong fuel odor is emitted and the engine runs too rich. On the other hand, regulator pressure that is too low results in poor engine performance or detonation can take place, due to the lean mixture.

CAUTION: Always relieve the fuel pressure before disconnecting a fuel system component to avoid gasoline spills that may cause a fire, resulting in personal injury and/or property damage.

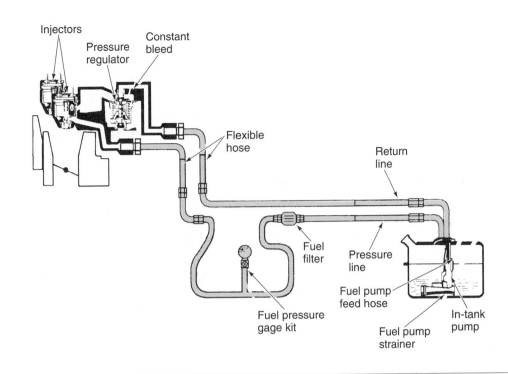

Figure 11-15 Fuel lines and fuel filter location for a TBI system. Note, the fuel pressure gage kit has been added to the illustration and is not part of the system. *(Courtesy of Chevrolet Motor Division—GMC)*

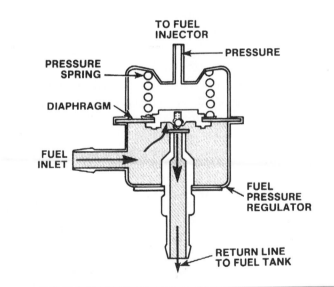

Figure 11-16 Operation of a diaphragm-operated fuel pressure regulator.

Throttle Body Internal Design and Operation

When fuel enters the throttle body fuel inlet, the fuel surrounds the injector or injectors at all times. Each injector is sealed into the throttle body with O-ring seals, which prevent fuel leakage around the injector at the top or bottom. Fuel is supplied from the injector through a passage to the pressure regulator. A diaphragm and valve assembly is mounted in this regulator and a diaphragm spring holds the valve closed. At a specific fuel pressure, the regulator diaphragm is forced upward to open the valve, and some excess fuel is returned to the fuel tank.

When the pressure regulator valve opens, fuel pressure decreases slightly, and the spring closes the regulator valve. This action causes the fuel pressure to increase and reopen the pressure regulator valve. In most TBI systems, the fuel pressure regulator controls fuel pressure at 10 psi to 25 psi (70 kPa to 172 kPa). The fuel pressure must be high enough to prevent fuel boiling in the TBI assembly. When the pressure on a liquid is increased, the boiling point is raised proportionally. If fuel boiling occurs in the TBI assembly, vapor and fuel are discharged from the injectors. The computer program assumes that the injectors are discharged liquid fuel. Vapor discharge from the injectors creates a lean air-fuel ration, which results in lack of engine power and acceleration stumbles.

Throttle Body Temperature Sensor

A throttle body temperature sensor is mounted in some Chrysler TBI assemblies (Figure 11-17). When the TBI assembly reaches the temperature at which some fuel boiling may occur, the throttle body temperature sensor signals the computer to provide a slightly richer air-fuel ratio. This action compensates for the vapor discharge from the injectors.

The fuel pressure is regulated at 14.5 psi (100 kPa) in Chrysler TBI assemblies with a temperature sensor. These systems were used from 1986 through 1990, and they are referred to as low pressure TBI (LPTBI). A black plastic rivet is located on the top of the pressure regulator in a LPTBI system. In 1991, Chrysler installed a pressure regulator on their TBI systems that controls the fuel pressure at 39.2 psi (270 kPa). These systems are called high pressure TBI (HPTBI). A white plastic rivet is located on top of these pressure regulators, and the TBI temperature sensor is no longer required with the higher fuel pressure.

Injector Internal Design and Electrical Connections

The plunger and valve seat are held downward by a spring. In this position, the seat closes the metering orifices in the end of the injector. Openings in the sides of the injector allow fuel to enter the cavity surrounding the injector tip. A mesh screen filter inside the injector openings removes dirt particles from the fuel. In some injectors, a diaphragm is located between the valve seat and the housing (Figure 11-18). The tip of the injector may contain up to six metering orifices, but some injectors have a single metering orifice. Injector design varies depending on the manufacturer.

Each injector contains two terminals, and an internal coil is connected across these terminals. A movable plunger is positioned in the center of the coil, and the lower end of the plunger has a

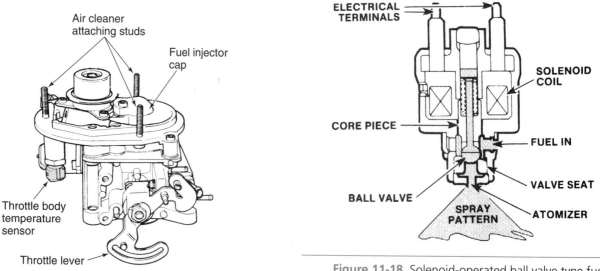

Figure 11-17 Throttle body temperature sensor. *(Courtesy of Chrysler Corporation)*

Figure 11-18 Solenoid-operated ball valve type fuel injector used in a TBI system. When electronically energized, the ball valve lifts off the valve seat, allowing fuel to spray into the throttle body housing.

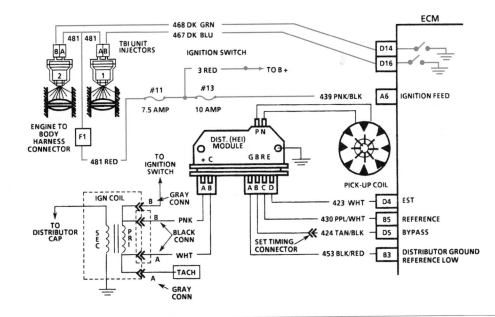

Figure 11-19 When the ignition switch is turned on, 12 volts are applied to one terminal on each injector. The other injector terminal is connected to and completed by the computer. *(Courtesy of Chevrolet Motor Division—GMC)*

tapered valve seat. When the ignition switch is turned on, 12 V are supplied to one of the injector terminals, and the other injector terminal is connected to the computer. When the computer grounds this terminal, current flows through the injector coil to ground in the computer (Figure 11-19).

When this action occurs, the injector coil magnetism moves the plunger and valve seat upward and fuel sprays out the injector orifices into the airstream above the throttle.

Pulse Width

The length of time that the computer grounds the injector is referred to as pulse width. Under most operating conditions, the computer provides the correct injector pulse width to maintain the stoichiometric air-fuel ratio. For example, the computer might ground the injector for 2 milliseconds at idle speed and 7 milliseconds at wide-open throttle to provide the stoichiometric air-fuel ratio. In many TBI systems, the computer grounds an injector each time a signal is received from the distributor pickup. This type of TBI system may be referred to as a synchronized system, because the injector pulses are synchronized with the pick-up signals. In a dual injector throttle body assembly, the computer grounds the injectors alternately under most operating conditions.

Air-Fuel Ratio Enrichment

When the coolant temperature sensor signal to the computer indicates that the engine coolant is cold, the computer increases the injector pulse width to provide a richer air-fuel ratio. This action eliminates the need for a conventional choke on a TBI assembly. The PCM supplies the proper air-fuel ratio and engine rpm when starting a cold engine. This eliminates the need for the driver to depress the accelerator pedal while starting the engine.

When a TBI-equipped engine is cold, the computer provides a very rich air-fuel ratio for faster starting. However, if the engine does not start because of an ignition defect, the engine becomes flooded quickly. Under this condition, excessive fuel may run past the piston rings into the crankcase. Therefore, when a cold TBI engine does not start, periods of long cranking should be avoided.

If the driver suspects that the air-fuel ratio is extremely rich, he or she may push the accelerator pedal to the wide-open position when starting a cold engine. Under these conditions, the

Starting a fuel injected engine without depressing the accelerator pedal may be called no-touch starting.

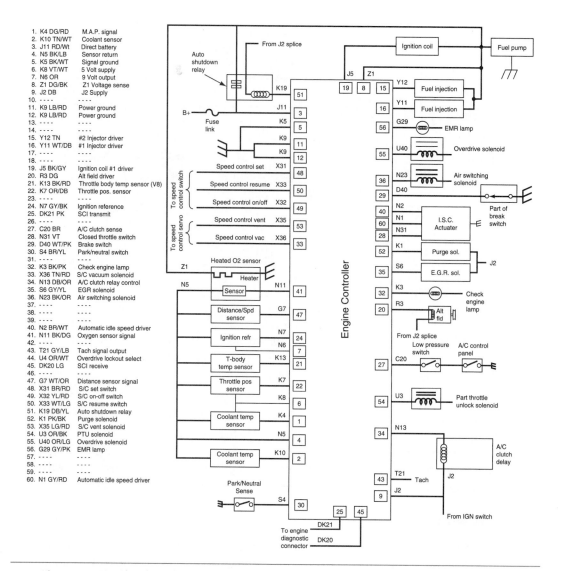

1. K4 DG/RD — M.A.P. signal
2. K10 TN/WT — Coolant sensor
3. J11 RD/Wt — Direct battery
4. N5 BK/LB — Sensor return
5. K5 BK/WT — Signal ground
6. K8 VT/WT — 5 Volt supply
7. N6 OR — 9 Volt output
8. Z1 DG/BK — Z1 Voltage sense
9. J2 DB — J2 Supply
10. ---- — ----
11. K9 LB/RD — Power ground
12. K9 LB/RD — Power ground
13. ---- — ----
14. ---- — ----
15. Y12 TN — #2 Injector driver
16. Y11 WT/DB — #1 Injector driver
17. ---- — ----
18. ---- — ----
19. J5 BK/GY — Ignition coil #1 driver
20. R3 DG — Alt field driver
21. K13 BK/RD — Throttle body temp sensor (V8)
22. K7 OR/DB — Throttle pos. sensor
23. ---- — ----
24. N7 GY/BK — Ignition reference
25. DK21 PK — SCI transmit
26. ---- — ----
27. C20 BR — A/C clutch sense
28. N31 VT — Closed throttle switch
29. D40 WT/PK — Brake switch
30. S4 BR/YL — Park/neutral switch
31. ---- — ----
32. K3 BK/PK — Check engine lamp
33. X36 TN/RD — S/C vacuum solenoid
34. N13 DB/OR — A/C clutch relay control
35. S6 GY/YL — EGR solenoid
36. N23 BK/OR — Air switching solenoid
37. ---- — ----
38. ---- — ----
39. ---- — ----
40. N2 BR/WT — Automatic idle speed driver
41. N11 BK/DG — Oxygen sensor signal
42. ---- — ----
43. T21 GY/LB — Tach signal output
44. U4 OR/WT — Overdrive lockout select
45. DK20 LG — SCI receive
46. ---- — ----
47. G7 WT/OR — Distance sensor signal
48. X31 BR/RD — S/C set switch
49. X32 YL/RD — S/C on-off switch
50. X33 WT/LG — S/C resume switch
51. K19 DB/YL — Auto shutdown relay
52. K1 PK/BK — Purge solenoid
53. X35 LG/RD — S/C vent solenoid
54. U3 OR/BK — PTU solenoid
55. U40 OR/LG — Overdrive solenoid
56. G29 GY/PK — EMR lamp
57. ---- — ----
58. ---- — ----
59. ---- — ----
60. N1 GY/RD — Automatic idle speed driver

Figure 11-20 Chrysler's dual TBI system that is used on 3.9L, 5.2L, and 5.9L engines. *(Courtesy of Chrysler Corporation)*

The clear flood mode provides a very lean air-fuel ratio to correct for an excessively rich condition during cold startup.

computer program provides a very lean air-fuel ratio of approximately 18:1. This may be referred to as s a **clear flood mode**. Under normal conditions, the driver should not push on the accelerator pedal at any time when starting an engine with TBI. When the engine is decelerated, the computer reduces injector pulse width in many TBI systems to provide a lean air-fuel ratio, which reduces emissions and improves fuel economy.

Typical Dual Throttle Body Injection System

Dual throttle body fuel injection systems have two injectors.

⚠️ **WARNING:** Never disconnect any computer system component with the ignition switch on. This action may damage computer system components.

The inputs in a Chrysler dual TBI system are (Figure 11-20):

1. Heated oxygen (O_2) sensor
2. Distance/speed sensor
3. Distributor pickup
4. Throttle body temperature sensor
5. Manifold absolute pressure (MAP) sensor

6. Coolant temperature sensor

7. Park/neutral switch

8. Cruise control switches

9. Brake switch

10. Ignition switch

11. A/C switch

The outputs on the Chrysler TBI system are:

1. Injectors

2. Spark advance

3. Automatic shutdown (ASD) relay

4. Alternator field

5. Cruise control vent and vacuum solenoids

6. Throttle kicker motor

7. Purge solenoid

8. EGR solenoid

9. Check engine lamp

10. Part throttle unlock solenoid

11. Up overdrive solenoid

12. Air switching solenoid

13. A/C clutch relay

14. Tachometer

As shown in the figure, the alternator field and the cruise control switches are connected to the PCM terminals. These connections indicate that the voltage regulator and the cruise control module are part of the PCM board.

If the A/C mode is selected on the instrument panel controls, and input signal is sent from the A/C control panel switch and the A/C low pressure switch to the PCM. When this signal is received, the PCM grounds the A/C relay winding, which closes the relay points and supplies voltage to the compressor clutch. If the engine is operating at wide-open throttle, the PCM will not ground the A/C relay winding.

The throttle kicker, or idle speed, motor is a reversible electric motor. The stern of this motor pushes against the throttle linkage. The PCM operates this motor to control the idle speed under all engine operating conditions (Figure 11-21).

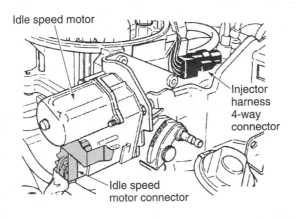

Figure 11-21 Idle speed control motor. *(Courtesy of Chrysler Corporation)*

Typical Single Throttle Body Injection System

WARNING: Never short across or ground any computer system terminals unless instructed to do so in the vehicle manufacturer's service manual. This action may damage computer system components.

The inputs on a General Motors 2.5L, four-cylinder, single TBI system are (Figures 11-22 and 11-23):

1. Coolant temperature sensor
2. Oxygen (O_2) sensor
3. Distributor reference
4. Power steering switch
5. Ignition on

Shop Manual

Chapter 11, page 479

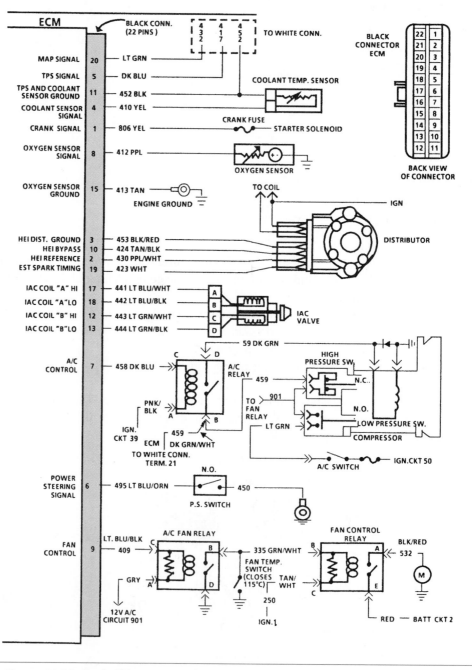

Figure 11-22 Single TBI system used on GM's 2.5L engine. *(Courtesy of Chevrolet Motor Division—GMC)*

350

6. Manifold absolute pressure (MAP) sensor
7. Throttle position sensor (TPS)
8. Vehicle speed sensor (VSS)
9. Park/neutral switch
10. Crank signal
11. A./C switch

The outputs on a 2.5L TBI system are:

1. Injector
2. Spark advance
3. Idle air control (IAC) valve

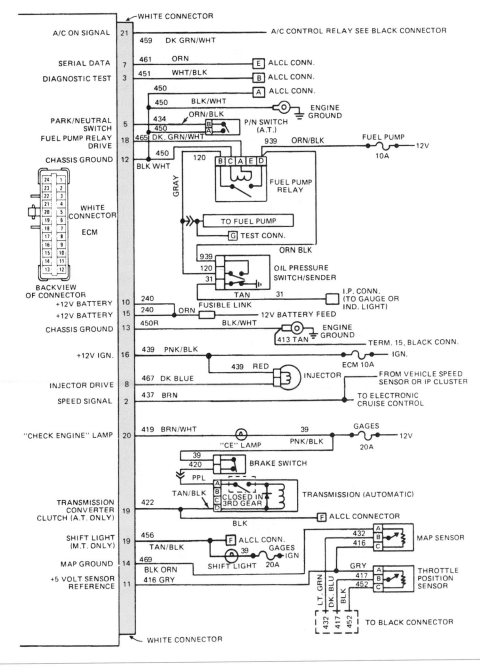

Figure 11-23 Single TBI system used on GM's 2.5L engine. *(Courtesy of Chevrolet Motor Division—GMC)*

4. A/C relay

5. A/C fan relay

6. Fuel pump relay

7. Torque converter clutch (TCC)

8. Malfunction indicator light (MIL)

9. Shift lamp, manual transaxles only

Port Fuel Injection

Port fuel injection (PFI) systems use one injector at each cylinder, They are mounted in the intake manifold near the cylinder head where they can inject a fine, atomized fuel mist as close as possible to the intake valve (Figure 11-24). Fuel lines run to each cylinder from a fuel manifold, usually referred to as a **fuel rail**. The fuel rail assembly on a PFI system of V6 and V8 engines usually consists of a left- and right-hand rail assembly. The two rails can be connected either by crossover and return fuel tubes or by a mechanical bracket arrangement. Since each cylinder has its own injector, fuel distribution is exactly equal. With little or no fuel to wet the manifold walls, there is no need for manifold heat or any early fuel evaporation system. Fuel does not collect in puddles at the base of the manifold.

The throttle body in a port fuel injection system controls the amount of air that enters the engine as well as the amount of vacuum in the manifold. It also houses and controls the idle air control motor and the throttle position sensor. The TPS enables the PCM to know where the throttle is positioned at all times.

The throttle body is a single cast aluminum housing with a single throttle blade attached to the throttle shaft. The TPS and the IAC valve/motor are also attached to the housing. The throttle shaft is controlled by the accelerator pedal. The throttle shaft extends the full length of the housing. The throttle bore controls the amount of incoming air that enters the air induction system. A small amount of coolant is also routed through a passage in the throttle body to prevent icing during cold weather.

Port systems require an additional control system that throttle body injection units do not require. While throttle body injectors are mounted above the throttle plates and are not affected by fluctuations in manifold vacuum, port system injectors have their tips located in the manifold where constant changes in vacuum would affect the amount of fuel injected (at a given pulse width). To compensate for these fluctuations, port injection systems are equipped with fuel pressure regulators (Figure 11-25) that sense manifold vacuum and continually adjust the fuel pressure to maintain a constant pressure drop across the injector tips at all times.

Port injection is a term that may be applied to any fuel injection system that has its injectors located in the intake ports.

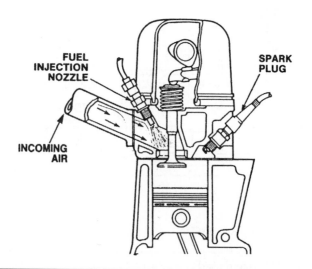

Figure 11-24 Port fuel injection systems use an injector at each cylinder.

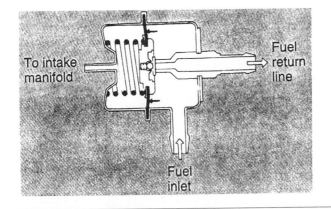

Figure 11-25 The fuel pressure regulator for a port fuel injection system.

There are many similarities in the MFI and SFI systems supplied by the various domestic and import vehicle manufacturers. For example, both MFI and SFI systems have injectors installed in the intake ports near the intake valve, and many of these systems share similar inputs and outputs. One of the major differences in MFI and SFI systems is the method of connecting the injectors to the computer. In SFI systems, each injector is connected individually into the computer and the computer grounds one injector at a time. In MFI systems, the injectors are grouped together in pairs or groups and these groups of injectors share a common wire to the computer. For example, on some four-cylinder engines, the injectors are connected in pairs, and each pair of injectors has a common connection to the computer. On some V6 engines, each group of three injectors has a common ground wire connected to the computer. Groups of four injectors share a common ground connection to the computer on some V8 engines.

Port Fuel Injection System Design

Basically the same electric in-tank fuel pumps and fuel pump circuits are found on TBI, MFI, and SFI systems. Some MFI and SFI systems, such as those on Ford trucks have a booster fuel pump on the frame rail in addition to the in-tank pump. A fuel filter is connected in the fuel line from the tank to the injectors. This filter may be under the vehicle or in the engine compartment. The fuel line from the filter is connected to a hollow fuel rail that is bolted to the intake manifold. The lower end of each port injector is sealed in the intake manifold with an O-ring seal, and a similar seal near the top of the injector seals the injector to the fuel rail. Most fuel rails have a Schrader valve. A pressure gauge may be connected to this valve when testing fuel pressure (Figure 11-26). A dust cap on the Schrader valve must be removed prior to gauge installation.

A Schrader valve is best described as being like an air valve on a tire.

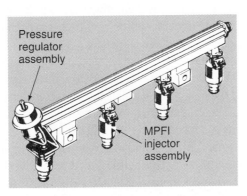

Figure 11-26 Fuel rail and injectors for a four-cylinder port injected engine. *(Courtesy of Oldsmobile Division—GMC)*

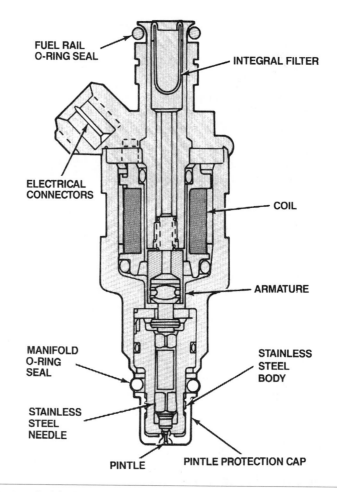

FUEL RAIL
O-RING SEAL

INTEGRAL FILTER

ELECTRICAL
CONNECTORS

COIL

ARMATURE

MANIFOLD
O-RING
SEAL

STAINLESS
STEEL
BODY

STAINLESS
STEEL
NEEDLE

PINTLE

PINTLE PROTECTION CAP

Figure 11-27 A typical fuel injector used in multiport fuel injection systems. *(Reprinted with the permission of Ford Motor Company)*

Each injector has a movable armature in the center of the injector, and a pintle with a tapered tip is positioned at the lower end of the armature. A spring pushes the armature and pintle downward so the pintle tip seats in the discharge orifice. The injector coil surrounds the armature, and the two ends of the winding are connected to the terminals on the side of the injector. An integral filter is located inside the top of the injector. When the ignition switch is turned on, voltage is supplied to one injector terminal and the other terminal is connected through the computer. Each time the computer completes the circuit from the injector winding to ground, current flows through the injector coil, and the coil magnetism moves the plunger and pintle upward. Under this condition, the pintle tip is unseated from the injector orifice, and fuel sprays out this orifice into the intake port (Figure 11-27).

The computer is programmed to ground the injectors well ahead of the actual intake valve openings so the intake ports are filled with fuel vapor before the intake valves open (Figure 11-28). In both SFI and MFI systems, the computer supplies the correct injector pulse width to provide the stoichiometric air-fuel ratio. The computer increases the injector pulse width to provide air-fuel ratio enrichment while starting a cold engine. A clear flood mode is also available in the computer in MFI and SFI systems. On some TBI, MFI, and SFI systems, if the ignition system is not firing, the computer stops operating the injectors. This action prevents severe flooding from long cranking periods while starting a cold engine. On many MFI and SFI systems, the computer decreases injector pulse width while the engine is decelerating to provide improved emission levels and fuel economy. On some of these systems, the computer stops operating the injectors while the engine is decelerating in a certain rpm range.

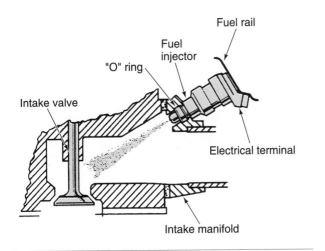

Figure 11-28 A port fuel injector sprays fuel into the intake port and fills the port with fuel vapors before the intake valve opens. *(Courtesy of Oldsmobile Division—GMC)*

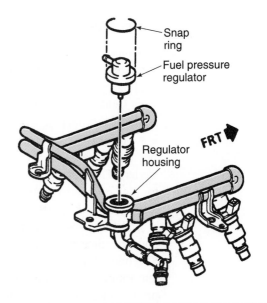

Figure 11-29 A one-piece, U-shaped fuel rail for a V6 engine. *(Courtesy of Oldsmobile Division—GMC)*

On many V6 and V8 engines, there is a fuel rail on each side of the intake manifold. On these engines, the fuel is supplied to one rail and a connecting hose carries fuel to the second rail. A pressure regulator is connected to the end of the fuel rail and excess fuel is returned from the regulator through a return line to the fuel tank. On some V6 engines, the fuel rail is U-shaped and each injector is mounted in this one-piece rail (Figure 11-29). Many fuel rails are made from steel or aluminum alloy. Plastic-type fuel rails have been installed recently on some cars. These fuel rails transfer less heat to the fuel and reduce the possibility of fuel boiling in the rail. On some V6 engines such as the General Motors 2.8L and 3.1L V6, a one-piece fuel rail is mounted under the top of the intake manifold. In these MFI systems, the top of the intake must be removed to gain access to the fuel rail and injectors (Figure 11-30).

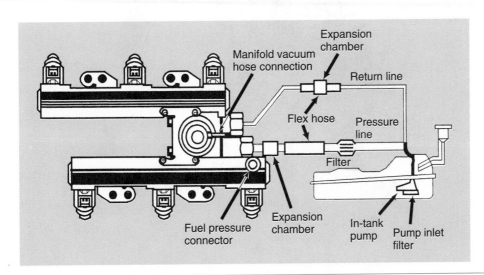

Figure 11-30 A one-piece fuel rail that is mounted under the top of the intake manifold. *(Courtesy of Oldsmobile Division—GMC)*

Cold Start Injector

A significant number of engines supplied by domestic and import manufacturers have a **cold start injector** and the injectors in each intake port. A pick-up pipe is connected from the fuel rail to the cold start injector, and the end of this injector is mounted in the intake manifold (Figure 11-31).

Unlike the intake port injectors that are operated by the PCM, the cold start injector is operated by a thermo time switch that senses coolant temperature. When the engine is cranked, voltage is supplied from the starter solenoid to one terminal on the cold start injector. If the coolant temperature is below 95°F (35°C), the thermo switch grounds the other cold start injector terminal. Under this condition, the cold start injector is energized while cranking the engine, and this injector pintle opens to spray fuel into the intake manifold in addition to the fuel injected by the injectors in the intake ports.

A bimetal switch in the thermo time switch is heated as current flows through the injector coil. The bimetal switch action opens the circuit through the thermo time switch in a maximum of 8 seconds. The actual time that the thermo time switch remains closed is determined by the coolant temperature. In this MFI system, the pulse width supplied by the PCM to the intake port injectors is programmed to operate with the cold start injector and supply the correct air-fuel ratio while cranking a cold engine.

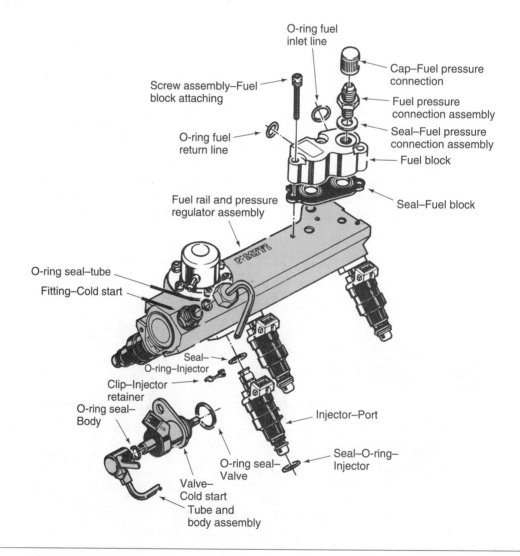

Figure 11-31 Cold start injector and fuel rail for a V6 engine. *(Courtesy of Chevrolet Motor Division—GMC)*

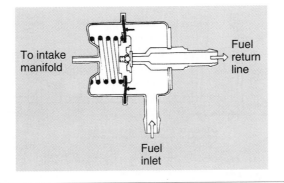

Figure 11-32 An internal pressure regulator. *(Courtesy of Chrysler Corporation)*

Pressure Regulators

The pressure regulator on MFI and SFI systems is similar to the regulator on TBI systems. A diaphragm and valve assembly is positioned in the center of the regulator, and a diaphragm spring seats the valve on the fuel outlet (Figure 11-32).

Shop Manual
Chapter 11, page 497

When fuel pressure reaches the regulator setting, the diaphragm moves against the spring tension and the valve opens. This action allows fuel to flow through the return line to the fuel tank. The fuel pressure drops slightly when the pressure regulator valve opens, and the spring closes the regulator valve. Under this condition, the fuel pressure increases and reopens the regulator valve.

A vacuum hose is connected from the intake manifold to the vacuum inlet on the pressure regulator. This hose supplies vacuum to the area where the diaphragm spring is located. This vacuum works with the fuel pressure to move the diaphragm and open the valve. When the engine is running at idle speed, high manifold vacuum is supplied to the pressure regulator. Under this condition, fuel pressure opens the regulator valve. If the engine is operating at wide-open throttle, a very low manifold vacuum is supplied to the pressure regulator. When this condition is present, the vacuum does not help to open the regulator valve and a higher fuel pressure is required to open the valve.

When the engine is idling, higher manifold vacuum is supplied to the injector tips and the injectors are discharging fuel into this vacuum. Under wide-open throttle conditions, the very low manifold vacuum is supplied to the injector tips. When this condition is present, the injectors are actually discharging fuel into a higher pressure compared to idle speed conditions because the very low manifold vacuum is closer to a positive pressure. If the fuel pressure remained constant at idle and wide-open throttle conditions, the injectors would discharge less fuel into the higher pressure in the intake manifold at wide-open throttle. The increase in fuel pressure supplied by the pressure regulator at wide-open throttle maintains the same pressure drop across the injectors at idle speed and wide-open throttle. When this same pressure drop is maintained, the change in pressure at the injector tips does not affect the amount of fuel discharged by the injectors.

Typical Multiport Fuel Injection System

In many Ford MFI systems on V8 engines, the injectors are connected in groups of four on the ground side. Each group shares a common ground wire to the powertrain control module (Figure 11-33). When the injectors are connected in groups of four on the ground side, a group of injectors is grounded by the computer every crankshaft revolution. The inputs are:

Shop Manual
Chapter 11, page 497

1. Heated exhaust gas oxygen (HEGO) sensor
2. Engine coolant temperature (ECT) sensor
3. Throttle position sensor (TPS)
4. Intake air temperature (IAT) sensor
5. Manifold absolute pressure (MAP) sensor
6. EGR valve position sensor

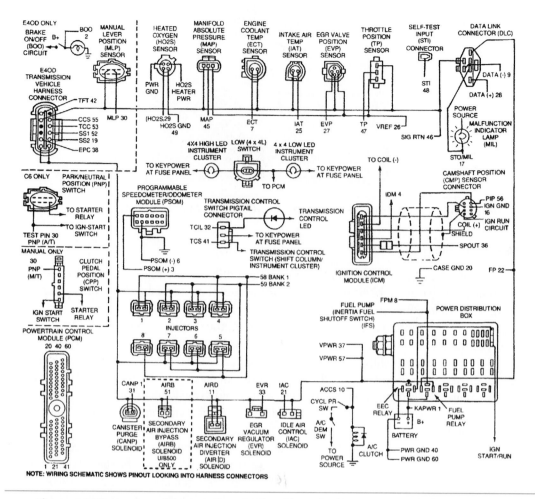

Figure 11-33 Inputs and outputs for a Ford multiport fuel injection system. *(Reprinted with the permission of Ford Motor Company)*

7. Brake on/off (BOO) switch
8. Manual lever position (MLP) sensor
9. Neutral/park switch
10. Self-test connector
11. Self-test input

The outputs on the 5.8L V8 MFI system are:

1. Injectors
2. Spark advance
3. Idle speed control (IAC) solenoid
4. EGR vacuum regulator (EVR) solenoid
5. Canister purge solenoid
6. Secondary air injection bypass (AIRB) solenoid
7. Secondary air injection diverter (AIRD) solenoid

Typical Import Multiport Fuel Injection System

Some Toyota Camry models are equipped with a multiport fuel injection system. In these systems, injectors one and three and injectors two and four share common ground wire connections to the PCM (Figure 11-34). The PCM grounds a pair of injectors once every crankshaft revolution.

Shop Manual
Chapter 11, page 497

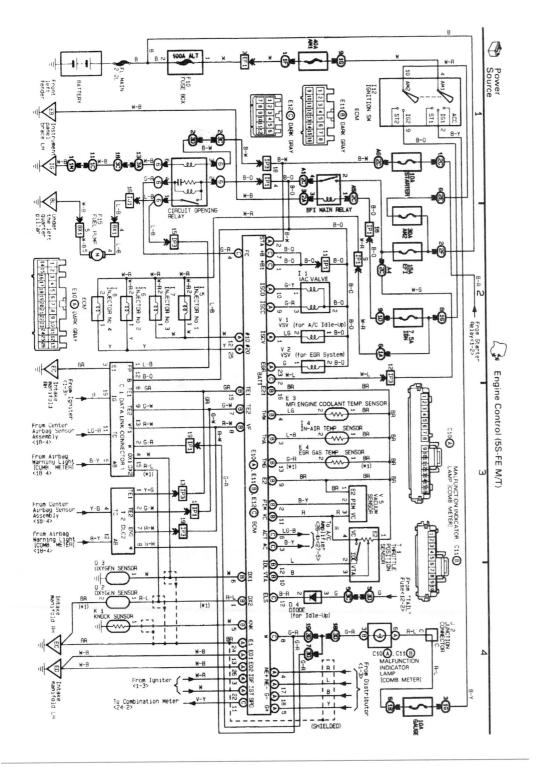

Figure 11-34 The multiport fuel injection system for a Toyota Camry.

The inputs in this MFI system are:

1. Dual oxygen sensors

2. Knock sensor

3. Throttle position sensor

4. Exhaust gas recirculation (EGR) gas temperature sensor

5. In-air temperature sensor

6. Engine coolant temperature sensor

7. Vacuum sensor

8. Distributor pickup

The outputs on this MFI system are:

1. Injectors

2. Spark advance

3. Vacuum switching valve (VSV) for EGR

4. VSV for idle up

5. Idle air control (IAC) valve

6. Malfunction indicator light (MIL)

7. Data link connector

The VSV solenoid for idle up is energized by the PCM under certain conditions, such as when the A/C is turned on. When the VSV solenoid is energized, it allows more airflow through the solenoid into the intake manifold to increase the idle speed. The **malfunction indicator lamp (MIL)** light is illuminated by the PCM if the PCM senses a defect in the system. During the diagnostic procedure, the MIL light flashes the fault codes.

A BIT OF HISTORY

Since the 1970s, fuel system technology has developed quickly from carburetors to computer-controlled carburetors and then to throttle body injection systems. Many of the throttle body injection systems have been replaced with multiport fuel injections systems, and a significant number of multiport injection systems have been changed to sequential fuel injection systems.

Sequential Fuel Injection Systems

Shop Manual
Chapter 11, page 497

While all port injection systems operate using an injector at each cylinder, they do not all fire the injectors in the same manner. This one statement best defines the difference between typical multiport injection systems and sequential fuel injection systems.

Sequential fuel injection (SFI) systems control each injector individually so that it is opened just before the intake valve opens. This means that the mixture is never static in the intake manifold and that adjustments to the mixture can be made almost instantaneously between the firing of one injector and the next. Sequential firing is the most accurate and desirable method of regulating port injection.

In port injection systems, the injectors are grouped together in pairs or groups, and these pairs or groups of injectors are turned on at the same time. When the injectors are split into two equal groups the groups are fired alternatively, with one group firing each engine revolution (Figure 11-35).

Since only two injectors are fired close to the time when the intake valve is about to open, the fuel charge for the remaining cylinders must stand in the intake manifold for varying periods of time. These periods of time are very short, therefore the standing of fuel in the intake manifold is not that great a disadvantage of MPI systems. At idle speeds this wait is about 150 milliseconds, and at higher speeds the time is much less. The primary advantage of the SFI is the ability to make instantaneous changes to the mixture.

In SFI systems, each injector is connected individually into the computer, and the computer completes the ground for each injector, one at a time. In MPI systems, the injectors are grouped and all injectors within the group share the same common ground wire.

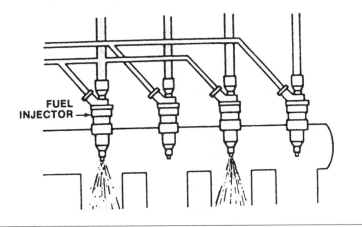

Figure 11-35 A grouped fire port injection system.

Some injection systems fire all of the injectors at the same time for every engine revolution. This type of system offers easy programming and relatively fast adjustment to the air-fuel mixture. The injectors are connected in parallel so the PCM sends out just one signal for all injectors. They all open and close at the same time. It simplifies the electronics without compromising injection efficiency. The amount of fuel required for each four-stroke cycle is divided in half and delivered in two injections, one for every 360 degrees of crankshaft rotation. The fact that the intake charge must still wait in the manifold for varying periods of time is the system's major drawback.

Typical Sequential Fuel Injection System

In a Chrysler SFI system on a 3.5L engine, each injector has a separate ground wire connected into the PCM (Figure 11-36). Many Chrysler engines previous to 1992 have multiport fuel injection (MFI) systems with the injectors connected in pairs on the ground side. Each pair of injectors shares a common ground wire into the PCM. Voltage is supplied through the ASD relay points to the injectors when the ignition switch is turned on, and a separate fuel pump relay supplies voltage to the fuel pump. This engine is equipped with an electronic ignition system, and the crank and cam sensors are inputs for this system. Since these inputs are connected to the PCM, the ignition module is contained in the PCM.

EI systems are ignition systems without a distributor.

Returnless Fuel System Pressure Regulators

Some later model Chrysler SFI systems are referred to as returnless systems. In these systems, the fuel pressure regulator and filter are mounted in the top of the assembly containing the fuel pump and fuel gauge sending unit in the fuel tank (Figure 11-37). The fuel line from the fuel rail under the hood is connected to the filter with a quick-disconnect fitting.

Fuel enters the filter through the fuel supply tube in the center of the regulator and filter assembly. Fuel pressure is applied against the regulator seat washer, which is seated by the seat control spring (Figure 11-38). At the specified regulator pressure, the seat is forced downward against the spring, and fuel flows past the seat into the cavity around the seat control spring. Fuel returns from this cavity to the fuel tank. When the pressure drops slightly, the seat closes again. With the returnless fuel system, only the fuel needed by the engine is filtered, thus allowing the use of a smaller fuel filter.

The inputs shown in the figure on the Chrysler 3.5L SFI system are:

1. Dual heated oxygen (O_2) sensor
2. Crank sensor
3. Cam sensor
4. Throttle position sensor (TPS)

Shop Manual
Chapter 11, page 497

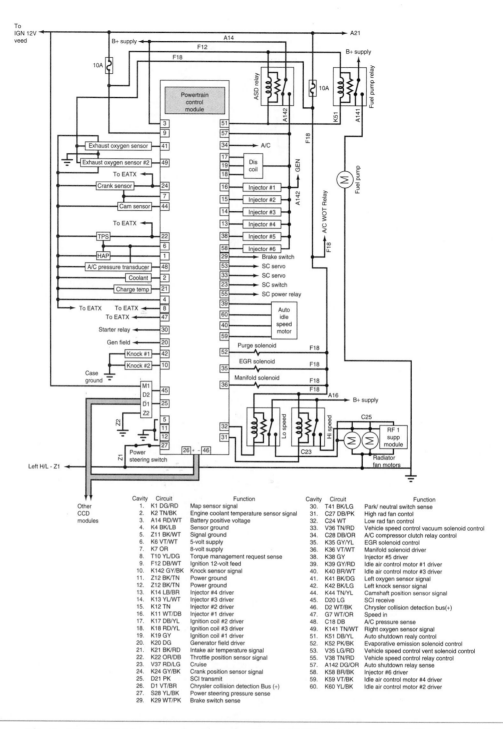

Cavity	Circuit	Function		Cavity	Circuit	Function
1.	K1 DG/RD	Map sensor signal		30.	T41 BK/LG	Park/ neutral switch sense
2.	K2 TN/BK	Engine coolant temperature sensor signal		31.	C27 DB/PK	High rad fan contol
3.	A14 RD/WT	Battery positive voltage		32.	C24 WT	Low rad fan control
4.	K4 BK/LB	Sensor ground		33.	V36 TN/RD	Vehicle speed control vacuum solenoid control
5.	Z11 BK/WT	Signal ground		34.	C28 DB/OR	A/C compressor clutch relay control
6.	K6 VT/WT	5-volt supply		35.	K35 GY/YL	EGR solenoid control
7.	K7 OR	8-volt supply		36.	K36 VT/WT	Manifold solenoid driver
8.	T10 YL/DG	Torque management request sense		38.	K38 GY	Injector #5 driver
9.	F12 DB/WT	Ignition 12-volt feed		39.	K39 GY/RD	Idle air control motor #1 driver
10.	K142 GY/BK	Knock sensor signal		40.	K40 BR/WT	Idle air control motor #3 driver
11.	Z12 BK/TN	Power ground		41.	K41 BK/DG	Left oxygen sensor signal
12.	Z12 BK/TN	Power ground		42.	K42 BK/LG	Left knock sensor signal
13.	K14 LB/BR	Injector #4 driver		44.	K44 TN/YL	Camshaft position sensor signal
14.	K13 YL/WT	Injector #3 driver		45.	D20 LG	SCI receive
15.	K12 TN	Injector #2 driver		46.	D2 WT/BK	Chrysler collision detection bus(+)
16.	K11 WT/DB	Injector #1 driver		47.	G7 WT/OR	Speed in
17.	K17 DB/YL	Ignition coil #2 driver		48.	C18 DB	A/C pressure sense
18.	K18 RD/YL	Ignition coil #3 driver		49.	K141 TN/WT	Right oxygen sensor signal
19.	K19 GY	Ignition coil #1 driver		51.	K51 DB/YL	Auto shutdown realy control
20.	K20 DG	Generator field driver		52.	K52 PK/BK	Evaporative emission solenoid control
21.	K21 BK/RD	Intake air temperature signal		53.	V35 LG/RD	Vehicle speed control vent solenoid control
22.	K22 OR/DB	Throttle position sensor signal		55.	V38 TN/RD	Vehicle speed control relay control
23.	V37 RD/LG	Cruise		57.	A142 DG/OR	Auto shutdown relay sense
24.	K24 GY/BK	Crank position sensor signal		58.	K58 BR/BK	Injector #6 driver
25.	D21 PK	SCI transmit		59.	K59 VT/BK	Idle air control motor #4 driver
26.	D1 VT/BR	Chrysler collision detection Bus (+)		60.	K60 YL/BK	Idle air control motor #2 driver
27.	S28 YL/BK	Power steering pressure sense				
29.	K29 WT/PK	Brake switch sense				

Figure 11-36 PCM inputs and outputs for Chrysler's SFI system used on 3.5L engines. *(Courtesy of Chrysler Corporation)*

5. Manifold absolute pressure (MAP) sensor

6. Coolant temperature sensor

7. Charge temperature sensor

8. Electronic automatic transaxle (EATX) computer

9. Starter relay

10. Generator field

11. Dual knock sensors

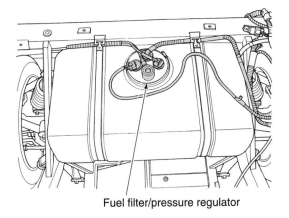

Fuel filter/pressure regulator

Figure 11-37 Returnless fuel system with the pressure regulator and filter mounted in the fuel tank with the fuel pump and fuel gauge sending unit. *(Courtesy of Chrysler Corporation)*

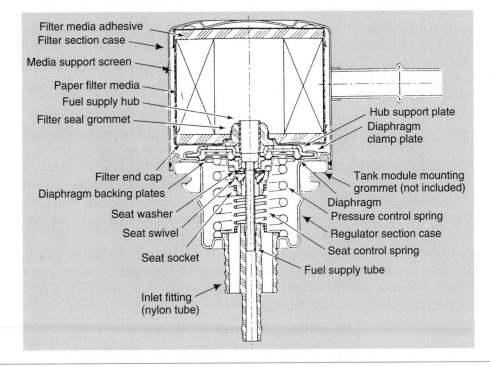

Figure 11-38 Pressure regulator and fuel filter from a returnless fuel system. *(Courtesy of Chrysler Corporation)*

12. Data links to other computers such as the EATX

13. Battery voltage

14. Power steering switch

15. Brake switch

16. Ignition switch

17. Cruise control switches

18. Park/neutral switch, starter relay

Many V6 and V8 engines now have dual oxygen sensors that provide improved control of the air-fuel ratio in each bank of cylinders. An oxygen sensor is usually mounted in each exhaust manifold. Dual knock sensors are located in many V6 and V8 engines to improve spark knock or detonation control. The knock sensors are positioned in each side of the block, or cylinder heads.

The outputs on the Chrysler 3.5L SFI system are:

1. Automatic shutdown (ASD) relay
2. Fuel pump relay
3. Ignition coil
4. Spark advance
5. Injectors
6. Cruise control
7. Automatic idle speed motor
8. Purge solenoid
9. EGR solenoid
10. Manifold solenoid
11. Dual radiator fan relays

The Chrysler SFI system has many similarities to the Chrysler TBI system. For example, the voltage regulator and the cruise control module are contained in the PCM board. The SFI system on the 3.5L engine has a low-speed and a high-speed cooling fan relay. At a specific coolant temperature, the PCM grounds the low-speed relay winding, which closes the relay points and supplies voltage to the fan motor. If the engine coolant temperature continues to increase, the PCM grounds the high-speed cooling fan relay winding, which closes the fan relay points and supplies voltage to the high-speed fan motor.

The manifold solenoid controls the vacuum supplied to the intake manifold tuning valve. This solenoid is mounted on the right shock tower and the manifold tuning valve is positioned near the center of the intake manifold (Figure 11-39). The manifold contains a pivoted butterfly valve that opens and closes to change the length of the intake manifold air passages. This butterfly valve is mounted on a shaft, and the outer end of the shaft is connected through a linkage to a diaphragm in a sealed vacuum chamber. A vacuum hose is connected from the outlet fitting on the manifold solenoid to the vacuum chamber in the manifold tuning valve. Another vacuum hose is connected from the inlet fitting on the manifold solenoid to the intake manifold.

One terminal on the manifold solenoid winding is connected to the ignition switch and the other terminal on this winding is connected to the PCM. While the engine is running at lower rpm, the PCM opens the manifold solenoid circuit. Under this condition, the solenoid shuts off the manifold vacuum to the intake manifold tuning valve, and the butterfly valve closes some of the air passages inside the intake manifold.

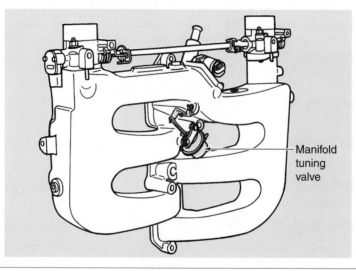

Figure 11-39 Intake manifold tuning valve for Chrysler's 3.5L engine. *(Courtesy of Chrysler Corporation)*

At higher engine rpm, the PCM ground the manifold solenoid winding and energizes the solenoid. This action opens the vacuum passage through the solenoid and supplies vacuum to the intake manifold tuning valve. Under this condition, the butterfly valve is moved so it opens additional air passages inside the intake manifold to improve air flow and increase engine horsepower and torque.

Typical Import Sequential Fuel Injection System

The Nissan electronic concentrated engine control system (ECCS) is an SFI system that has many of the same inputs and outputs as the other systems. The inputs shown in the wiring diagram for the ECCS are (Figure 11-40):

Shop Manual
Chapter 11, page 497

1. Crank angle sensor

2. Water temperature sensor

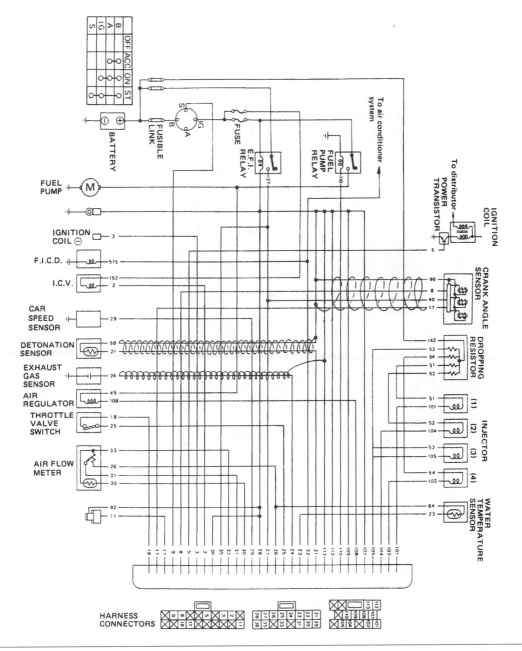

Figure 11-40 Inputs and outputs in Nissan's ECCS system. *(Courtesy of Nissan Motors)*

3. Car speed sensor
4. Detonation sensor
5. Exhaust gas sensor
6. Throttle valve switch
7. Air flow meter

A barometric pressure sensor is located in the PCM. This sensor is not serviced separately. This sensor sends a signal to the PCM in relation to barometric pressure as it varies with altitude and climatic conditions. The crank angle sensor is located in the distributor. It sends engine rpm and crankshaft position signals to the PCM (Figure 11-41).

System outputs as shown in the wiring diagram for the ECCS are:

1. Injectors
2. Spark advance

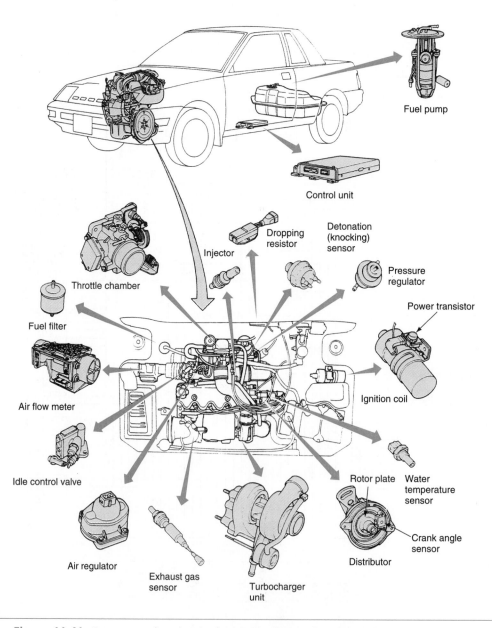

Figure 11-41 Component location in the ECCS system. *(Courtesy of Nissan Motors)*

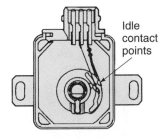

Figure 11-42 Typical idle switch. *(Courtesy of Nissan Motors)*

3. Air regulator
4. Idle control valve (ICV)
5. Fast idle control device (FICD)
6. Fuel pump relay
7. Electronic fuel injection (EFI) relay

The system also uses some components that are not commonly found on other systems, such as the dropping resistor assembly which contains a resistor connected in series with each injector. These resistors protect the injectors from sudden voltage changes and provide constant injector operation. The system uses a vane-type mass air flow sensor. The throttle valve switch is mounted in the throttle chamber and the switch contacts are closed when the throttle is in the idle position (Figure 11-42). When the throttle is opened from the idle position, the switch contacts open. The throttle valve switch signal informs the PCM when the throttle is in the idle position.

The idle control valve (ICV) solenoid and the fast idle control device (FICD) solenoid are mounted on the intake manifold. When the idle speed drops below a specific rpm, the PCM energized the ICV solenoid, and additional air is bypassed through this solenoid into the intake manifold to increase the idle speed. The FICD solenoid is energized when the A/C is on, and air flows past this solenoid into the intake manifold to maintain idle speed and compensate for the compressor load on the engine. The ICV and FICD assemblies contain an idle adjust screw to set the specified idle rpm (Figure 11-43).

The operation of the fuel pump relay and in-tank fuel pump is similar to other fuel pump circuits. When the ignition switch is turned on, voltage is supplied to the EFI relay winding and current flows through this winding and the PCM to ground. This action closes the EFI relay points, and voltage is supplied through these points to the PCM and the crank angle sensor.

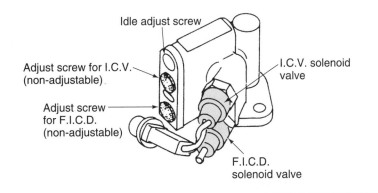

Figure 11-43 Idle control valve solenoid and fast idle control device solenoid valve. *(Courtesy of Nissan Motors)*

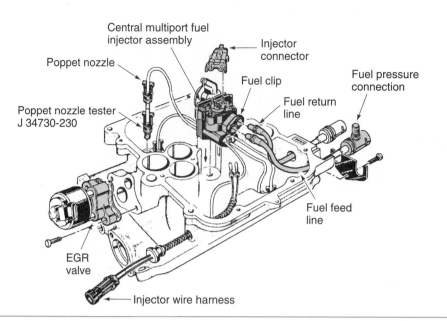

Central multiport fuel
injector assembly

Injector
connector

Poppet nozzle

Fuel clip

Fuel pressure
connection

Poppet nozzle tester
J 34730-230

Fuel return
line

Fuel feed
line

EGR
valve

Injector wire harness

Figure 11-44 Central multiport fuel injection components in the lower half of the intake manifold. *(Courtesy of Chevrolet Motor Division—GMC)*

Shop Manual
Chapter 11, page 497

Central Multiport Fuel Injection (CMFI)

GM recently introduced a new EFI design that incorporates many features of a TBI system with those of port injection. This design first appeared on GM's 4.3L V6 engine. The **central multiport fuel injection (CMFI)** system uses one injector to control the fuel flow to six individual poppet nozzles (Figure 11-44). The CMFI injector assembly consists of a fuel metering body, pressure regulator, one fuel injector, six poppet nozzles with nylon fuel tubes, and a gasket seal. The injector distributes metered fuel through a six hole distribution gasket. The gasket seals the injector to the six lines to the nozzles.

Each nozzle contains a check ball and extension spring that regulates fuel flow. The poppet nozzle opens when high pressure is exerted on the check ball. This action allows the nozzles to feed individual cylinders with atomized fuel.

Pressure Regulator

The pressure regulator is mounted with the central injector. Since this regulator is mounted inside the intake manifold, vacuum from the intake is supplied through an opening in the regulator cover to the regulator diaphragm. The regulator spring pushes downward on the diaphragm and closes the valve. Fuel pressure from the in-tank fuel pump pushes the diaphragm upward and opens the valve, which allows fuel to flow through this valve and the return line to the fuel tank (Figure 11-45).

The pressure regulator is designed to regulate fuel pressure at 54 to 64 psi (370 to 440 kPa), which is higher than many port fuel injection systems. Higher pressure is required in the CMFI system to prevent fuel vaporization from the extra heat encountered with the CMFI assembly, poppet nozzles, and lines mounted inside the intake manifold. The pressure regulator operates the same as the regulators explained previously in this chapter.

Injector Design and Operation

A pivoted armature is mounted under the injector winding in the central injector. The lower side of this armature acts as a valve that covers the six outlet ports to the nylon tubes and poppet nozzles. A supply of fuel at a constant pressure surrounds the injector armature while the ignition switch is on. Each time the PCM grounds the injector winding, the armature is lifted upward, which operates the injector ports. Under this condition, fuel is forced from the nylon tubes to the poppet nozzles (Figure 11-46).

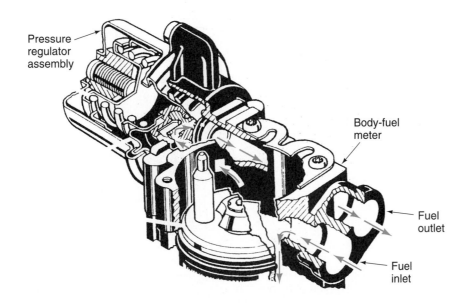

Figure 11-45 Pressure regulator for a CMFI system. *(Courtesy of Chevrolet Motor Division— GMC)*

The amount of fuel delivered by the central injector is determined by the length of time that the PCM keeps the injector winding grounded. This time is referred to as pulse width. When the PCM opens the injector ground circuit, the injector spring pushes the armature downward and closes the injector ports. The injector winding has low resistance, and the PCM operates the injector with a peak-and-hold current. When the PCM grounds the injector winding, the current flow in this circuit increases rapidly to 4 amperes. When the current flow reaches this value, a current-limiting

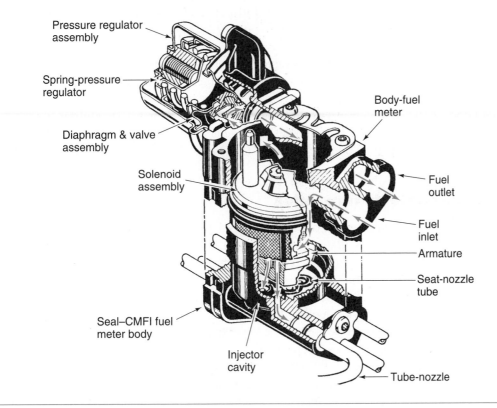

Figure 11-46 Central injector operation. *(Courtesy of Chevrolet Motor Division—GMC)*

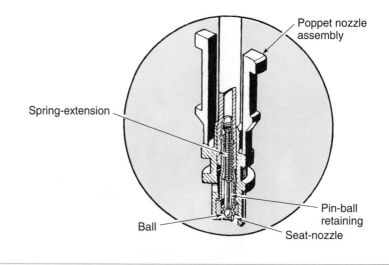

Figure 11-47 Internal design of a poppet nozzle. *(Courtesy of Chevrolet Motor Division—GMC)*

circuit in the PCM limits the current flow to 1 ampere for the remainder of the injector pulse width. The peak-and-hold function provides faster injector armature opening and closing.

Poppet Nozzles

The **poppet nozzles** are snapped into opening in the lower half of the intake manifold, and the tip of each nozzle directs fuel into an intake port. Each port nozzle contains a valve with a check ball seat in the tip of the nozzle. A spring holds the valve and check ball seat in the closed position. When fuel pressure is applied from the central injector through the nylon lines to the poppet nozzles, this pressure forces the valve and check ball seat open against spring pressure. The poppet nozzles open when the fuel pressure exceeds 37 to 43 psi (254 to 296 kPa), and the fuel sprays from these nozzles into the intake ports (Figure 11-47).

When fuel pressure drops below this value, the poppet nozzles close. Under this condition, approximately 40 psi (276 kPa) fuel pressure remains in the nylon lines and poppet nozzles. This pressure prevents fuel vaporization in the nylon lines and nozzles during hot engine operation or hot soak periods. If a leak occurs in a nylon or other CMFI component, fuel drains from the bottom of the intake manifold through two drain holes to the center cylinder intake ports. The in-tank fuel pump, fuel filter, lines, and fuel pump circuit used with the CMFI system are similar to those used with SFI and MFI systems.

Intake Manifold Tuning Valve

The IMTV makes the intake manifold an "active" manifold; that is, it responds to changing driver demands.

The two-piece aluminum intake manifold has an integral throttle body. An **intake manifold tuning valve (IMTV)** assembly is mounted in the top of the intake manifold. The IMTV assembly contains an electric solenoid that operates a rectangular-shaped valve inside the manifold. Two zip tubes are connected from the throttle body to dual plenums in the upper half of the intake manifold (Figure 11-48). Each plenum feeds three tuned intake runners.

The IMTV rectangular valve is mounted in an opening between the two halves of the upper intake manifold. While the engine is operating at lower rpm, the IMTV remains closed, and airflow in the two halves of the upper intake is separated.

When the engine rpm reaches 3,025 to 4,650 and the throttle opening exceeds 36%, the PCM energizes the IMTV solenoid. This action opens the rectangular valve, and the two halves of the intake are now joined through this valve opening. Airflow in the intake increases to improve horsepower and torque.

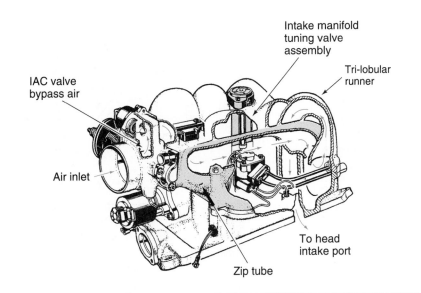

Figure 11-48 Intake manifold tuning valve with the zip tubes in the intake manifold. *(Courtesy of Chevrolet Motor Division—GMC)*

One end of the IMTV relay winding is connected to the ignition switch and the other end to the PCM. When the engine operating conditions require IMTV operation, the PCM grounds the IMTV relay winding. This action closes the relay points. Current flows through the IMTV relay points to the IMTV solenoid winding in the intake manifold to open the rectangular valve (Figure 11-49).

Idle Speed Control Systems

Idle speed control is a function of the PCM. Based on operating conditions and inputs from various sensors, the PCM regulates the idle speed to control emissions.

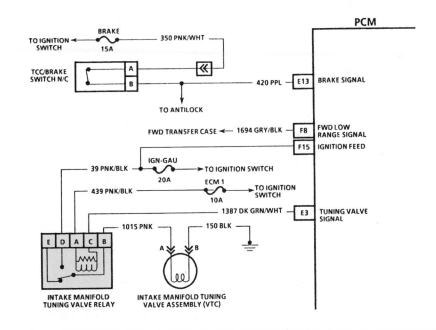

Figure 11-49 IMTV relay and solenoid wiring diagram. *(Courtesy of Chevrolet Motor Division—GMC)*

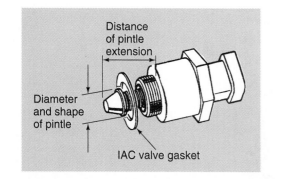

Figure 11-50 IAC BPA motor that threads into the TBI assembly. *(Courtesy of Chevrolet Motor Division—GMC)*

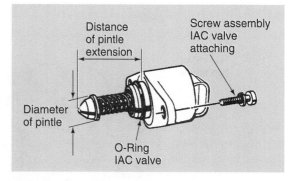

Figure 11-51 IAC BPA motor that bolts onto the TBI assembly. *(Courtesy of Chevrolet Motor Division—GMC)*

Idle Air Control Bypass Air (IAC BPA) Motors

The IAC is used on many EFI systems, as well as computer-controlled carburetors.

Shop Manual
Chapter 11, page 507

Most multiport fuel injection (MFI) and sequential fuel injection (SFI) systems have an idle air control bypass (IAC BPA) motor in the throttle body to control idle speed. This type of IAC BPA motor is also used on some TBI assemblies. In any of these systems, the IAC BPA motor is threaded or bolted into the throttle body assembly (Figures 11-50 and 11-51).

The IAC BPA motor contains a reversible DC motor that is operated by the PCM. A tapered valve is located on the end of the IAC motor stem with air passages above and below the throttle connected to this tapered valve.

The position of this valve controls idle speed by regulating the amount of air bypassing the throttle. If the motor opens the tapered valve and allows more air to bypass the throttle, engine speed is increased. Decreasing the amount of air bypassing the throttle reduces idle speed. Four wires from the motor are connected to the PCM. On many systems, the PCM operates the motor to move the plunger and tapered valve inward and outward in steps, or counts.

⚠ **WARNING:** When installing an IAC BPA motor, if the motor stem is extended more than specified, the motor will be damaged. Use hand pressure to retract the plunger to the specified length prior to motor installation.

IAC BPA Valves

Some MFI or SFI systems have a flap-type IAC BPA valve that regulates the amount of air bypassing the throttle to control idle speed. The PCM opens and closes this valve to regulate the airflow past the throttle. Three wires are connected from the valve to the PCM.

Other MFI and SFI systems have an IAC BPA valve used with a starting air valve and a fast idle thermo switch (Figure 11-52). Two wires are connected from the IAC BPA valve winding to the PCM (Figure 11-53). When the PCM energizes the valves winding, the valve shaft is moved by the coil magnetism. This action opens the valve airflow from the air cleaner through the valve into the intake manifold. If the PCM de-energizes the valve's winding, the shaft spring pushes the valve closed (Figure 11-54). The PCM pulses the valve open and closed to regulate airflow past the valve intake manifold and control idle speed.

Fast Idle Thermo Valve

The fast idle thermo valve is positioned under the throttle body (Figure 11-55). A thermowax element is located in the bottom of this valve. When the engine is cold, the thermowax contracts, moving the valve plunger to open the air bypass valve (Figure 11-56). This valve opening allows air to flow past the valve into the intake manifold, which increases engine speed. As the engine warms

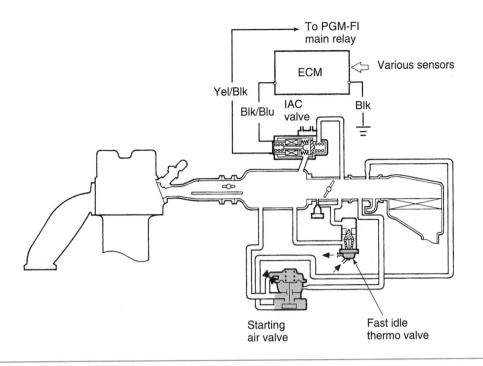

Figure 11-52 IAC BPA valve used with a starting air valve and fast idle thermo valve. *(Courtesy of American Honda Motor Company, Inc.)*

up, the thermowax expands and gradually closes the valve, which slowly reduces engine speed. When the engine reaches normal operating temperature, the thermowax expands enough to close the valve and stop the air flow past the valve. The PCM is programmed to operate the IAC BPA valve with the fast idle thermo valve to provide the proper engine idle speed at all engine temperatures.

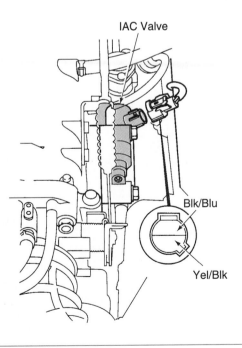

Figure 11-53 IAC BPA valve wiring and mounting. *(Courtesy of American Honda Motor Company, Inc.)*

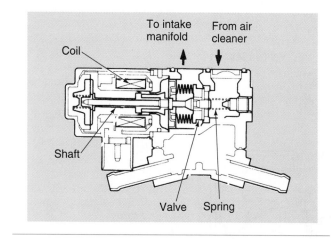

Figure 11-54 Internal design of an IAC BPA valve. *(Courtesy of American Honda Motor Company, Inc.)*

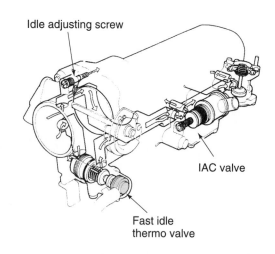

Figure 11-55 Fast idle thermo valve location. *(Courtesy of American Honda Motor Company, Inc.)*

Starting Air Valve

The starting air valve contains a diaphragm in the upper part of the valve. A stem is connected from this diaphragm to a valve in the lower part of the valve (Figure 11-57). A spring above the diaphragm forces the diaphragm downward and opens the valve. Intake manifold vacuum is supplied to the area above the diaphragm.

When the engine is cranking, the intake manifold vacuum will not move the diaphragm in the starting air valve. Under this condition, airflows past the valve into the intake manifold to provide the proper air-fuel ratio for starting. Once the engine starts, the manifold vacuum moves the diaphragm upward and closes the valve, which stops the air flow past the valve. This valve remains closed under all engine operating conditions. The PCM is programmed to operate the injectors and supply the proper air-fuel ratio while starting with air flowing through the starting air valve.

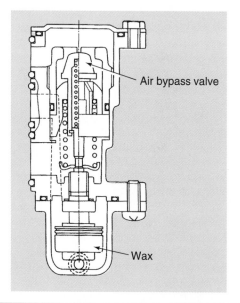

Figure 11-56 Internal design of a fast idle thermo valve. *(Courtesy of American Honda Motor Company, Inc.)*

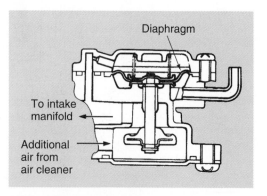

Figure 11-57 Starting air valve. *(Courtesy of American Honda Motor Company, Inc.)*

Summary

- There are three types of electronic fuel injection systems: throttle body, port injection, and central multiport injection. In the throttle body injection system, fuel is delivered to a central point. In the port injection system, there is one injector at each cylinder. Central multiport is a mixture of both throttle body and port injection.

- Port injection systems use one of four firing systems: grounded single fire, grouped double fire, simultaneous double fire, or sequential fire.

- The electronic fuel injection system includes a fuel delivery system, system sensors, electronic control unit, and fuel injectors.

- The volume air flow sensor and mass air flow sensor determine the amount of air entering the engine. The MAP sensor measures changes in the intake manifold pressure that results from changes in engine load and speed.

- The heart of the fuel injection system is the electronic control unit. The PCM receives signals from all the system sensors, processes them, and transmits programmed electrical pulses to the fuel injectors.

- Two types of fuel injectors are currently in use: top feed and bottom feed. Top-feed injectors are used in port injection systems. Bottom-feed injectors are used in throttle body injection systems.

- Two methods are used to control idle speed on engines equipped with fuel injection: an auxiliary air valve and an idle speed solenoid.

- Compared to cast iron or aluminum intake manifolds, glass fiber-reinforced nylon resin intake manifolds save weight and improve fuel economy and hot start performance.

- Many of the same input sensors are used on computer-controlled carburetor systems and EFI systems.

- In a speed density EFI system, the computer uses the MAP or MAF and engine rpm inputs to calculate the amount of air entering the engine. The computer then calculates the required amount of fuel to go with the air entering the engine.

- In any EFI system, the fuel pressure must be high enough to prevent fuel boiling.

- In an EFI system, the computer supplies the proper air-fuel ratio by controlling injector pulse width.

- In an EFI system, the computer increases injector pulse width to provide air-fuel ratio enrichment while starting a cold engine.

- Most computers provide a clear flood mode if a cold engine becomes flooded. This mode is activated by pressing the gas pedal to the floor while cranking the cold engine.

- In an SFI system, each injector has an individual ground wire connected to the computer.

- In an MFI system, the injectors are connected together in pairs or groups on the ground side.

- Compared to steel or aluminum alloy fuel rails, plastic-type fuel rails transfer less heat to the fuel and reduce the possibility of fuel boiling in the fuel rail.

- The cold start injector is operated by a thermo time switch, and this injector sprays additional fuel into the intake manifold while cranking a cold engine.

- The pressure regulator maintains the specified fuel system pressure and returns excess fuel to the fuel tank.

- In a returnless fuel system, the pressure regulator and filter assembly is mounted with the fuel pump and gauge sending unit assembly on top of the fuel tank. This pressure regulator returns fuel directly into the fuel tank.

Terms to Know

Central multiport fuel injection (CMFI)

Clear flood mode

Cold start injector

Electronic fuel injection (EFI)

Fuel pressure regulator

Fuel rail

Idle air control (IAC)

Intake manifold tuning valve (IMTV)

Malfunction indicator lamp (MIL)

Mass air flow sensor

Multiport injection (MPI)

Pintle

Poppet nozzles

Port fuel injection (PFI)

Sequential fuel injection (SFI)

Speed density

Thermo time switch

Throttle body injection (TBI)

❑ A central multiport injection system has one central injector and a poppet nozzle in each intake port. The central injector is operated by the PCM, and the poppet nozzles are operated by fuel pressure.

❑ An intake manifold tuning valve (IMTV) controls the air passages inside the intake manifold to provide improved airflow in the manifold.

❑ An IAC BPA motor controls idle speed by controlling the amount of air bypassing the throttle.

❑ An IAC BPA valve opens and closes the air bypass passage around the throttle to control idle speed.

❑ A fast idle thermo valve is operated by a thermowax element that expands and contracts as it is heated and cooled.

❑ A starting air valve is vacuum-operated, and this valve allows additional airflow into the intake manifold while starting.

Review Questions

Short Answer Essays

1. Describe how an IAC BPA motor controls idle speed.

2. Explain the purpose of the TPS input in a speed-density fuel injection system.

3. Describe the advantages of MFI and SFI systems compared to carburetor fuel systems.

4. Describe the purpose of a throttle body temperature signal on a TBI system.

5. Explain how the computer controls the air-fuel ratio on an EFI system.

6. Explain why a choke is not required on an EFI system.

7. Explain the major differences between throttle body fuel injection and port fuel injection systems.

8. What is meant by sequential firing of fuel injectors?

9. Describe the purpose of a manifold absolute pressure (MAP) sensor.

10. Explain the basic operation of a CMFI system.

Fill-in-the-Blanks

1. The length of time that an injector is energized is called _____

2. When the IAC BPA motor increases the amount of air bypassing the throttle, engine rpm is _____.

3. The computer determines the air entering the engine from the _____ and _____ input signals in a speed-density EFI system.

4. Compared to TBI systems, MFI and SFI systems require _____ fuel pressure.

5. In EFI systems, the fuel pressure must be high enough to prevent _____ _____.

6. The computer program assumes that only liquid fuel is available at the injector and that a specified _____ _____ is available at the injector.

7. On an SFI system, each injector has an individual _____ _____ connected to the computer.

8. If the injector pulse width is increased on EFI systems, the air-fuel ratio becomes _____.

9. When an engine is idling, the pressure regulator provides _____ fuel pressure compared to the fuel pressure at wide-open throttle.

10. In a central multiport fuel injection (CMFI) system, the air-fuel ratio is determined by the pulse width on the _____ _____ _____.

ASE Style Review Questions

1. While discussing EFI systems:
   Technician A says the PCM provides the proper air-fuel ratio by controlling the fuel pressure.
   Technician B says the PCM provides the proper air-fuel ratio by controlling injector pulse width.
   Who is correct?
   A. A only   **C.** Both A and B
   B. B only   **D.** Neither A nor B

2. While discussing cold start injector systems:
   Technician A says the cold start injector is operated by a thermo time switch.
   Technician B says the cold start injector is operated by the PCM.
   Who is correct?
   A. A only   **C.** Both A and B
   B. B only   **D.** Neither A nor B

3. While discussing IAC BPA valves:
   Technician A says the PCM opens and closes the IAC BPA valve to control the amount of air bypassing the throttle.
   Technician B says on some systems the IAC BPA valve is used with a fast idle thermo valve and a starting air valve.
   Who is correct?
   A. A only   **C.** Both A and B
   B. B only   **D.** Neither A nor B

4. While discussing electronic fuel injection principles:
   Technician A says the computer uses the TPS and ECT signals to determine the air entering the engine in a speed-density system.
   Technician B says the computer uses the TPS and oxygen sensor signals to determine the air entering the engine in a speed-density system.
   Who is correct?
   A. A only   **C.** Both A and B
   B. B only   **D.** Neither A nor B

5. While discussing TBI and MFI systems:
   Technician A says that in a TBI or MFI system higher-than-normal fuel pressure causes a lean air-fuel ratio.
   Technician B says that in these systems higher-than-normal fuel pressure causes a rich air-fuel ratio.
   Who is correct?
   A. A only   **C.** Both A and B
   B. B only   **D.** Neither A nor B

6. While discussing fuel pressure regulators:
   Technician A says that the pressure regulator in an SFI system maintains the same fuel pressure regardless of throttle opening.
   Technician B says that the manifold vacuum connection to the pressure regulator in an SFI system causes higher fuel pressure at wide-open throttle.
   Who is correct?
   A. A only   **C.** Both A and B
   B. B only   **D.** Neither A nor B

7. While discussing returnless fuel systems:
   Technician A says in a returnless fuel system the pressure regulator is mounted on the fuel rail.
   Technician B says in this type of fuel system the filter and pressure regulator are combined in one unit.
   Who is correct?
   A. A only   **C.** Both A and B
   B. B only   **D.** Neither A nor B

8. While discussing fuel boiling in the fuel rail:

   Technician A says fuel boiling in the fuel rail causes a lean air-fuel ratio.

   Technician B says the computer will compensate for the improper air-fuel ratio caused by the fuel boiling in the fuel rail.

   Who is correct?

   A. A only   **C.** Both A and B

   B. B only   **D.** Neither A nor B

9. While discussing central multiport injection (CMFI):

   Technician A says the poppet nozzles are opened by the computer.

   Technician B says the poppet nozzles are opened by fuel pressure.

   Who is correct?

   A. A only   **C.** Both A and B

   B. B only   **D.** Neither A nor B

Distributor Ignition Systems

Upon completion and review of this chapter, you should be able to:

❏ Describe the operation of distributor-based ignition systems.

❏ Explain the purpose of the electronic control unit.

❏ Describe the various types of spark timing systems, including electronic switching systems and their related engine position sensors.

❏ Describe the operation of the various switching devices used in distributors.

❏ Describe the major differences between a Dura-Spark II and a Dura-Spark I ignition system.

❏ Explain the term variable dwell as it relates to a high energy ignition (HEI) system.

❏ Describe the differences in operation of a thick film integrated system and a high energy ignition system.

❏ Explain the basic operation of a computer-controlled ignition system.

❏ Explain how the fuel injection system may rely on components of the ignition system.

Introduction

One of the requirements for an efficient running engine is the correct amount of heat delivered into the cylinders at the right time. This requirement is the responsibility of the ignition system. The ignition system supplies properly timed high-voltage surges to the spark plugs. These voltage surges cause combustion inside the cylinder. For each cylinder in an engine, the ignition system has three main jobs. First, it must generate an electrical spark that has enough heat to ignite the air-fuel mixture in the combustion chamber. Secondly, it must maintain that spark long enough to allow for the combustion of all the air and fuel in the cylinders. Lastly, it must deliver the spark to each cylinder so that combustion can begin at the right time during the compression stroke of each cylinder.

The job of the ignition system is not easy. When it fails to provide the correct amount of heat at the correct time, exhaust emissions increase and engine performance and fuel economy decrease. Through the years, the ignition system has become more precise, reliable, and durable.

From the fully mechanical breaker point system, ignition technology progressed to basic electronic triggering and switching devices. The electronic switching components are normally inside a separate housing known as an electronic control module or control unit (Figure 12-1). The original electronic ignitions still relied on mechanical and vacuum advance mechanisms in the distributor.

In order to have an efficient running engine there must be the correct amount of air mixed with the correct amount of fuel, in a sealed container, shocked by the correct amount of heat at the right time.

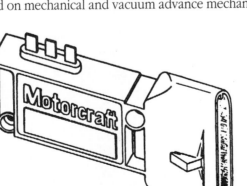

B4046-A

Figure 12-1 A control module for a distributor ignition system. *(Reprinted with the permission of Ford Motor Company)*

As technology advanced, many manufacturers expanded the ability of the ignition control modules. For example, by tying a manifold vacuum sensor into the ignition module circuitry, the module could now detect when the engine was under heavy load and retard the timing automatically. Similar add-on sensors and circuits were designed to control spark knock, start-up emissions, and altitude compensation. The expansion of the control module's duties led to the development of computer-controlled ignition systems.

Computer-controlled ignition systems offer continuous spark timing control through a network of engine sensors and a central microprocessor. Based on the inputs it receives, the central microprocessor or computer makes decisions regarding spark timing and sends signals to the ignition module to fire the spark plugs according to those inputs and according to the programs in its memory. Computer-controlled ignition systems may or may not use a distributor to distribute secondary voltage to the spark plugs.

Distributor Ignition System Components

All ignition systems share a number of common components. Some, such as the battery and ignition switch perform simple functions. The battery supplies low-voltage current to the ignition primary circuit. The current flows when the ignition switch is in the start or the run position. Full-battery voltage is always present at the ignition switch, as if it were directly connected to the battery.

Ignition Coil

A laminated soft iron core is positioned at the center of the ignition coil, and an insulated primary winding is wound around this core. The primary winding has approximately 200 turns of heavier wire, and the two ends of this winding are connected to the primary terminals on top of the coil tower (Figure 12-2). These terminals are usually identified with positive and negative symbols. An enamel-type insulation prevents the primary turns from touching each other, and paper insulation is positioned between the layers of turns.

A typical secondary coil winding may contain 22,000 turns of very fine wire that is wound inside the primary winding. A similar insulation method is used on the secondary and primary windings. The ends of the secondary winding are usually connected to one of the primary terminals and the high tension terminal in the coil tower. When the winding and core assembly is mounted in the coil container, the core rests on a ceramic insulating cup in the bottom of the container. Metal sheathing is placed around the outside of the coil windings in the container, and this

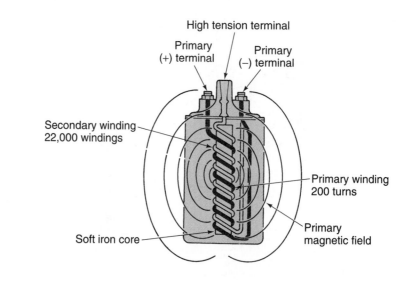

Figure 12-2 Ignition coil design. *(Courtesy of Chrysler Corporation)*

In the SAE J1930 terminology, the term distributor ignition (DI) replaces all previous terms for electronically controlled distributor-type ignition systems.

Shop Manual
Chapter 12, page 525

Some manufacturers use a transformer, called an igniter, in place of a coil. An igniter is very much like an ignition coil.

sheathing concentrates the magnetic field on the outside of the windings. A sealing washer is positioned on the container under the tower, and the container is crimped over the tower in the manufacturing process. The coil is filled with oil through the screw hole in the high tension terminal before this screw is installed. When the coil is oil-filled, air space and moisture formation are eliminated, and the oil helps to cool the coil.

Ignition Switch and Resistor

Some ignition systems have a ballast resistor connected in series between the ignition switch and the coil positive primary terminal. This resistor supplies the correct amount of voltage and current to the coil. In some ignition systems, a calibrated resistance wire is used in place of the ballast resistor. Many ignition systems at the present time have eliminated the resistor. These systems supply 12 V directly to the coil.

The ignition resistor may be called a ballast resistor.

There are some electronic ignition systems that do not require a ballast resistor. For instance, some control units directly regulate the current flow through the primary of the coil. Hall effect systems do not require ballast resistors either. The signal voltage is not changed by the speed of the distributor as it is in an inductive magnetic signal generating system.

Triggering Devices

Some distributor ignition (DI) systems have the primary triggering device located in the distributor, while other systems use triggering devices external to the distributor. Regardless of location, these triggering devices keep the ignition system synchronized with the position of the engine's pistons. Many distributors were fitted with a pick-up coil bolted to the pick-up plate. The two ends of the pick-up winding are connected to the lead wires extending from the distributor (Figure 12-3). These pick-up leads are connected to the ignition electronic control unit.

A reluctor is pressed onto the distributor shaft and a roll pin prevents reluctor rotation on the shaft. The reluctor has a high point for each engine cylinder. The reluctor high points rotate past the pick-up coil with a small gap. Voltage signals are produced by the pick-up coil as the high points move toward and away from it.

Electronic Control Unit

The electronic control unit (ECU) contains a circuit board with many electronic components. A power transistor is mounted on a heat sink on the front of some ECUs, and a sealing compound is placed on the back of the ECU, which prevents servicing of this component. These control modules are mounted either inside or outside the distributor assembly. The size and shape of the modules also varies with manufacturer.

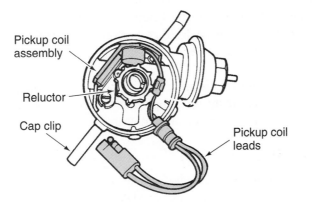

Figure 12-3 Pick-up unit mounted inside a distributor. *(Courtesy of Chrysler Corporation)*

Distributor

The reluctor and distributor shaft assembly rotates on bushings in the aluminum distributor housing. A roll pin extends through a retainer and the distributor shaft to retain the shaft in the distributor. Another roll pin retains the drive gear to the lower end of the shaft. On many engines, this drive gear is meshed with the camshaft gear to drive the distributor. Some distributors have an offset slot on the end of the distributor shaft that meshes with a matching slot in a gear driven from the camshaft. The gear size is designed to drive the distributor at the same speed as the camshaft, which rotates at one-half the speed of the crankshaft in a four-stroke cycle engine.

If the distributor has advance mechanisms, the centrifugal advance is sometimes mounted under the pick-up plate and the vacuum advance is positioned on the side of the distributor. Most engines manufactured in recent years have computer-controlled spark advance, and the distributor advance mechanisms are eliminated.

As its name implies, a distributor mechanically distributes the spark so that it arrives at the right time during the compression stroke of each cylinder. The distributor's shaft, rotor, and cap perform this function.

The distributor cap and rotor receive high voltage from the secondary winding via a high-tension wire. The voltage enters the distributor cap through the coil tower. The rotor then sends the voltage from the coil tower to the spark plug electrodes inside the distributor cap. The distributor cap is made from silicone plastic or similar material that offers protection from chemical attack. The coil tower contains a carbon insert that carries the voltage from the high-tension coil lead to the raised portion of the electrode on the rotor. Spaced evenly around the coil tower are the spark plug electrodes and towers for each spark plug.

The coil or center electrode inside the distributor cap may be called the center button.

An air gap of a few thousandths of an inch exists between the tip of the rotor electrode and the spark plug electrode inside the cap. This gap is necessary in order to prevent the two electrodes from making contact. The rotor is positioned on top of the distributor shaft. A metal strip on the top of the rotor makes contact with the center distributor cap terminal, and the outer end of the strip rotates past the cap terminals as it rotates. This action completes the circuit between the ignition coil and the individual spark plugs according to the firing order.

The rotor is positioned on top of the distributor shaft and a projection inside the rotor fits into a slot in the shaft to ensure that the rotor can be installed only in one position. A metal strip on top of the rotor makes contact with the center distributor cap terminal. The outer end of this strip rotates past the cap terminals as the shaft rotates (Figure 12-4). The alignment of the rotor's strip with a cap terminal allows current flow from the coil secondary winding to the appropriate spark plug wire.

Spark Plugs

Every type of ignition system uses spark plugs. The spark plugs provide the crucial air gap across which the high-voltage current from the coil flows across in the form of an arc. The three main parts of a spark plug are the steel core, the ceramic core, or insulator, which acts as a heat conductor; and a pair of electrodes, one insulated in the core and the other grounded on the shell. The shell holds the ceramic core and electrodes in a gas-tight assembly and has the threads needed for plug installation in the engine (Figure 12-5).

An ignition cable connects the secondary to the top of the plug. Current flows through the center of the plug and arcs from the tip of the center electrode to the ground electrode. The resulting spark ignites the air-fuel mixture in the combustion chamber.

The spark plug electrodes provide gaps inside each combustion chamber across which the secondary current flows to ignite the air-fuel mixture in the combustion chambers.

Primary Circuit Operation

The primary circuit of a DI system is controlled electronically by a triggering device and the electronic control unit that contains some type of switching device.

When the ignition switch is in the on position, current from the battery flows through the ignition switch and primary circuit resistor to the primary winding of the ignition coil. From here it

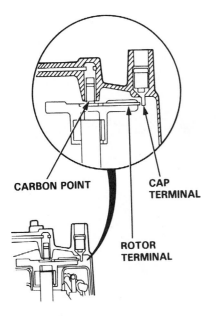

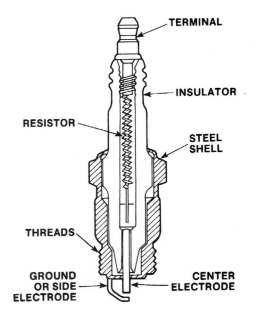

Figure 12-4 Relationship of a rotor and distributor cap. *(Courtesy of American Honda Motor Company, Inc.)*

Figure 12-5 A typical spark plug.

passes through some type of switching device and back to ground. The switching device is controlled by the triggering device. The current flow in the ignition coil's primary winding creates a magnetic field. The switching device or control module interrupts this current flow at predetermined times. When it does, the magnetic field in the primary winding collapses. This collapse generates a high-voltage surge in the secondary winding of the ignition coil. The secondary circuit of the system begins at this point.

Dwell time is the length of time, in degrees of distributor rotation, that current is flowing in the primary circuit.

Secondary Circuit Operation

In a DI system, high voltage from the secondary winding passes through an ignition cable running from the coil to the distributor. The distributor then distributes the high voltage to the individual spark plugs through a set of ignition cables. The cables are arranged in the distributor cap according to the firing order of the engine. The rotor is driven by the distributor shaft; it rotates and completes the electrical path from the secondary winding of the coil to the individual spark plugs. The distributor delivers the spark to match the compression stroke of the piston. The distributor assembly may also have the capability of advancing or retarding ignition timing.

The typical amount of secondary coil voltage required to jump the spark plug gap is 10,000 V. Most coils have a maximum secondary voltage of more than 25,000 V. The difference between the required voltage and the maximum available voltage is referred to as secondary reserve voltage. This reserve voltage is necessary to compensate for high cylinder pressures and increased secondary resistances as the spark plug gap increases through use. The maximum available voltage must always exceed the required firing voltage or ignition misfire will occur. If there is an insufficient amount of voltage available to push current across the gap, the spark plug will not fire.

Distributor Timing Advance Units

As stated, early electronic ignition systems changed the timing mechanically just like breaker point systems. At idle, the firing of the spark plug usually occurs just before the piston reaches top dead center. At higher engine speeds however, the spark must be delivered to the cylinder much earlier in the cycle to achieve maximum power from the air-fuel mixture, since the engine is moving

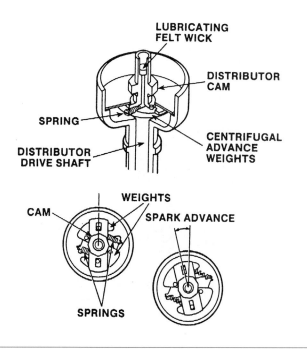

Figure 12-6 Typical centrifugal advance mechanism.

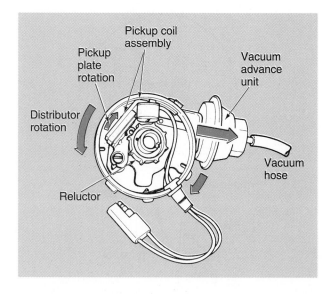

Figure 12-7 Action of a vacuum advance unit. *(Courtesy of Chrysler Corporation)*

through the cycle more quickly. To change the timing of the spark in relation to engine speed, the centrifugal advance mechanism is used (Figure 12-6).

This mechanism consists of a set of weights and springs connected to the distributor shaft and a distributor armature assembly. During idle speeds, the springs keep the weights in place and the armature and distributor shaft rotate as one assembly. When speed increases, centrifugal force causes the weights to slowly move out against the tension of the springs. This allows the armature assembly to move ahead in relation to the distributor shaft rotation. The ignition's triggering device is mounted to the armature assembly. Therefore, as the assembly moves ahead, ignition timing becomes more advanced. As engine speed decreases, the springs pull the weights back in. This action returns the ignition timing to its base setting.

During part-throttle engine operation, high vacuum is present in the intake manifold. To get the most power and the best fuel economy from the engine, the plugs must fire even earlier during the compression stroke than is provided by a centrifugal advance mechanism. This additional advance is provided for by the vacuum advance unit.

When there is a great load on the engine its manifold vacuum is very low.

A spring-loaded diaphragm (Figure 12-7) inside a metal housing is connected to the movable plate on which the pick-up coil is mounted. Vacuum is applied to one side of the diaphragm in the housing chamber while the other side of the diaphragm is open to the atmosphere. Any increase in vacuum allows atmospheric pressure to push the diaphragm. In turn, this causes the movable plate to rotate. The more vacuum present on one side of the diaphragm, the more atmospheric pressure is able to cause a change in timing. The rotation of the movable plate moves the pick-up coil so the armature develops a signal earlier. These units are also equipped with a spring that retards the timing as vacuum decreases.

Triggering Devices

The primary circuit is controlled by a negative-positive-negative (NPN) transistor. The transistor's emitter is connected to ground. The collector is connected to the negative (-) terminal of the coil. When the triggering device supplies a small amount of current to the base of the switching transistor, the collector and emitter allows current to build up in the coil primary circuit. When the current to the base is interrupted by the switching device, the collector and emitter no longer allows current flow in the primary.

Engine Position Sensors

The time when the primary circuit must be opened and closed is related to the position of the pistons and the crankshaft. Therefore, the position of the crankshaft is used to control the flow of current to the base of the switching transistor.

A number of different types of sensors are used to monitor the position of the crankshaft and control the flow of current to the base of the transistor. These engine position sensors and generators serve as triggering devices and include magnetic pulse generators, metal detection sensors, Hall effect sensors, and photoelectric sensors. These are all described in some detail in Chapter 5.

All four types of sensors can be mounted in the distributor, which is turned by the camshaft. Magnetic pulse generators and Hall effect sensors can also be located on the crankshaft. Both Hall effect sensors and magnetic pulse generators can also be used as camshaft reference sensors to identify which cylinder is the next one to fire.

Magnetic Pulse Generator

Shop Manual
Chapter 12, page 535

The magnetic pulse or PM generator operates on basic electromagnetic principles. As the disc teeth approach the pick-up coil, they repel the magnetic field, forcing it to concentrate around the pick-up coil (Figure 12-8). Once the tooth passes by the pick-up coil, the magnetic field is free to expand or unconcentrate until the next tooth on the disc approaches. Approaching teeth concentrate the magnetic lines of force, while passing teeth allow them to expand. This pulsation of the magnetic field causes the lines of magnetic force to cut across the winding in the pick-up coil, inducing a small amount of AC voltage that is sent to the switching device in the primary circuit.

When a disc tooth is directly in line with the pick-up coil, the magnetic field is not expanding or contracting. Since there is no movement or change in the field, voltage at this precise movement drops to zero. At this point, the switching device inside the ignition module reacts to the zero voltage signal by turning the ignition's primary circuit current off.

As soon as the tooth rotates past the pick-up coil, the magnetic field expands again and another voltage signal is induced. The only difference is that the polarity of the charge is reversed. Negative becomes positive or positive becomes negative. Upon sensing this change in voltage, the switching device turns the primary circuit back on and the process begins all over.

Metal Detection Sensors

Metal detection sensors work much like a magnetic pulse generator with one major difference, the pick-up coil of a metal detection sensor does not have a permanent magnet. Instead, the pick-up coil is an electromagnet. A low level of current is supplied to the coil by an electronic control unit,

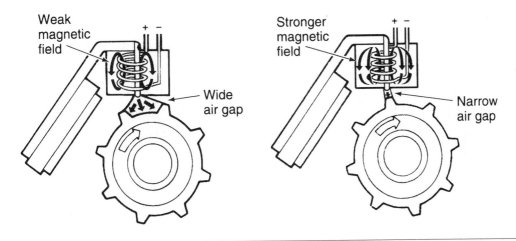

Figure 12-8 Action of a magnetic pick-up unit. *(Courtesy of Chrysler Corporation)*

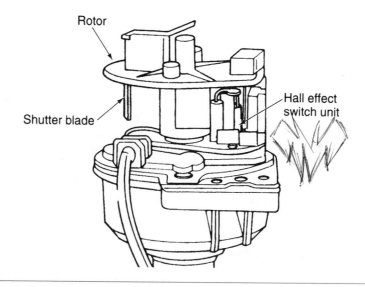

Figure 12-9 Hall effect switch located inside the distributor. *(Courtesy of Chrysler Corporation)*

inducing a weak magnetic field around the coil. As the reluctor on the distributor shaft rotates, the trigger teeth pass very close to the coil. As the teeth pass in and out of the coil's magnetic field, the magnetic field builds and collapses, producing a corresponding change in the coil's voltage. The voltage changes are monitored by the control unit to determine crankshaft position.

Hall Effect Sensor

The Hall effect switch is the most commonly used engine position sensor (Figure 12-9). The Hall effect sensor produces an accurate voltage signal throughout the entire rpm range of the engine. It also produces a square wave signal that is more compatible with the digital signals required by computers.

A Hall switch is based on the Hall effect principle which states: If a current is allowed to flow through a thin conducting material, and that material is exposed to a magnetic field, voltage is produced.

The Hall switch is described as being "on" any time the Hall layer is exposed to a magnetic field and a Hall voltage is being produced (Figure 12-10). A shutter wheel performs the same function as the timing disc on magnetic pulse generators. The only difference is the shutter wheel creates a magnetic shunt that changes the field strength through the Hall element. When a vane of the shutter wheel is positioned between the magnet and Hall element, the metallic vane blocks the

Shop Manual
Chapter 12, page 535

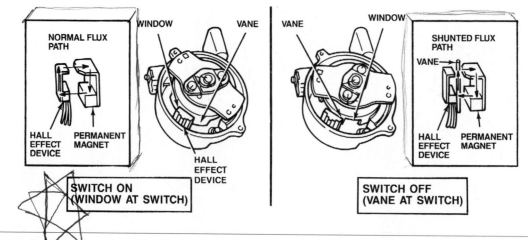

Figure 12-10 Operation of a Hall effect switch. *(Reprinted with the permission of Ford Motor Company)*

magnetic field and keeps it from permeating the Hall layer. As a result, Hall output voltage is low. Conversely, when a window rotates into the air gap, the magnetic field is able to penetrate the Hall layer, which in turn pushes the Hall voltage to its maximum range.

Photoelectric Sensor

A fourth type of crankshaft position sensor is the photoelectric sensor. The parts of this sensor include a light-emitting diode (LED), a light sensitive phototransistor, and a slotted disc called a light beam interrupter.

The slotted disc is attached to the distributor shaft. The LED and the photo cell are situated over and under the disc opposite of each other. As the slotted disc rotates between the LED and photo cell, light from the LED shines through the slots. The intermittent flashes of light are translated into voltage pulses by the photo cell.

DI Systems with an External Ignition Module

Through the years there have been many different designs of DI systems, all operate in basically the same way but are configured differently. The systems described in this section represent the different designs used by manufacturers. These designs are based on the location of the ECU. Systems are either mounted in the module away from the distributor, on the distributor, or in the distributor.

Shop Manual
Chapter 12, page 527

The armature has also been called the stator or reluctor.

Dura-Spark Ignition Systems

Ford Motor Company used two generations of Dura-Spark ignition systems. The second design (Dura-Spark II) is based on the first (Dura-Spark I). The Dura-Spark II (Figure 12-11) had the ECU mounted away from the distributor, typically on a fender wall. The distributor is fitted with a cen-

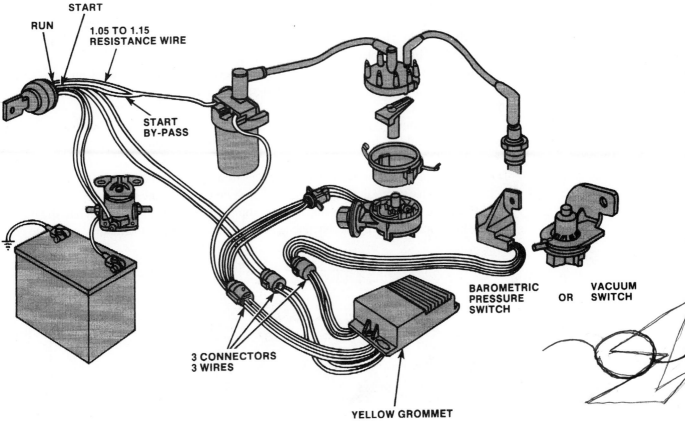

Figure 12-11 Dura-Spark II ignition system. *(Reprinted with the permission of Ford Motor Company)*

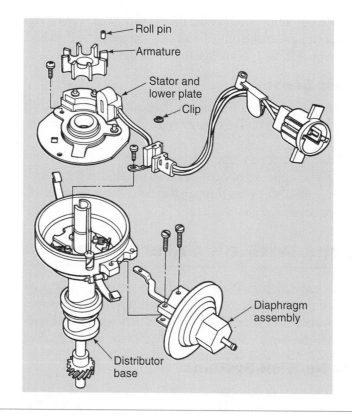

Figure 12-12 Dura-Spark II distributor assembly. *(Reprinted with the permission of Ford Motor Company)*

trifugal advance assembly and a vacuum advance unit. An armature is mounted to the distributor shaft with a roll pin. The armature has a high point for each cylinder. The magnetic pick-up coil is mounted to the breaker plate (Figure 12-12).

In this system, a resistance wire is connected between the ignition switch and the positive primary coil terminal. A bypass wire is also connected from the ignition switch to the positive coil terminal (Figure 12-13).

When the engine is running, 9.6 V are supplied through the resistance wire to the coil. While the engine is cranking, full battery voltage is supplied to the coil through the bypass wire. This increases primary current flow and allows the coil to provide high voltage to ensure good starting. The negative primary coil terminal is referred to as a distributor electronic control (dec) or tachometer (tach) terminal. This terminal is connected to the ignition module.

When the ignition switch is on, 12 V are supplied from the ignition switch to the module. A white wire supplies 12 V from a separate ignition switch terminal to the module while the engine is

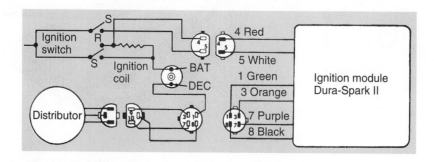

Figure 12-13 Schematic of a Dura-Spark II ignition system. *(Reprinted with the permission of Ford Motor Company)*

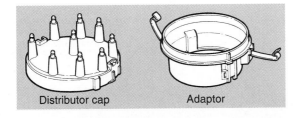

Figure 12-14 Two-piece distributor cap for a Dura-Spark II ignition system. *(Reprinted with the permission of Ford Motor Company)*

cranked. The distributor pick-up coil is connected through two wires to the module. A wire is also connected from the distributor housing to the module. This wire supplies a ground connection from the module to the pick-up plate; therefore, the module does not need to be grounded where it is mounted.

An armature is mounted on the distributor shaft with a roll pin. The armature has a high point for each engine cylinder. Rivets are used to attach the pick-up coil to the plate, which indicates that the pick-up gap is not adjustable. The Dura-Spark II distributor has conventional centrifugal and vacuum advances.

A unique feature of this ignition system is the design of the distributor. The cap is a two-piece unit. An adaptor, the lower portion of the cap, is positioned on top of the distributor housing. Its upper diameter of the adaptor is larger than the lower. This increased diameter allows for a larger distributor cap (Figure 12-14). The larger diameter cap places the spark plug terminals further apart. This helps to prevent cross-firing.

Dura-Spark I Ignition Systems

Dura-Spark I ignition systems are similar to Dura-Spark II systems except for these differences:

1. The primary circuit does not use a resistance wire, rather an ordinary wire is connected from the ignition switch to the positive primary coil terminal (Figure 12-15).

2. The primary coil winding resistance is lower (0.75 ohm compared to 1 to 2 ohms).

3. The module has a variable dwell circuit and this module is not interchangeable with a Dura-Spark II module.

DI Systems with Module Mounted on Distributor

In Ford's **thick film integrated (TFI)** ignition system, the resistor wire between the ignition switch and the positive primary coil is discontinued and an ordinary wire is connected between these

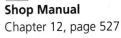
Shop Manual
Chapter 12, page 527

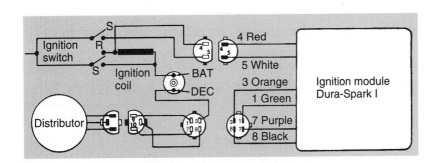

Figure 12-15 Schematic of a Dura-Spark I ignition system. *(Reprinted with the permission of Ford Motor Company)*

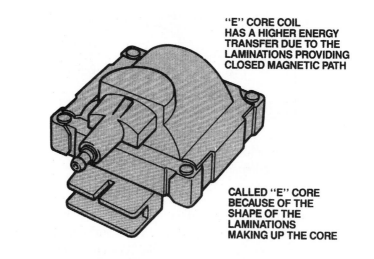

"E" CORE COIL HAS A HIGHER ENERGY TRANSFER DUE TO THE LAMINATIONS PROVIDING CLOSED MAGNETIC PATH

CALLED "E" CORE BECAUSE OF THE SHAPE OF THE LAMINATIONS MAKING UP THE CORE

Figure 12-16 Typical E-core ignition coil. *(Reprinted with the permission of Ford Motor Company)*

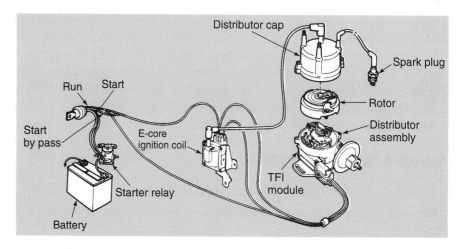

Distributor cap

Spark plug

Run Start

Start by pass

Start

E-core ignition coil

Rotor

Distributor assembly

TFI module

Starter relay

Battery

Figure 12-17 TFI ignition system. *(Reprinted with the permission of Ford Motor Company)*

E-core coils are set in epoxy when they are made. This is why they are sometimes called Epoxy coils.

Shop Manual
Chapter 12, page 527

components. This ignition system has an E-core coil (Figure 12-16). The negative primary coil terminal is referred to as a tach terminal. This terminal is connected to the module that is bolted on the side of the distributor housing (Figure 12-17). Mounting the module onto the side of the distributor eliminates the need to run wires from the control module to the distributor.

A wire is connected from the ignition switch start terminal to the module. The module-to-pick-up terminals extend through the distributor housing, and three pick-up lead wires are connected to these terminals (Figure 12-18). Conventional centrifugal and vacuum advances are used in the TFI distributor. The TFI ignition system operation and distributor advance operation are similar to the systems explained previously.

DI Systems with an Internal Ignition Module

Perhaps the best example of a DI system with the ignition module inside the distributor is the GM **High Energy Ignition (HEI)** system. Some HEI units also contain the ignition coil, others have the coil remotely mounted away from the distributor. Some early HEI designs have centrifugal and vacuum advance units, while others utilize electronic spark timing.

In a high energy ignition (HEI) system, the pick-up coil surrounds the distributor shaft, and a flat magnetic plate is bolted between the pick-up coil and the pole piece (Figure 12-19). A timer

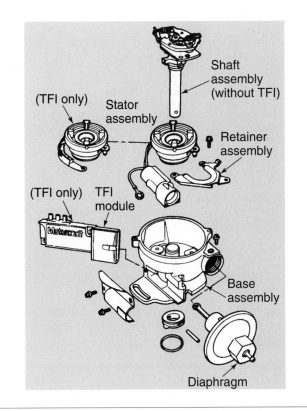

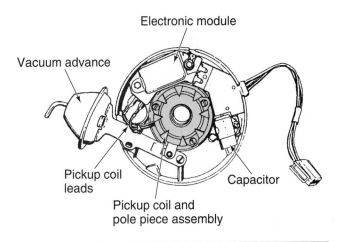

Figure 12-19 HEI component location. *(Courtesy of Oldsmobile Division—GMC)*

Figure 12-18 TFI distributor assembly. *(Reprinted with the permission of Ford Motor Company)*

core that has one high point for each engine cylinder is attached to the distributor shaft. The number of timer core high points matches the number of teeth on the pole piece. This design allows the **timer core** high points to be aligned with the pole piece teeth at the same time.

The dual pick-up lead wires are connected to the HEI module, which is bolted to the distributor housing. Reversal of the pick-up wires on the module is impossible because the wire terminals are different sizes. Heat dissipating grease must be placed on the module mounting surface to prevent excessive module heat. A capacitor is connected from the module voltage supply terminal to ground on the distributor housing (Figure 12-20).

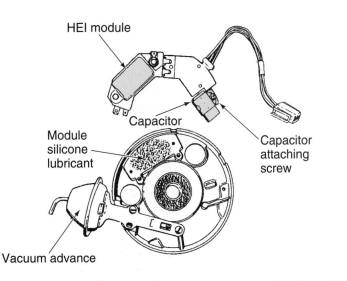

Figure 12-20 HEI control module mounting. *(Courtesy of Oldsmobile Division—GMC)*

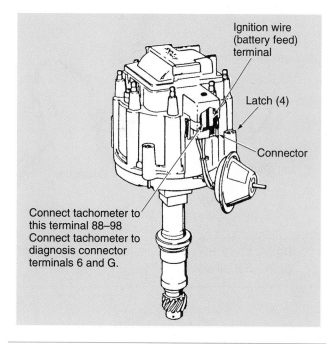

Connect tachometer to this terminal 88–98 Connect tachometer to diagnosis connector terminals 6 and G.

Figure 12-21 HEI distributor terminal identification. *(Courtesy of Oldsmobile Division—GMC)*

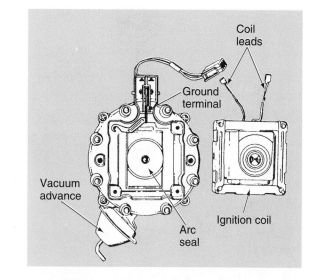

Figure 12-22 Location of ignition coil when it is part of the distributor assembly. *(Courtesy of Oldsmobile Division—GMC)*

The coil battery terminal is connected directly to the ignition switch, and the coil tachometer (tach) terminal is connected to the module (Figure 12-21). A wire also extends from the coil battery terminal to the module. In many HEI systems with an integral coil, a ground wire is connected between the coil frame and the distributor housing to dissipate induced voltages in the coil frame (Figure 12-22).

HEI coils are similar to Ford TFI ignition E-core coils as they both have windings set in epoxy and a laminated iron frame surrounding the windings. The HEI centrifugal advance mechanism is mounted under the rotor, and the rotor is bolted to the centrifugal advance plate.

HEI System Operation

When the ignition switch is on and the distributor shaft is not turning, the module opens the primary ignition circuit. As the engine is cranked and the timer core high points approach alignment with the pole piece teeth, a positive voltage is induced in the pick-up coil. This voltage signal causes the module to close the primary circuit, and current begins to flow through the primary circuit. Under this condition, the magnetic field expands around the coil windings.

At the instant of alignment between the timer core high points and pole piece teeth, the pick-up coil voltage drops to zero. As these high points move out of alignment, a negative voltage is induced in the pick-up coil. This voltage signal to the module causes the module to open the primary circuit. When this action occurs, the magnetic field collapses across the ignition coil windings, and the high induced secondary voltage forces current through the secondary circuit and across the spark plug gap.

HEI modules have a **variable dwell** feature, which closes the primary circuit sooner as engine speed increases. In an eight-cylinder distributor, there are 45° of distributor shaft rotation between the timer core high points. At idle speed, the module closes the primary circuit for 15° and opens the circuit for 30°. If the engine is operating at high speed, the module may close the primary circuit for 32° and open the circuit for 13°. This dwell increase in relation to engine speed provides a stronger magnetic field in the coil and improved secondary voltage at high speed. Since dwell is a function of the module, there is no dwell adjustment.

DI Systems with Computer-Controlled Ignition Timing

Computer-controlled ignition systems control the primary circuit and distribute the firing voltages in the same manner as other types of electronic ignition systems. The main difference between the systems is the elimination of any mechanical or vacuum advance devices from the distributor in the computer-controlled systems. In these systems, the distributor's sole purpose is to generate the primary circuit's switching signal and distribute the secondary voltage to the spark plugs. Timing advance is controlled by a microprocessor, or computer. In fact, some of these systems have even removed the primary switching function from the distributor by using a crankshaft position sensor. The distributor's only job then is to distribute secondary voltage to the spark plugs.

Spark timing on these systems is controlled by a computer that continuously varies ignition timing to obtain optimum air-fuel combustion. The computer monitors the engine operating parameters with sensors. Based on this input, the computer signals an ignition module to collapse the primary circuit, allowing the secondary circuit to fire the spark plugs (Figure 12-23).

Timing control is selected by the computer's program. During engine starting, computer control is bypassed and the mechanical setting of the distributor controls spark timing. Once the engine is started and running, spark timing is controlled by the computer. This scheme or **strategy** allows the engine to start regardless of whether the electronic control system is functioning properly or not.

The goal of computerized spark timing is to produce maximum engine power, top fuel efficiency, and minimum emissions levels during all types of operating conditions. The computer does this by continuously adjusting ignition timing. The computer determines the best spark timing based on certain engine operating conditions such as crankshaft position, engine speed, throttle position, engine coolant temperature, and initial and operating manifold or barometric pressure. Once the computer receives input from these and other sensors, it compares the existing operating conditions to information permanently stored or programmed into its memory. The computer matches the existing conditions to a set of conditions stored in its memory, determines proper timing setting, and sends a signal to the ignition module to fire the plugs.

The computer continuously monitors existing conditions, adjusting timing to match what its memory tells it is the ideal setting for those conditions. It can do this very quickly, making thousands of decisions in a single second. The control computer typically has the following types of information permanently programmed into it.

Speed-related Spark Control. As engine speed increases to a particular point, there is a need for more advanced timing. As the engine slows, the timing should be retarded or have less advance. The computer bases speed-related spark advance decisions on engine speed and signals from the TPS.

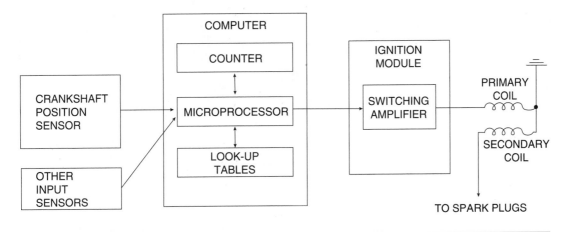

Figure 12-23 Typical flowchart for a computer-controlled ignition system.

Load-related Spark Control. This is used to improve power and fuel economy during acceleration and heavy load conditions. The computer defines the load and the ideal spark advance by processing information from the TPS, MAP, and engine speed sensors. Typically, the more load on an engine, the less spark advance is needed.

Warm-up Spark Advance. This is used when the engine is cold, since a greater amount of advance is required while the engine warms up.

Special Spark Advance. This is used to improve fuel economy during steady driving conditions. During constant speed and load conditions, the engine will be more efficient with much advance timing.

Spark Advance due to Barometric Pressure. This is used when barometric pressure exceeds a preset calibrated value.

All of this information is looked at by the computer to determine the ideal spark timing for all conditions. The calibrated or programmed information in the computer is contained in what is called software **look-up tables**.

Ignition timing can also work in conjunction with the electronic fuel control system to provide emission control, optimum fuel economy, and improved driveability. They are all dependent on spark advance. An example of this type of system is used by Chrysler.

Chrysler's system has two Hall effect switches in the distributor when the engine is equipped with port fuel injection (Figure 12-24). In some units, the pick-up unit used for ignition triggering is located above the pick-up plate in the distributor and is referred to as the reference pickup. The second pick-up unit is positioned below the plate. A ring with two notches is attached to the distributor shaft and rotates through the lower pick-up unit. This lower pickup is called the synchronizer (SYNC) pickup.

In other designs, the two pick-up units are mounted below the pick-up plate and one set of blades rotates through both Hall effect units (Figure 12-25). The shutter blade representing the number one cylinder has a large opening in the center of the blade. When this blade rotates through the SYNC pickup, a different signal is produced compared to the other blades. This number one blade signal informs the PCM when to activate the injectors.

Shop Manual
Chapter 12, page 543

WARNING: Some vehicle manufacturers do not recommend removing or disabling spark plug wires with the engine running for more than ten seconds. Such action may cause catalytic converter overheating. Always consult the vehicle manufacturer's service manual regarding the removal of spark plug wires with the engine running.

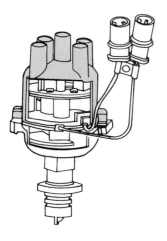

Figure 12-24 Distributor assembly with a reference and SYNC pickup. *(Courtesy of Chrysler Corporation)*

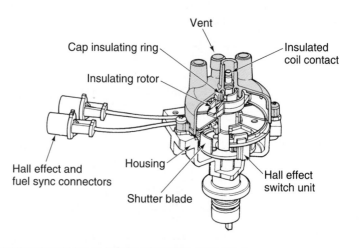

Figure 12-25 Distributor assembly with two Hall effect switches below the pick-up plate. *(Courtesy of Chrysler Corporation)*

In Chrysler's system, the Hall effect switch and the distributor plate are a one-piece assembly, and the shutter blades are attached to the rotor. The shutter blades are insulated from the secondary electrode of the rotor. When the ignition switch is on, 9.2 V are supplied by the PCM to the distributor pick-up. A ground wire extends from the Hall effect pickup to the PCM. The Hall effect pick-up signal is sent through a wire to the PCM.

Each time a shutter blade moves into the Hall effect pick-up, the pick-up signal changes from 0 V to 8 V. This signal always occurs at the same crankshaft position, such as 10° BTDC on the compression stroke. The 0 V-to-8 V Hall effect pick-up signal informs the PCM of crankshaft speed and piston position. When the 0 V-to-8 V signal is received, the PCM scans the sensor inputs and calculates the exact spark advance required for the next cylinder firing.

This calculation is completed in a few milliseconds; then the PCM opens the primary ignition circuit and fires the next spark plug in the firing order at the correct instant to provide the precise spark advance required by the engine.

There are two Hall effect pickups in distributors used with Chrysler port fuel injection, whereas the distributors used with throttle body injection have a single pickup. When the SYNC pick-up signal is received by the PCM every one-half distributor shaft revolution, or once every crankshaft revolution, the PCM grounds two injectors to inject fuel into two cylinders.

Distributors with Optical-Type Pickups

Shop Manual
Chapter 12, page 536

The 3.0L V6 engine available in some Chrysler products has a distributor fitted with an optical pick-up assembly with two light-emitting diodes and two photo diodes. A thin plate attached to the distributor shaft rotates between the LEDs above the plate and the photo diodes below the plate (Figure 12-26). This plate contains six equally spaced slots, which rotate directly below the inner LED and photo diode.

The inner LED and photo diode act as the reference pickup. As in Hall effect pick-up systems, the reference pickup in the optical distributor provides a crankshaft position and speed signal to the PCM. When the ignition switch is on, the PCM supplies voltage to the optical pickup that causes the LEDs to emit light. If a solid part of the plate is under the reference LED, this light does not shine on the photo diode. Under this condition, the photo diode does not conduct current and the reference voltage signal to the PCM is 5 V. As a reference slot moves under the LED, the light shines on the photo diode and this diode conducts current. When this action occurs, the reference voltage signal to the PCM is 0 V.

The outer LED, photo diode, and row of slots perform a similar function to the SYNC pickup in a distributor with Hall effect pickups. The outer row of slots is closely spaced, and the width between each slot represents 2° of crankshaft rotation. On the outer row there is one area where

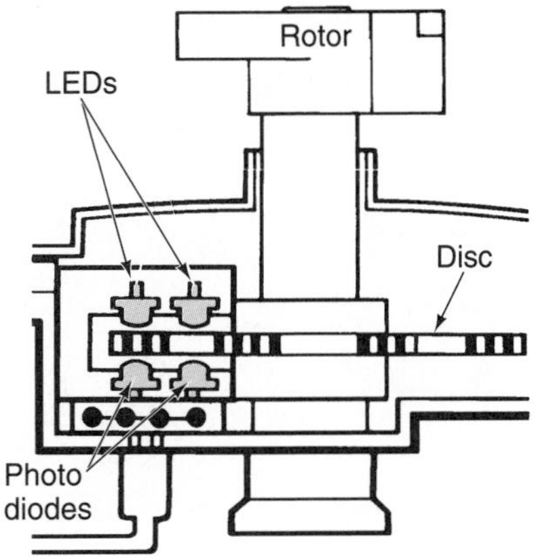

Figure 12-26 Location of LEDs and photo diodes. *(Courtesy of Chrysler Corporation)*

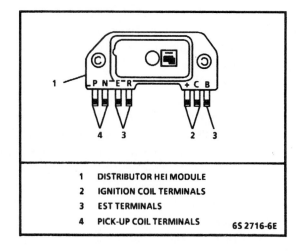

1 DISTRIBUTOR HEI MODULE
2 IGNITION COIL TERMINALS
3 EST TERMINALS
4 PICK-UP COIL TERMINALS

6S 2716-6E

Figure 12-27 A seven-terminal GM HEI system with EST. *(Courtesy of Chevrolet Motor Division—GMC)*

the slots are missing. When this blank area rotates under the LED, a different SYNC voltage signal is produced which informs the PCM regarding the number one piston position. The PCM uses this signal for injector control. As the outer row of slots rotates under the outer LED, the SYNC voltage signal to the PCM cycles from 0 V to 5 V. The reference pick-up signal informs the PCM when each piston is a specific number of degrees before TDC on the compression stroke.

When this signal is received, the PCM scans the inputs and calculates the spark advance required by the engine. The SYNC sensor signals always keep the PCM informed regarding the exact crankshaft position. The PCM opens the primary ignition circuit and fires the next spark plug in the firing order to provide the calculated spark advance.

GM's HEI with Electronic Spark Timing

Shop Manual
Chapter 12, page 543

The ground terminal on some HEI/EST modules is referred to as the reference low terminal.

A seven- (Figure 12-27) or eight-terminal module is used in some General Motors distributors with computer-controlled spark advance and fuel injection. Two of the module terminals are connected to the coil primary terminals, and two other module terminals are connected to the pick-up coil. The other four module terminals are connected through a four-wire harness to the PCM. These four wires are identified as bypass, **electronic spark timing (EST)**, ground, and reference wires (Figure 12-28).

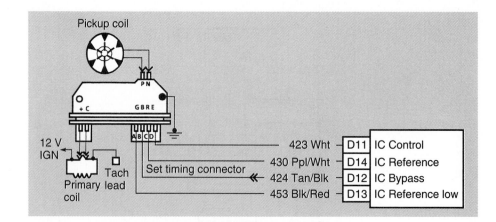

Figure 12-28 Schematic of an HEI system with EST. *(Courtesy of Chevrolet Motor Division—GMC)*

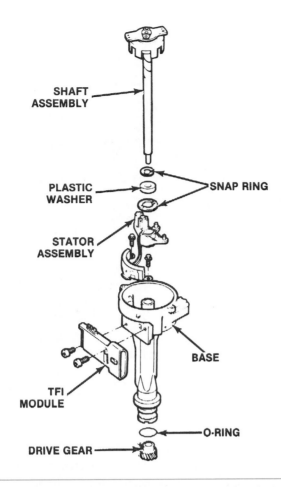

SHAFT
ASSEMBLY

PLASTIC
WASHER

SNAP RING

STATOR
ASSEMBLY

TFI
MODULE

BASE

O-RING

DRIVE GEAR

Figure 12-29 A typical Ford universal distributor. *(Reprinted with the permission of Ford Motor Company)*

When the engine is starting, the pick-up coil signal goes directly to the module. Immediately after the engine starts, the PCM sends a 5-V signal through the bypass wire to the module. This signal switches the module circuit and forces the pick-up signal to travel through the reference wire to the PCM. Crankshaft position and speed information are obtained from the pick-up signal to the PCM.

The PCM scans the input sensors and then sends a signal on the EST wire to the module. This signal commands the module to open the primary circuit and fire the next spark plug at the right instant to provide the precise spark advance indicated by the input signals.

Ford's TFI-IV System

In some DI systems with computer-controlled spark advance, the module is mounted externally from the distributor. It may be mounted away from the distributor or mounted to it. Ford Motor Company's TFI-IV system is such a system (Figure 12-29). This system is similar in appearance to the TFI system described earlier. The TFI-IV is only used with Ford's EEC-IV system (Figure 12-30).

The distributor contains the Profile Ignition Pickup (PIP), an octane rod, and a Hall effect vane switch stator. During the period of time the Hall effect device is turned on and off, a digital voltage pulse is produced. The pulse is used by the EEC-IV electronics for crankshaft position and the calculation of the required ignition timing. Ignition timing required for a particular operating condition is determined by inputs from various sensors which are correlated with values in the computer's memory.

The PIP signal is an indication of crankshaft position and engine speed. This signal is fed into the TFI-IV module and to the PCM. This signal, along with many others, allows the PCM to accurately calculate ignition timing needs for the conditions. The PCM produces a signal "Spout" and

Shop Manual
Chapter 12, page 543

EEC-IV stands for Electronic Engine Control, fourth generation.

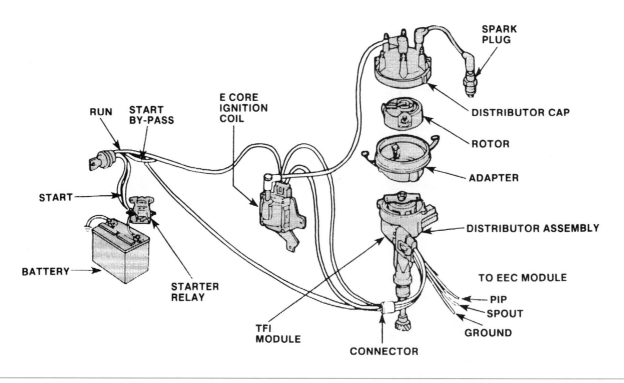

Figure 12-30 TFI-IV ignition system. *(Reprinted with the permission of Ford Motor Company)*

Spout simply stands for Spark out.

sends it to the TFI-IV module. The module compares the spout signal to the PIP signal then controls the activity of the ignition coil and the firing of the spark plugs (Figure 12-31).

The distribution of spark to the cylinders is accomplished in the same way as in other distributors, utilizing a distributor cap, rotor, and spark plug wires. The octane rod inside the distributor allows for fixed octane adjustments. The adjustment is made by replacing the standard (normal octane fuel) zero-degree retard rod with a 3- or 6-degree (lower octane) retard rod.

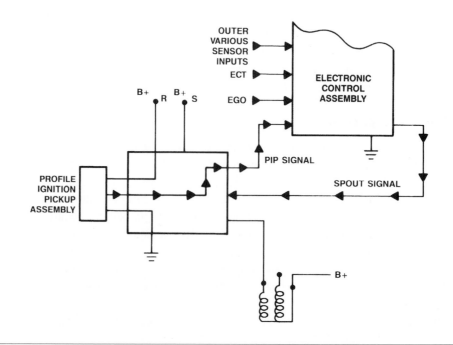

Figure 12-31 PIP and SPOUT signals. *(Reprinted with the permission of Ford Motor Company)*

Summary

❏ The electronic control unit (ECU), or module, opens and closes the primary circuit.

❏ The distributor must distribute the secondary current flow from the coil secondary winding to the spark plug wires.

❏ The distributor may house the switching device plus centrifugal or vacuum timing advance mechanisms. Some systems locate the switching device outside the distributor housing.

❏ Ignition timing is directly related to the position of the crankshaft. Magnetic pulse generators and Hall effect sensors are the most widely used engine position sensors. They generate an electrical signal at certain times during crankshaft rotation which triggers the switching device to control ignition timing.

❏ DI systems may have a module built into the computer, mounted on the distributor, or mounted externally from the distributor.

❏ Thick film integrated (TFI) ignition and high energy ignition (HEI) systems have the module mounted on the distributor housing.

❏ High energy ignition (HEI) systems may have a coil integral with the distributor cap or an external coil.

❏ In some distributors, the Hall effect switch is replaced with two light-emitting diodes (LEDs) and photo diodes, which produce voltage signals as a plate with two rows of slots rotates between the LEDs and photo diodes.

❏ In some distributors, one pickup is used for ignition triggering and a second pickup is used for injector sequencing.

❏ Computer-controlled ignition eliminates centrifugal and vacuum timing mechanisms. The computer receives input from numerous sensors. Based on this data, the computer determines the optimum firing time and signals an ignition module to activate the secondary circuit at the precise time needed.

Terms to Know

Electronic spark timing (EST)

High energy ignition (HEI)

Look-up tables

Strategy

Thick film integrated (TFI)

Timer core

Variable dwell

Review Questions

Short Answer Essay

1. Explain how a distributor distributes high voltage to the individual spark plugs.

2. Name four different types of triggering units commonly used in ignition systems.

3. Explain why Ford uses a two-piece distributor cap in their Dura-Spark II system.

4. Describe the variable dwell feature of some DI systems.

5. Explain the components and operation of a magnetic pulse generator.

6. Describe the responsibilities of an ignition system.

7. Explain the purpose of the SYNC pickup in an electronic ignition system with computer-controlled spark advance and two distributor pickups.

8. Explain the purpose of the reference pick-up signal in an optical-type distributor with computer-controlled spark advance.

9. Describe what is contained in the look-up tables of a computer.

10. Give some examples of information that a computer uses to determine the ideal amount of ignition advance for a particular operating condition.

Fill-in-the-Blanks

1. Typically, the more load on an engine, the _____ spark advance is needed.

2. The magnetic field surrounding the pick-up coil in a magnetic pulse generator moves when the _____.

3. The purpose of the electronic control unit (ECU), or module, is to _____ and _____ the primary circuit.

4. The goal of computerized spark timing is to produce _____ power, _____ fuel efficiency, and _____ emission levels.

5. The rotor in a distributor rotates at the same speed as the engine's _____.

6. The ignition coil of HEI units are either located _____ or _____ of the distributor.

7. Chrysler distributors may have _____ Hall effect units if the engine is equipped with port fuel injection.

8. _____ _____ _____ and _____ - _____ sensors are the most widely used engine position sensors.

9. In systems with variable dwell, dwell normally _____ with an increase in engine speed.

10. In an electronic ignition system, a _____ actually controls the primary coil current.

ASE Style Review Questions

1. While discussing pick-up coil operation:
Technician A says the pick-up coil magnetic field strength is increasing when the reluctor high point is directly aligned with the pick-up coil.
Technician B says the pick-up coil magnetic field strength is increasing when the reluctor high point moves out of alignment with the pickup.
Who is correct?
  A. A only
  B. B only
  C. Both A and B
  D. Neither A nor B

2. *Technician A* says a magnetic pulse generator is equipped with a permanent magnet.
Technician B says a Hall effect switch is equipped with a permanent magnet.
Who is correct?
  A. A only
  B. B only
  C. Both A and B
  D. Neither A nor B

3. While discussing PM generators:
Technician A says the pick-up coil does not produce a voltage signal when a reluctor tooth approaches the coil.
Technician B says the pick-up coil does not produce a voltage signal when a reluctor tooth moves away from the coil.
Who is correct?
  A. A only
  B. B only
  C. Both A and B
  D. Neither A nor B

4. While discussing engine position sensors:
Technician A says a metal detection sensor needs to have its voltage signal amplified, inverted, and shaped into a clean square wave.
Technician B says a Hall effect sensor needs to have its voltage signal amplified, inverted, and shaped into a clean square wave signal.
Who is correct?
  A. A only
  B. B only
  C. Both A and B
  D. Neither A nor B

5. While discussing ignition systems:
   Technician A says an ignition system must supply high voltage surges to the spark plugs.
   Technician B says the system must maintain the spark long enough to burn all of the air-fuel mixture in the cylinder.
   Who is correct?
   A. A only   **C.** Both A and B
   B. B only   **D.** Neither A nor B

6. While discussing ignition timing requirements:
   Technician A says more advanced timing is desired when the engine is under a heavy load.
   Technician B says more advanced timing is desired when the engine is running at high engine speeds.
   Who is correct?
   A. A only   **C.** Both A and B
   B. B only   **D.** Neither A nor B

7. While discussing primary ignition circuit resistors:
   Technician A says some systems have one to make sure the correct amount of voltage and current is available to the primary coil.
   Technician B says these resistors must be used with electronic triggering devices, such as a Hall effect switch.
   Who is correct?
   A. A only   **C.** Both A and B
   B. B only   **D.** Neither A nor B

8. While discussing ignition coils:
   Technician A says the primary winding is made of thicker wire than the secondary.
   Technician B says the secondary winding is made up of many more turns of wire than the primary.
   Who is correct?
   A. A only   **C.** Both A and B
   B. B only   **D.** Neither A nor B

9. While discussing ignition timing:
   Technician A says changes in barometric pressure require changes in ignition timing.
   Technician B says engine temperature has little effect on the required amount of spark advance.
   Who is correct?
   A. A only   **C.** Both A and B
   B. B only   **D.** Neither A nor B

10. While discussing different ignition system designs:
    Technician A says all HEI systems have conventional centrifugal and vacuum advance units.
    Technician B says all Ford TFI ignition systems have computer-controlled ignition timing.
    Who is correct?
    A. A only   **C.** Both A and B
    B. B only   **D.** Neither A nor B

Electronic Ignition Systems

Upon completion and review of this chapter, you should be able to:

❑ Describe the advantages of electronic ignition (EI) systems.

❑ Describe the operation of distributorless ignition systems.

❑ Describe the coil secondary-to-spark plug wiring connections on an EI system, including an explanation of how the spark plugs fire.

❑ Explain the coil and injector sequencing on a Chrysler EI system while cranking the engine.

❑ Explain why some EI systems may be called slow-start systems.

❑ Explain the purpose of the SYNC sensor signal in an EI system with a combined crankshaft and SYNC sensor.

❑ Explain the operation of the 3X and 18X crankshaft sensor signals during engine starting on a fast-start EI system.

❑ Describe the design and location of the reluctor ring and magnetic sensor on an EI system.

Introduction

Distributorless ignition systems are commonly referred to as DIS.

Electronic ignition (EI) systems (Figure 13-1), have no distributor, rather spark distribution is controlled by an electronic control unit and/or the vehicle's computer. Instead of a single ignition coil for all cylinders, each cylinder may have its own ignition coil, or two cylinders may share one coil.

Figure 13-1 A supercharged Buick 3800 V6 engine with an electronic ignition (EI) system. *(Courtesy of Buick Motor Division—GMC)*

The coils are wired directly to the spark plug they control. An ignition control module, tied into the vehicle's computer control system, controls the firing order and the spark timing and advance. This module is typically located under the coil assembly.

In many EI systems, a crank sensor located at the front of the crankshaft is used to trigger the ignition system. When a distributor is used in the ignition system, the distributor drive gear, shaft, and bushings are subject to wear. Worn distributor components cause erratic ignition timing and spark advance, which result in reduced economy and performance plus increased exhaust emissions. Since the distributor is eliminated in EI systems, ignition timing remains more stable over the life of the engine, which means improved economy and performance with reduced emissions.

There are many advantages of a **distributorless ignition system (DIS)** over one that uses a distributor. Here are some of the more important ones:

In the SAE J1930 terminology, the term electronic ignition (EI) replaces all previous terms for distributorless ignition systems.

- ❏ Fewer moving parts, therefore less friction and wear.
- ❏ Flexibility in mounting location. This is important with today's smaller engine compartments.
- ❏ Less required maintenance; there is no rotor or distributor cap to service.
- ❏ Reduced radio frequency interference because there is no rotor to cap gap.
- ❏ Elimination of a common cause of ignition misfire, the buildup of water and ozone/nitric acid in the distributor cap.
- ❏ Elimination of mechanical timing adjustments.
- ❏ Has no mechanical load on the engine to operate.
- ❏ Increased available time for coil saturation.
- ❏ Increased time between firings, which allows the coil to cool more.

In a distributorless ignition system, each end of the secondary winding of a coil is connected to a spark plug. These two spark plugs are on companion cylinders and are at top dead center (TDC) at the same time. One of them is on the compression stroke and the other is on its exhaust stroke. When a coil fires its two spark plugs, the plug that fires into the exhaust stroke is the waste plug.

Since the polarity of the primary and secondary windings are fixed, one plug always fires in a forward direction and the other in a reverse direction. The voltage dropped across each plug is determined by its polarity and cylinder pressure. Therefore the spark plug firing into the exhaust requires much less voltage than the one firing into the compressed gases. The voltage demand to fire both plugs in series is greater than the required voltage to fire one plug. This is not a problem in these systems; there is plenty of voltage available because of the longer coil saturation times due to the longer intervals between firings.

Distributorless ignition systems have had many names prior to SAE's J1930 list of acceptable names. Some of the commonly used names were: EDIS, DIS, C3I, and IDI.

Basic Components

A BIT OF HISTORY

A very crude but effective ignition system with individual coils for each cylinder was used on the Model T Ford. This system was based on a magneto that rotated with the engine. The magneto sent timed, low voltage signals to a wooden box that contained four coils. Each of these coils were connected to a spark plug. The pulses from the magneto caused the magnetic field in the coils to build and collapse, firing the spark plugs according to a firing order.

Electronic ignition systems electronically perform the functions of a distributor. They control spark timing and advance in the same manner as the computer-controlled ignition systems. Yet EI is a step

Figure 13-2 A coil pack for a six-cylinder EI system.

beyond the computer-controlled system because it also distributes spark electronically instead of mechanically. The distributor is completely eliminated from these systems. Distributorless systems use multiple coils and modules to provide and distribute high secondary voltages directly from the coil to the plug.

The computer, ignition module, and position sensors combine to control spark timing and advance. The computer collects and processes information to determine the ideal amount of spark advance for the operating conditions. The ignition module uses crank/cam sensor data to control the timing of the primary circuit in the coils. The ignition module synchronizes the coils' firing sequence in relation to crankshaft position and firing order of the engine. Therefore, the ignition module takes the place of the distributor.

Depending on the exact EI system, the ignition coils can be serviced as a complete unit or separately. The coil assembly (Figure 13-2) is typically called a **coil pack** and is comprised of two or more individual coils.

On those EI systems that use one coil per spark plug, the electronic ignition module determines when each spark plug should fire and controls the on/off time of each plug's coil.

The systems with a coil for every two spark plugs also use an electronic ignition module, but they use the **waste spark** method of spark distribution. Each end of the coil's secondary winding is attached to a spark plug. Each coil is connected to a pair of spark plugs. When the field collapses in the coil, voltage is sent to both spark plugs that are attached to the coil. In all V6s, the paired cylinders are 1 and 4, 2 and 5, and 3 and 6 (or 4 and 1 and 3 and 2 on four-cylinder engines). With this arrangement, one cylinder of each pair is on its compression stroke while the other is on the exhaust stroke. Both cylinders get spark simultaneously, but only one spark generates power while the other is wasted out the exhaust. During the next revolution, the roles are reversed.

Due to the way the secondary coils are wired, when the induced voltage cuts across the primary and secondary windings of the coil, one plug fires in the normal direction—positive center electrode to negative side electrode—and the other plug fires just the reverse-side to center electrode (Figure 13-3). Both plugs fire simultaneously, completing the series circuit. Each plug in the set always fires the same way, regardless of the stroke the piston is on.

The coil is able to overcome the increased voltage requirements caused by reversed polarity and still fire two plugs simultaneously because each coil is capable of producing very high voltages. There is very little resistance across the plug gap on exhaust, so that plug requires very little voltage to fire, thereby providing its mate (the plug that is on compression) with plenty of available voltage.

> ⚠ **WARNING:** Most vehicle manufacturers do not recommend removing a spark plug wire completely with the engine running on an EI system. This action may damage the insulation somewhere in the secondary ignition circuit, such as in the coil or spark plug wires. It is permissible to crank, or start, the engine with the proper test spark plug connected to a spark plug wire on an EI system.

Shop Manual
Chapter 13, page 559

About 30% more voltage is required to fire a spark plug with reversed polarity.

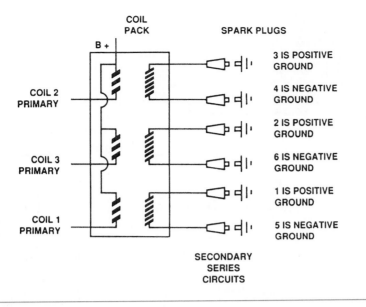

COIL PACK

SPARK PLUGS

B +

COIL 2 PRIMARY

COIL 3 PRIMARY

COIL 1 PRIMARY

3 IS POSITIVE GROUND

4 IS NEGATIVE GROUND

2 IS POSITIVE GROUND

6 IS NEGATIVE GROUND

1 IS POSITIVE GROUND

5 IS NEGATIVE GROUND

SECONDARY SERIES CIRCUITS

Figure 13-3 Polarity of the spark plugs in an EI system. *(Reprinted with the permission of Ford Motor Company)*

Primary current is controlled by transistors in the control module. There is one switching transistor for each ignition coil in the system. The transistors complete the ground circuit for the primary, thereby allowing for a dwell period. When primary current flow is interrupted, secondary voltage is induced in the coil and the coil's spark plug(s) fire. The timing and sequencing of ignition coil action is determined by the control module and input from a triggering device.

The control module is also responsible for limiting the dwell time. In EI systems there is time between plug firings to saturate the coil. Achieving maximum current flow through the coil is great if the system needs the high voltage that may be available. If the high voltage is not needed, the high current is not needed and the heat it produces is not desired. Therefore the control module is programmed to only allow total coil saturation when the very high voltage is needed or the need for it is anticipated.

A few engines have an EI system in which the coils are mounted directly over the spark plugs so that no wiring between the coils and plugs is necessary. On other systems, the coil packs are mounted remote from the spark plugs (Figure 13-4). High-tension secondary wires carry high-voltage current from the coils to the plugs.

A few EI systems have one coil per cylinder with two spark plugs per cylinder. During starting only one plug is fired. Once the engine is running the other plug also fires. One spark plug is located on the intake side of the combustion chamber while the other is located on the exhaust side. Two coil packs are used, one for the plugs on the intake side and the other for the plugs on the exhaust side. These systems are called **dual plug** systems. During dual plug operation, the two coil packs are synchronized so that each cylinder's two plugs fire at the same time. The coils fire two spark plugs at the same time. Therefore on a four-cylinder engine, four spark plugs are fired at a time, two during the compression stroke of the cylinder and two during the exhaust stroke of another cylinder.

EI System Operation

Shop Manual
Chapter 13, page 552

From a general operating standpoint, most distributorless ignition systems are similar. However, there are variations in the way different distributorless systems obtain a timing reference in regard to crankshaft and camshaft position.

Some engines use separate Hall effect sensors to monitor crankshaft and camshaft position for the control of ignition and fuel injection firing orders. The crankshaft pulley has interrupter

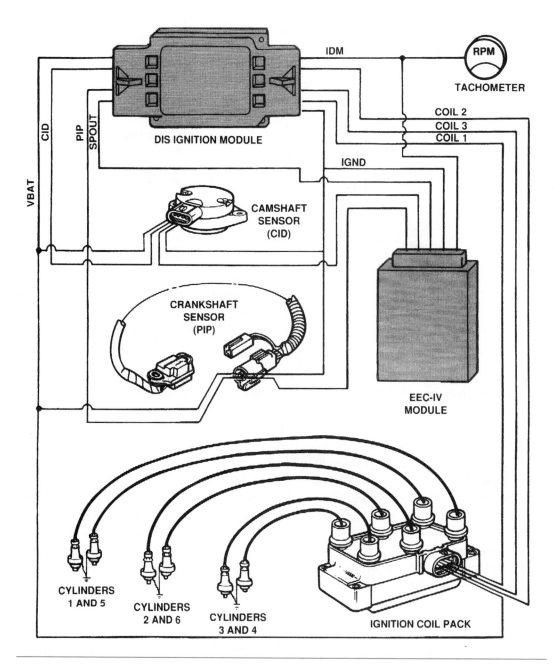

Figure 13-4 Typical EI system. *(Reprinted with the permission of Ford Motor Company)*

rings that equal in number to half of the cylinders of the engine. The resultant signal informs the powertrain control module (PCM) as to when to fire the plugs. The camshaft sensor helps the computer determine when the number one piston is at TDC on the compression stroke.

Defining the different types of EI systems used by manufacturers focuses on the location and type of sensors used. There are other differences, such as the construction of the coil pack, wherein some are a sealed assembly and others have individually mounted ignition coils. Some EI systems have a camshaft sensor mounted in the opening where the distributor was mounted. The camshaft sensor ring has one notch and produces a leading edge and trailing edge signal once per camshaft revolution. These systems also use a crankshaft sensor. Both the camshaft and crankshaft sensors are Hall effect sensors.

Some systems have the camshaft sensor mounted in the front of the timing chain cover. A magnet on the camshaft gear rotates past the inner end of the camshaft sensor and produces a signal for each camshaft revolution.

Other systems use a dual crankshaft sensor located behind the crankshaft pulley. When this type of sensor is used, there are two interrupter rings on the back of the pulley that rotate through the Hall effect switches at the dual crankshaft sensor. The inner ring with three equally spaced blades rotate through the inner Hall effect switch, whereas the outer ring with one opening rotates through the outer Hall effect.

In this dual sensor, the inner sensor provides three leading edge signals and the outer sensor produces one leading edge during one complete revolution of the crankshaft. The outer sensor is the SYNC sensor.

The examples given so far depend on two revolutions of the crankshaft to inform the PCM as to when the number one cylinder is ready. These systems are referred to as **slow-start** systems because the engine must crank through two crankshaft revolutions before ignition begins.

The **fast start** electronic ignition system used in GM's Northstar system uses two crankshaft position sensors. A reluctor ring with 24 evenly spaced notches and 8 unevenly spaced notches is cast onto the center of the crankshaft.

The signals from the two sensors are sent to the ignition control module. This module counts the number of signals from one of the sensors that are between the other sensor signals to sequence the ignition coils properly. This allows the ignition system to begin firing the spark plugs within 180° of crankshaft rotation while starting the engine. This system allows for much quicker starting than other EI systems that require the crankshaft to rotate one or two times before the coils are sequenced.

Finally, some engines use a magnetic pulse generator. The timing wheel is cast on the crankshaft and has machined slots on it. If the engine has six cylinders, there will be seven slots, six of which are spaced exactly 60° apart and the seventh notch is located 10° from the number six notch and is used to synchronize the coil firing sequence in relation to crankshaft position. The same triggering wheel can be and is used on four-cylinder engines. The computer only needs to be programmed to interpret the signals differently than on a six-cylinder engine.

The magnetic sensor, which protrudes into the side of the block, generates a small AC voltage each time one of the machined slots passes by. By counting the time between pulses, the ignition module picks out the unevenly spaced seventh slot, which starts the calculation of the ignition coil sequencing. Once its counting is synchronized with the crankshaft, the module is programmed to accept the AC voltage signals of the select notches for firing purposes.

The development and spreading popularity of EI is the result of reduced emissions, improved fuel economy, and increased component reliability brought about by these systems.

EI offers advantages in production costs and maintenance considerations. By removing the distributor, the manufacturers realize a substantial savings in ignition parts and related machining costs. By eliminating the distributor, they also do away with cracked caps, eroded carbon buttons, burned-through rotors, moisture misfiring, base timing adjustments, and other problems.

CAUTION: Since EI systems have considerably higher maximum secondary voltage compared to point-type or electronic ignition systems, greater electrical shocks are obtained from EI systems. Although such shocks may not be directly harmful to the human body, they may cause you to jump or react suddenly, which could result in personal injury. For example, when you jump suddenly as a result of EI electrical shock, you may hit your head on the vehicle hood or push your hand into a rotating cooling fan.

EI Systems with the Cam Sensor in the Distributor Opening

Shop Manual
Chapter 13, page 574

Locating the cam sensor in the opening previously occupied by the distributor merely takes advantage of the bore and gear that was already present. Seeing that the distributor was driven by the camshaft at camshaft speed, driving a camshaft position sensor by the same mechanism just made

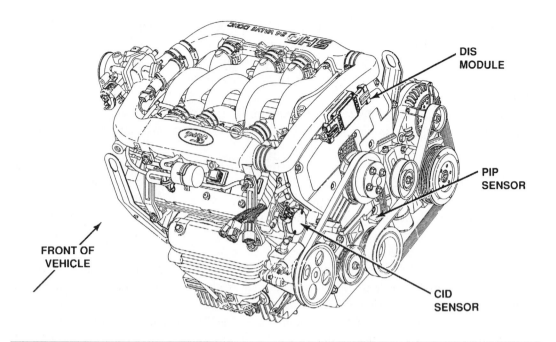

Figure 13-5 Location of EI components. The cylinder identification (CID) is a camshaft position sensor located where a distributor might be. *(Reprinted with the permission of Ford Motor Company)*

sense. This modification really made sense when older engine designs were modified for distributorless ignition. A few new engine designs place the cam sensor where the distributor would have been (Figure 13-5).

In GM's Type 1 EI system, used on the 3.8L turbocharged sequential fuel injection (SFI) engine, the camshaft sensor is mounted in the distributor opening (Figure 13-6). The camshaft sensor ring has one notch; thus, it produces a leading edge and trailing edge signal once per camshaft revolution, or once every two crankshaft revolutions.

This design has a crankshaft sensor mounted behind the crankshaft pulley on the front of the engine block. The camshaft and crankshaft sensors both contain Hall effect switches. An inter-

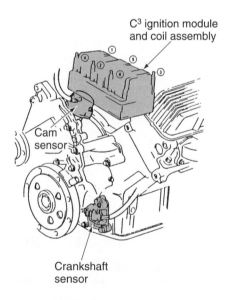

Figure 13-6 EI components on a 3.8L turbocharged SFI engine. *(Courtesy of Buick Motor Division—GMC)*

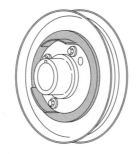

Figure 13-7 Crankshaft pulley with an interrupter ring. *(Courtesy of Buick Motor Division—GMC)*

rupter ring with three equally spaced blades is positioned on the back of the crankshaft pulley. These blades rotate through the Hall effect switch in the crankshaft sensor (Figure 13-7).

As the crankshaft rotates and the interrupter passes in and out of the Hall effect switch, the switch turns the module reference voltage on and off. The three signals are identical and the control module cannot distinguish which of these signals to assign to a particular coil.

The signal from the cam sensor gives the module the information it needs to assign the signals from the crankshaft sensors to the appropriate coils. The camshaft sensor synchronizes the crankshaft sensor signals with the position of the number one cylinder. From there the module can energize the coils according to the firing order of the engine. Once the engine has started, the camshaft signal serves no purpose.

Ford uses a similar system on some of their engines. This system is similar to the TFI-IV system, except it does not use a distributor. A profile ignition pickup (PIP) sensor is used to monitor crankshaft position and engine speed. A cylinder identification (CID) sensor is used to identify a specific point in piston number one's travel. The EEC-IV module determines spark advance by using the PIP signal to establish base timing. A SPOUT signal is sent to the ignition (DIS) module to fire the coil and to control dwell time. Dwell time is controlled by the EEC module. CID is important to the module; it needs to know which coil to fire.

The CID (Figure 13-8) is a Hall effect switch assembly placed in the distributor bore and driven by the camshaft. The CID is a camshaft position sensor. The PIP is a Hall effect crankshaft posi-

The reference voltage from the Hall effect switch actually moves from a high voltage (6-8 volts) to low voltage (0–0.5 volt).

The SPOUT signal is the spark out signal.

Shop Manual
Chapter 13, page 568

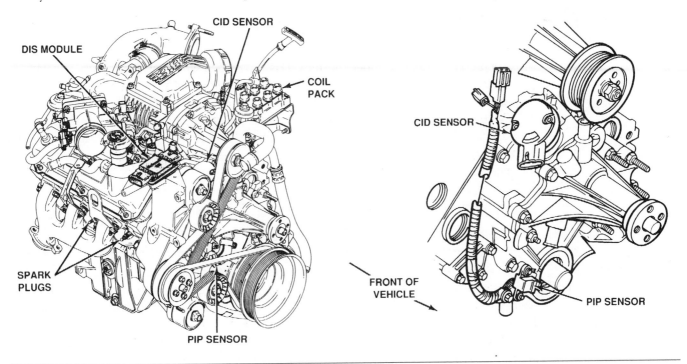

Figure 13-8 Camshaft sensor located in distributor opening. *(Reprinted with the permission of Ford Motor Company)*

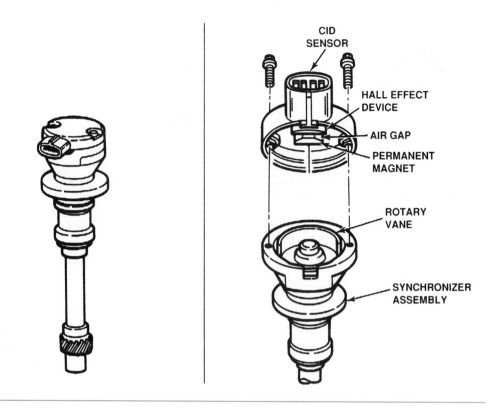

Figure 13-9 Camshaft sensor construction. *(Reprinted with the permission of Ford Motor Company)*

tion sensor that produces a digital signal. The PIP sends a high signal when the Hall effect switch vane shunts the magnetic lines of flux. It sends a low signal when a window allows the flux lines to cut through the Hall unit. On six-cylinder engines, the PIP produces three high signals for every crankshaft revolution. The leading edge of PIP always occurs at 10° BTDC. Therefore base timing is set at 10° BTDC.

The leading edge of the CID (Figure 13-9) signal to the module occurs when piston number one is at 26° ATDC on its compression stroke. The trailing edge always occurs at 26° ATDC on the intake stroke of piston number one. Input from the CID is also used to synchronize fuel injection timing. The difference between the PIP and CID signals is that the CID produces one signal for every two revolutions of the crankshaft or one signal for every revolution of the camshaft (Figure 13-10).

When the engine is cranking, the module waits for a transition in the CID signal, either from low to high or high to low. At that point, it prepares coil number two for firing. Once the transition is complete, the module is in synch with the engine and is ready to fire the coils in accordance with the engine's firing order.

EI Systems with the Cam Sensor in the Timing Chain Cover

Since the spark plug firing on the exhaust stroke has no effect on engine operation, the term "waste spark" may be applied to these systems.

The camshaft sensor in GM's 3.8L non-turbocharged SFI V6 engine is positioned in the front of the timing gear cover (Figure 13-11). A magnet on the camshaft gear rotates past the inner end of the camshaft sensor and produces a signal for each camshaft revolution, or every two crankshaft revolutions.

The secondary coil windings are connected to the external coil terminals and two spark plug wires are connected to these terminals. In these systems, the secondary voltage and current move from the coil secondary winding down through the electrodes in one spark plug and up through

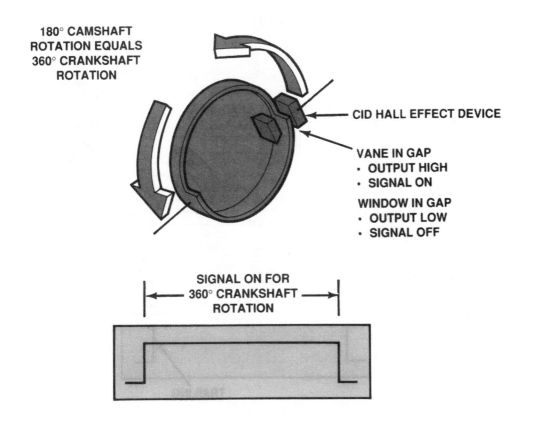

180° CAMSHAFT ROTATION EQUALS 360° CRANKSHAFT ROTATION

CID HALL EFFECT DEVICE

VANE IN GAP
• OUTPUT HIGH
• SIGNAL ON

WINDOW IN GAP
• OUTPUT LOW
• SIGNAL OFF

SIGNAL ON FOR 360° CRANKSHAFT ROTATION

Figure 13-10 Camshaft sensor operation and resulting signal. *(Reprinted with the permission of Ford Motor Company)*

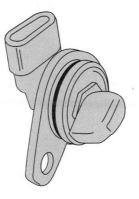

Figure 13-11 Camshaft sensor for a Buick 3.8L non-turbocharged SFI engine. *(Courtesy of Buick Motor Division—GMC)*

the electrodes in the other spark plug. One spark plug fires while the cylinder is on the compression stroke, and the other spark plug fires with the cylinder on the exhaust stroke.

Each coil primary winding is connected into the coil module. When the ignition switch is turned on, 12 V is supplied through a pair of fuses to coil module terminals N and P. Three wires from camshaft and crankshaft sensors are connected to the coil module, and five wires are connected between the powertrain control module and the coil module (Figure 13-12).

On the 3.8L turbocharged SFI engines and 3.8L non-turbocharged SFI V6 engines, the firing order is 1-6-5-4-3-2. On these engines, spark plugs 1-4, 6-3, and 5-2 are paired together on the coil assembly. When a trailing edge camshaft sensor signal is received during initial starting, the coil

Shop Manual
Chapter 13, page 574

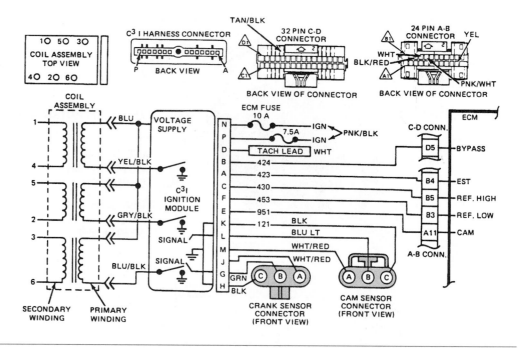

Figure 13-12 Schematic for the EI system on a Buick 3.8L turbocharged SFI engine. *(Courtesy of Buick Motor Division—GMC)*

These EI systems may be referred to as slow-start systems, because a cam sensor signal must be sent to the coil module before the module sequences the coils and begins firing the spark plugs. This signal may require two crankshaft revolutions.

module prepares to fire the coil connected to spark plugs 5-2. After the camshaft sensor signal is received, the next trailing edge crankshaft sensor signal turns on the primary circuit of the 5-2 coil, and the next leading edge crankshaft sensor signal informs the coil module to open the primary circuit of the 5-2 coil (Figure 13-13). When this coil fires, one of these cylinders is always on the compression stroke, and the other cylinder is on the exhaust stroke. After the 5-2 coil firing, the coil module fires the 1-4 coil and the 6-3 coil in sequence. This firing sequence provides the correct firing order.

On an SFI engine, the PCM grounds each injector individually. The cam sensor signal is also used for injector sequencing on the 3.8L turbocharged and non-turbocharged SFI engines. This cam sensor signal is sent from the cam sensor through the coil module to the PCM. The PCM grounds each injector in the intake port when the piston for that cylinder is at 70° before TDC on the intake stroke.

When a crankshaft sensor failure occurs, the engine does not start. If the camshaft sensor signal becomes defective with the engine running, the engine continues to run, but the PCM reverts to multiport fuel injection without the camshaft signal information. Under this condition, engine performance and economy decrease and emission levels may increase. When the engine is shut off with a defective cam sensor, it will not restart.

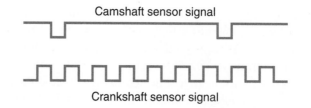

Figure 13-13 Relationship between camshaft and crankshaft sensor signals on a Buick 3.8L SFI engine. *(Courtesy of Buick Motor Division—GMC)*

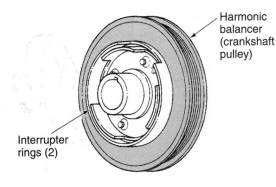

Harmonic
balancer
(crankshaft
pulley)

Interrupter
rings (2)

Figure 13-14 Crankshaft pulley vanes and dual crank-shaft sensor for GM's 3.3L and 3300 engines. *(Courtesy of Oldsmobile Division—GMC)*

Figure 13-15 Dual crankshaft sensor for GM's 3.3L and 3300 engines. *(Courtesy of Oldsmobile Division—GMC)*

EI Systems with a Dual Crankshaft Sensor

GM's 3.0L V6 and the 3300 V6 multiport fuel injection engines have a dual crankshaft sensor located behind the crankshaft pulley on the front of the engine block. When this type of sensor is used, there are two interrupter rings on the back of the crankshaft pulley that rotate through the Hall effect switches in the dual crankshaft sensor (Figure 13-14). The inner ring with three equally spaced blades rotates through the inner Hall effect switch, whereas the outer ring with one opening rotates through the outer Hall effect switch in the dual sensor (Figure 13-15). In this dual sensor, the inner sensor provides three leading edge signals per crankshaft revolution, and the outer sensor produces one leading edge signal during this time period. This outer sensor is referred to as a synchronizer (SYNC) sensor. The signal from this sensor informs the coil module regarding crankshaft position. The SYNC sensor signal occurs once per crankshaft revolution and this signal is synchronized with the inner crankshaft sensor signal to fire the 6-3 coil.

The primary and secondary coil winding connections and the coil module-to-PCM wiring on the 3.3L, or 3300 V6 engine, are much the same as the wiring connections on the 3.8L turbocharged engine. The main difference in the EI systems on these engines is in the sensor wiring. Four wires extend from the dual crankshaft sensor to the coil module (Figure 13-16).

Shop Manual
Chapter 13, page 574

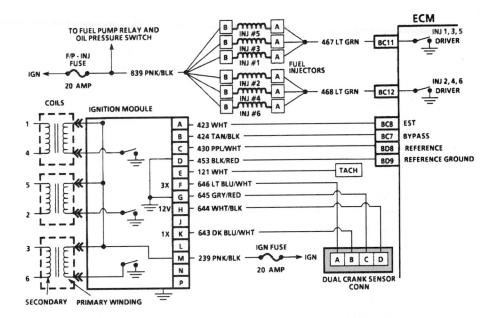

Figure 13-16 Schematic for the EI system used on GM's 3.3L and 3300 engines. *(Courtesy of Oldsmobile Division—GMC)*

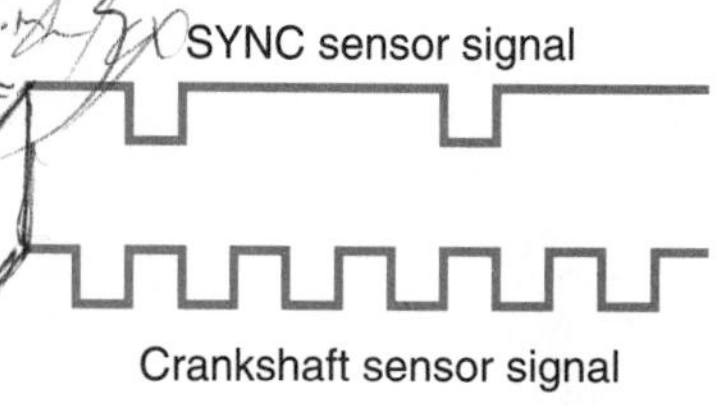

SYNC sensor signal

Crankshaft sensor signal

Figure 13-17 Relationship between the inner crankshaft sensor signals and the outer SYNC signals. *(Courtesy of Oldsmobile Division—GMC)*

System Operation

During engine starting on the 3.3L or 3300 MFI engine, the leading edge SYNC sensor signal informs the coil module to sequence the primary of the 5-2 coil. The module grounds the 5-2 coil primary winding when the next trailing edge crankshaft sensor signal is received. The next leading edge inner crankshaft sensor signal informs the coil module to open the primary circuit of coil 5-2 (Figure 13-17). The coil module closes and opens the primary circuits on coils 1-4 and 6-3 in response to the next inner crankshaft sensor signals. Therefore, the firing order of 1-6-5-4-3-2 is obtained from this coil sequencing. Once the engine is running, the coil module sequences the coils properly without using the SYNC sensor signal.

Signals from the SYNC sensor are only required for coil sequencing during engine starting. Since the inner crankshaft sensor blades are spaced 60° apart and the length of each blade is 60° of crankshaft rotation, a leading edge inner crankshaft sensor signal is received every 120° of crankshaft rotation. Each time one of these leading edge signals occurs during initial starting, the coil module opens a coil primary circuit and fires a pair of spark plugs. While the engine is starting, the bypass signal from the PCM to the coil module remains at 0 V, and leading edge crankshaft sensor signals occur at 10° BTDC on the compression stroke. The basic timing is 10° BTDC, and the PCM does not provide spark advance while the engine is starting. The operation of these EI systems during engine starting may be referred to as a **bypass mode**. Since the SYNC signal is used for coil sequencing, it is sent to the coil module and this signal is not sent to the PCM.

Once the engine starts, the PCM sends a 5-V signal through the bypass wire to the coil module, and the system enters the electronic spark timing (EST) mode. When this signal is received, the inner crankshaft sensor signal is sent through the reference high wire to the PCM. The inner crankshaft sensor signal provides crankshaft position and speed information to the PCM. On the 3.3L, 3300, 3.8L turbocharged SFI, and 3.8L non-turbocharged SFI engines, the crankshaft sensor signal occurs when each piston is at 10° BTDC on the compression stroke. When the PCM receives a crankshaft sensor signal, the PCM scans all the sensor inputs and then sends a voltage signal through the electronic spark timing (EST) wire to the coil module. This EST signal commands the coil module to open a coil primary circuit and fire a coil and a pair of spark plugs to provide the precise spark advance required by the engine. If either the crankshaft or SYNC signals fail, the engine will not start. On some 3.0L V6 engines, repeated attempts to start the engine with a defective SYNC sensor signal may allow the engine to start.

Fast-Start EI Systems

While starting the engine, fast-start EI systems start firing the spark plugs in 180 degrees or less of crankshaft rotation.

A fast-start system is used on GM's 3800 engine, an updated version of the 3.8L engine. This fast-start EI system has a dual crankshaft sensor at the front of the crankshaft. The cam sensor is mounted in the timing gear cover as on previous 3.8L SFI engines. Two Hall effect switches are located in the dual crankshaft sensor, and two matching interrupter rings are attached to the back of the crankshaft pulley (Figure 13-18). The inner ring on the crankshaft pulley has three blades of unequal lengths with unequal spaces between the blades. These blades have 110°, 100°, and 90° of crankshaft rotation each, and the spaces between the blades are 30°, 20°, and 10° each. On the

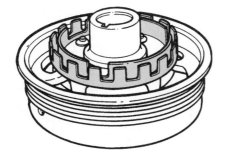

Figure 13-18 Interrupter rings on a crankshaft pulley. *(Courtesy of Oldsmobile Division—GMC)*

Shop Manual
Chapter 13, page 578

outer ring, there are 18 blades of equal length with equal spaces between the blades. The signal from the inner Hall effect switch is referred to as the 3X signal, while the outer Hall effect switch is called the 18X signal. These signals are sent from the dual crankshaft sensor to the coil module. The coil module and coil assembly in this fast-start system are similar to the components found on other systems, but they are not interchangeable.

The leading edge 3X signals are spaced 120° apart, and each pair of cylinders that is connected to the same coil reaches TDC on the compression stroke 75° after the 3X leading edge signal. Cylinders 1-4, 3-6, and 2-5 are paired together by the ignition coils. Since the leading edge 3X signals are spaced 120° apart, spark plug firings still occur at the correct intervals. An 18X leading edge signal occurs once every 20° of crankshaft rotation, or 18 times during one crankshaft revolution. This sensor also provides a trailing edge signal the same number of times per crankshaft revolution. Therefore, leading edge and trailing edge 18X signals occur every 10° of crankshaft rotation.

The coil module monitors the 18X signal in relation to the 3X signal to provide coil sequencing while the engine is starting. For example, during the 10-degree window on the 3X interrupter blade, one trailing edge 18X signal is received. While the 20-degree window in the 18X interrupter blade rotates through the Hall effect switch, one leading edge and one trailing edge 18X signal are produced. During the 30-degree window rotation on the 3X interrupter blade, two trailing edge and one leading edge 18X signals are sent to the coil module (Figure 13-19).

The coil module knows which coil to sequence from the number of 18X signals received during each 3X window rotation. For example, when two 18X signals are received, the coil module is programmed to sequence coil 3-6 next in the firing sequence. Within 120° of crankshaft rotation, the coil module can identify which coil to sequence, and thus start firing the spark plugs. Therefore, the system fires the spark plugs with less crankshaft rotation during initial starting than the previous slow-start systems.

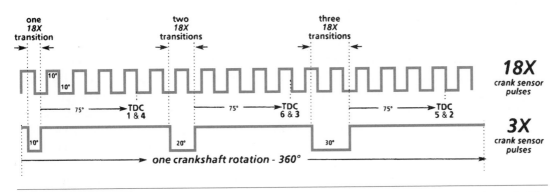

Figure 13-19 3X and 18X crankshaft sensor signals. *(Courtesy of Oldsmobile Division—GMC)*

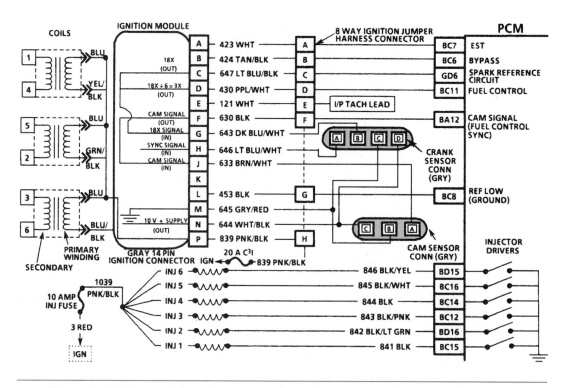

Figure 13-20 Schematic for a GM type 1 EI fast-start system. *(Courtesy of Oldsmobile Division—GMC)*

Once the engine is running, a 5-V signal is sent from the PCM through the bypass wire to the coil module. When this signal is received, the system switches to the EST mode, and the PCM uses the 18X signal for crankshaft position and speed information. The 18X signal may be referred to as a high resolution signal (Figure 13-20).

If the 18X signal is not present, the engine will not start. When the 3X signal fails with the engine running, the engine continues to run but will refuse to restart.

In this system, the cam sensor signal is used for injector sequencing, but it is not required for coil sequencing. If the cam sensor signal fails, the PCM logic begins sequencing the injectors after two cranking revolutions. There is a one in six chance that the PCM logic will ground the injectors in the normal sequence. When the PCM logic does not ground the injectors in the normal sequence, the engine hesitates on acceleration.

Fast-Start EI Systems with Two Crankshaft Sensors

Shop Manual
Chapter 13, page 578

The fast-start electronic ignition system used in GM's Northstar system uses two crankshaft position sensors. A reluctor ring with 24 evenly spaced notches and 8 unevenly spaced notches is cast onto the center of the crankshaft (Figure 13-21).

When the reluctor ring rotates past the magnetic-type sensors, each sensor produces 32 high and low voltage signals per crankshaft revolution. The "A" sensor is positioned in the upper crankcase, and the "B" sensor is positioned in the lower crankcase. Since the A sensor is above the B sensor, the signal from the A sensor occurs 27° before the B sensor signal (Figure 13-22).

The signals from the two sensors are sent to the ignition control module (ICM). This module counts the number of B sensor signals between the A sensor signals to sequence the ignition coils properly. There can be zero, one, or two B sensor signals between the A signals. When starting the engine, the module begins counting B sensor signals between the A signals as soon as the module senses zero B signals between sensor signals. After the module senses four B signals, the module

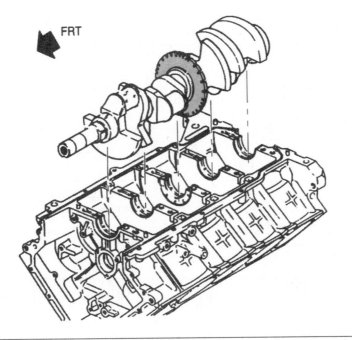

Figure 13-21 Crankshaft with reluctor for a Northstar engine. *(Courtesy of Cadillac Motor Division—GMC)*

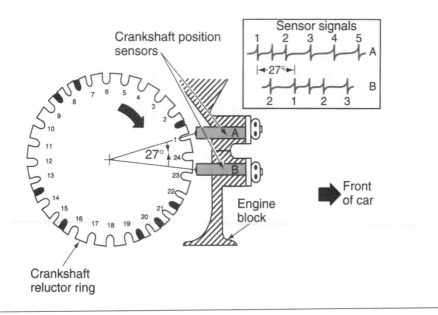

Figure 13-22 A and B sensors in a Northstar engine. *(Courtesy of Cadillac Motor Division—GMC)*

sequences the coils properly (Figure 13-23). This allows the ignition system to begin firing the spark plugs within 180° of crankshaft rotation while starting the engine. This system allows for much quicker starting than other EI systems that require the crankshaft to rotate one or two times before the coils are sequenced.

System Operation

Once the engine starts, the PCM switches from the bypass mode to the ignition control mode. The ICM uses the A and B crankshaft sensor signals to determine the exact crankshaft position and rpm, and the ICM relays this information to the PCM with 4X reference and 24X voltage signals. The

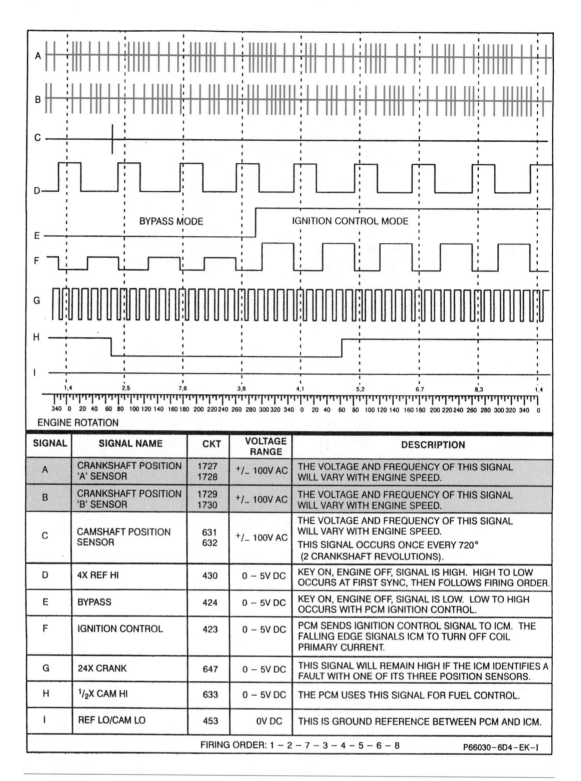

SIGNAL	SIGNAL NAME	CKT	VOLTAGE RANGE	DESCRIPTION
A	CRANKSHAFT POSITION 'A' SENSOR	1727 1728	+/− 100V AC	THE VOLTAGE AND FREQUENCY OF THIS SIGNAL WILL VARY WITH ENGINE SPEED.
B	CRANKSHAFT POSITION 'B' SENSOR	1729 1730	+/− 100V AC	THE VOLTAGE AND FREQUENCY OF THIS SIGNAL WILL VARY WITH ENGINE SPEED.
C	CAMSHAFT POSITION SENSOR	631 632	+/− 100V AC	THE VOLTAGE AND FREQUENCY OF THIS SIGNAL WILL VARY WITH ENGINE SPEED. THIS SIGNAL OCCURS ONCE EVERY 720° (2 CRANKSHAFT REVOLUTIONS).
D	4X REF HI	430	0 − 5V DC	KEY ON, ENGINE OFF, SIGNAL IS HIGH. HIGH TO LOW OCCURS AT FIRST SYNC, THEN FOLLOWS FIRING ORDER.
E	BYPASS	424	0 − 5V DC	KEY ON, ENGINE OFF, SIGNAL IS LOW. LOW TO HIGH OCCURS WITH PCM IGNITION CONTROL.
F	IGNITION CONTROL	423	0 − 5V DC	PCM SENDS IGNITION CONTROL SIGNAL TO ICM. THE FALLING EDGE SIGNALS ICM TO TURN OFF COIL PRIMARY CURRENT.
G	24X CRANK	647	0 − 5V DC	THIS SIGNAL WILL REMAIN HIGH IF THE ICM IDENTIFIES A FAULT WITH ONE OF ITS THREE POSITION SENSORS.
H	$^1/_2$X CAM HI	633	0 − 5V DC	THE PCM USES THIS SIGNAL FOR FUEL CONTROL.
I	REF LO/CAM LO	453	0V DC	THIS IS GROUND REFERENCE BETWEEN PCM AND ICM.

FIRING ORDER: 1 − 2 − 7 − 3 − 4 − 5 − 6 − 8   P66030−6D4−EK−I

Figure 13-23 Operation of the EI fast-start system used on Northstar engines. *(Courtesy of Cadillac Motor Division—GMC)*

PCM scans all the other input sensor signals and calculates the precise spark advance required by the engine. The PCM sends an ignition control signal to the ICM that informs the ICM to turn off the appropriate coil primary at the proper time to supply the correct spark advance. The reference cam low wire is a ground wire between the ICM and the PCM (Figure 13-24). If the engine detonates, the knock sensor signal informs the PCM to reduce the spark advance.

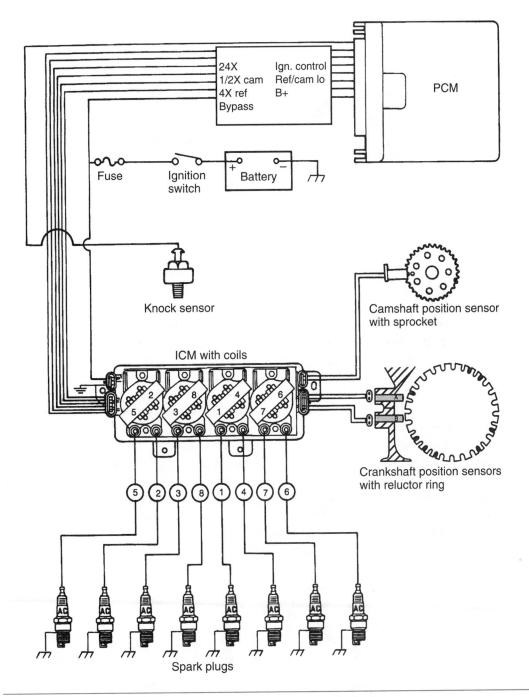

Figure 13-24 Complete EI fast-start system with A and B sensors. *(Courtesy of Cadillac Motor Division—GMC)*

The camshaft position sensor is located in the rear cylinder bank in front of the exhaust camshaft sprocket. A reluctor pin in the sprocket rotates past the sensor, and this sensor produces one high and one low voltage signal every camshaft revolution, or every two crankshaft revolutions (Figure 13-25). The PCM uses the camshaft position sensor signal to sequence the injectors properly.

EI Systems with a Crankshaft Reluctor Ring

Electronic ignition systems are used on GM's 2.0L and 2.5L four-cylinder engines and 2.8L and 3.1L V6 engines. The main difference between these EI systems and the previously explained EI systems

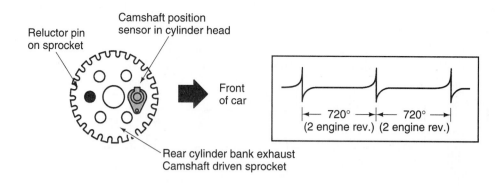

Figure 13-25 Crankshaft position sensor. *(Courtesy of Cadillac Motor Division—GMC)*

Shop Manual

Chapter 13, page 574

is in the crankshaft sensor. These other systems have a notched reluctor ring positioned near the center of the crankshaft (Figure 13-26). This ring is permanently cast on the crankshaft. A magnetic sensor containing a permanent magnet and a winding is mounted in an opening in the engine block. The tip of this sensor is 0.050 in. (1.27 mm) from the reluctor ring outer surface. This gap between the magnetic sensor tip and the reluctor ring is not adjustable. The magnetic sensor is retained in the engine block with a bolt and clamp. The coil assembly and coil module are similar to those used on other slow-start and fast-start EI systems.

On the 2.0L four-cylinder engine and the 2.8L and 3.1L V6 engines, the coil and module assembly is positioned on one side of the engine block and the magnetic sensor is located in the opposite side of the block. The magnetic sensor is mounted on the back of the coil module on the 2.5L four-cylinder engine. Therefore, the magnetic sensor is not visible until the coil and module assembly is removed from the engine block.

The reluctor ring has seven notches on four-cylinder or V6 engines. Six of these notches are spaced 60 degrees apart, and the seventh notch is positioned 10 degrees from the sixth notch. A signal from the seventh notch is referred to as a SYNC signal, which is used by the coil module for coil sequencing. On four-cylinder engines, the coil module is programmed to recognize the SYNC notch and count notch 1. When notch 2 passes the sensor tip, the coil module opens the 2-3 coil primary circuit and fires spark plugs 2 and 3. After this event, the coil module counts notches 3 and

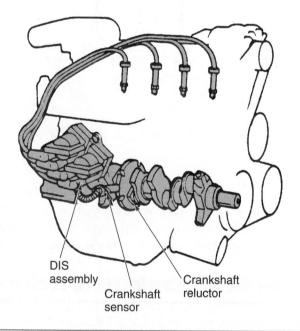

Figure 13-26 EI system on a four-cylinder engine. *(Courtesy of Oldsmobile Division—GMC)*

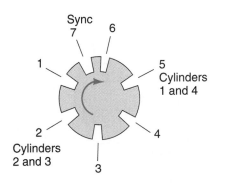

Figure 13-27 Reluctor ring and firing order for a 4-cylinder engine.

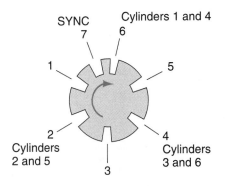

Figure 13-28 Reluctor ring and firing order for a V6 engine.

4, and then the module opens the primary circuit of the 1-4 coil when notch 5 rotates past the sensor. This module action fires spark plugs 1 and 4. The coil module counts notches 6 and 7, and the cycle starts repeating (Figure 13-27). During engine starting, the 2-3 coil fires before the 1-4 coil.

On the V6 engines, the coil module recognizes the SYNC notch and counts notch 1. When the signal from notch 2 is received, the coil module opens the primary circuit of the 2-5 coil, which fires spark plugs 2 and 5 (Figure 13-28). Notch 3 is counted next, and then the signal from notch 4 is used to open the primary circuit of the 3-6 coil and fire spark plugs 3 and 6. As the sequence continues, notch 5 is counted, and then the signal from notch 6 is used to open the primary circuit of the 1-4 coil and fire spark plugs 1 and 4. The cycle repeats as the crankshaft continues to rotate. During engine starting, the 2-5 coil fires first.

When the engine is running, a low voltage signal is sent via the reference wire from the coil module to the PCM when the notch that is 60 degrees ahead of when the cylinder event notch passes the sensor. When the cylinder event notch passes the sensor, the reference signal changes to high voltage. This reference signal informs the PCM regarding crankshaft position and speed. After a reference signal is received, the PCM scans the inputs and determines the spark advance required by the engine. When this calculation is completed in a few milliseconds, the PCM sends an EST signal to the coil module to open the primary circuit of the next coil at the correct instant to provide the required spark advance (Figure 13-29). The PCM can affect the spark advance during engine starting on these EI systems. If an engine with EI has multiport fuel injection or throttle body injection, a cam sensor signal is not required for injector sequencing.

Ford's EI System with a Crankshaft Reluctor Ring

Ford uses a similar system as GM, however the reluctor ring has many more slots. The crankshaft sensor for their 4.6L V8 engine is a variable reluctance sensor which is triggered by a 36 minus 1 (or 35) tooth trigger wheel located inside the front cover of the engine (Figure 13-30). The sensor provides two types of information: crankshaft position and engine speed.

The system utilizes four ignition coils, two in each housing (Figure 13-31). Each coil terminal has two cylinder numbers on them. This allows for interchangeability.

The trigger wheel has a tooth every 10 degrees with one tooth missing. When the part of the wheel that is missing a tooth passes by the sensor, there is a longer than normal pause between signals from the sensor. The ignition control module recognizes this and is able to identify this long pause as the location of piston number one (Figure 13-32).

Chrysler's EI System with a Crankshaft Reluctor Ring

Again, most systems that use a crankshaft reluctor ring are quite similar in construction and operation. Chrysler's is like the others but uses a different number of teeth on the reluctor, a camshaft sensor, and a camshaft reluctor, therefore the signals received by the control module are also different.

Shop Manual
Chapter 13, page 568

Trigger wheel is another name for a sensor reluctor.

Shop Manual
Chapter 13, page 565

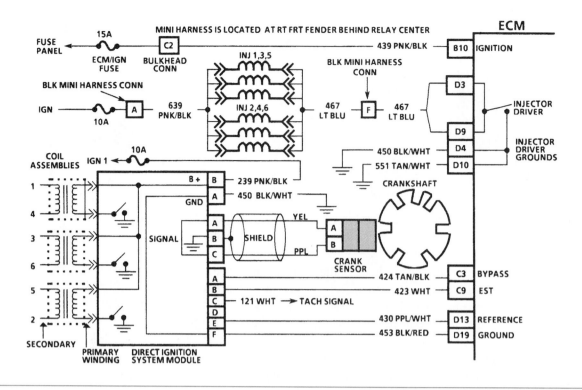

Figure 13-29 Schematic of an EI system for a V6 engine. *(Courtesy of Oldsmobile Division—GMC)*

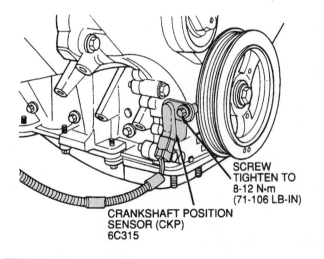

Figure 13-30 Location of crankshaft position sensor on a Ford 4.6L engine. *(Reprinted with the permission of Ford Motor Company)*

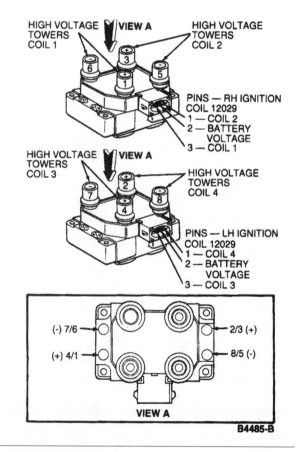

Figure 13-31 Ignition coils for a Ford 4.6L engine. *(Reprinted with the permission of Ford Motor Company)*

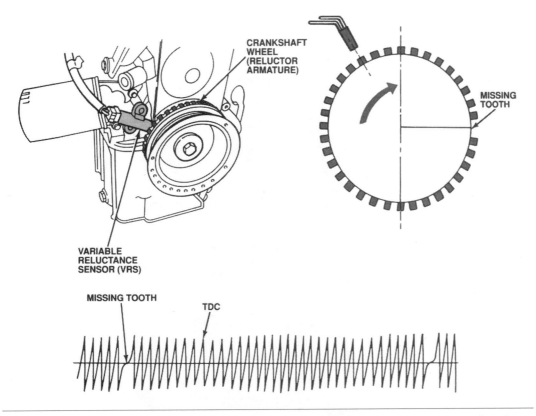

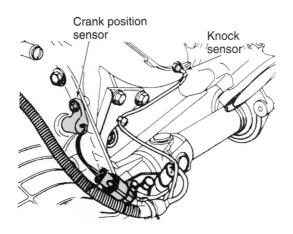

The crankshaft timing sensor is mounted in an opening in the transaxle bell housing (Figure 13-33). The inner end of this sensor is positioned near a series of notches and slots that are integral with the transaxle drive plate (Figure 13-34).

A bolt retains the crankshaft timing sensor to the bell housing. A group of four slots is located on the transaxle drive plate for each pair of engine cylinders. Thus, a total of 12 slots is positioned around the drive plate. The slots in each group are positioned 20 degrees apart. When the slots on the transaxle drive plate rotate past the crankshaft timing sensor, the voltage signal from the sensor changes from 0 V to 5 V. This varying voltage signal informs the PCM regarding crankshaft position and speed. The PCM calculates spark advance from this signal. The PCM also uses the crankshaft

Figure 13-33 Crankshaft position sensor mounted in a transaxle. *(Courtesy of Chrysler Corporation)*

Figure 13-34 Crankshaft timing sensor with the transaxle drive plate. *(Courtesy of Chrysler Corporation)*

timing sensor signal along with other inputs to determine air-fuel ratio. Base timing is determined by the signal from the last slot in each group of slots.

The camshaft reference sensor is mounted in the top of the timing gear cover (Figure 13-35). A notched ring on the camshaft gear rotates past the end of the camshaft reference sensor. This ring contains two single slots, two double slots, and a triple slot (Figure 13-36).

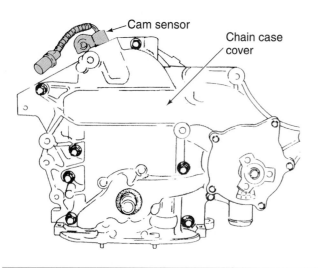

Figure 13-35 Camshaft reference sensor mounting in timing cover. *(Courtesy of Chrysler Corporation)*

Figure 13-36 Notched ring on camshaft gear. *(Courtesy of Chrysler Corporation)*

When a camshaft gear notch rotates past the camshaft reference sensor, the signal from the sensor changes from 0 V to 5 V. The single, double, and triple notches provide different voltage signals from the camshaft reference sensor as they rotate past the sensor. These signals are sent to the PCM. Since the ignition module is part of the PCM board, an external module is not used on these systems. The PCM determines the exact camshaft and crankshaft position from the camshaft reference sensor signals, and the PCM uses these signals to sequence the coil primary windings and each pair of injectors at the correct instant.

When the engine starts cranking, the spark plugs fire and the injectors discharge fuel within one crankshaft revolution. The PCM determines when to sequence the coils and injectors from the camshaft reference sensor signals. If the camshaft reference sensor or the crankshaft timing sensor is defective, the engine will not start. The spark plug wires from coil number one are connected to cylinders one and four, whereas the spark plug wires from coil number two go to cylinders two and five, and the spark plug wires on coil number three are attached to cylinders three and six. The cylinder firing order for the 3.3L V6 engine is 1-2-3-4-5-6.

Once the engine is started, the PCM knows the exact crankshaft position and speed from the camshaft reference sensor and crankshaft timing sensor signals. The leading edges of the slots in the transaxle drive plate rotate past the crankshaft timing sensor at 9°, 29°, 49°, and 69° BTDC. When the single camshaft reference sensor slot rotates past the sensor, one high and one low digital signal are received by the PCM. When these signals are received, the PCM is programmed to fire the number two coil next, which is connected to spark plugs 2 and 5 (Figure 13-37).

When the engine is cranking, all spark plug firings are at 9° BTDC on the compression stroke. The PCM also sequences injectors 2 and 3 when the single slot rotates past the camshaft reference sensor.

When the double notch rotates past the camshaft reference sensor, the PCM is programmed to fire coil 3 connected to spark plugs 3 and 6, and sequence injectors 1 and 6. When the triple notch rotates past the camshaft reference sensor, the PCM is programmed to fire coil 1 connected to spark plugs 1 and 4, and sequence injectors 4 and 5.

Since Chrysler engines are now equipped with sequential fuel injection, the PCM grounds each injector individually, but the proper injector sequencing is determined from the camshaft reference sensor signals.

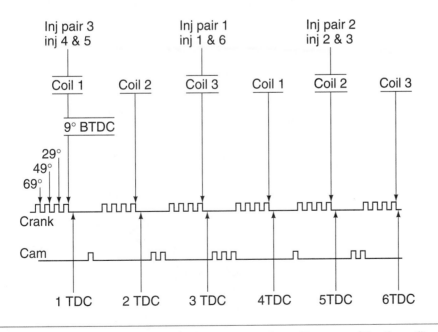

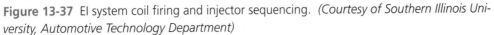

Figure 13-37 EI system coil firing and injector sequencing. *(Courtesy of Southern Illinois University, Automotive Technology Department)*

EI Systems with Coils Connected Directly to Spark Plugs

Shop Manual
Chapter 13, page 582

These EI systems have the same reluctor ring and magnetic sensor as other EI systems. However, on these EI systems, the spark plug wires are eliminated and the coil secondary terminals are connected directly to the spark plugs. Since the spark plug wires are eliminated, the chance of high voltage leaks is reduced. This EI system is used on GM's 2.3L Quad 4 engine. The ignition module and coils are mounted under a plate that is positioned between the cam covers on top of the engine (Figure 13-38). Each coil is connected to a pair of spark plugs. The spark plug pairs are 1-4 and 2-3. Since the coil module contacts the mounting plate, it acts as a heat shield. The mounting plate must be grounded to the engine to provide ignition operation.

The operation and electrical connections for these EI systems are similar to the EI system operations explained previously (Figure 13-39). An added feature of this system is the retarding of ignition timing for a few milliseconds during transmission shifting to lower the engine torque load on the transmission clutches.

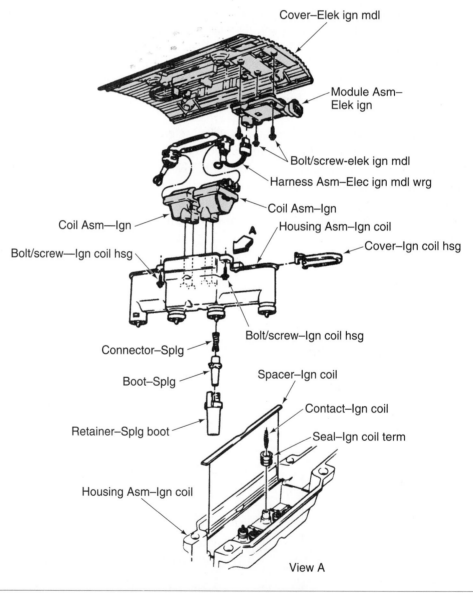

Figure 13-38 The coil module, coil assemblies, and cover plate for an EI system with the plugs connected directly to the ignition coils. *(Courtesy of Oldsmobile Division—GMC)*

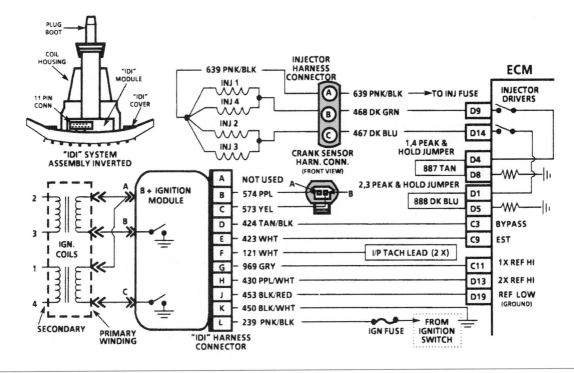

Figure 13-39 Schematic for an EI system with the plugs connected directly to the ignition coils. *(Courtesy of Oldsmobile Division—GMC)*

Dual Spark Plug EI Systems

Shop Manual
Chapter 13, page 568

Some EI systems fire two spark plugs at each cylinder, these systems are known as dual plug (DP) systems. Commonly found on four-cylinder engines, the system (Figure 13-40) is comprised of a dual plug inhibit circuit, eight spark plugs, and two coil packs (each one containing two coils), as

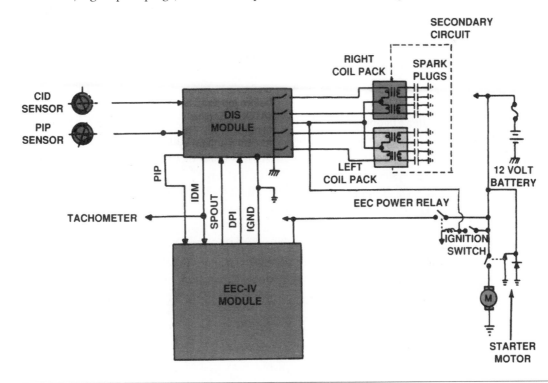

Figure 13-40 Schematic of Ford's DP EI system. *(Reprinted with the permission of Ford Motor Company)*

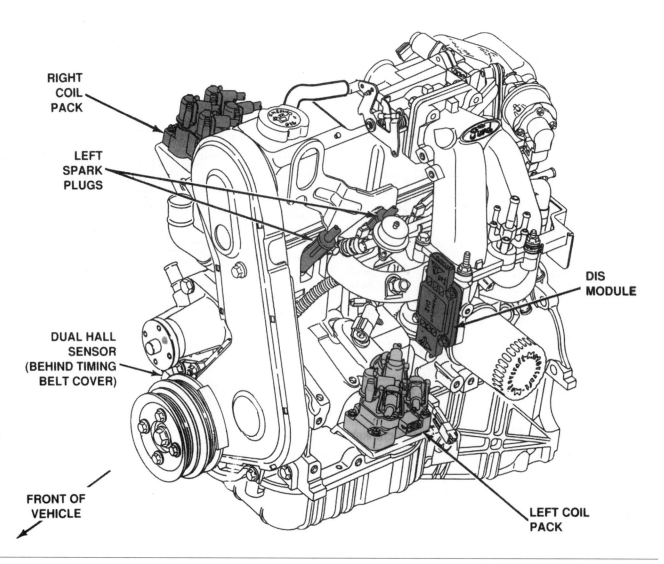

**RIGHT
COIL
PACK**

**LEFT
SPARK
PLUGS**

**DUAL HALL
SENSOR
(BEHIND TIMING
BELT COVER)**

**FRONT OF
VEHICLE**

**DIS
MODULE**

**LEFT COIL
PACK**

Figure 13-41 Location of components in Ford's DP system. *(Reprinted with the permission of Ford Motor Company)*

well as other EI components. The spark plugs are placed across from each other in the cylinder. The coil packs are placed on either side of the engine and are connected to the plugs on that side (Figure 13-41). Only those components that are unique to this system will be discussed here, as the basics of the system are similar to other EI systems.

Ford's DP system uses a dual Hall effect switch (Figure 13-42) that contains the CID and PIP sensors. The dual switch unit has one permanent magnet, two Hall devices, and a rotating assembly with two separate vanes. The vane assembly is part of the crankshaft pulley assembly. The CID and PIP signals are used in the same way as other Ford EI systems.

During engine start-up, the dual plug mode of operation is inhibited and only the spark plugs on the right side are fired. Once the engine is running, both sets of plugs fire. Each coil fires two plugs, which are wired in series. One spark plug in each cylinder always fires from the center electrode to the ground electrode and its companion plug fires from the ground to the center electrode (Figure 13-43). Because of the way the coils are wired, the two spark plugs in a cylinder fire in opposite directions (Figure 13-44).

The ground electrode of a spark plug is also called the side electrode.

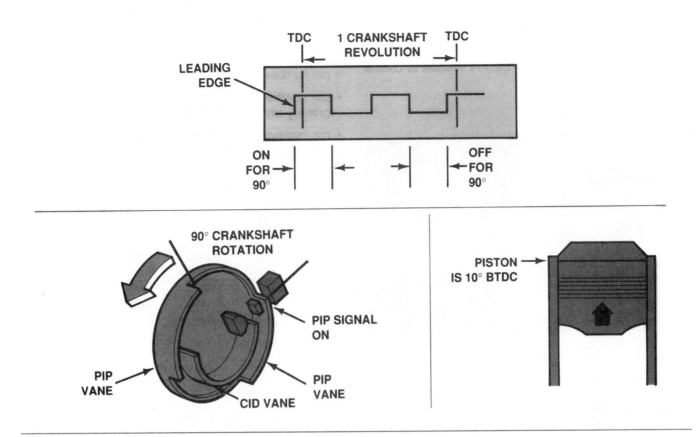

Figure 13-42 PIP signal. *(Reprinted with the permission of Ford Motor Company)*

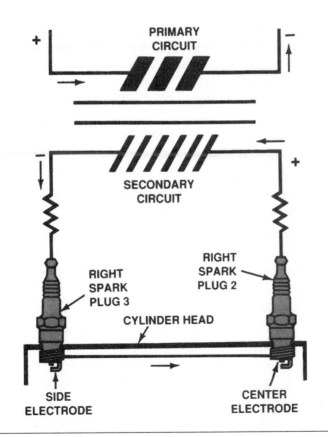

Figure 13-43 Spark plug circuit and firing in an EI system. *(Reprinted with the permission of Ford Motor Company)*

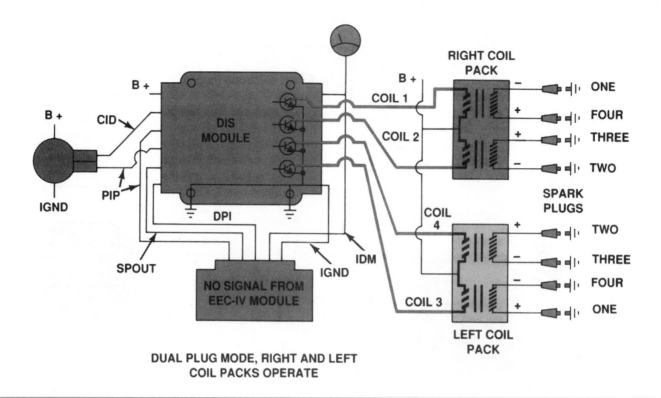

RIGHT COIL PACK

B +
COIL 1
COIL 2

- ONE
+ FOUR
+ THREE
- TWO

B +
CID
PIP
IGND
DIS MODULE
DPI
SPOUT
IGND
IDM
NO SIGNAL FROM EEC-IV MODULE

COIL 4
COIL 3

SPARK PLUGS

+ TWO
- THREE
- FOUR
+ ONE

LEFT COIL PACK

DUAL PLUG MODE, RIGHT AND LEFT COIL PACKS OPERATE

Figure 13-44 Notice the polarity of the spark plugs in the same cylinder in Ford's DP system. *(Reprinted with the permission of Ford Motor Company)*

Summary

Terms to Know

Bypass mode

Coil Pack

Distributorless ignition system (DIS)

Dual plug

Fast-start

Slow-start

Waste spark

❏ Direct ignition systems eliminate the distributor. Each spark plug, or in some cases, pair of spark plugs, has its (their) own ignition coil. Primary circuit switching and timing control is done using a special ignition module tied into the vehicle control computer.

❏ EI systems provide longer spark duration at the spark plug electrodes than conventional electronic ignition systems, and this helps to fire leaner air-fuel ratios in today's engines.

❏ In an EI system, the coil fires two spark plugs in series and the current flows down through one spark plug and up through the other spark plug.

❏ In an EI system, while a pair of spark plugs is firing, one of the cylinders is on the compression stroke and the other cylinder in the pair is on the exhaust stroke.

❏ In some EI systems, the camshaft sensor signal informs the computer when to sequence the coils and fuel injectors.

❏ In some EI systems, the crankshaft sensor signal provides engine speed and crankshaft position information to the computer.

❏ In some EI systems, the camshaft sensor is positioned in the former distributor opening.

❏ Some EI systems have a combined crankshaft and SYNC sensor at the front of the crankshaft.

❏ Some EI systems may be called slow-start systems because as many as two crankshaft revolutions are required before the ignition system begins firing.

❏ Some EI systems are called fast-start systems because the spark plugs begin firing within 120 degrees of crankshaft rotation.

❏ A notched reluctor ring on the crankshaft is used to produce voltage signals in a magnetic sensor in some EI systems.

❏ In other EI systems, the coils are mounted directly on top of the spark plugs, and the spark plug wires are eliminated.

Review Questions

Short Answer Essay

1. Explain how the plugs fire in a two-plug-per-coil EI system.

2. What is the purpose of the 18X crankshaft sensor?

3. What is the difference between a slow-start and a fast-start EI system?

4. How do EI ignition systems differ from conventional electronic ignition systems?

5. What does the term "waste spark" mean?

6. Why do EI systems have a primary current limiting function?

7. Why are four spark plugs fired at the same time in Ford's DP EI system?

8. Explain why a distributorless ignition system has more than one ignition coil.

9. Name five advantages of an EI system over a DI system.

10. Describe the pickup and triggering mechanism on an EI system with a reluctor ring mounted near the center of the crankshaft.

Fill-in-the Blanks

1. A camshaft position sensor is typically used to monitor the _____ of _____ and the crankshaft sensor is used to monitor the _____ of the _____.

2. The camshaft sensor ring has one notch, therefore the sensor produces one _____ _____ signal and one_____ _____ signal for each revolution of the camshaft.

3. The engine's camshaft rotates at _____ the speed of the crankshaft.

4. An EI coil pack normally consists of two or more _____ _____ and a(n) _____ _____ .

5. Camshaft sensors can be mounted in the _____ _____ or in the _____ _____ _____.

6. Slow-start EI systems may require _____ crankshaft revolutions before they will start.

7. EI systems help to prevent cylinder misfiring because these systems have a longer _____ _____.

8. If the crankshaft sensor signal is defective, the engine will not _____.

9. In an EI system with the coils and module mounted under a plate on top of the engine, the secondary _____ _____ wires are not required.

10. In an EI system with the coils and module mounted under a plate on top of the engine, the _____ _____ must be grounded to the engine.

ASE Style Review Questions

1. While discussing the advantages of EI systems:
   Technician A says EI systems help to prevent misfiring of lean air-fuel ratios.
   Technician B says EI systems provide more stable control of ignition timing and spark advance compared to a distributor.
   Who is correct?
   A. A only   C. Both A and B
   B. B only   D. Neither A nor B

2. While discussing electronic ignition (EI) systems:
   Technician A says when a pair of spark plugs are firing, one of the cylinders is on the exhaust stroke and the other cylinder is on the power stroke.
   Technician B says each pair of spark plugs is fired in series.
   Who is correct?
   A. A only   C. Both A and B
   B. B only   D. Neither A nor B

3. *Technician A* says EI systems typically use one coil for two spark plugs.
   Technician B says some EI systems rely on a waste spark system to fire the spark plug.
   Who is correct?
   A. A only   C. Both A and B
   B. B only   D. Neither A nor B

4. While discussing EI systems:
   Technician A says one spark plug fires from the center to the ground electrode and its companion plug fires from the ground electrode to the center electrode.
   Technician B says the pair of plugs that are fired by one coil are connected together by a secondary cable.
   Who is correct?
   A. A only   C. Both A and B
   B. B only   D. Neither A nor B

5. While discussing EI systems:
   Technician A says an advantage of EI over DI is the longer time available for dwell.
   Technician B says an advantage of EI over DI is the longer period of time between coil firings.
   Who is correct?
   A. A only   C. Both A and B
   B. B only   D. Neither A nor B

6. *Technician A* says a dual crankshaft sensor monitors crankshaft and camshaft location.
   Technician B says many EI systems have a camshaft sensor in addition to the crankshaft sensor.
   Who is correct?
   A. A only   C. Both A and B
   B. B only   D. Neither A nor B

7. *Technician A* says the camshaft sensor is used to synchronize the crankshaft sensor with the position of piston number one.
   Technician B says the camshaft sensor is used to synchronize the firing of the fuel injectors in a sequentially fuel injected engine.
   Who is correct?
   A. A only   C. Both A and B
   B. B only   D. Neither A nor B

8. While discussing Ford EI systems:
   Technician A says the CID sensor monitors the position of the crankshaft.
   Technician B says the PIP and CID are redundant and monitor the same thing.
   Who is correct?
   A. A only   C. Both A and B
   B. B only   D. Neither A nor B

9. While discussing GM's EI system for a 3.8L turbocharged SFI engine:
   Technician A says the engine won't start if the camshaft sensor is faulty.
   Technician B says if the camshaft sensor becomes faulty while the engine is running, the engine will continue to run but will run poorly.
   Who is correct?
   A. A only   C. Both A and B
   B. B only   D. Neither A nor B

10. While discussing the fast-start EI system used with the Northstar engine:
    Technician A says the A and B sensors are redundant and only the B sensor is critical to engine starting.
    Technician B says the signal from the A sensor occurs 27 degrees before the B signals.
    Who is correct?
    A. A only   C. Both A and B
    B. B only   D. Neither A nor B

On Board Diagnostic (OBD) II Systems

Upon completion and review of this chapter, you should be able to:

❏ Explain the reasons for OBD II.

❏ Describe the primary provisions of OBD II.

❏ Explain the requirements to illuminate the malfunction indicator light in an On Board Diagnostic II system.

❏ Briefly describe the monitored systems in an OBD II system.

❏ Describe the main hardware differences between an OBD II system and other systems.

❏ Explain the advantage of a EEPROM compared to a PROM.

❏ Describe an OBD II warm-up cycle.

❏ Explain trip and drive cycle in an OBD II system.

❏ Describe how engine misfire is detected in an OBD II system.

❏ Describe the differences between a type A and B misfire.

❏ Describe the purpose of having two oxygen sensors in an exhaust system.

❏ List five things that were mandated by OBD II.

❏ Briefly describe what the comprehensive component monitor looks at.

Introduction

The On Board Diagnostic II (OBD II) systems were developed in response to the federal government's and the state of California's emission control system monitoring standards for all automotive manufacturers. The main goal of OBD II was to detect when engine or system wear or when component failure caused exhaust emissions to increase by 50% or more. OBD II also called for standard service procedures without the use of dedicated special tools. To accomplish these goals, manufacturers needed to change many aspects of their electronic engine control systems. According to the guidelines of OBD II, all vehicles must have:

❏ A universal diagnostic test connector, known as the Data Link Connector (DLC), with dedicated pin assignments.

❏ A standard location for the DLC. It must be under the dash on the driver's side of the vehicle and must be visible (SAE standard J1962).

❏ A standard list of Diagnostic Trouble Codes (DTCs), SAE's standard J2012.

❏ A standard communication protocol (SAE standard J1850).

❏ The use of common scan tools on all vehicle makes and modes (SAE standard J1979).

❏ Common diagnostic test modes (SAE standard J2190).

❏ Vehicle identification must be automatically transmitted to the scan tool.

❏ Stored trouble codes must be able to be cleared from the computer's memory with the scan tool.

❏ The ability to record, and store in memory, a snapshot of the operating conditions that existed when a fault occurred.

❏ The ability to store a code whenever something goes wrong and affects exhaust quality.

❏ A standard glossary of terms, acronyms, and definitions must be used for all components in the electronic control systems (SAE standard J1930).

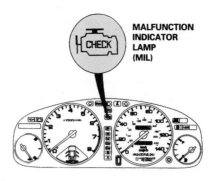

Figure 14-1 A standard MIL. *(Courtesy of American Honda Motor Company, Inc.)*

The OBD II systems must illuminate the malfunction indicator lamp (MIL) (Figure 14-1) if the vehicle conditions would allow emissions to exceed 1.5 times the allowable standard for that model year based on a Federal Test Procedure (FTP). When a component or strategy failure allows emissions to exceed this level, the MIL is illuminated to inform the driver of a problem and a diagnostic trouble code is stored in the powertrain control module (PCM).

Besides enhancements to the computer's capacities, some additional hardware is required to monitor the emissions performance closely enough to fulfill the tighter constraints and beyond merely keeping track of component failures. In most cases, this hardware consists of an additional heated oxygen sensor down the exhaust stream from the catalytic converter (Figure 14-2), upgrading specific connectors and components to last the mandated 100,000 miles or 10 years, in some cases a more precise crankshaft or camshaft position sensor (to detect misfires), and a new standardized 16-pin DLC.

OBD II Introduction and Implementation

A BIT OF HISTORY

The first manufacturer to have a fully compliant OBD II system was Toyota in 1994. Toyota's strategies are very much like those of the domestic car manufacturers. They also included some of their own diagnostic routines and DTCs.

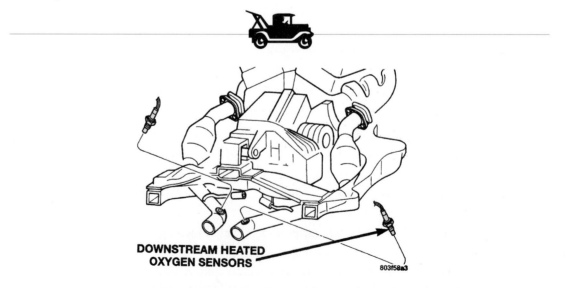

DOWNSTREAM HEATED OXYGEN SENSORS

803f58a3

Figure 14-2 Downstream oxygen sensors. *(Courtesy of Chrysler Corporation)*

The following is an outline of the development of computerized engine control systems leading to OBD II compliance for the domestic manufacturers:

Chrysler

1981-1983	Lean Burn Computer
1983-1987	Logic Module/Power Module System
1988-1989	Single Module Engine Controller
1990-1995	Single Board Engine Controller
1995-	OBD II

Ford

1978-1979	EEC I
1979	EEC II
1980-1984	EEC III
1980-1991	MCU
1984-1995	EEC IV
1994-	EEC V (OBD II)

General Motors

1975-1979	Cadillac EFI
1980	C-4
1981-1990	CCC (Carburetor controls)
1982-1996	EFI/PFI
1994-1996	OBD II Phase In
1996	OBD II

Vehicle manufacturers installed OBD II systems on some 1994 models. For example, Ford installed their EEC V system on the 3.8L Mustang and the 4.6L Thunderbird and Cougar. In 1995, the OBD II system was also installed on Lincoln Town Cars and Continentals, Grand Marquis, Crown Victorias, Rangers, and Windstars. Since OBD II systems were not mandated until 1996, some of the 1994 and 1995 OBD II systems are partial systems that do not have all of the monitors required for a complete OBD II system.

The EEC V system hardware includes the PCM, sensors, switches, actuators, solenoids, wires, terminals, and the constant control relay module. The software is the program in the PCM that the PCM uses to control the outputs. The major difference between EEC V with OBD II and previous systems is the software containing the monitoring strategies inside the PCM.

The main hardware difference in the EEC V system is the addition of a downstream **heated oxygen sensor** (HO_2S) mounted downstream from the catalytic converter. The conventional HO_2S is mounted in the exhaust manifold. The HO_2S are identified by their position in relation to the number one cylinder and their location relative to the converters. Sensors on the same side of the vehicle as the number one cylinder have a 1 prefix, and sensors mounted downstream from the converter have a 2 suffix (Figure 14-3).

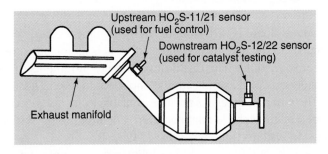

Figure 14-3 OBD II system with heated oxygen sensors in the exhaust manifold and after the catalytic converter. *(Reprinted with the permission of Ford Motor Company)*

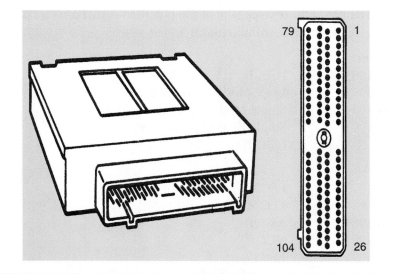

Figure 14-4 EEC V PCM and wiring terminal with 104 terminal pins. *(Reprinted with the permission of Ford Motor Company)*

The EEC V PCM has 104 terminals compared to the previous 60 terminals on the EEC IV PCM (Figure 14-4). The additional terminals are required for the extra inputs and outputs in the EEC V system with OBD II. The EEC V PCM has smaller diameter pins and the wires have thinner high-temperature insulation, which allows more wires to occupy the same space. Some of the PCM terminals are goldplated, and the connectors on the HO_2S and throttle position sensor have improved seals to keep out dirt and moisture.

Instead of a fixed, unalterable programmable read only memory (PROM), the PCMs are equipped with an **electronically erasable programmable read only memory (EEPROM)** to store a large amount of information. The EEPROM is soldered into the PCM and is not replaceable. The EEPROM stores data without the need for a continuing source of electrical power.

Ford's SBDS is a state-of-the-art diagnostic machine found in Ford dealerships.

The EEPROM is an integrated circuit that contains the program used by the PCM to provide powertrain control. It is possible to erase and reprogram the EEPROM without removing this chip from the computer. When a modification to the PCM operating strategy is required, it is no longer necessary to replace the PCM. The EEPROM may be reprogrammed through the DLC using the Service Bay Diagnostic System (SBDS).

For example, if the vehicle calibrations are updated for a specific car model sold in California, the SBDS may be used to erase the EEPROM. After the erasing procedure, the EEPROM is reprogrammed with the updated information. A red-colored CD-ROM that is sent to all SBDS machine locations each month must be used to complete the reprogramming procedure. PCM recalibrations must be directed by a service bulletin or recall letter.

Monitoring Capabilities

The California Air Resources Board (CARB) found that by the time a computer or emission system component failure occurs and the malfunction indicator light is illuminated, the vehicle emissions have been excessive for some time. The CARB developed requirements to monitor the performance of emission systems, as well as indicating component failure. These requirements were accepted by the EPA. The monitoring results must be available to service personnel without special test equipment marketed by the vehicle manufacturer. The monitoring system for engine, computer system, and emission system equipment is called OBD II.

Computer systems without OBD II have the ability to detect component and system failure. Computer systems with OBD II are capable of monitoring the ability of systems and components to maintain low emission levels.

Computer systems with OBD II capabilities are similar to previous systems except for the monitoring systems and the monitoring strategies in the PCM, which are extensive. New refinements are frequently incorporated into the PCM and other system components as improved technology is developed. Monitors included in OBD II are:

1. Catalyst efficiency monitor
2. Engine misfire monitor
3. Fuel system monitor
4. Heated exhaust gas oxygen sensor monitor
5. Exhaust gas recirculation monitor
6. Evaporative system monitor
7. Secondary air injection monitor
8. Comprehensive component monitor

OBD II systems will perform certain tests on various subsystems of the engine management system. OBD II is designed to turn on the MIL when the vehicle has any failure that could potentially cause the vehicle to exceed its designed emission standard by a factor of 1.5. The system does that by the use of a monitor. If one or more monitored systems are out limit, then the MIL turns on to indicate a problem. The various monitors are:

Catalyst Efficiency Monitor

OBD II vehicles use a minimum of two oxygen sensors (O_2S). One of these is used for feedback to the PCM for fuel control and the other, located at the rear of the catalytic converter, gives an indication of the efficiency of the converter. If the converter is operating properly, the signal from the pre-catalyst O_2 will have oscillations while the post-catalyst O_2 will be relatively flat (Figure 14-5). Once the signal from the rear sensor approaches that of the front sensor, the MIL comes on and a DTC is set.

Shop Manual
Chapter 14, page 607

The downstream HO_2Ss have additional protection to prevent the collection of condensation on the ceramic. The internal heater is not turned on until the engine coolant temperature (ECT) sensor signal indicates a warmed up engine. This action prevents cracking of the ceramic. Gold-plated pins and sockets are used in the HO_2Ss, and the downstream and upstream sensors have different wiring harness connectors.

A catalytic converter stores oxygen during lean engine operation and gives up this stored oxygen during rich operation to burn up excessive hydrocarbons. Catalytic converter efficiency is measured by monitoring the oxygen storage capacity of the converter during closed loop operation.

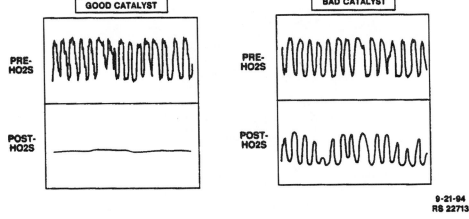

Figure 14-5 Oxygen sensor signal for a good and bad catalytic converter. *(Courtesy of Oldsmobile Division—GMC)*

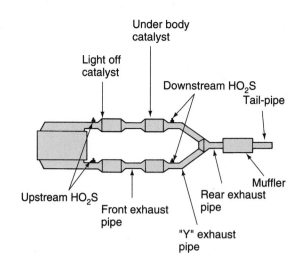

Figure 14-6 Catalyst efficiency monitoring system. (*Reprinted with the permission of Ford Motor Company*)

When the catalytic converter is storing oxygen properly, the downstream HO_2Ss provide low-frequency voltage signals. If the catalytic converter is not storing oxygen properly, the voltage signal frequency increases on the downstream HO_2Ss until the frequency of the downstream HO_2Ss approaches the frequency of the upstream HO_2Ss (Figure 14-6). When the downstream HO_2Ss voltage signals reach a certain frequency, a DTC is set in the PCM memory. If the fault occurs on three drive cycles, the MIL light is illuminated.

Misfire Monitor

Shop Manual
Chapter 14, page 616

If a cylinder misfires, unburned hydrocarbons (HC) are exhausted from the cylinder, and these excessive HC emissions enter the catalytic converter. When the catalytic converter changes these excessive HC emissions to carbon dioxide and water, the catalytic converter is overheated and the honeycomb in the converter may melt together into a solid mass. If this occurs, the converter is no longer efficient in reducing emissions.

Cylinder misfire monitoring requires measuring the contribution of each cylinder to engine power. The misfire monitoring system uses a highly accurate crankshaft angle measurement to measure the crankshaft acceleration each time a cylinder fires. A high data rate crankshaft sensor is required for this function (Figure 14-7). The PCM monitors the crankshaft acceleration time for

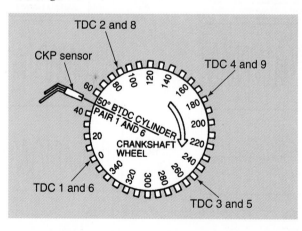

Figure 14-7 High data rate crankshaft sensor used for misfire detection. (*Reprinted with the permission of Ford Motor Company*)

each cylinder firing. If a cylinder is contributing normal power, a specific crankshaft acceleration time occurs. When a cylinder misfires, the cylinder does not contribute to engine power, and crankshaft acceleration for that cylinder is slowed.

Most OBD II systems allow a random misfire rate of about 2% before a misfire is flagged as a fault. It is important to note that this monitor only looks at the speed of acceleration of the crankshaft during a cylinder's firing stroke. It cannot determine if the problem is fuel-, ignition-, or mechanical-related. Misfire is categorized as type A, B, or C. **Type A misfire** could cause immediate catalyst damage. **Type B misfire** could cause emissions of 1.5 times the design standard, and type C could cause an inspection/maintenance (I/M) failure. When there is a type A misfire, the MIL will flash. If there is a type B or C misfire, the MIL will turn on but will not flash.

The misfire monitoring sequence includes an adaptive feature compensating for variations in engine characteristics caused by manufacturing tolerances and component wear (Figure 14-8). It also has the adaptive capability to allow vibration at different engine speeds and loads. When an individual cylinder's contribution to engine speed falls below a certain threshold, the misfire monitoring sequence calculates the vibration, tolerance, and load factors before setting a misfire code.

▲ **WARNING:** Never short across or ground the terminals of a computer system unless instructed to do so by the manufacturer. Grounding or shorting the terminals may cause damage to the system and set DTCs in the PCM's memory.

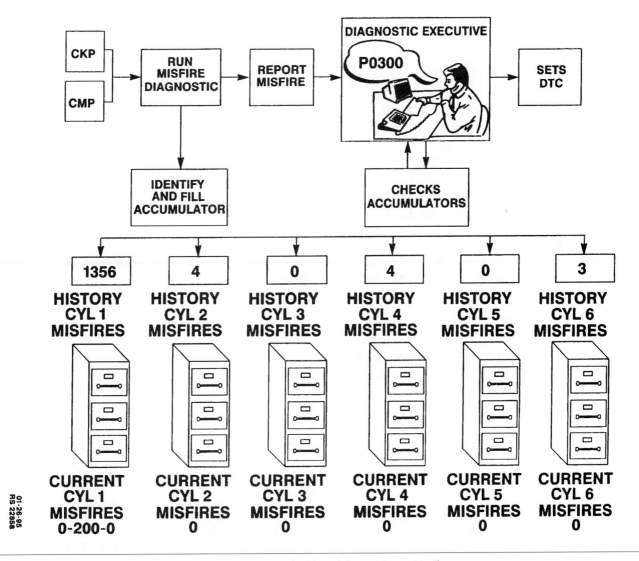

Figure 14-8 Engine misfire file system. *(Courtesy of Oldsmobile Division—GMC)*

Type A Misfires

Type A and type B engine misfires are detected by the misfire monitor. When detecting a type A misfire, the monitor checks cylinder misfiring over a 200-rpm period. If cylinder misfires are between 2% and 20%, the monitor considers the misfiring to be excessive. Under this condition, the PCM may shut off the fuel to the misfiring cylinder or cylinders to limit catalytic converter heat. The PCM may turn off two injectors at the same time on misfiring cylinders. When the engine is operating under heavy load, the PCM will not turn off the injectors on misfiring cylinders.

If the misfire monitor detects a type A cylinder misfire and the PCM does not shut off the injector or injectors, the MIL light begins flashing. When the misfire monitor detects a type A cylinder misfire, and the PCM shuts off an injector or injectors, the MIL is illuminated continually.

A misfire means a lack of combustion in at least one cylinder for at least one combustion event. A misfire pumps unburned fuels through the exhaust. Although the converter can handle an occasional sample of raw fuel, too much fuel to the converter can overheat and destroy it.

Type B Misfires

To detect a type B cylinder misfire, the misfire monitor checks cylinder misfiring over a 1,000-rpm period. If cylinder misfiring exceeds 2% to 3% during this period, the monitor considers the misfiring to be excessive. This amount of cylinder misfiring may not overheat the catalytic converter, but it may cause excessive emission levels. When a type B misfire is detected, a pending DTC is set in the PCM memory. If this fault is detected on a second consecutive drive cycle, the MIL is illuminated.

Fuel System Monitoring

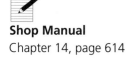

Shop Manual
Chapter 14, page 614

The **fuel system monitor** checks **short-term fuel trim** (**SFT**) and **long-term fuel trim** (**LFT**) while the PCM is operating in closed loop. Fuel trim is similar to idle air control in that the system looks at the present state of this indicator compared to the desired range. On GM vehicles, this was known as Block Integrator and Block Learn. When a fuel system problem causes the PCM to make fuel trim corrections for an excessive time period, the fuel system monitor sets a DTC and illuminates the MIL if the fault occurs on two consecutive drive cycles. The fuel system monitor operates continually when the PCM is in closed loop. The fuel system monitor does not involve any new hardware.

Heated Oxygen Sensor Monitor

Shop Manual
Chapter 14, page 616

The system also monitors lean to rich and rich to lean time responses. This test can pick up a lazy O_2 sensor that cannot switch fast enough to keep proper control of the air-fuel mixture in the system. These sensors are the heated type and the amount of time before activity of the sensor signal is present is an indication of whether it is functional or not. Some systems use current flow to indicate if the heater is working or not.

All of the system's HO_2Ss are monitored once per drive cycle, but the heated oxygen sensor monitor provides separate tests for the upstream and downstream sensors. The heated oxygen sensor monitor checks the voltage signal frequency of the upstream HO_2S. Excessive time between signal voltage frequency indicates a faulty sensor. At certain times, the heated oxygen sensor monitor varies the fuel delivery and checks for HO_2Ss response. A slow response in the sensor voltage signal frequency indicates a faulty sensor. The sensor signal is also monitored for excessive voltage.

The heated oxygen sensor monitor also checks the frequency of the rear HO_2S sensor signals, and checks these sensor signals for excessively high voltage. If the monitor does not detect signal voltage frequency within a specific range, the rear HO_2Ss are considered faulty. The heated oxygen sensor monitor will command the PCM to vary the air-fuel ratio to check the rear HO_2S response.

On many V-type engines, there are two primary or upstream O_2 sensors, one for each cylinder bank. These individual sensors report the air-fuel mixture and combustion for the cylinders on that side of the engine.

EGR System Monitoring

The exhaust gas recirculation (EGR) monitors use several different strategies to determine if the system is operating properly. Some monitor the temperature within the EGR passages. A high temperature indicates that exhaust gas is present. Other systems look at the manifold absolute pressure (MAP) signal, energize the EGR valve, and look for corresponding change in vacuum levels.

The EGR system may contain a delta pressure feedback EGR (DPFE) sensor. An orifice is located under the EGR valve, and small exhaust pressure hoses are connected from each side of this orifice to the DPFE sensor. During the EGR monitor, the PCM first checks the DPFE signal. If this sensor signal is within the normal range, the monitor proceeds with the tests.

With the engine idling and the EGR valve closed, the PCM checks for pressure difference at the two pressure hoses connected to the DPFE sensor. When the EGR valve is closed and there is no EGR flow, the pressure should be the same at both pipes. If the pressure is different at these two hoses, the EGR valve is stuck open.

The PCM commands the EGR valve to open and then checks the pressure at the two exhaust hoses connected to the DPFE sensor. With the EGR valve open and EGR flow through the orifice, there should be higher pressure at the upstream hose than at the downstream hose (Figure 14-9).

The PCM checks the EGR flow by checking the DPFE signal value against an expected DPFE value for the engine operating conditions at steady throttle within a specific rpm range. If a fault is detected in any of the EGR monitor tests, a DTC is set in the PCM memory. If the fault occurs during two drive cycles, the MIL is illuminated. The EGR monitor operates once per OBD II trip.

Shop Manual
Chapter 14, page 616

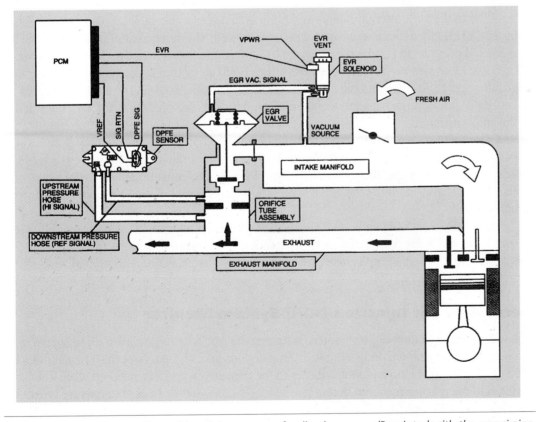

Figure 14-9 EGR system with a delta pressure feedback sensor. *(Reprinted with the permission of Ford Motor Company)*

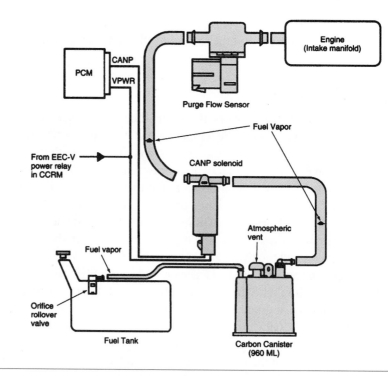

Figure 14-10 EVAP system with a purge flow sensor. *(Reprinted with the permission of Ford Motor Company)*

Evaporative (EVAP) Emission System Monitor

Shop Manual
Chapter 14, page 616

In addition to the various components and failures that could affect the tailpipe emissions on a vehicle, OBD II monitors the evaporative systems as well. The system tests the ability of the fuel tank to hold pressure and of the purge system to vent the gas fumes from the charcoal canister.

Some EVAP systems have a purge flow sensor (PFS) connected to the vacuum hose between the canister purge solenoid and the intake manifold (Figure 14-10). The PCM monitors the PFS signal once per drive cycle to determine if there is vapor flow or no vapor flow through the solenoid to the intake manifold.

Other EVAP systems have a vapor management valve connected in the vacuum hose between the canister and the intake manifold (Figure 14-11). The vapor management valve is a normally closed valve. The PCM operates the valve to control vapor flow from the canister to the intake manifold. The PCM also monitors the valve's operation to determine if the EVAP system is purging vapors properly.

Since 1996, some GM vehicles have an enhanced evaporative system monitor. This system detects leaks and restrictions in the EVAP system. A newly designed fuel tank filler cap is used on this system. In these enhanced systems, an evaporative system leak or a missing fuel tank cap will cause the MIL to turn on.

Secondary Air Injection (AIR) System Monitor

Shop Manual
Chapter 14, page 616

The AIR system operation can be verified by turning the AIR system on to inject air upstream of the oxygen sensor while monitoring its signal. Many designs inject air into the exhaust manifold when the engine is in open loop and switch the air to the converter when it is in closed loop. If the air is diverted to the exhaust manifold during closed loop, the O_2 sensor thinks the mixture is lean and the signal should drop.

On some vehicles, the AIR system is monitored with passive and active tests. During the passive test, the voltage of the pre-catalyst HO_2S is monitored from startup to closed loop operation.

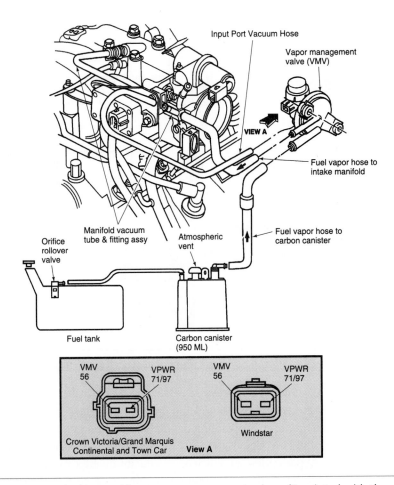

Input Port Vacuum Hose

Vapor management
valve (VMV)

VIEW A

Fuel vapor hose to
intake manifold

Manifold vacuum
tube & fitting assy

Orifice
rollover
valve

Atmospheric
vent

Fuel vapor hose to
carbon canister

Fuel tank

Carbon canister
(950 ML)

VMV
56   VPWR
         71/97

VMV
56   VPWR
         71/97

Crown Victoria/Grand Marquis
Continental and Town Car   View A

Windstar

Figure 14-11 EVAP system with a vapor management valve. *(Reprinted with the permission of Ford Motor Company)*

The AIR pump is normally on during this time. Once the HO_2S is warm enough to produce a voltage signal, the voltage should be low if the AIR pump is delivering air to the exhaust manifold. The secondary AIR monitor will indicate a pass if the HO_2S voltage is low at this time. The passive test also looks for a higher HO_2S voltage when the AIR flow to the exhaust manifold is turned off by the PCM. When the AIR system passes the passive test, no further testing is done. If the AIR system fails the passive test or if the test is inconclusive, the AIR monitor in the PCM proceeds with the active test.

During the active test, the PCM cycles the AIR flow to the exhaust manifold on and off during closed loop operation and monitors the pre-catalyst HO_2S voltage and the short-term fuel trim value. When the AIR flow to the exhaust manifold is turned on, the sensor's voltage should decrease and the short-term fuel trim should indicate a richer condition. The secondary AIR system monitor illuminates the MIL and stores a DTC in the PCM's memory if the AIR system fails the active test on two consecutive trips.

Some Ford vehicles have an electric air pump system. In this system, the air pump is controlled by a solid-state relay. The relay is operated by a signal from the PCM. An air-injection bypass solenoid is also operated by the PCM. This solenoid supplies vacuum to dual air diverter valves (Figure 14-12).

When the engine is started, the PCM signals the relay to start the air pump. This module supplies the high current required for air pump operation. The air pump may provide a 10-second delay in pump operation after the engine is started. The PCM also energizes the air-injection bypass solenoid. When this solenoid is energized, it supplies vacuum to the dual air diverter valves. This action opens the normally closed air diverter valves. Air from the pump is now delivered to the exhaust manifold. The purpose of the air pump is to oxidize HC and CO in the exhaust manifolds

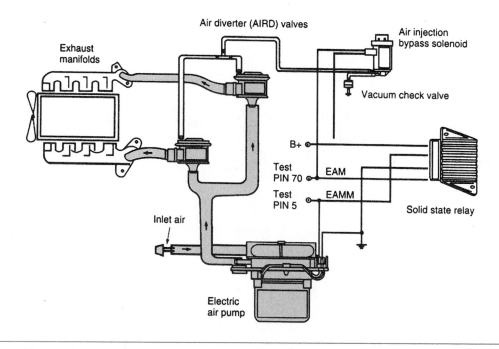

Figure 14-12 An electric air pump system. *(Reprinted with the permission of Ford Motor Company)*

for 20 to 120 seconds after the engine has started and until the catalytic converter is working properly. The length of time the air pump is operating depends on the temperature of the engine. Once the catalyst is warmed up, the PCM signals the relay to shut down the air pump. The PCM also de-energizes the air injection bypass solenoid which allows the air diverter valves to close.

The PCM monitors the relay and the air pump to determine if secondary air is present. This PCM monitor for the air pump system functions once per drive cycle. When a malfunction occurs in the air pump system on two consecutive drive cycles, a DTC is stored and the MIL is turned on. If the malfunction corrects itself, the MIL is turned off after three consecutive drive cycles in which the fault is not present.

Comprehensive Monitor

Shop Manual
Chapter 14, page 616

The system looks at any electronic input that could affect emissions. The strategy is to look for opens and shorts or input signal values that are out of the normal range. It also looks to see if the actuators have their intended effect on the system and to monitor other abnormalities.

The **comprehensive component monitor** (CCM) uses two strategies to monitor inputs and two strategies to monitor outputs. One strategy for monitoring inputs involves checking certain inputs for electrical defects and out-of-range values by checking the input signals at the analog-to-digital converter. The input signals monitored in this way are (Figure 14-13):

1. Rear HO_2S inputs

2. HO_2S inputs

3. Mass air flow sensor

4. Manual lever position sensor

5. Throttle position sensor

6. Engine coolant temperature sensor

7. Intake air temperature sensor

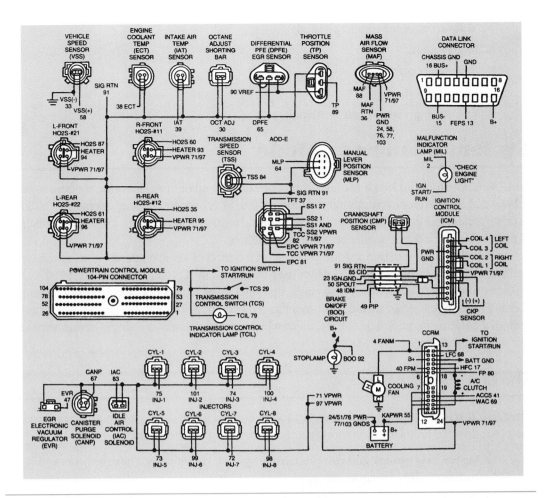

The CCM checks frequency signal inputs by performing rationality checks. During a rationality check, the monitor uses other sensor readings and calculations to determine if a sensor reading is proper for the present conditions. The CCM checks these inputs with rationality checks:

1. Crankshaft position sensor
2. Output shaft speed sensor
3. Ignition diagnostic monitor
4. Camshaft position sensor
5. Vehicle speed sensor

The PCM output that controls the idle air control motor is monitored by checking the idle speed demanded by the inputs against the closed loop idle speed correction supplied by the PCM to the idle air control (IAC) motor.

The output state monitor in the CCM checks most of the outputs by monitoring the voltage of each output solenoid, relay, or actuator at the output driver in the PCM. If the output is off, this voltage should be high. This voltage is pulled low when the output is on.

Monitored outputs include:

1. Wide-open throttle A/C cutoff
2. Shift solenoid 1
3. Shift solenoid 2

4. Torque converter clutch solenoid

   5. HO$_2$S heaters

   6. High fan control

   7. Fan control

   8. Electronic pressure control solenoid

System Readiness Mode

Shop Manual
Chapter 14, page 606

All OBD II scan tools include a readiness function showing all of the monitoring sequences on the vehicle and the status of each: complete or incomplete. If vehicle travel time, operating conditions, or other parameters were insufficient for a monitoring sequence to complete a test, the scanner will indicate which monitoring sequence is not yet complete.

OBD II standards define a warm-up cycle as a period of vehicle operation, after the engine had been turned off, in which the coolant rises by at least 40°C and reaches at least 160°C. Most DTCs are automatically erased after 40 warm-up cycles if the fault is not detected during that time. Some manufacturers retain erased DTCs in a flagged condition. This can be useful if the technician notices a pattern of component failure, all of which may be related to a single intermittent cause like low fuel pressure.

OBD II Trip

A trip in an OBD II system refers to starting and driving the vehicle until five monitors are completed.

The OBD II **trip** consists of an engine start following an engine off period, with enough vehicle travel to allow the following monitoring sequences to complete their tests:

❏ Misfire, fuel system, and comprehensive system components. These are checked continuously throughout the trip.

❏ EGR. This test requires a series of idle speed operations, acceleration, and deceleration to satisfy the conditions needed for completion.

❏ HO$_2$S. This test requires a steady speed for about 20 seconds at speeds between 20 and 45 mph after warmup to be complete (Figure 14-14).

The trip display is provided on a scan tester. Some scan testers display YES when the five monitors are completed in a trip. When the five monitors are not completed during a trip, the scan tester displays NO in the trip display.

Depending on the monitor, the system tests the component or system once per trip. Trips are based on the driving cycle that the vehicle experiences during the FTP. Basically the cycle includes operation during warmup, cruise, acceleration, and deceleration with certain time requirements for each mode of operation.

Drive Cycle

An OBD II **drive cycle** (Figure 14-15) consists of an engine start and vehicle operation that brings the vehicle into closed loop and includes whatever specific operating conditions are necessary either to initiate and complete a specific monitoring sequence or to verify a symptom or verify a repair. A monitoring sequence is an operational strategy designed to test the operation of a specific system, function, or component.

A minimum drive cycle is from engine startup and vehicle operation until after the PCM enters closed loop. To complete a drive cycle, all five trip monitors must be completed followed by the catalyst monitor. The catalyst monitor must be completed after the other five monitors are completed in a trip. A steady throttle opening between 40 to 60 mph (64 to 96 kph) for 80 seconds is required to complete the catalyst efficiency monitor.

DRIVE INSTRUCTIONS OVER TIME

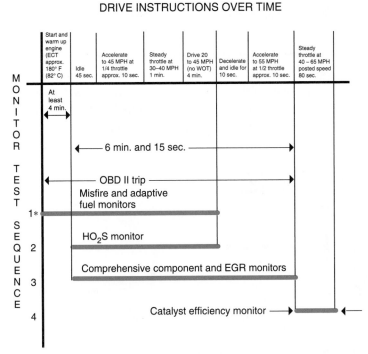

* Since the Misfire, Adaptive Fuel, EGR (requiring idles and accelerations) and Comprehensive Monitors are continuously checked by the OBD II system, the test sequence may vary on each vehicle due to outside ambient temperature, engine/vehicle performance temperature and driving conditions.

Figure 14-14 OBD trip cycle. *(Reprinted with the permission of Ford Motor Company)*

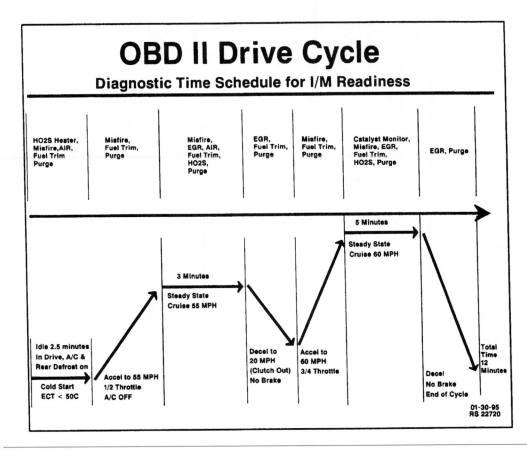

Figure 14-15 An OBD II drive cycle. *(Courtesy of Oldsmobile Division—GMC)*

Ford's Electronic Engine Control V (EEC V) System

One of the most electronically advanced Ford production vehicles is the Lincoln Continental. The EEC V application, which consists of the PCM and ten other modules, uses a DLC as the universal link to system diagnosis. The major difference on this vehicle is that the modules communicate (Figure 14-16) with each other through the use of one twisted pair of wires called the MCN (multiplexing communications network). This network is accessible through the DLC.

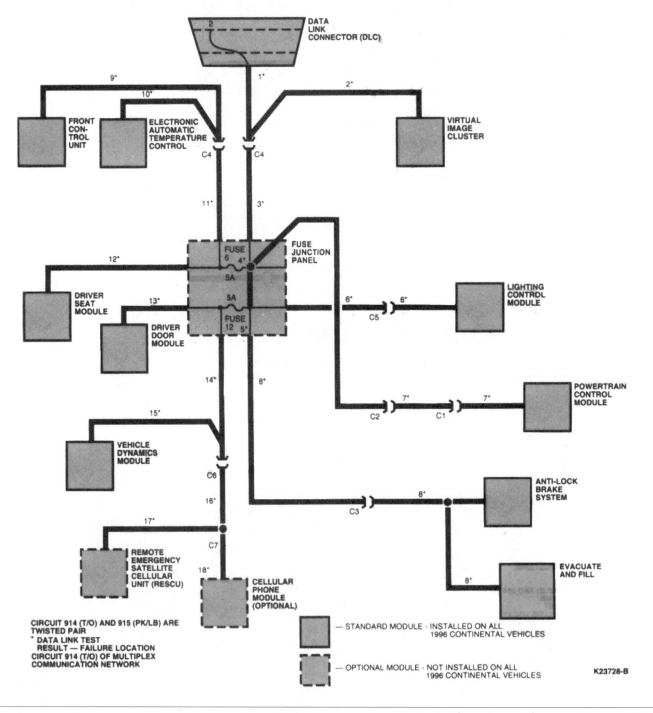

Figure 14-16 The multiplex communication network circuit on a Lincoln Continental. *(Reprinted with the permission of Ford Motor Company)*

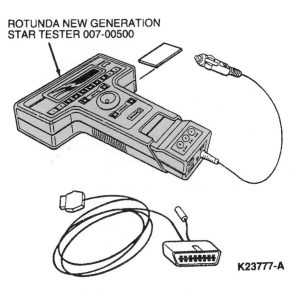

ROTUNDA NEW GENERATION
STAR TESTER 007-00500

K23777-A

Figure 14-17 Ford's New Generation STAR tester. *(Reprinted with the permission of Ford Motor Company)*

The DLC pin designations dictated by the SAE J1962 document allow for some flexibility between vehicle manufacturers. While certain pin assignments must remain uniform, the use of the ISO protocol is up to the manufacturer. Ford chose to designate pin numbers 2 and 10 to communicate OBD II and Standard Corporate Protocol (SCP) information to their New Generation STAR (NGS) tester (Figure 14-17).

On the EEC V, the standard corporate protocol (SCP) data links are the carrier for the MCN. If there is a problem with one of the modules in the system, a technician can monitor the module that controls the component or system. The technician can also control the actuators through bi-directional communications between the scan tool and the module being tested. This enables the technician to view PIDs (Parameter Identifications) related to the component he has commanded to turn on, through the scan tool.

Ford's Short-term fuel trim (SFT) and Long-term fuel trim (LFT) strategies monitor the oxygen sensor signal. The information gathered by the PCM is used to make adjustments to the fuel control calculations.

Should the exhaust from a vehicle indicate a rich condition, the O_2 sensor signal prompts the computer to subtract fuel. The reverse condition is an engine running lean in which the PCM adds fuel. The computer on GM vehicles relates fuel strategy to the scan tool through binary messages sent as a series of zeroes and ones. Since zero holds a place, and the computer has the ability to count to 255, half of that amount is 128. The technician uses the reference number 128 as the midpoint of the fuel strategy when the PCM is operating in closed loop. SFT control on a GM vehicle includes: numbers higher than 128 indicate fuel is being added, numbers lower than 128 indicate fuel is being subtracted, and the constant crossing above and below the reference point of 128 indicates proper system operation.

On a Ford vehicle, zero is the midpoint of the fuel strategy when the computer is in closed loop. Ford's fuel cells are illustrated as a percentage: numbers without a minus sign indicate fuel is being added and numbers with a minus sign indicate fuel is being subtracted. The constant change or crossing above and below the zero line indicate proper system operation. If the SFT readings are constantly on either side of the zero line, the engine is not operating efficiently.

The GM long-term fuel trim strategy is displayed on the scan tool in the same way as the SFT. Long-term fuel trim represents the PCM learning driver habits, engine variables, and road conditions. If the number displayed on the scan tool is above 128, the computer has learned to compensate for a lean exhaust. If the number is below 128, the computer has learned to compensate for a rich exhaust.

The Ford LFT strategy is displayed as a percentage on the scan tool in the same way as the SFT. LFT is the computer learning the driver's habits, engine variables, and road conditions. Numbers without a minus sign indicate the computer has compensated for a lean exhaust. Numbers with a minus sign indicate the PCM has compensated for a rich exhaust. As the LFT learns to compensate for an exhaust that is rich or lean, the SFT returns to a value that crosses the zero reference point. If the engine's condition is too far toward either the lean or rich side, the LFT will not compensate and a DTC will be set.

Test Connector

OBD II and the Society of Automotive Engineers (SAE) standards require the DLC to be mounted in the passenger compartment out of sight of vehicle passengers. The DLC must be a 16-terminal connector with 9 terminals defined by the SAE (Figure 14-18).

The standard DLC (Figure 14-19) is a 16-pin connector. The same pins are used for the same information, regardless of the vehicle's make, model, and year. The connector is D-shaped and has guide keys which allow the scan tool to only be installed one way. Using a standard connector design and by designating the pins allows data retrieval with any scan tool designed for OBD II. Some European vehicles meet OBD II standards by providing the designated DLC along with their own connector for their own scan tool.

The DLC is designed only for scan tool use. You cannot jump across any of the terminals to display codes on an instrument panel or other indicator lamp. The MIL is only used to inform the driver that the vehicle should be serviced soon. It also informs a technician that the computer has set a trouble code.

The DLC must be easily accessible while sitting in the driver's seat (Figure 14-20). All DLCs must be located somewhere between the left end of the instrument panel and a position 300 millimeters to the right of the center. The DLC cannot be hidden behind panels and must be accessible

When a vehicle has a 16-pin DLC, this does not necessarily mean that the vehicle is equipped with OBD II.

Cavity	General Assignment
1	Ignition Control
2	BUS (+) SCP
3	Discretionary (Not Used)
4	Chassis Ground
5	Signal Ground (SIG RTN)
6	Discretionary (Not Used)
7	K Line of ISO 9141
8	Discretionary (Not Used)
9	Discretionary (Not Used)
10	BUS (-) SCP
11	Discretionary (Not Used)
12	Discretionary (Not Used)
13	FEPS (Flash EEPROM)
14	Discretionary (Not Used)
15	L Line of ISO 9141
16	Battery Power

Figure 14-18 DLC terminal identification. *(Reprinted with the permission of Ford Motor Company)*

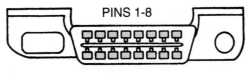

PINS 1-8

PINS 9-16

Figure 14-19 A 16-pin DLC. *(Reprinted with the permission of Ford Motor Company)*

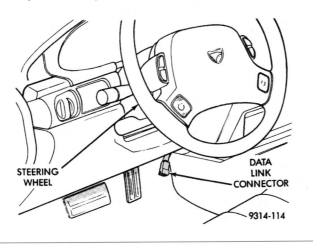

STEERING WHEEL

DATA LINK CONNECTOR

9314-114

Figure 14-20 Designated location for the DLC. *(Courtesy of Chrysler Corporation)*

without tools. Any generic scan tool can be connected to the DLC and can access the diagnostic data stream.

The connector pins are arranged in two rows and are numbered consecutively. Seven of the sixteen pins have been assigned by the OBD II standard. The remaining nine pins can be used by the individual manufacturers to meet their needs and desires.

Malfunction Indicator Lamp Operation

An OBD II system continuously monitors the entire emissions system, switches on a MIL if something goes wrong, and stores a fault code in the PCM when it detects a problem. The codes are well defined and can lead a technician to the problem. A scan tool must be used to access and interpret emission-related DTCs regardless of the make and model of the vehicle (Figure 14-21).

Shop Manual
Chapter 14, page 607

DATA LINK CONNECTOR (16P)

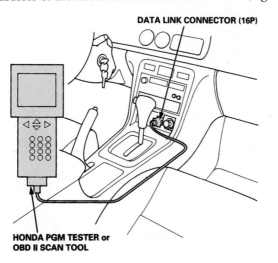

HONDA PGM TESTER or
OBD II SCAN TOOL

Figure 14-21 Scan tool connected to the DLC. *(Courtesy of American Honda Motor Company, Inc.)*

451

Many of the trouble codes from an OBD II system will mean the same thing regardless of manufacturer. However, some of the codes will pertain only to a particular system or will mean something different with each system. The DTC is a five-character code with both letters and numbers (Figure 14-22). This is called the alphanumeric system. The first character of the code is a letter. This defines the system where the code was set. Currently there are four possible first character codes: B for body, C for chassis, P for powertrain, and U for undefined. The U-codes are designated for future use.

The second character is a number. This defines the code as being a mandated code or a special manufacturer code. This number will either be a 0, 1, 2, or 3. A 0 code means that the fault is defined or mandated by OBD II. A "1" code means that the code is manufacturer specific. Codes of "2" or "3" are designated for future use.

The third through fifth characters are numbers. These describe the fault. The third character of a powertrain code tells you where the fault occurred. The remaining two characters describe the exact condition that set the code.

When the same fault has been detected during two drive cycles, a DTC is stored in the PCM memory. If a misfire occurs, the misfire monitor will store a DTC immediately in the PCM memory, depending on the type of misfire.

If a misfire that threatens engine or catalyst damage occurs, the misfire monitor flashes the MIL on the first occurrence of the misfire. A fault detected by the catalyst monitor must occur on three drive cycles before the MIL is illuminated.

The SAE J2012 standards specify that all DTCs will have a five-digit alphanumeric numbering and lettering system. The following prefixes indicate the general area to which the DTC belongs:

1. P — power train
2. B — body
3. C — chassis

The first number in the DTC indicates who is responsible for the DTC definition.

1. 0 — SAE
2. 1 — manufacturer

The third digit in the DTC indicates the subgroup to which the DTC belongs. The possible subgroups are:

0 — Total system
1 — Fuel-air control
2 — Fuel-air control
3 — Ignition system misfire
4 — Auxiliary emission controls
5 — Idle speed control
6 — PCM and I/O
7 — Transmission
8 — Non-EEC power train

The fourth and fifth digits indicate the specific area where the trouble exists. Code P1711 has this interpretation:

P — Power train DTC
1 — Manufacturer-defined code
7 — Transmission subgroup
11 — Transmission oil temperature (TOT) sensor and related circuit

Figure 14-22 OBD II DTC interpretation chart.

For the misfire and fuel system monitors, if the fault does not occur on three consecutive drive cycles under similar conditions, the MIL is turned off. The system defines similar conditions as:

1. Engine speed within 375 rpm compared to when the fault was detected.

2. Engine load within 10% compared to when the fault was detected.

3. Engine warm-up state or coolant temperature must match the temperature when the fault was detected.

For the catalyst efficiency, HO_2S, EGR, and comprehensive component monitors, the MIL is turned off if the same fault does not reappear for three consecutive drive cycles. When the fault is no longer present and the MIL is turned off, the DTC is erased after 40 engine warm-up cycles. A technician may use a scan tester to erase DTCs immediately.

A pending DTC is a code representing a fault that has occurred, but that has not occurred enough times to illuminate the MIL. Some scan testers are capable of reading pending DTCs with the continuous DTCs.

Data Links

Each manufacturer must use the same protocol between the PCM and its sensors and actuators. The same protocol must be used to send diagnostic information to the scan tool through the DLC.

A variety of protocol data links are connected to the DLC on OBD II systems. For example, some General Motors vehicles have universal asynchronous receive and transmit (UART) data links connected to the DLC (Figure 14-23). When computers or the scan tool are not communicating on the UART data links, the at-rest voltage is 5 V. When a computer or scan tester is transmitting data on the UART bus wires, the voltage is pulled to ground level. Data is transmitted at 8,192 bits per second on the UART system. This data has a fixed pulse width.

Other GM vehicles have class 2 data links connected to the DLC (Figure 14-24). Data transmission speed is 10.4 kilobits per second on class 2 data links. Most GM OBD II systems have class 2 data links. This data link system meets the SAE J1850 standard for serial data transmission. The at-rest voltage is 0 V on class 2 data links while the transmission voltage is 7 V. A variable pulse width is used to transmit data in class 2 systems. When two devices are attempting to transmit data at the same time, the data is automatically prioritized so only the high-priority data is allowed to continue. The low-priority data must wait until the high-priority transmission is completed. When a GM OBD II system has a VCM (Variable Communication Monitor) and class 2 data links, the VCM is referred to as VCM-A. On those systems without class 2 data links, the computer is called a VCM. Some VCMs are used on OBD II systems.

Some Ford vehicles have international standards organization (ISO) data links connected to the 16-pin DLC. These data links transmit data from the antilock brake system module, driver seat module, lighting control module, and driver door module to the DLC.

Standard corporate protocol (SCP) data links are also connected to the 16-pin DLC on some Ford vehicles. These data links meet the SAE J1850 standard. Data from the PCM and electronic automatic temperature control is transmitted on the SCP data links. The computers connected to ISO or SCP data links may vary on the year, make, and model of the vehicle.

Some Ford cars, such as the Town Car, Grand Marquis, and Crown Victoria, have audio corporate protocol (ACP) data links, which transmit data between the audio system components. Since this system has internal self-diagnostic capabilities and does not require the use of a scan tool, the ACP data links are not connected to the 16-pin DLC.

Data links may be referred to as bus wires.

A **protocol** is an agreed upon digital code that the PCM uses to communicate with a scan tool.

Test Modes

All OBD II systems have the same basic test modes. These test modes must be accessible with an OBD II scan tool. Mode 1 is the **Parameter Identification (PID)** mode. It allows access to certain

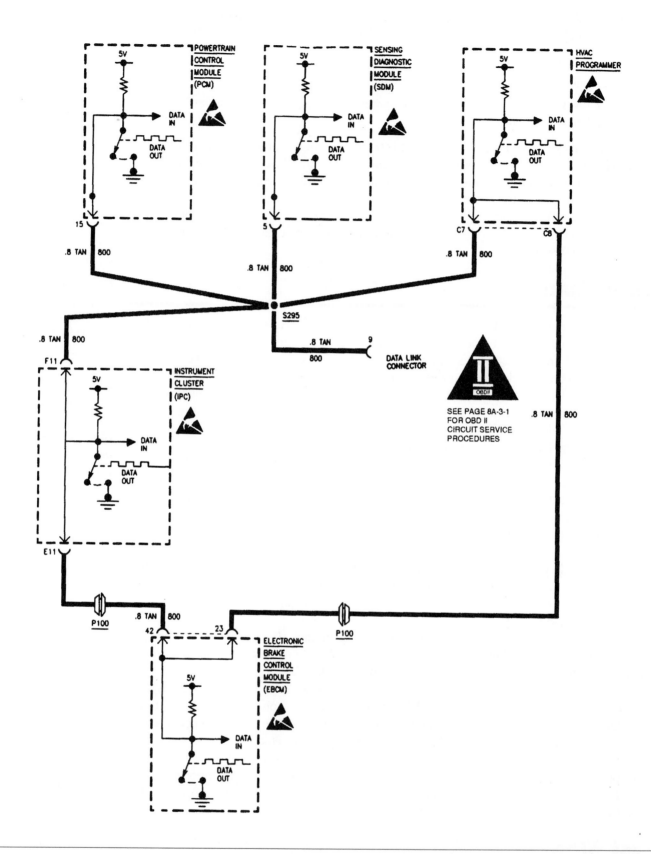

Figure 14-23 GM's UART communication system. *(Courtesy of Oldsmobile Division—GMC)*

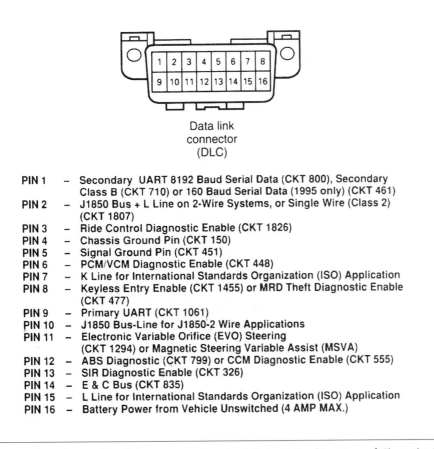

Data link
connector
(DLC)

PIN 1 — Secondary UART 8192 Baud Serial Data (CKT 800), Secondary
Class B (CKT 710) or 160 Baud Serial Data (1995 only) (CKT 461)
PIN 2 — J1850 Bus + L Line on 2-Wire Systems, or Single Wire (Class 2)
(CKT 1807)
PIN 3 — Ride Control Diagnostic Enable (CKT 1826)
PIN 4 — Chassis Ground Pin (CKT 150)
PIN 5 — Signal Ground Pin (CKT 451)
PIN 6 — PCM/VCM Diagnostic Enable (CKT 448)
PIN 7 — K Line for International Standards Organization (ISO) Application
PIN 8 — Keyless Entry Enable (CKT 1455) or MRD Theft Diagnostic Enable
(CKT 477)
PIN 9 — Primary UART (CKT 1061)
PIN 10 — J1850 Bus-Line for J1850-2 Wire Applications
PIN 11 — Electronic Variable Orifice (EVO) Steering
(CKT 1294) or Magnetic Steering Variable Assist (MSVA)
PIN 12 — ABS Diagnostic (CKT 799) or CCM Diagnostic Enable (CKT 555)
PIN 13 — SIR Diagnostic Enable (CKT 326)
PIN 14 — E & C Bus (CKT 835)
PIN 15 — L Line for International Standards Organization (ISO) Application
PIN 16 — Battery Power from Vehicle Unswitched (4 AMP MAX.)

Figure 14-24 Class 2 data links connected to the 16-pin DLC. *(Courtesy of Chevrolet Motor Division—GMC)*

data values, analog and digital inputs and outputs, calculated values, and system status information. Some of the PID values will be manufacturer specific, others are common to all vehicles.

Mode 2 is the Freeze Frame Data Access mode. This mode permits access to emission-related data values from specific generic PIDs. These values represent the operating conditions at the time the fault was recognized and logged into memory as a DTC. Once a DTC and a set of freeze frame data are stored in memory, they will stay in memory even if other emission-related DTCs are stored. The number of these sets of freeze frames that can be stored are limited. On 1996 GM vehicles, the possible number of stored sets is five.

There is one type of failure that is an exception to this rule—misfire. Fuel system misfires will overwrite any other type of data except for other fuel system misfire data. This data can only be removed with a scan tool. When a scan tool is used to erase a DTC, it automatically erases all freeze frame data associated with the events that lead to that DTC.

Mode 3 permits scan tools to obtain stored DTCs. The information is transmitted from the PCM to the scan tool following an OBD II Mode 3 request. Either the DTC, its descriptive text, or both will be displayed on the scan tool.

The PCM reset mode, Mode 4, allows the scan tool to clear all emission-related diagnostic information from its memory. Once the PCM has been reset, the PCM stores an inspection maintenance readiness code until all OBD II system monitors or components have been tested to satisfy an OBD trip cycle without any other faults occurring. Specific conditions must be met before the requirements for a trip are satisfied.

Mode 5 is the oxygen sensor monitoring test. This mode gives the oxygen sensor fault limits and the actual oxygen sensor outputs during the test cycle. The test cycle includes specific operating conditions that must be met to complete the test. This information helps determine the effectiveness of the catalytic converter.

Mode 6 is the output state mode (OTM) which allows a technician to activate or deactivate the system's actuators through the scan tool. When the OTM is engaged, the actuators can be controlled without affecting the radiator fans. The fans are controlled separately, this gives a pure look at the effectiveness and action of the outputs.

Snapshots

Shop Manual
Chapter 14, page 609

The primary purpose of OBD II is to make the diagnosis of emissions and driveability problems simple and uniform in the future. No longer will it be necessary to learn entirely new systems from each manufacturer. The basic plan is to allow technicians to diagnose any vehicle with the same diagnostic tools. Manufacturers can introduce special diagnostic tools or capacities for their own systems, providing that standard scan tools, along with DMMs and lab scopes can analyze the system. These special tools can have additional capabilities beyond those given in the standards. One of the mandated capabilities is the **freeze frame** or snapshot feature. This is the ability of the system to record data from all of its sensors and actuators at a time when the system turns on the MIL. GM expanded this capability to include "failure reports" which does the same thing as the snapshot, but also includes any fault stored in memory, not just those related to the emissions-related circuits.

The basic advantage of the snapshot feature is the ability to look at the existing conditions when a code was set. This will be especially valuable for diagnosing intermittent problems. Whenever a code is set, a record of all related activities will be stored in memory. This allows the technician to look at the action of sensors and actuators when the code was set. This helps identify the cause of the problem.

OBD Terms

This new terminology is commonly called J1930 terminology because they conform to the SAE standard J1930.

All vehicle manufacturers must use the same names and acronyms for all electric and electronic systems related to the engine and emission control systems. Previously, there were many names for the same component. Now all similar components will be referred to with the same name. Beginning with the 1993 model year, all service information was required to use the new terms. Figure 14-25 is a list of the previous names followed by the currently acceptable names and acronyms according to J1930.

J1930 Terminology List

NOTE: Certain Ford Component names have been changed in this Service Manual to conform to Society of Automotive Engineers (SAE) directive J1930.

SAE J1930 standardizes automotive component names for all vehicle manufacturers.

New Term	New Acronyms/ Abbreviations	Old Acronyms/ Term
Accelerator Pedal	AP	– Accelerator
Air Cleaner	ACL	– Air Cleaner
Air Conditioning	A/C	– A/C – Air Conditioning
Barometric Pressure	BARO	– BP – Barometric Pressure
Battery Positive Voltage	B+	– BATT+ – Battery Positive
Camshaft Position	CMP	– Camshaft Sensor
Carburetor	CARB	– CARB Carburetor
Continuous Fuel Injection	CFI	– Continuous Fuel Injection
Charge Air Cooler	CAC	– After Cooler – Intercooler
Closed Loop	CL	– EEC
Closed Throttle Position	CTP	– CTP – Closed Throttle Position
Clutch Pedal Position	CPP	– CES – CIS – Clutch Engage Switch – Clutch Interlock Switch
Continuous Trap Oxidizer	CTOX	– CTO
Crankshaft Position	CKP	– CPS – VRS – Variable Reluctance Sensor
Data Link Connector	DLC	– Self-Test Connector
Diagnostic Test Mode	DTM	– Self-Test Mode
Diagnostic Trouble Code	DTC	– Self-Test Code
Distributor Ignition	DI	– CBD – DS – TFI – Closed Bowl Distrbutor – Duraspark Ignition – Thick Film Ignition
Early Fuel Evaporation	EFE	– EFE – Early Fuel Evaporation

Figure 14-25 A list of SAE J1930 terms.

J1930 Terminology List

New Term	New Acronyms/ Abbreviations	Old Acronyms/ Term
Electrically Erasable Programmable Read Only Memory	EEPROM	– E2PROM
Electronic Ignition	EI	– DIS – EDIS – Distributorless Ignition System – Electronic Distributorless Ignition System
Engine Coolant Level	ECL	– Engine Coolant Level
Engine Coolant Temperature	ECT	– ECT – Engine Coolant Temperature
Engine Control Module	ECM	– ECM – Engine Control Module
Engine Speed	RPM	– RPM – Revolutions Per Minute
Erasable Programmable Read Only Memory	EPROM	– EPROM – Erasable Programmable Read Only Memory
Evaporative Emission	EVAP	– EVP Sensor – EVR Solenoid
Exhaust Gas Recirculation	EGR	– EGR – Exhaust Gas Recirculation
Fan Control	FC	– EDF – Electro-Drive Fan
Flash Electrically Erasable Programmable Read Only Memory	FEEPROM	– FEEPROM – Flash Electrically Erasable Programmable Read Only Memory
Flash Erasable Programmable Read Only Memory	FEPROM	– FEPROM – Flash Erasable Programmable Read Only Memory
Flexible Fuel	FF	– FCS – FFS – FFV – Fuel Compensation Sensor – Flex Fuel Sensor
Fourth Gear	4GR	– Fourth Gear
Fuel Pump	FP	– FP – Fuel Pump
Generator	GEN	– ALT – Alternator
Ground	GND	– GND – Ground
Heated Oxygen Sensor	HO2S	– HEGO – Heated Exhaust Gas Oxygen Sensor
Idle Air Control	IAC	– IAC – Idle Air Bypass Control

Figure 14-25 A list of SAE J1930 terms. *(continued)*

J1930 Terminology List

New Term	New Acronyms/ Abbreviations	Old Acronyms/ Term
Idle Speed Control	ISC	– Idle Speed Control
Ignition Control Module	ICM	– DIS Module – EDIS Module – TFI Module
Indirect Fuel Injection	IFI	– IDFI – Indirect Fuel Injection
Inertia Fuel Shutoff	IFS	– Inertia Switch
Intake Air Temperature	IAT	– ACT – Air Charge Temperature
Knock Sensor	KS	– KS – Knock Sensor
Malfunction Indicator Lamp	MIL	– CEL – "CHECK ENGINE" Light – "SERVICE ENGINE SOON" Light
Manifold Absolute Pressure	MAP	– MAP – Manifold Absolute Pressure
Manifold Differential Pressure	MDP	– MDP – Manifold Differential Pressure
Manifold Surface Temperature	MST	– MST – Manifold Surface Temperature
Manifold Vacuum Zone	MVZ	– MVZ – Manifold Vacuum Zone
Mass Air Flow	MAF	– MAF – Mass Air Flow
Mixture Control	MC	– Mixture Control
Multiport Fuel Injection	MFI	– EFI – Electronic Fuel Injection
Non-Volatile Random Access Memory	NVRAM	– NVM – Non-Volatile Memory
On-Board Diagnostic	OBD	– Self-Test – On-Board Diagnostic
Open Loop	OL	– OL – Open Loop
Oxidation Catalytic Converter	OC	– COC – Conventional Oxidation Catalyst
Oxygen Sensor	O2S	– EGO
Park/Neutral Position	PNP	– NDS – NGS – TSN – Neutral Drive Switch – Neutral Gear Switch – Transmission Select Neutral
Periodic Trap Oxidizer	PTOX	– PTOX – Periodic Trap Oxidizer

Figure 14-25 A list of SAE J1930 terms. *(continued)*

J1930 Terminology List

New Term	New Acronyms/ Abbreviations	Old Acronyms/ Term
Power Steering Pressure	PSP	– PSPS – Power Steering Pressure Switch
Powertrain Control Module	PCM	– ECA – ECM – ECU – EEC Processor – Engine Control Assembly – Engine Control Module – Engine Control Unit
Programmable Read Only Memory	PROM	– PROM – Programmable Read Only Memory
Pulsed Secondary Air Injection	PAIR	– MPA – PA – Thermactor II – Managed Pulse Air – Pulse Air
Random Access Memory	RAM	– RAM – Random Access Memory
Read Only Memory	ROM	– ROM – Read Only Memory
Relay Module	RM	– RM – Relay Module RM
Scan Tool	ST	– GST – NGS – Generic Scan Tool – New Generation STAR Tester – Enhanced Scan Tool OBD II ST
Secondary Air Injection	AIR	– AM – CT – MTA – Air Management – Conventional Thermactor – Managed Thermactor Air – Thermactor
Sequential Multiport Fuel Injection	SFI	– SEFI – Sequential Electronic Fuel Injection
Service Reminder Indicator	SRI	– SRI – Service Reminder Indicator
Smoke Puff Limiter	SPL	– SPL – Smoke Puff Limiter
Supercharger	SC	– SC – Supercharger
Supercharger Bypass	SCB	– SCB – Supercharger Bypass
System Readiness Test[1]	SRT[1]	— —

Figure 14-25 A list of SAE J1930 terms. *(continued)*

J1930 Terminology List

New Term	New Acronyms/ Abbreviations	Old Acronyms/ Term
Thermal Vacuum Valve	TVV	– Thermal Vacuum Switch
Third Gear	3GR	– Third Gear
Three Way Catalytic Converter	TWC	– TWC – Three Way Catalytic Converter
Three Way + Oxidation Catalytic Converter	TWC+OC	– TWC & COC – Dual Bed – Three Way Catalyst and Conventional Oxidation Catalyst
Throttle Body	TB	– TB – Throttle Body
Throttle Body Fuel Injection	TBI	– CFI – Central Fuel Injection – EFI
Throttle Position	TP	– TP – Throttle Position
Torque Converter Clutch	TCC	– CCC – CCO – MCCC – Converter Clutch Control – Converter Clutch Override – Modulated Converter Clutch Control
Transmission Control Module	TCM	– 4EAT Module
Transmission Range	TR	– PRNDL
Turbocharger	TC	– TC – Turbocharger
Vehicle Speed Sensor	VSS	– VSS –Vehicle Speed Sensor
Voltage Regulator	VR	– VR – Voltage Regulator
Volume Air Flow	VAF	– VAF – Volume Air Flow
Warm-Up Oxidation Catalytic Converter	WU-OC	– WV-OC – Warm-up Oxidation Catalytic Converter
Warm-Up Three Way Catalytic Converter	WU-TWC	– WU-TWC – Warm-up Three Way Catalytic Converter
Wide Open Throttle	WOT	– Full Throttle – WOT – Wide Open Throttle

Figure 14-25 A list of SAE J1930 terms. *(continued)*

Summary

Terms to Know

Catalyst efficiency
monitor

Comprehensive
component monitor

Drive cycle

Electronically erasable
programmable read
only memory (EEPROM)

Engine misfire monitor

Freeze frame

Fuel system monitor

Heated oxygen sensor
monitor

Long-term fuel trim
(LFT)

Parameter
identification (PID)

Protocol

Short-term fuel trim
(SFT)

Snapshot

Trip

Type A misfire

Type B misfire

❏ An OBD II system has many monitors to check system operation, and the MIL is illuminated if vehicle emissions exceed 1.5 times the allowable standard for that model year.

❏ According to the guidelines of OBD II, all vehicles must have a universal diagnostic test connector, a standard location for the DLC, a standard list of Diagnostic Trouble Codes, a standard communication protocol, common use of scan tools on all vehicle makes and models, common diagnostic test modes, the ability to record and store in memory a snapshot of the operating conditions that existed when a fault occurred, and a standard glossary of terms, acronyms, and definitions that must be used for all components in the electronic control systems.

❏ Monitors included in OBD II are: catalyst efficiency, engine misfire, fuel system, heated exhaust gas oxygen sensor, EGR, EVAP, secondary air injection, and comprehensive component monitors.

❏ OBD II vehicles use a minimum of two oxygen sensors. One of these is used for feedback to the PCM for fuel control and the other gives an indication of the efficiency of the converter.

❏ Cylinder misfire monitoring requires measuring the contribution of each cylinder to engine power.

❏ Fuel system monitoring checks short-term fuel trim and long-term fuel trim while the PCM is operating in closed loop.

❏ Heated oxygen sensor monitor checks lean-to-rich and rich-to-lean time responses.

❏ The EGR monitors use several different strategies to determine if the system is operating properly.

❏ OBD II monitors the evaporative system's ability of the fuel tank to hold pressure and of the purge system to vent the gas fumes from the charcoal canister.

❏ The AIR system operation can be verified by turning the AIR system on to inject air upstream of the oxygen sensor while monitoring its signal.

❏ The comprehensive monitor looks at any electronic input that could affect emissions.

❏ An OBD II drive cycle includes whatever specific operating conditions are necessary either to initiate and complete a specific monitoring sequence or to verify a symptom or verify a repair.

❏ The DLC must be a 16-terminal connector with 12 terminals defined by the SAE.

❏ An OBD II system continuously monitors the entire emissions system, switches on a MIL if something goes wrong, and stores a fault code in the PCM when it detects a problem.

❏ Each manufacturer must use the same protocol between the PCM and its sensors and actuators. The same protocol must be used to send diagnostic information to the scan tool through the DLC.

❏ All OBD II systems have the same basic test modes.

❏ One of the mandated capabilities is the "freeze frame" or snapshot feature, which gives the ability to record data from all of its sensors and actuators at a time when the system turns on the MIL.

❏ Compared to previous systems, the main difference in an OBD II system is in the software contained in the PCM.

❏ An EEPROM may be erased and reprogrammed with the proper equipment without removing the chip.

Review Questions

Short Answer Essays

1. Describe the main hardware differences between an OBD II system and other systems.

2. Explain the advantage of a flash EEPROM compared to a PROM.

3. Describe an OBD II warm-up cycle.

4. Explain trip and drive cycle in an OBD II system.

5. Describe how engine misfire is detected in an OBD II system.

6. Describe the differences between an A misfire and a B misfire.

7. Describe the purpose of having two oxygen sensors in an exhaust system.

8. List five things that were mandated by OBD II.

9. Briefly describe what the comprehensive component monitor looks at.

10. Describe briefly five of the monitors in an OBD II system.

Fill-in-the-Blanks

1. Some scan tools require a(n) _____ adapter to provide OBD II DLC capabilities.

2. OBD II regulations required that all vehicles produced after _____ be equipped with these new systems, however some manufacturers started to equip their vehicles with this equipment in _____.

3. OBD II regulations require the DLC to be mounted in the _____ compartment.

4. The downstream HO_2S monitors _____ efficiency.

5. Type B engine misfires are excessive if the misfiring exceeds _____ to _____ percent in a _____ rpm period.

6. The _____ monitor system checks the action of the canister purge system.

7. The _____ monitor system has a(n) _____ and _____ test to check the efficiency of the air injection system.

8. The fuel monitor checks _____ _____ fuel trim and _____ _____ fuel trim.

9. OBD stands for _____ _____ _____.

10. In an OBD II system, the MIL is illuminated if the emission levels exceed _____ times the standard for that model year.

ASE Style Review Questions

1. While discussing OBD II systems:
   Technician A says the PCM illuminates the MIL if a defect causes emissions levels to exceed 2.5 times the emission standards for that model year vehicle.
   Technician B says if a misfire condition threatens engine or catalyst damage, the PCM flashes the MIL.
   Who is correct?
   - **A.** A only
   - **B.** B only
   - **C.** Both A and B
   - **D.** Neither A nor B

2. While discussing the catalyst efficiency monitor:
   Technician A says if the catalytic converter is not reducing emissions properly, the voltage frequency increases on the downstream HO_2S.
   Technician B says if a fault occurs in the catalyst monitor system on three drive cycles, the MIL will be illuminated.
   Who is correct?
   - **A.** A only
   - **B.** B only
   - **C.** Both A and B
   - **D.** Neither A nor B

3. While discussing the misfire monitor:
   Technician A says while detecting type A misfires, the monitor checks cylinder misfiring over a 500 rpm period.
   Technician B says while detecting type B misfires, the monitor checks cylinder misfires over a 1,000 rpm period.
   Who is correct?
   - **A.** A only
   - **B.** B only
   - **C.** Both A and B
   - **D.** Neither A nor B

4. While discussing OBD II systems:
   Technician A says these systems have two heated oxygen sensors downstream from the catalytic converters to monitor converter operation.
   Technician B says the PCM checks exhaust temperature to monitor ignition misfiring.
   Who is correct?
   - **A.** A only
   - **B.** B only
   - **C.** Both A and B
   - **D.** Neither A nor B

5. While discussing OBD II guidelines:
   Technician A says they call for a standard scan tool to be used on all vehicles regardless of make and model.
   Technician B says they call for a standard location of the DLC.
   Who is correct?
   - **A.** A only
   - **B.** B only
   - **C.** Both A and B
   - **D.** Neither A nor B

6. While discussing OBD II O_2 sensors:
   Technician A says the downstream O_2 sensor signal should show clean toggles from rich to lean.
   Technician B says the upstream O_2 sensor signal should show clean toggles from rich to lean.
   Who is correct?
   - **A.** A only
   - **B.** B only
   - **C.** Both A and B
   - **D.** Neither A nor B

7. While discussing EVAP system monitoring:
   Technician A says it tests the ability of the system to hold a vacuum.
   Technician B says it tests the system's ability to vent gas fumes from the charcoal canister.
   Who is correct?
   - **A.** A only
   - **B.** B only
   - **C.** Both A and B
   - **D.** Neither A nor B

8. While discussing monitoring systems:
   Technician A says the fuel system monitor checks the short-term and long-term fuel trim.
   Technician B says the heated oxygen sensor monitoring system checks lean-to-rich and rich-to-lean response times.
   Who is correct?
   - **A.** A only
   - **B.** B only
   - **C.** Both A and B
   - **D.** Neither A nor B

9. While discussing the comprehensive monitoring system:
   Technician A says it tests various input circuits.
   Technician B says it tests various output circuits.
   Who is correct?
   - **A.** A only
   - **B.** B only
   - **C.** Both A and B
   - **D.** Neither A nor B

10. While discussing the DLC in OBD II systems:
    Technician A says it must be located in the engine compartment.
    Technician B says it has at least 16 terminals.
    Who is correct?
    - **A.** A only
    - **B.** B only
    - **C.** Both A and B
    - **D.** Neither A nor B

Related Systems

Upon completion and review of this chapter, you should be able to:

❑ List the basic systems that make up an automobile and name their major components and functions.

❑ Describe the various clutch components and their functions.

❑ Explain the design characteristics of the gears used in manual transmission and transaxles.

❑ Explain the fundamentals of torque multiplication and overdrive.

❑ Explain the basic design and operation of standard and lockup torque converters.

❑ Explain the function and operation of a differential and drive axles.

❑ Describe how an automotive air conditioning system operates.

❑ Understand how cruise or speed control operates and the differences of various systems.

❑ Consider the use and value of engine cooling fans.

❑ Identify the major components of a typical drum brake and describe their functions.

❑ Describe the components of a hydraulic brake system and their operation.

❑ Briefly describe the operation of drum and disc brakes.

❑ Describe the main purposes of the steering and suspension systems.

Introduction

Although the main emphasis of this book is on the engine and its systems, there are other systems that affect the way a vehicle or engine runs, or appears to run. The engine is the power source for a car or truck; poor driveability can result from problems other than the engine. To better understand this, let's define driveability. Good driveabilty requires the following conditions:

1. The engine must start quickly.

2. The engine must idle smoothly.

3. The engine's idle speed must be constant to prevent stalling or racing of the engine.

4. The engine must accelerate smoothly and without hesitation.

5. The engine must run smoothly at all speeds.

6. Normal amounts of fuel should be used by the engine.

7. Exhaust emissions should be at a minimum and there should be no noticeable smoke from the tailpipe.

Most of these conditions do only concern the engine. Some are affected by other systems. For example, if a car has a flat tire it will not accelerate smoothly or quickly. This may seem ridiculously simple, but you would be surprised at how often simple problems are the cause of driveability problems.

In this chapter we will look at the components and systems that affect driveability. Most of these systems have the purpose of moving the car down the road, comfortably and safely.

The engine provides the power to drive the wheels of the vehicle. All automobile engines, both gasoline and diesel, are classified as internal combustion engines because the combustion or burning that creates energy takes place inside the engine. The combustion process is the burning of an air and fuel mixture. As a result of combustion, large amounts of pressure are generated in the engine. This pressure or energy is used to power the car. The engine must be built strong enough to hold the pressure and temperatures formed by combustion.

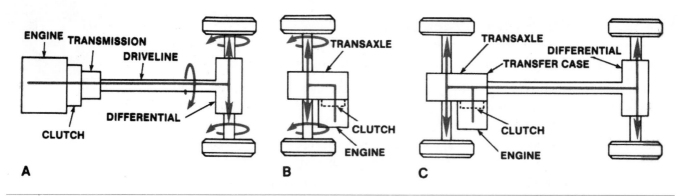

Figure 15-1 Typical drivetrains (A) rear-wheel drive, (B) front-wheel drive, and (C) four-wheel drive.

The drive shaft assembly is called the vehicle's driveline.

The drivetrain is made up of all components that transfer power from the engine to the driving wheels of the vehicle. The exact components used in a vehicle's drivetrain depend on whether the vehicle is equipped with rear-wheel drive, front-wheel drive, or four-wheel drive (Figure 15-1).

Power flow through the drivetrain of a rear-wheel-drive vehicle passes through the clutch or torque converter, manual or automatic transmission, and the driveline. Then it goes through the rear differential, the rear-driving axles, and onto the rear wheels.

Power flow through the drivetrain of a front-wheel-drive vehicle passes through the clutch or torque converter, manual or automatic transmission, then it goes through a front differential, the driving axles, and onto the front wheels.

Four-wheel drive, or all-wheel drive, vehicles combine features of both rear- and front-wheel-drive systems so that power can be delivered to all wheels either on a permanent or on-demand basis.

Clutches

Shop Manual
Chapter 15, page 633

Clutches are used on vehicles with manual transmissions or transaxles. The clutch is used to mechanically connect the engine's flywheel to the transmission or transaxle input shaft (Figure 15- 2). It does this through the use of a special friction plate that is splined to the input shaft. When the clutch is engaged, the friction plate contacts the flywheel, transferring power through the plate to the input shaft.

When stopping, starting, and shifting from one gear to the next, the clutch is disengaged by pushing down on the clutch pedal. This moves the clutch plate away from the flywheel. Power flow to the transmission stops. The driver can then shift gears without damaging the transmission or transaxle. Allowing the clutch pedal to come up re-engages the clutch. This allows power to flow from the engine through the transmission.

All manual transmissions require a clutch to engage or disengage the transmission. If the vehicle had no clutch and the engine was always connected to the transmission, the engine would stop every time the vehicle was brought to a stop. The clutch allows the engine to idle while the vehicle is stopped. It also allows for easy shifting between gears.

The clutch engages the transmission gradually by allowing a certain amount of slippage between the input and the output shafts on the clutch. The basic principle of engaging a clutch is demonstrated in Figure 15-3. The flywheel and the pressure plate are the drive or driving members of the clutch. The driven member connected to the transmission input shaft is the **clutch disc**, also called the friction disc. As long as the clutch is disengaged (clutch pedal depressed), the drive members turn independently of the driven member, and the engine is disconnected from the transmission. However, when the clutch is engaged (clutch pedal released), the pressure plate moves in the direction of the arrows and the clutch disc is bound between the two revolving drive members and forced to turn at the same speed.

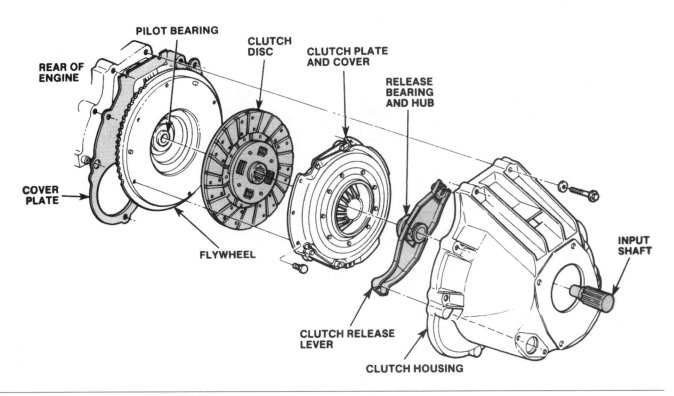

Figure 15-2 Parts of a clutch assembly. *(Reprinted with the permission of Ford Motor Company)*

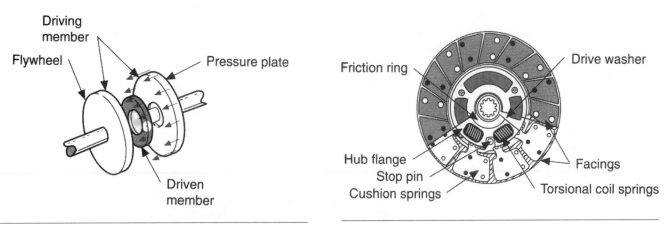

Figure 15-3 When the clutch is engaged, the driven member is squeezed between the two driving members. The transmission is connected to the driven member. *(Courtesy of General Motors Corporation, Service Technology Group)*

Figure 15-4 Parts of a clutch disc. *(Courtesy of General Motors Corporation, Service Technology Group)*

The **flywheel** is the main driving member of the clutch. The rear surface of the flywheel is a friction surface machined very flat to assure smooth clutch engagement.

A bore in the center of the flywheel and crankshaft holds the **pilot bushing**, which supports the front end of the transmission input shaft and maintains alignment with the engine's crankshaft. Sometimes a ball or roller needle bearing is used instead of a pilot bushing.

The clutch disc receives the driving motion from the flywheel and pressure plate assembly and transmits that motion to the transmission input shaft. The parts of a clutch disc are shown in Figure 15-4.

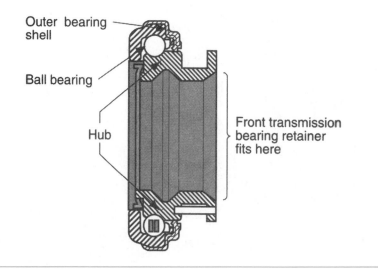

Figure 15-5 Typical clutch release bearing. *(Courtesy of General Motors Corporation, Service Technology Group)*

The clutch disc is designed to absorb such things as crankshaft vibration, abrupt clutch engagement, and driveline shock. Torsional coil springs allow the disc to rotate slightly in relation to the pressure plate while they absorb the torque forces. The number and tension of these springs is determined by engine torque and vehicle weight. Stop pins limit this torsional movement to approximately 3/8 inch.

The purpose of the pressure plate assembly is two-fold. First, it must squeeze the clutch disc onto the flywheel with sufficient force to transmit engine torque efficiently. Second, it must move away from the clutch disc so that the clutch disc can stop rotating, even though the flywheel and pressure plate continue to rotate.

The **clutch release bearing** is usually a sealed, prelubricated ball bearing (Figure 15-5). Its function is to smoothly and quietly move the pressure plate release levers or diaphragm spring through the engagement and disengagement process.

The release bearing is mounted on an iron casting called a hub, which slides on a hollow shaft at the front of the transmission housing. This hollow shaft, shown in Figure 15-6, is part of the transmission bearing retainer.

The clutch release bearing is commonly referred to as the throw out bearing.

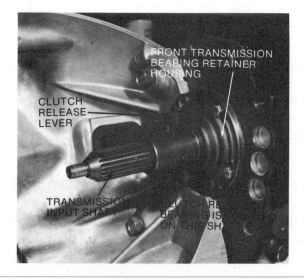

Figure 15-6 The clutch release bearing assembly slides on the hollow shaft of the front transmission bearing retainer housing.

To disengage the clutch, the release bearing is moved forward on its shaft by the **clutch fork**. As the release bearing contacts the release levers or diaphragm spring of the pressure plate assembly, it begins to rotate with the rotating pressure plate assembly. As the release bearing continues forward, the clutch disc is disengaged from the pressure plate and flywheel.

To engage the clutch, the release bearing slides to the rear of the shaft. The pressure plate moves forward and traps the clutch disc against the flywheel to transmit engine torque to the transmission input shaft. Once the clutch is fully engaged, the release bearing is normally stationary.

The clutch linkage is a series of parts that connects the clutch pedal to the clutch fork. It is through the clutch linkage that the operator controls the engagement and disengagement of the clutch assembly smoothly and with little effort.

Clutch linkage can be mechanical or hydraulic. Mechanical clutch linkage can be divided into two types: shaft and lever linkage and cable linkage. The shaft and lever clutch linkage consists of the various shafts, levers, adjustable rods, and pivots that transmit clutch pedal motion to the clutch fork. A rod connects the clutch pedal to the lever and shaft assembly. When the upper lever is moved by the clutch pedal, the shaft rotates and moves the lower lever, which is connected to a pushrod that is attached to the clutch fork.

A cable linkage can perform the same controlling action as the shaft and lever linkage, but with fewer parts. The clutch cable system does not take up much room. It also has the advantage of flexible installation so it can be routed around the power brake and steering units. These advantages help to make it the most commonly used clutch linkage.

Frequently, the clutch assembly is controlled by a hydraulic system (Figure 15-7). In the hydraulic clutch linkage system, hydraulic (liquid) pressure transmits motion from one sealed cylinder to another through a hydraulic line. Like the cable linkage assembly, the hydraulic linkage is compact and flexible. In addition, the hydraulic pressure developed by the master cylinder decreases required pedal effort and provides a precise method of controlling clutch operation. Brake fluid is commonly used as the hydraulic fluid in hydraulic clutch systems.

When the clutch pedal is depressed, the movement of the piston develops hydraulic pressure that is displaced from the master cylinder, through a tube, into the slave cylinder. The slave cylinder piston movement is transmitted to the clutch fork, which disengages the clutch.

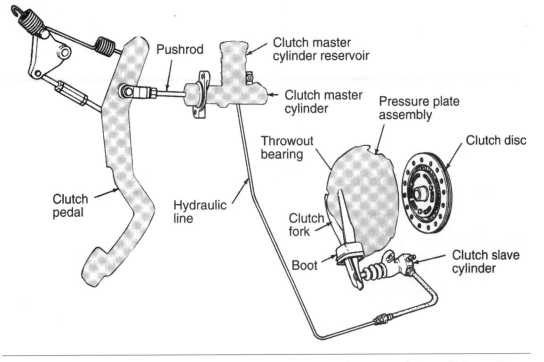

Figure 15-7 Typical hydraulic clutch linkage arrangement. *(Courtesy of General Motors Corporation, Service Technology Group)*

When the clutch pedal is released, the piston is forced back to the engaged position by the master cylinder piston return spring. External springs move the slave cylinder pushrod and piston back to the engaged position. Fluid pressure returns through the hydraulic tubing to the master cylinder assembly. There is no hydraulic pressure in the system when the clutch assembly is in the engaged position.

Manual Transmissions

Shop Manual
Chapter 15, page 635

The transmission or transaxle is a vital link in the powertrain of any modern vehicle. The purpose of the transmission or transaxle is to use gears of various sizes to give the engine a mechanical advantage over the driving wheels. During normal operating conditions, power from the engine is transferred through the engaged clutch to the input shaft of the transmission or transaxle. Gears in the transmission or transaxle housing alter the torque and speed of this power input before passing it on to other components in the powertrain. Without the mechanical advantage the gearing provides, an engine can generate only limited torque at low speeds. Without sufficient torque, moving a vehicle from a standing start would be impossible.

In any engine, the crankshaft always rotates in the same direction. If the engine transmitted its power directly to the drive axles, the wheels could be driven only in one direction. Instead, the transmission or transaxle provides the gearing needed to reverse direction so the vehicle can be driven backward.

Vehicles propelled by the rear wheels normally use a transmission. Transmission gearing is located within an aluminum or iron casting called the transmission case assembly (Figure 15-8). The transmission case assembly is attached to the rear of the engine, which is normally located in the front of the vehicle. A drive shaft links the output shaft of the transmission with the differential and drive axles located in a separate housing at the rear of the vehicle (Figure 15-9). The differential splits the driveline power and redirects it to the two rear drive axles, which then pass it onto the wheels. For many years, rear-wheel drive systems were the conventional method of propelling a vehicle.

Front-wheel drive vehicles are propelled by the front wheels. For this reason, they must use a drive design different from that of a rear-wheel drive vehicle. The transaxle is the special power transfer unit commonly used on front-wheel drive vehicles. A transaxle combines the transmission gearing, differential, and drive axle connections into a single case aluminum housing located in front of the vehicle (Figure 15-10). This design offers many advantages. One major advantage is the good traction on slippery roads due to the weight of the powertrain components being directly over the driving axles of the vehicle.

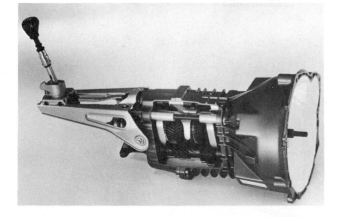

Figure 15-8 Typical five-speed manual transmission with the mainshaft (speed) gears, countershaft cluster gears, and shaft forks visible through the case cutaway. *(Reprinted with the permission of Ford Motor Company)*

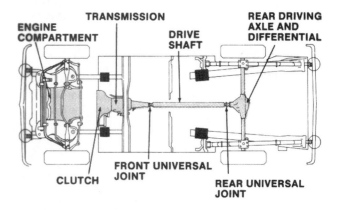

Figure 15-9 Location of typical rear-wheel drive powertrain components. *(Courtesy of Chrysler Corporation)*

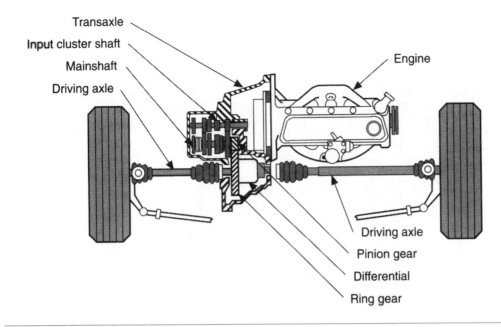

Figure 15-10 Location of typical front-wheel drive powertrain components.

Four-wheel drive vehicles typically use a transmission and transfer case. The transfer case mounts on the side or back of the transmission. A chain or gear drive inside the transfer case receives power from the transmission and transfers it to two separate drive shafts. One drive shaft connects to a differential on the front drive axle. The other drive shaft connects to a differential on the rear drive axle.

Most manual transmissions and transaxles are constant mesh, fully synchronized units. Constant mesh means that regardless of the vehicle being stationary or moving, the gears within the unit are constantly in mesh. Fully synchronized means that the unit uses a mechanism of brass rings and clutches to bring rotating shafts and gears to the same speed before shifts occur. This promotes smooth shifting. In a vehicle equipped with a four-speed manual shaft transmission or transaxle, all four forward gears are synchronized. Reverse gearing may or may not be synchronized, depending on the type of transmission or transaxle.

Torque Converters

An automatic transmission eliminates the use of a mechanical clutch and shift lever. In place of a clutch, it uses a fluid coupling called a **torque converter** to transfer power from the engine's flywheel to the transmission input shaft. The torque converter allows for smooth transfer of power at all engine speeds (Figure 15-11).

The torque converter operates through hydraulic force provided by automatic transmission fluid. The torque converter changes or multiplies the twisting motion of the engine crankshaft and directs it through the transmission.

The torque converter automatically engages and disengages power from the engine to the transmission in relation to engine rpm. With the engine running at the correct idle speed, there is not enough fluid flow for power transfer through the torque converter. As engine speed is increased, the added fluid flow creates sufficient force to transmit engine power through the torque converter assembly to the transmission.

A standard torque converter consists of three elements (Figure 15-12): the impeller, the stator assembly, and the turbine.

The **impeller** assembly is the input (drive) member. It receives power from the engine. The **turbine** is the output (driven) member. It is applied to the forward clutch of the transmission and

Shop Manual
Chapter 15, page 635

A torque converter's impeller is also called the pump assembly.

Torque converters are commonly abbreviated as T/Cs.

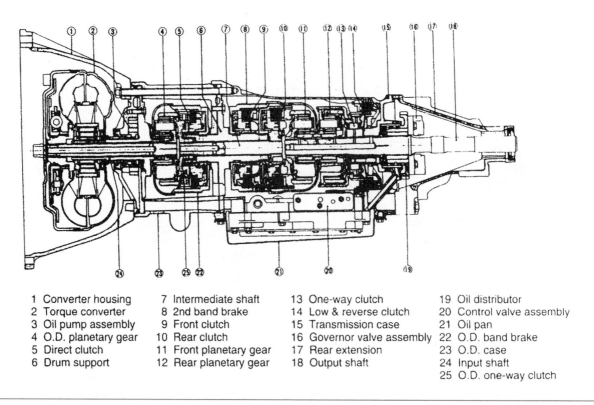

1 Converter housing	7 Intermediate shaft	13 One-way clutch	19 Oil distributor
2 Torque converter	8 2nd band brake	14 Low & reverse clutch	20 Control valve assembly
3 Oil pump assembly	9 Front clutch	15 Transmission case	21 Oil pan
4 O.D. planetary gear	10 Rear clutch	16 Governor valve assembly	22 O.D. band brake
5 Direct clutch	11 Front planetary gear	17 Rear extension	23 O.D. case
6 Drum support	12 Rear planetary gear	18 Output shaft	24 Input shaft
			25 O.D. one-way clutch

Figure 15-11 Typical torque converter and automatic transmission. *(Courtesy of General Motors Corporation, Service Technology Group)*

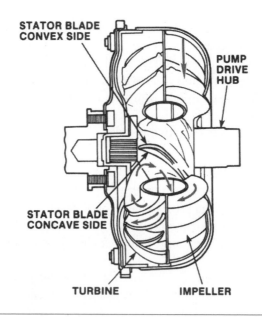

Figure 15-12 A torque converter's major internal parts are its impeller, turbine, and stator. *(Reprinted with the permission of Ford Motor Company)*

to the turbine shaft assembly. The **stator** assembly is the reaction member or torque multiplier. The stator is supported on a roller race, which operates as an overrunning clutch and permits the stator to rotate freely in one direction and lock up in the opposite direction.

Transmission oil is used as the medium to transfer energy in the T/C. Figure 15-13A illustrates the T/C impeller or pump at rest. Figure 15-13B shows it being driven. As the pump impeller rotates, centrifugal force throws the oil outward and upward due to the curved shape of the impeller housing.

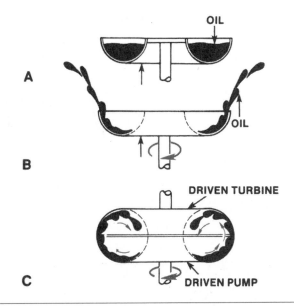

Figure 15-13 Fluid travel inside the torque converter: (A) fluid at rest in impeller/pump, (B) fluid thrown up and outward by spinning pump, and (C) fluid flow harnessed by turbine and redirected back into the pump.

The faster the impeller rotates, the greater the centrifugal force becomes. In Figure 15-13B, the oil is simply flying out of the housing and is not producing any work. To harness some of this energy, the turbine assembly is mounted on top of the impeller (Figure 15-13C). Now the oil thrown outward and upward from the impeller strikes the curved vanes of the turbine, causing the turbine to rotate. An oil pump driven by the converter shell and the engine continually delivers oil under pressure into the T/C through a hollow shaft at the center axis of the rotating torque converter assembly. A seal prevents the loss of fluid from the system.

With the transmission in gear and the engine at idle, the vehicle can be held stationary by applying the brakes. Since the impeller is driven by engine speed, it turns slowly creating little centrifugal force within the torque converter. Therefore, little or no power is transferred to the transmission.

When the throttle is opened, engine speed, impeller speed, and the amount of centrifugal force generated in the torque converter increase dramatically. Oil is then directed against the turbine blades, which transfer power to the turbine shaft and transmission.

Types of Oil Flow

Two types of oil flow take place inside the torque converter: rotary and vortex flow (Figure 15-14). Rotary oil flow is the oil flow around the circumference of the torque converter caused by the rotation of the torque converter on its axis. Vortex oil flow is the oil flow occurring from the impeller to the turbine and back to the impeller.

Figure 15-15 also shows the oil flow pattern as the speed of the turbine approaches the speed of the impeller. This is known as the **coupling point**. The turbine and the impeller are running at essentially the same speed. They cannot run at exactly the same speed due to slippage between them. The only way they can turn at exactly the same speed is by using a lockup clutch to mechanically tie them together. Torque converter multiplication can only occur when the impeller is rotating faster than the turbine.

As the vehicle begins to move, the stator stays in its stationary or locked position because of the difference between the impeller and turbine speeds.

As vehicle road speed increases, turbine speed increases until it approaches impeller speed. Oil exiting the turbine vanes strikes the back face of the stator, causing the stator to rotate in the same direction as the turbine and impeller.

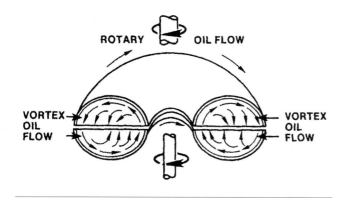

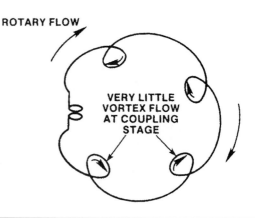

Figure 15-14 Rotary and vortex oil flow in the torque converter.

Figure 15-15 Rotary flow is at its greatest at the coupling stage.

If the vehicle slows, engine speed also slows along with turbine speed. This decrease in turbine speed allows the oil flow to change direction. It now strikes the front face of the stator vanes, halting the turning stator and attempting to rotate it in the opposite direction.

As this happens, the stator is locked in position. In a stationary position, the stator now redirects the oil exiting the turbine so that torque is again multiplied.

Lockup Torque Converters

A lockup torque converter eliminates the 10% slip that takes place between the impeller and turbine at the coupling stage of operation. The engagement of a clutch between the engine crankshaft and the turbine assembly has the advantage of improving fuel economy and reducing torque converter operational heat and engine speed.

The lockup torque converter clutch assembly is controlled by the powertrain control module (PCM). When the computer receives electronic signals from the different sensors confirming the requirements for lockup have been met, lockup clutch engagement begins. These sensors include an engine coolant sensor, vehicle speed sensor, engine vacuum sensor, and throttle position sensor.

The system operates in the following manner. The engine operates for more than five minutes and the engine coolant temperature sensor reports 150°F. Engagement could take place if all the other sensors agree. However, the vehicle operates in congested traffic at speeds varying from 15 to 35 mph and the converted clutch engagement speed is approximately 40 mph. Under these operating conditions the vehicle speed sensor reports that the vehicle speed is too low for clutch converter engagement. In addition, the throttle position sensor reports to the computer the unsteady up and down movement of the throttle. The computer interprets this as a reason not to engage the converter clutch. The brake switch also opens periodically. Thus, the computer does not energize the clutch solenoid to engage the converter clutch.

If the vehicle breaks out of congested traffic and is traveling at a steady higher speed, the speed sensor reports that the vehicle is at a speed higher than converter clutch engagement speed. The throttle position sensor reports that the throttle is in a steady position in favor of engagement, and the brake switch is closed because the driver's foot is not on the brake pedal. When the computer scans all the sensors and determines that all sensor and switch signals favor engagement, it energizes the converter clutch solenoid and the converter clutch is engaged for lockup operation.

Automatic Transmissions

Shop Manual
Chapter 15, page 636

Many rear-wheel drive and four-wheel drive vehicles are equipped with automatic transmissions. Automatic transaxles, which combine an automatic transmission and final drive assembly in a single unit, are used on front-wheel drive, all-wheel drive, and some rear-wheel drive vehicles (Figure 15-16).

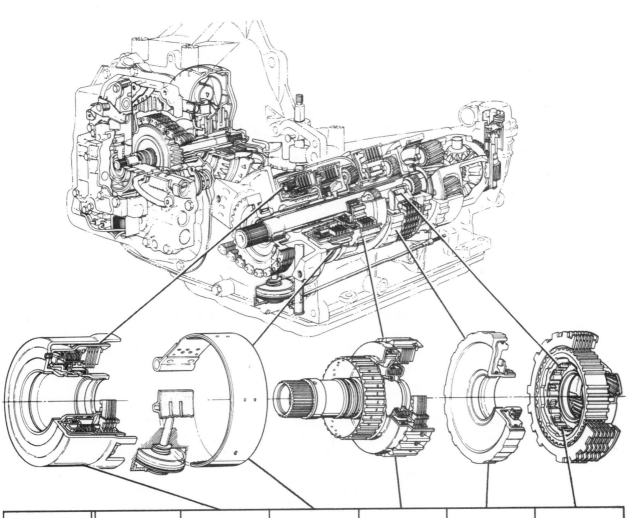

RANGE	GEAR	DIRECT CLUTCH	INTERMEDIATE BAND	FORWARD CLUTCH	LO-REVERSE CLUTCH	ROLLER CLUTCH
P-N						
D	1st			APPLIED		HOLDING
	2nd		APPLIED	APPLIED		
	3rd	APPLIED		APPLIED		
2	1st			APPLIED		HOLDING
	2nd		APPLIED	APPLIED		
1	1st			APPLIED	APPLIED	HOLDING
R	REVERSE	APPLIED			APPLIED	

Figure 15-16 Interior view of a typical transaxle.

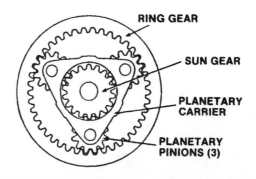

RING GEAR

SUN GEAR

PLANETARY CARRIER

PLANETARY PINIONS (3)

Figure 15-17 Planetary gear configuration is similar to the solar system, with the sun gear surrounded by the planetary pinion gears. The ring gear surrounds the complete gearset.

An automatic transmission or transaxle selects gear ratios according to engine speed, powertrain load, vehicle speed, and other operating factors. Little effort is needed on the part of the driver because both upshifts and downshifts occur automatically. A driver-operated clutch is not needed to change gears, and the vehicle can be brought to a stop without shifting to neutral. This is a great convenience, particularly in stop-and-go traffic. The driver can also manually select a lower forward gear, reverse, neutral, or park. Depending on the forward range selected, the transmission can provide engine braking during deceleration.

Until recently, all automatic transmissions were controlled by hydraulics. However, many new systems now feature computer-controlled operation of the torque converter and transmission. Based on input data supplied by electronic sensors and switches, the computer sets the torque converter's operating mode, controls the transmission's shifting sequence, and in some cases regulates transmission oil pressure.

All automatic transmissions rely on planetary gearsets to transfer power and multiply engine torque to the drive axle. Compound gearsets combine two simple planetary gearsets so that load can be spread over a greater number of teeth for strength and also to obtain the largest number of gear ratios possible in a compact area.

A simple planetary gearset consists of three parts: a sun gear, a carrier with planetary pinions mounted to it, and an internally toothed ring gear or annulus. The sun gear is located in the center of the assembly (Figure 15-17). It can be either a spur or helical gear design. It meshes with the teeth of the planetary pinion gears. Planetary pinion gears are small gears fitted into a framework called the planetary carrier. The planetary carrier is designed with a shaft for each of the planetary pinion gears.

The planetary pinions surround the sun gear's center axis and they themselves are surrounded by the annulus or ring gear, which is the largest part of the simple gearset. The ring gear acts like a band to hold the entire gearset together and provide great strength to the unit.

Any one of the three members can be used as the driving or input member. At the same time, another member might be kept from rotating and thus becomes the held or stationary member. The third member then becomes the driven or output member. Depending on which member is the driver, which is held, and which is driven, either a torque increase or a speed increase is produced by the planetary gearset. Output direction can also be reversed through various combinations.

Figure 15-18 summarizes the basic laws of simple planetary gears. It indicates the resultant speed, torque, and direction of the various combinations available. Also, remember that when an external-to-external gear tooth set is in mesh, there is a change in the direction of rotation at the output. When an external gear tooth is in mesh with an internal gear, the output rotation for both gears is the same.

The planetary pinion gears are called planetary pinions for short.

476

Sun Gear	Carrier	Ring Gear	Speed	Torque	Direction
1. Input	Output	Held	Maximum reduction	Increase	Same as input
2. Held	Output	Input	Minimum reduction	Increase	Same as input
3. Output	Input	Held	Maximum increase	Reduction	Same as input
4. Held	Input	Output	Minimum increase	Reduction	Same as input
5. Input	Held	Output	Reduction	Increase	Reverse of input
6. Output	Held	Input	Increase	Reduction	Reverse of input
7. When any two members are held together, speed and direction are the same as input. Direct 1:1 drive occurs.					
8. When no member is held or locked together, output cannot occur. The result is a neutral condition.					

Figure 15-18 Laws of simple planetary gear.

Planetary Gear Controls

Certain parts of the planetary gear train must be held while others must be driven to provide the needed torque multiplication and direction for vehicle operation. Planetary gear controls is the general term used to describe transmission bands, servos, and clutches.

A **band** is a braking assembly positioned around a stationary or rotating drum. The band brings a drum to a stop by wrapping itself around the drum and holding it. The band is hydraulically applied by a servo assembly. Connected to the drum is a member of the planetary gear train. The purpose of a band is to control the planetary gear train by holding the drum and connecting planetary gear member stationary.

In contrast to a band, which can only hold a planetary gear member, transmission clutches, either overrunning or multiple-disc, are capable of both holding and driving members. A **multiple-disc clutch** uses a series of hollow friction discs to transmit torque or apply braking force. The discs have internal teeth that are sized and shaped to mesh with splines on the clutch assembly hub. In turn, this hub is connected to a planetary gear train component so gearset members receive the desired braking or transfer force when the clutch is applied or released.

Multiple-disc clutches have a large drum-shaped housing that can be either a separate casting or part of the existing transmission housing (Figure 15-19). This drum housing holds all other clutch components: the cylinder, hub, piston, piston return springs, seals, pressure plate, **clutch pack**, and snap rings.

A clutch pack contains plain steel and friction discs.

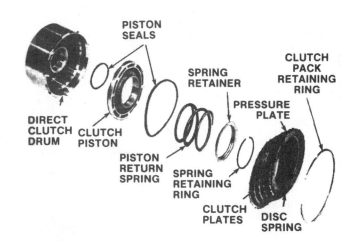

Figure 15-19 Exploded view of multiple-disc clutch assembly. *(Reprinted with the permission of Ford Motor Company)*

Hydraulic Systems

A hydraulic system uses a liquid to perform work. In an automatic transmission, this liquid is automatic transmission fluid (ATF). An automatic transmission uses ATF fluid pressure to control the action of the planetary gearsets. This fluid pressure is regulated and directed to change gears automatically through the use of various pressure regulators and control valves.

The **valve body** can be best understood as the control center of an automatic transmission (Figure 15-20). The purpose of the valve body is to sense the load on the vehicle's engine and drivetrain and the operator's driving requirements.

Many very precisely machined holes are located in the valve body to accommodate the various valves. The purpose of a valve is to start, stop, or direct and regulate fluid flow. The movement of the valves engages and disengages the gears. The choice of gear is determined by the driver and the placement of the shift lever. The automatic shifts are determined by vehicle speed and load.

Electronic Controls

Shifting in an automatic transmission is controlled by a hydraulic system. Most late-model automatic transmissions use electronics to control shifting with the hydraulic system. In a hydraulic system, an intricate network of valves and other components use hydraulic pressure to control the operation of planetary gearsets. These gearsets generate the three or four forward speeds, neutral, park, and reverse gears normally found on automatic transmissions. Newer electronic shifting systems use electric solenoids to control shifting mechanisms. Electronic shifting is precise and can be varied to

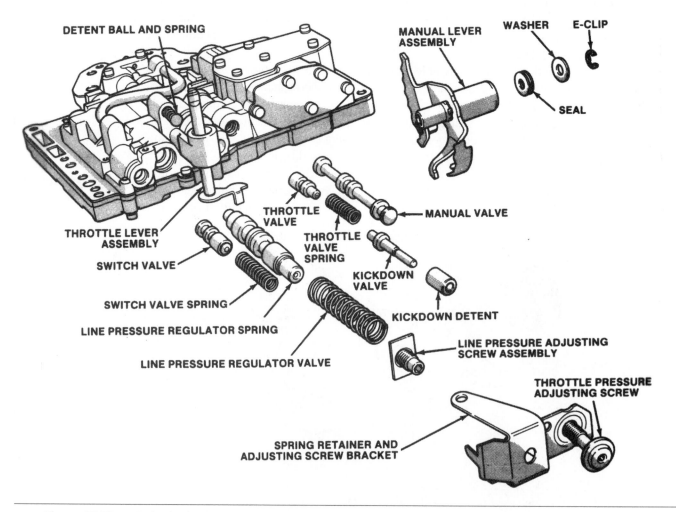

Figure 15-20 Typical valve body assembly. *(Courtesy of Chrysler Corporation)*

suit certain operating conditions. Electronic control is superior because information about the engine, fuel, ignition, vacuum, and operating temperature is fed into the computer so that shifting and lockup are closely monitored to take place at exactly the right time.

When shifting of the automatic transmission is controlled by a computer, input signals for engine and road speed, manifold vacuum, engine operating temperature, gear selection, throttle position, and other factors are fed to the computer. The computer produces output signals that activate relays, which in turn, operate electrical solenoid valves. Instead of hydraulic pressures and springs, the motion of the solenoids controls the position of the valves. When activated, a solenoid moves its valve to control fluid pressure in a valve body.

In order for the computer in an electronic control system to determine when to start a gear change, it must be able to refer to **shift schedules** that it has stored in its memory. A shift schedule contains the actual shift points to be used by the computer according to the input data it receives from the sensors. Shift schedule logic chooses the proper shift schedule for the current conditions of the transmission. It uses the shift schedule to select the appropriate gear, then determines the correct shift schedule or pattern that should be followed.

The first input a computer looks at to determine the correct shift logic is the position of the gear shift lever. All shift schedules are based on the gear selected by the driver. The choices of shift schedules are limited by the type and size of engine that is coupled to the automatic transmission. Each engine/transmission combination has a different set of shift schedules. These schedules are coded by selector lever position and current gear range, and use throttle angle and vehicle speed as primary determining factors. The computer also looks at different temperature, load, and engine operation inputs for more information.

The shift schedules set the conditions that need to be met for a change in gears. Since the computer frequently reviews the input information, it can make quick adjustments to the schedule if needed and as needed. The result of the computer's processing of this information and commanding outcomes according to a logically program is optimum shifting of the automatic transmission. This results in improved fuel economy and overall performance.

The electronic control systems used by the manufacturers differ with the various transmission models and the engines they are attached to. The components in each system and the overall operation of the system also varies with the different transmissions, however, all operate in a similar fashion and use basically the same parts.

Driveline

Drivelines are used on rear-wheel drive vehicles and four-wheel drive vehicles. They connect the output shaft of the transmission with the gearing in the rear axle housing (Figure 15-21), on rear-wheel drive vehicles. They are also used to connect the output shaft to the front and rear drive axles on a four-wheel drive vehicle.

Shop Manual
Chapter 15, page 640

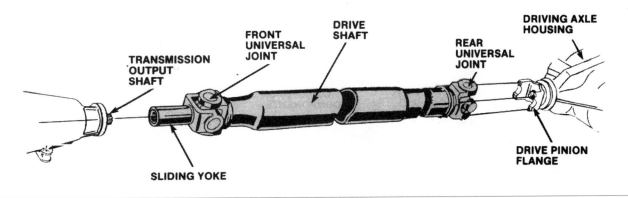

Figure 15-21 Output power from the transmission is connected to the differential in the drive axle housing by a drive shaft.

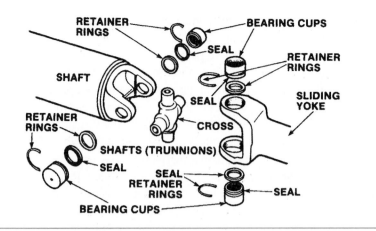

Figure 15-22 Exploded view of Cardan universal joint.

A driveline consists of a hollow drive or propeller shaft that is connected to the transmission and drive axle differential by universal joints. The drive shaft is nothing more than an extension of the transmission output shaft. The drive shaft transfers engine torque from the transmission to the rear driving axle.

The universal or U-joint allows two rotating shafts to operate at a slight angle to each other. Although simple in appearance (Figure 15-22), the universal joint is more intricate than it seems. This is because its natural action is to speed up and slow down twice in each revolution when operating at an angle. The rate of speed varies, depending upon the steepness of the U-joint angle.

The universal joint operating angle is derived by taking the difference between the transmission installation angle and the drive shaft installation angle. For instance, in Figure 15-23, a true horizontal centerline is drawn through the transmission. In order to align better with the rear axle, the transmission is installed at an angle 5 degrees off the true horizontal line. The drive shaft installation angle is 8 degrees off the true horizontal. The difference between the transmission installation angle and the drive shaft installation angle is 3 degrees. Therefore, the universal joint operating angle is 3 degrees. When the universal joint is operating at an angle, the driven yoke speeds up and slows down twice during each drive shaft revolution. This acceleration and deceleration of the universal joint is known as speed variation or fluctuation.

The speed changes are not normally visible during rotation. They might be felt as torsional vibrations due to improper installation, steep or unequal operating angles, and high speeds.

Speed variations must be canceled at exactly the same point in drive shaft rotation. The two driving yokes at opposite ends of the drive shaft must be at the same point of rotation.

Vibrations can be reduced by using canceling angles (Figure 15-24). Carefully examine the illustration, and note that the operating angle at the front of the drive shaft is offset by the one at the rear of the drive shaft. When the front universal joint accelerates, causing a vibration, the rear universal joint decelerates, causing a vibration. The vibrations created by the two joints oppose and dampen the vibrations from one to the other. The use of canceling angles provides a smoother drive shaft operation.

The drive shafts on front-wheel drive vehicles are correctly called axle or half shafts.

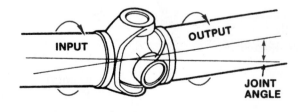

Figure 15-23 U-joint action.

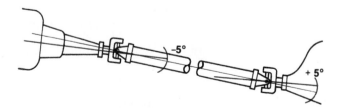

Figure 15-24 Canceling angles reduce vibrations.

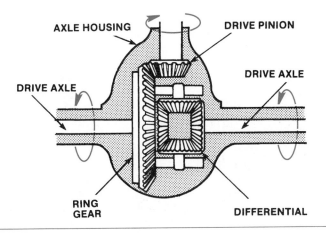

Figure 15-25 Rear differential components.

Differential

On rear-wheel drive vehicles, the drive shaft turns perpendicular to the forward motion of the vehicle. The differential gearing in the rear axle housing is designed to turn the direction of the power so that it can be used to drive the wheels of the vehicle. The power flows into the differential, where it changes direction, then to the rear axles and wheels (Figure 15-25).

The differential also performs two other important jobs. It multiplies the torque of the power it receives from the drive shaft by providing a final gear reduction. Also, it divides this power between the left and right driving axles and wheels in such a way that a differential wheel speed is possible. This means one wheel can turn faster than the other when going around turns. All vehicles use a differential to provide an additional gear reduction (torque increase) above and beyond what the transmission or transaxle gearing can produce. This is known as the **final drive gear**. In a transmission equipped vehicle, the differential gearing is located in the rear axle housing. In a transaxle, however, the final reduction is produced by the final drive gears housed in the transaxle case.

Driving Axles

Driving axles are solid steel shafts that transfer differential torque to the driving wheels. A separate axle shaft is used for each driving wheel. The driving axles and part of the differential are enclosed in an axle housing that protects and supports these parts.

Each driving axle is connected to the side gears in the differential. The inner or differential ends of the axles are splined to fit into the side gears. As the side gears are turned, the axles to which they are splined turn at the same speed.

At their outer or wheel ends, the axles are attached to the driving wheels. For attachment to a wheel, the outer end of each axle has a flange mounted to it. A flange is a rim for attaching one part to another part. To hold the wheel in place against the flange, studs are used. Studs are threaded shafts, resembling bolts without heads. One end of the stud is screwed or pressed into the flange. The wheel fits over the studs and a nut, called the lug nut, is tightened over the open end of the stud. This holds the wheel in place.

The inner end of each axle is supported by the differential carrier. The outer end of the axle shaft is supported by a bearing inside the axle housing. A bearing supports and holds a rotating part in place. This bearing, called the axle bearing, allows the axle to rotate smoothly inside the axle housing.

Heating and Air Conditioning

Shop Manual
Chapter 15, page 641

An automotive air conditioning (A/C) system is a closed pressurized system. It consists of a compressor, condenser, receiver/dryer or accumulator, expansion valve or orifice tube, and an evaporator (Figure 15-26).

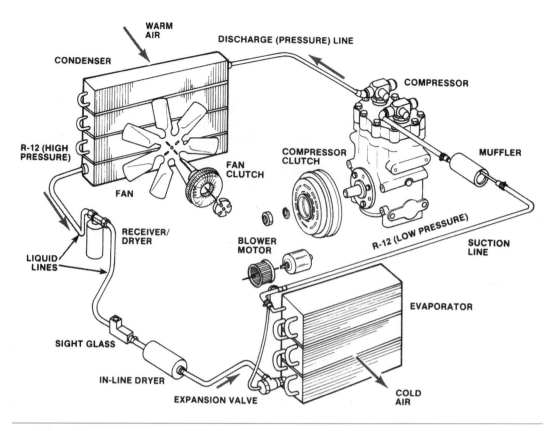

Figure 15-26 Major and secondary components of a typical air conditioning system.

In a basic air conditioning system, the heat is absorbed and transferred in the following steps (Figure 15-27):

1. Refrigerant leaves the compressor as a high-pressure, high-temperature vapor.

2. By removing heat via the condenser, the vapor becomes a high-pressure, low-temperature liquid.

3. Moisture and contaminants are removed by the receiver/dryer, where the cleaned refrigerant is stored until it is needed.

4. The expansion valve converts the high-pressure liquid changes into a low-pressure liquid by controlling its flow into the evaporator.

5. Heat is absorbed from the air inside the passenger compartment by the low-pressure, low-temperature refrigerant, causing the liquid to vaporize.

6. The refrigerant returns to the compressor as a low-pressure, higher-temperature vapor.

The compressor is the heart of the automotive air conditioning system. It separates the high-pressure and low-pressure sides of the system. The primary purpose of the unit is to draw the low-pressure vapor from the evaporator and compress this vapor into high-temperature, high-pressure vapor. This action results in the refrigerant having a higher temperature than surrounding air, and enables the condenser to condense the vapor back to a liquid. The secondary purpose of the compressor is to circulate or pump the refrigerant through the condenser under the different pressures required for proper operation.

In a **cycling clutch** system, the compressor is run intermittently through controlling the application and release of its clutch by a thermostatic switch. The thermostatic switch senses the evaporator's outlet air temperature, through a capillary tube that is part of the switch assembly. With a high sensing temperature, the thermostatic switch is closed and the compressor clutch is energized. As the evaporator outlet temperature drops to a preset level, the thermostatic switch

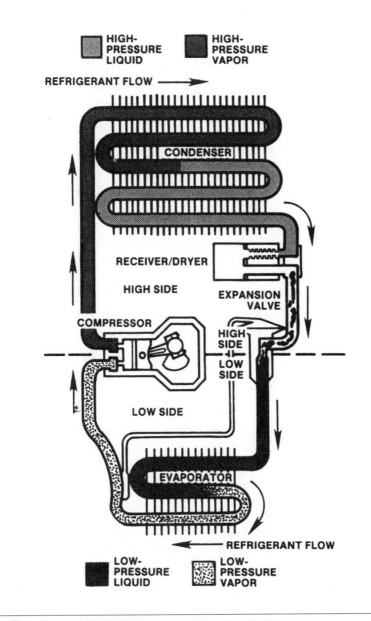

HIGH-PRESSURE LIQUID

HIGH-PRESSURE VAPOR

REFRIGERANT FLOW ⟶

CONDENSER

RECEIVER/DRYER

HIGH SIDE

EXPANSION VALVE

COMPRESSOR

HIGH SIDE

LOW SIDE

LOW SIDE

EVAPORATOR

⟵ REFRIGERANT FLOW

LOW-PRESSURE LIQUID

LOW-PRESSURE VAPOR

Figure 15-27 Basic refrigerant flow cycle.

opens the circuit to the compressor clutch. The compressor then ceases to operate until such time as the evaporator temperature rises above the switch setting. From this on and off operation is derived the term cycling clutch.

When the temperature of the evaporator approaches the freezing point (or the low setting of the switch), the thermostatic switch opens the circuit and disengages the compressor clutch. The compressor remains inoperative until the evaporator temperature rises to the preset temperature, at which time the switch closes and compressor operation resumes.

Engine Cooling Fans

With the advent of transverse-mounted engines, electrical cooling fans found their way under the hood of the modern automobile. Today most new cars are equipped with an electric cooling fan. They offer advantages over the mechanical cooling fans because of their ability to move large amounts of air independent of engine speed. Their circuitry is very simple, especially on a vehicle that is not air conditioned.

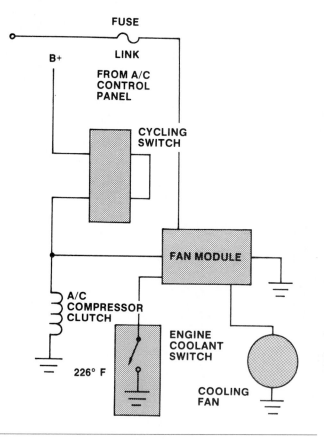

Figure 15-28 An A/C system in which the compressor and fan are cycled at the same time.

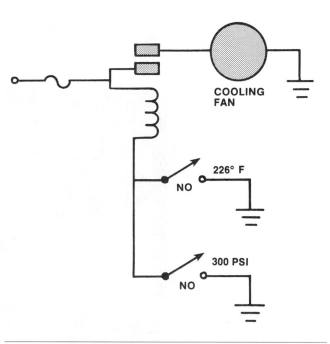

Figure 15-29 Cooling fan circuit with pressure switching.

Air conditioned vehicles, especially those with small engines, usually have additional circuitry to ensure that the cooling fan comes on when the compressor cycles on. Airflow through the condenser must be present for A/C to function correctly. With the condenser mounted in front of the radiator, the logical method of ensuring airflow is to turn the cooling fan on as the A/C compressor cycles on. Most of the manufacturers follow one of two methods. Either they cycle the compressor and the fan at the same time as Figure 15-28 shows, or they cycle the fan on when A/C high side pressure reaches a predetermined level (Figure 15-29). In either case, this ensures that airflow through the condenser keeps both the A/C and the engine cool.

Speed Control Systems

Shop Manual
Chapter 15, page 642

Cruise or speed control systems are designed to allow the driver to maintain a constant speed (usually about 30 mph) without having to apply continual foot pressure on the accelerator pedal. Selected cruise speeds are easily maintained and speed can be easily changed. Several override systems also allow the vehicle to be accelerated, slowed, or stopped. Because of the constant changes and improvements in technology, each cruise control system may be considerably different. There are several types that are used, including the non-resume type, the resume type, and the electronic type.

When engaged, the cruise control components set the throttle position to the desired speed. The speed is maintained unless heavy loads and steep hills interfere. The cruise control is disengaged whenever the brake pedal is depressed. The common speed or cruise control system components function in the following manner:

❏ The cruise control switch is located on the end of the turn signal or near the center or sides of the steering wheel. There are usually several functions on the switch, including off-on, resume, and engage buttons. The switch is different for resume and non-resume systems.

❏ The transducer is a device that controls the speed of the vehicle. When the transducer is engaged, it senses vehicle speed and controls a vacuum source (usually the manifold). The vacuum source is used to maintain a certain position on a servo. The speed control is sensed from the lower cable and casing assembly attached to the transmission.

❏ The servo unit is connected to the throttle by a rod or linkage, a head chain, or a bowden cable. The servo unit maintains the desired car speed by receiving a controlled amount of vacuum from the transducer. The variation in vacuum changes the position of the throttle. When a vacuum is applied, the servo spring is compressed and the throttle is positioned correctly. When the vacuum is released, the servo spring is relaxed and the system is not operating.

❏ There are two brake-activated switches. They are operated by the position of the brake. When the brake pedal is depressed, the brake release switch disengages the system. A vacuum release valve is also used to disengage the system when the brake pedal is depressed.

Figure 15-30 shows an electrical and vacuum circuit diagram. The system operates by controlling vacuum to the servo through various solenoids and switches. Cruise control can also be obtained by using electronic components rather than mechanical components. Depending upon the vehicle manufacturer, several additional components may be used.

The electronic control module is used to control the servo unit. The servo unit is again used to control the vacuum which in turn controls the throttle. The vehicle speed sensor (VSS) buffer amplifier is used to monitor or sense vehicle speed. The signal created is sent to the electronic control module. A generator speed sensor may also be used in conjunction with the VSS. The clutch switch is used on vehicles with manual transmissions to disengage the cruise control when the clutch is depressed. The accumulator is used as a vacuum storage tank on vehicles that have low vacuum during heavy load and high road speed.

Figure 15-31 shows how electronic cruise control components work together. The throttle position is controlled by the **servo unit**. The servo unit uses vacuum working against a spring pressure to operate an internal diaphragm. The servo unit vacuum circuit is controlled electronically by the controller.

The controller has several inputs that help determine how it will affect the servo. These inputs include a brake release switch (clutch release switch), a speedometer, buffer amplifier, or generator speed sensor, and a turn signal mode switch or speed control on the steering wheel (signal to control the cruise control).

Brake Systems

Automobiles are stopped by activating the brake system (Figure 15-32). Brakes, which are located at each wheel, utilize friction to slow and stop the automobile.

The brakes are activated when the vehicle operator depresses a brake pedal. The brake pedal is connected to a plunger in a *master cylinder*, which is filled with hydraulic fluid. When the brake pedal is depressed, a force is put onto the hydraulic fluid in the master cylinder. The force is increased by the master cylinder and transferred through brake hoses and lines to the four brake assemblies.

Two types of brakes are used on automobiles: disc brakes and drum brakes. Many automobiles use a combination of the two types; disc brakes at the front wheels and drum brakes at the rear wheels.

Most vehicles have power-assisted brakes. A brake booster typically uses manifold vacuum to increase the pressure applied to the plunger in the master cylinder. This lessens the amount of

Shop Manual
Chapter 15, page 642

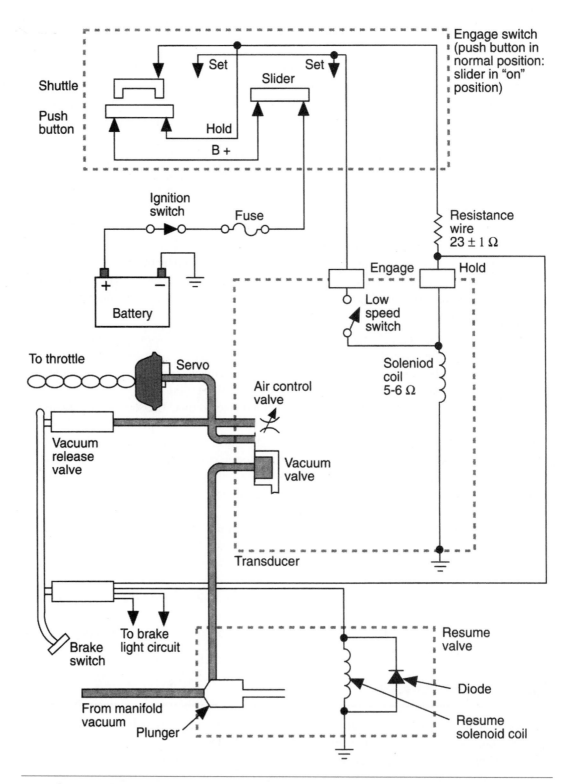

Figure 15-30 A cruise control circuit with vacuum and electrical systems. *(Courtesy of General Motors Corporation, Service Technology Group)*

pressure that must be applied to the brake pedal by the operator and increases the responsiveness of the brake system.

The brake system is designed to slow and halt the motion of a vehicle. To do that, various components within a hydraulic brake system must convert the momentum of the vehicle into heat. They do so by using friction.

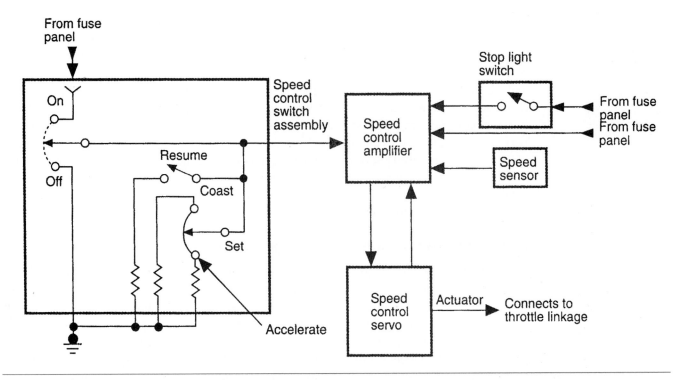

Figure 15-31 Electronic cruise control uses an electronic control module (controller) to operate a servo that controls the position of the throttle. *(Courtesy of General Motors Corporation, Service Technology Group)*

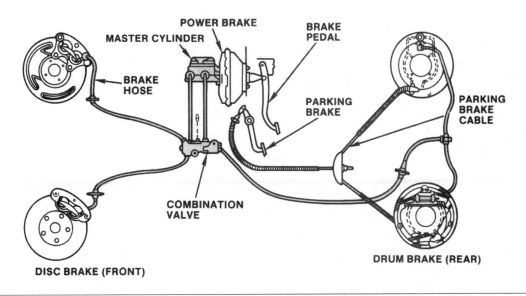

Figure 15-32 Many automobiles feature a combination of drum and disc brakes.

As the brakes on a moving automobile are actuated, rough-textured pads or shoes are pressed against rotating parts of the vehicle—either rotors or drums. The kinetic energy, or momentum, of the vehicle is then converted into heat energy by the kinetic friction of rubbing surfaces and the car or truck slows down.

When the vehicle comes to a stop, it is held in place by static friction. The friction between the surfaces of the brakes, as well as the friction between the tires and the road, resist any movement. To overcome the static friction that holds the car motionless, the brakes are released. The heat energy of combustion in the engine crankcase is converted into kinetic energy by the transmission and drivetrain, and the vehicle moves.

Disc brake rotors are commonly called discs.

Static friction also plays an important part in controlling a moving vehicle. The rotating tires grip the road, and the static friction between these two surfaces enables the driver to control the speed and direction of the car. When the brakes are applied, the kinetic friction of the rubbing brake components slows the rotation of the tires. This increases the static friction between the tires and the road, decreasing the motion of the car. If the kinetic or sliding friction of the brake components overcomes the static friction between the tires and road, the wheels lock up and the car begins to skid. Static friction then exists between the components in the brakes and kinetic friction between the skidding tires and the road—the car is out of control. Obviously, the most effective braking effort is achieved just below the brake component kinetic friction levels that result in wheel lockup. This is the role antilock braking systems play in modern vehicles. By electronically pumping the brakes on and off many times each second, antilock systems keep kinetic friction below the static friction between the tires and road.

Hydraulic Brake System Components

A drum brake assembly consists of a cast-iron drum, which is bolted to and rotates with the vehicle's wheel, and a fixed backing plate to which are attached the shoes and other components—wheel cylinders, automatic adjusters, and linkages (Figure 15-33). Additionally, there might be some extra hardware for parking brakes. The shoes are surfaced with frictional linings, which contact the inside of the drum when the brakes are applied. The shoes are forced outward, by pistons located inside the wheel cylinder. They are actuated by hydraulic pressure. As the drum rubs against the shoes, the energy of the moving drum is transformed into heat. This heat energy is passed into the atmosphere. When the brake pedal is released, hydraulic pressure drops, and the pistons are pulled back to their unapplied position by return springs.

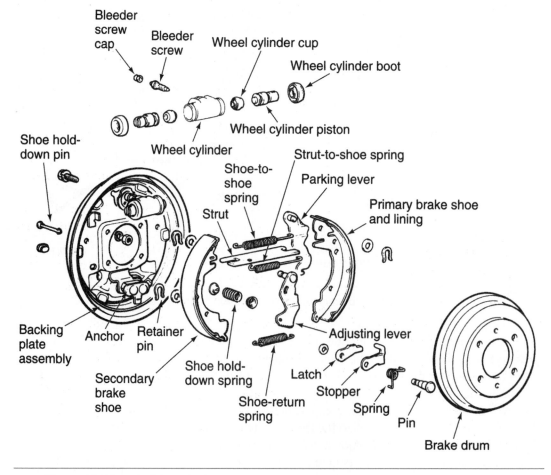

Figure 15-33 Typical drum brake.

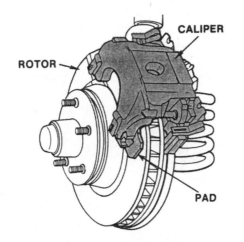

Figure 15-34 A typical disc brake.

Disc brakes resemble the brakes on a bicycle: the friction elements are in the form of pads, which are squeezed or clamped about the edge of a rotating wheel. With automotive disc brakes, this wheel is a *rotor* inboard of the vehicle wheel (Figure 15-34). The rotor is made of cast iron. Since the pads clamp against both sides of it, both sides are machined smooth. The pads are attached to metal shoes, which are actuated by pistons, the same as with drum brakes. The pistons are contained within a caliper assembly, a housing that wraps around the edge of the rotor. The caliper is kept from rotating by way of bolts holding it to the car's suspension framework.

The caliper is a housing containing the pistons and related seals, springs, and boots as well as the cylinders and fluid passages necessary to force the friction linings or pads against the rotor. The caliper resembles a hand in the way it wraps around the edge of the rotor. It is attached to the steering knuckle. Some models employ light spring pressure to keep the pads close against the rotor. In other caliper designs this is achieved by a unique type of seal that allows the piston to be pushed out the necessary amount, then retracts it just enough to pull the pad off the rotor.

Unlike shoes in a drum brake, the pads act perpendicular to the rotation of the disc when the brakes are applied. This effect is different from that produced in a brake drum, where frictional drag actually pulls the shoe into the drum. Disc brakes are said to be non-energized, and so require more force to achieve the same braking effort. For this reason, they are ordinarily used in conjunction with a power brake unit.

> Some brake discs are manufactured with two separate discs joined together by a finned center section. These discs are called ventilated brake rotors.

Suspension and Steering Systems

The suspension system on the automobile includes such components as the springs, shock absorbers, MacPherson struts, torsion bars, axles, and connecting linkages. These components are designed to support the body and frame, the engine, and the drivelines. Without these systems, the comfort and ease of driving the vehicle would be reduced. Figure 15-35 illustrates some of the components that are used in a suspension system.

Springs and torsion bars are used to support the axles of the vehicle. The two types of springs commonly used are the coil spring and the leaf spring. Torsion bars are made of long spring steel rods. One end of the rod is connected to the frame, while the other end is connected to the movable parts of the axles. As the axles move up and down, the rod twists and acts as a spring.

Shock absorbers slow down the upward and downward movement of the springs. This is necessary to limit the car's reaction to a bump in the road.

The steering system allows the driver to control the direction of the vehicle. A steering system includes the steering wheel, **steering gear**, steering shaft, and **steering linkage**.

Shop Manual
Chapter 15, page 644

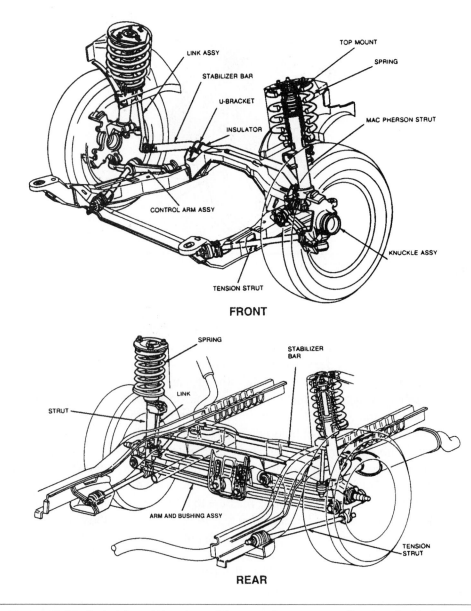

LINK ASSY
STABILIZER BAR
U-BRACKET
INSULATOR
CONTROL ARM ASSY
TENSION STRUT
TOP MOUNT
SPRING
MAC PHERSON STRUT
KNUCKLE ASSY

FRONT

SPRING
STABILIZER BAR
STRUT
LINK
ARM AND BUSHING ASSY
TENSION STRUT

REAR

Figure 15-35 Typical front (A) and rear (B) suspension systems. *(Reprinted with the permission of Ford Motor Company)*

There are two basic types of steering systems used in today's vehicles: a *rack-and-pinion* and *recirculating ball* systems (Figure 15-36). The rack-and-pinion system is commonly used in passenger cars. The recirculating ball system is normally used only on heavy vehicles, such as large pickup trucks, station wagons, and full-size luxury cars.

Steering gears provide a gear reduction to make changing the direction of the wheels easier. On all but a few subcompact and compact car models, the steering gear is also power assisted to ease the effort of turning the wheels. In a power-assisted system, a pump provides hydraulic fluid under pressure to the steering gear. A spool valve directs fluid to one side or the other of the steering gear to assist the operator in turning the wheels.

Principles of Wheel Alignment

Wheel alignment allows the wheels to roll without scuffing, dragging, or slipping on different types of road conditions. This gives greater safety in driving, easier steering, longer tire life, reduction in fuel consumption, and less strain on the parts that make up the front end of the vehicle.

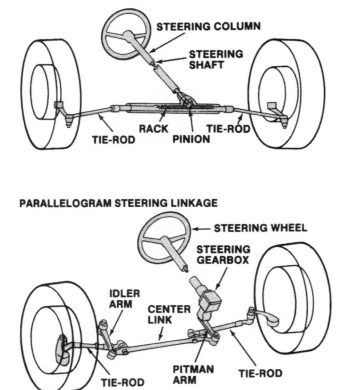

RACK-AND-PINION STEERING LINKAGE

STEERING COLUMN

STEERING SHAFT

TIE-ROD

RACK

PINION

TIE-ROD

TIE-ROD

PARALLELOGRAM STEERING LINKAGE

STEERING WHEEL

STEERING GEARBOX

IDLER ARM

CENTER LINK

PITMAN ARM

TIE-ROD

TIE-ROD

Figure 15-36 Common steering systems.

There is a multitude of angles and specifications that the automotive manufacturers must consider when designing a car. The multiple functions of the suspension system complicate things a great deal for design engineers. They must take into account more than basic geometry. Durability, maintenance, tire wear, available space, and production cost are all critical elements. Most elements contain a degree of compromise in order to satisfy the minimum requirements of each.

Most technicians do not need to be concerned with all of this. All they need to do is restore the vehicle to the condition the design engineer specified. To do this, however, the technician must be totally familiar with the purpose of basic alignment angles.

The alignment angles are designed in the vehicle to properly locate the vehicle's weight on moving parts and to facilitate steering. If these angles are not correct, the vehicle is misaligned. The effects of misalignment are given in Table 15–1. It is important to remember that alignment angles, when specified in the text, are those specific angles that should exist when the system is being measured under a given set of conditions. During regular performance, these angles change as the traveling surface and vehicle driving forces change.

Caster is the angle of the steering axis of a wheel from the vertical, as viewed from the side of the vehicle. The forward or rearward tilt from the vertical line (Figure 15-37) illustrates caster. Caster is the first angle adjusted during an alignment. Tilting the wheel forward is negative caster. Tilting backward is positive caster.

Caster is designed to provide steering stability. The caster angle for each wheel on an axle should be equal. Unequal caster angles cause the vehicle to steer toward the side with less caster. Too much negative caster can cause the vehicle to have sensitive steering at high speeds. The vehicle might wander as a result of negative caster.

Camber is the angle represented by the tilt of either the front or rear wheels inward or outward from the vertical as viewed from the front of the car (Figure 15-38). Camber is designed into

Table 15-1 EFFECTS OF INCORRECT ALIGNMENT

Problem	Effect
Incorrect camber setting	Tire wear Ball joint/wheel bearing wear Pull to side of most positive/least negative camber
Too much positive caster	Hard steering Excessive road shock Wheel shimmy
Too much negative caster	Wander Weave Instability at high speeds

Unequal caster	Pull to side most
	   negative/least positive
	   caster

Incorrect SAI	Instability Poor return Pull to side of lesser inclination Hard steering
Incorrect toe setting	Tire wear
Incorrect turning radius	Tire wear Squeal in turns

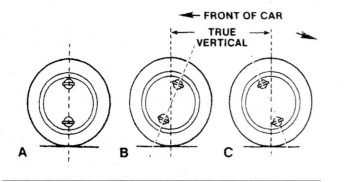

Figure 15-37 Three types of caster: (A) zero, (B) positive, and (C) negative.

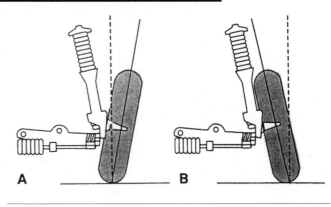

Figure 15-38 (A) Positive and (B) negative camber.

the vehicle to compensate for road crown, passenger weight, and vehicle weight. Camber is usually set equally for each wheel. Equal camber means each wheel is tilted outward or inward the same amount. Unequal camber causes tire wear and causes the vehicle to steer toward the side that is more positive.

Toe is the distance comparison between the leading edge and trailing edge of the front tires. If the leading edge distance is less, then there is toe-in. If it is greater, there is toe-out (Figure 15-39).

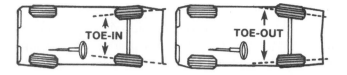

Figure 15-39 Typical rear toe condition.

Toe is critical as a tire-wearing angle. Wheels that do not track straight ahead have to drag as they travel forward. Excessive toe measurements (in or out) cause a sawtooth edge on the tread surface from dragging the tire sideways.

A main consideration in any alignment is to make sure the vehicle runs straight down the road, with the rear tires tracking directly behind the front tires when the steering wheel is in the straight-ahead position. The geometric centerline of the vehicle should parallel the road direction. This is the case when rear toe is parallel to the vehicle's geometric centerline in the straight-ahead position. If rear toe does not parallel the vehicle centerline, a thrust direction to the left or right is created (Figure 15-40). This difference of rear toe from the geometric centerline is called the **thrust angle**. The vehicle tends to travel in the direction of the thrust line, rather than straight ahead.

Steering axle inclination (SAI) locates the vehicle weight to the inside or outside of the vertical centerline of the tire. The SAI is the angle between true vertical and a line drawn between the steering pivots as viewed from the front of the vehicle. It is an engineering angle designed to project the weight of the vehicle to the road surface for stability. The SAI helps the vehicle's steering system return to straight ahead after a turn.

Turning radius or cornering angle is the amount of toe-out present in turns. As a car goes around a corner, the inside tire must travel in a smaller radius circle than the outside tire. This is accomplished by designing the steering geometry to turn the inside wheel sharper than the outside wheel. The result can be seen as toe-out in turns. This eliminates tire scrubbing on the road surface by keeping the tires pointed in the direction they have to move.

All vehicles are built around a geometric centerline that runs through the center of the chassis from the back to the front. The thrust line is the direction the rear axle would travel if unaffected by the front wheels. This condition is also called tracking. An ideal alignment has all four wheels parallel with the centerline, making the thrust line parallel with the centerline. However, the rear-wheel thrust line of a vehicle might not always be parallel to the actual centerline of the vehicle, so the angle of the thrust line must be checked first.

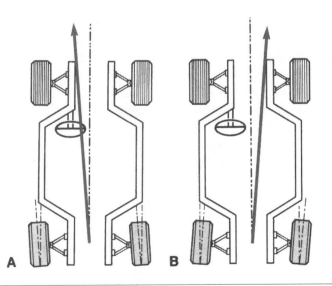

Figure 15-40 (A) Left and (B) right thrust direction.

Correct tracking refers to a situation with all suspension and wheels in their correct location and condition and aligned so that the rear wheels follow directly behind the front wheels while moving in a straight line. For this to occur, all wheels must be parallel with one another and axle and spindle lines must be at 90-degree angles to the vehicle centerline. Simply stated, all four wheels should form a perfect rectangle.

Rear Alignment

A car with a perfect front alignment can still experience poor handling and premature tire wear—particularly on front-wheel drive cars and cars with independent rear suspensions—if the rear suspension is misaligned. Approximately 80% of today's vehicles not only have front-end alignment specifications but also require rear-wheel alignment.

Like front camber, rear camber affects both tire wear and handling. The ideal situation is to have zero running camber on all four wheels to keep the tread in full contact with the road for optimum traction and handling. Camber is not a static angle. It changes as the suspension moves up and down. Camber also changes as the vehicle is loaded and the suspension sags under the weight.

Besides wearing the tires unevenly across the tread, uneven side-to-side camber (as when one wheel leans in and the other does not) creates a steering pull just like it does when the camber readings on the front wheels do not match. It is like leaning on a bicycle. A vehicle always pulls toward a wheel with the most positive camber. If the mismatch is at the rear wheels, the rear axle pulls toward the side with the greatest amount of positive camber. If the rear axle pulls to the right, the front of the car drifts to the left—and the result is a steering pull even though the front wheels may be perfectly aligned.

Rear toe, like front toe, is a critical tire wear angle. If toed-in or toed-out, the rear tires scuff just like the front ones. Either condition can also contribute to steering instability as well as reduced braking effectiveness. (Keep this in mind with antilock brake systems.)

Like camber, rear toe is not a static alignment angle. It changes as the suspension goes through jounce and rebound. It also changes in response to rolling resistance and the application of engine torque. With four-wheel drive vehicles, the front wheels tend to toe-in under power while the rear wheels toe-out in response to rolling resistance and suspension compliance. With rear-wheel drive vehicles, the opposite happens: the front wheels toe-out while the rear wheels on an independent suspension try to toe-in as they push the vehicle ahead.

Wheels and Tires

Shop Manual
Chapter 15, page 648

A vehicle's tires, wheels, and suspension and steering systems provide the contact between the driver and the road. They allow the driver to safely maneuver the vehicle in all types of conditions as well as ride in comfort and security. Tire design has improved dramatically during the past few years. Modern tires require increased attention to achieve their full potential of extended service and correct ride control. Tire wear that is uneven or premature is usually a good indicator of problems in the steering and suspension system. Tires become not only a good diagnostic aid to a technician, but they also can be clear evidence to the customer for the need to service the front end.

The primary purpose of tires is to provide traction. Tires also help the suspension absorb road shocks, but this is a side benefit. They must perform under a variety of conditions. The road might be wet or dry; paved with asphalt, concrete, or gravel; or there might be no road at all. The car might be traveling slowly on a straight road, or moving quickly through curves or over hills. All of these conditions call for special requirements that must be present, at least to some degree, in all tires.

In addition to providing good traction, tires are also designed to carry the weight of the vehicle, to withstand side thrust over varying speeds and conditions, and to transfer braking and driving torque to the road.

There are many different designs of tires available today (Figure 15-41), the most common of which is the **radial ply**. Radial ply tires have body cords that extend from bead to bead at an angle of

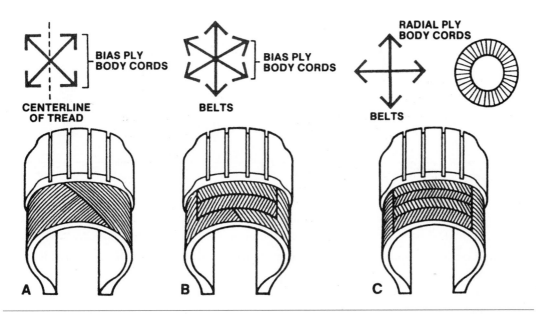

Figure 15-41 Three types of tire construction: (A) bias ply; (B) bias belted; and (C) radial ply.

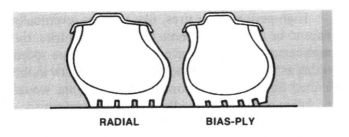

Figure 15-42 A radial tire's highly flexible sidewalls give maximum tread contact area during fast, hard turning.

about 90 degrees "radial" to the tire circumferential centerline—plus two or more layers of relatively inflexible belts under the tread. This construction of various combinations of rayon, nylon, fiberglass, and steel gives greater strength to the tread area and flexibility to the sidewall (Figure 15-42). The belts restrict tread motion during contact with the road, thus improving tread life and traction. Radial ply tires also offer greater fuel economy, increased skid resistance, and more positive braking.

Although the newer synthetics are being used more frequently in radial tires, steel is still the most popular belt material. Bias ply and belted bias are available in all cord materials mentioned earlier, except amarid and kevlar. Non-radial belts are usually of the same material as the sidewalls.

Combining Tire Types

As a general rule, tires should be replaced with the same size designation or an approved optional size as recommended by the auto or tire manufacturer. In addition to following the vehicle manufacturer's recommendations for tire size, type, inflation pressures, and rotation patterns, the following points should be observed:

1. Never mix size or construction types on the same axle.
2. Tires on the same axle should be of approximately equal tread depth.
3. All tires on station wagons and all other vehicles used for trailer towing should be of the same size, type, and load rating.
4. On some vehicles, the use of radial tires might hinder ride quality and control due to the design of the suspension system.

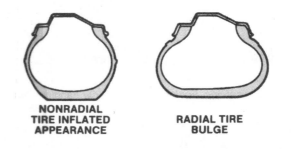

NONRADIAL
TIRE INFLATED
APPEARANCE

RADIAL TIRE
BULGE

Figure 15-43 Effects of inflation on tires.

5. New tires should be installed in pairs on the same axle. When replacing only one tire, it should be paired with the tire having the most tread to equalize braking traction.

6. If radial tires are used on the car, combine them with radial snow tires on the driving wheels.

7. Snow tires should be of a size and type equivalent to the other tires on the vehicle. Otherwise the safety and handling of the vehicle might be adversely affected.

Tire Care

To maximize tire performance, inspect for signs of improper inflation and uneven wear, which can indicate a need for balancing, rotation, or front suspension alignment. Tires should also be checked frequently for cuts, stone bruises, abrasions, blisters, and for objects that might have become imbedded in the tread. More frequent inspections are recommended when rapid or extreme temperature changes occur, or where road surfaces are rough or occasionally littered with debris.

A properly inflated tire (Figure 15-43) gives the best tire life, riding comfort, handling stability, and gas mileage for normal driving conditions. Too little air pressure can result in tire squeal, hard steering, excessive tire heat, abnormal tire wear, and increased fuel consumption by as much as 10%. An under-inflated tire shows maximum wear on the outside edges of the tread. There is little or no wear in the center. Conversely, an overinflated tire shows its wear in the center of the tread and little wear on the outside edges. A higher tire inflation pressure than recommended can cause a hard ride, tire bruising, and rapid wear at the center of the tire.

Tire/Wheel Balance

Proper front-end alignment allows the tires to roll straight without excessive tread wear. The wheels can go out of alignment from striking raised objects or pot holes. Misalignment subjects the tires to uneven and/or irregular wear. An out-of-balance condition can also cause increased wear on the ball joints, as well as deterioration of shock absorbers and other suspension components.

Should an inspection show uneven or irregular tire wear, wheel alignment and balance service is a must. Wheel balancing distributes weights along the wheel rim which counteract heavy spots in the wheels and tires and allow them to roll smoothly without vibration. There are two types of wheel imbalance: static and dynamic.

Static balance is the equal distribution of weight around the wheel. Wheels that are statically unbalanced cause a bouncing action called **wheel tramp** (Figure 15-44). This condition eventually causes uneven tire wear. As the name implies, static balance is balancing a wheel at rest. This is done by adding a compensating weight. Static balance is achieved when the wheel does not rotate by itself regardless of the position in which it is placed on its axis. A statically unbalanced wheel tends to rotate by itself until the heavy portion is down.

Dynamic balance is the equal distribution of weight on each side of the centerline. When the tire spins there is no tendency for the assembly to move from side to side. Wheels that are dynamically unbalanced can cause **wheel shimmy** and a wear pattern (Figure 15-45). Dynamic

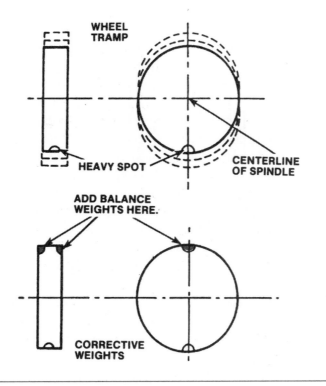

Figure 15-44 Static unbalance causes wheel tramp.

balance, simply stated, is balancing a wheel in motion. Once a wheel starts to rotate and is in motion, the static weights try to reach the true plane of rotation of the wheel because of the action of centrifugal force. In an attempt to reach the true plane of rotation when there is an imbalance, the static weights force the spindle to one side.

At 180 degrees of wheel rotation, static weights kick the spindle in the opposite direction. The resultant side thrusts cause the wheel assembly to wobble or wiggle. When severe enough, as already mentioned, it causes vibration and front-wheel shimmy.

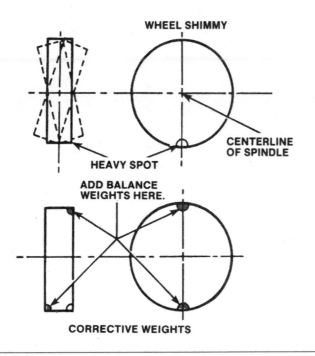

Figure 15-45 Dynamic unbalance causes wheel shimmy.

Summary

❏ The clutch, located between the transmission and the engine, provides a mechanical coupling between the engine flywheel and the transmission's input shaft. All manual transmissions and transaxles require a clutch.

❏ The flywheel, an important part of the engine, is also the main driving member of the clutch.

❏ The clutch disc receives the driving motion from the flywheel and pressure plate assembly and transmits that motion to the transmission input shaft.

❏ The two-fold purpose of the pressure plate assembly is to squeeze the clutch disc onto the flywheel and to move away from the clutch disc so that the disc can stop rotating.

❏ The clutch release bearing, also called a throwout bearing, smoothly and quietly moves the pressure plate release levers or diaphragm spring through the engagement and disengagement processes.

❏ The clutch fork moves the release bearing and hub back and forth. It is controlled by the clutch pedal and linkage.

❏ Clutch linkage can be mechanical or hydraulic. Mechanical linkage is divided into two types: shaft and lever linkage and cable linkage.

❏ A transmission or transaxle uses meshed gears of various sizes to give the engine a mechanical advantage over its driving wheels.

❏ Transaxles contain the gear train plus the differential gearing needed to produce the final gear ratios. Transaxles are commonly used on front-wheel drive vehicles.

❏ Transmissions are normally used on rear-wheel drive vehicles.

❏ All vehicles use a gearset in a differential to provide additional gear reduction (torque increase) above and beyond what the transmission or transaxle gearing can produce.

❏ Modern automatic transmissions use a computer to match the demand for acceleration with engine speed, wheel speed, and load conditions. It then chooses the proper gear ratio and, if necessary, initiates a gear change.

❏ The torque converter is a fluid clutch used to transfer engine torque from the engine to the transmission. It automatically engages and disengages power transfer from the engine to the transmission in relation to engine rpm. It consists of three elements: the impeller (input), turbine (output), and stator (torque multiplier). Two types of oil flow take place inside the torque converter: rotary and vortex flow. An overrunning clutch keeps the stator assembly from rotating in one direction and permits overrunning when turned in the opposite direction.

❏ A lockup torque converter eliminates the 10% slip that takes place between the impeller and turbine at the coupling stage of operation. There are two types: centrifugal lockup clutch and the more popular piston lockup clutch.

❏ Planetary gearsets transfer power and generate torque from the engine to the drive axle. Compound gearsets combine two simple planetary gearsets so that load can be spread over a greater number of teeth for strength and also to obtain the largest number of gear ratios possible in a compact area. A simple planetary gearset consists of a sun gear, a carrier with planetary pinions mounted to it, and an internally toothed ring gear.

❏ Planetary gear controls include transmission bands, servos, and clutches. A band is a braking assembly positioned around a drum. There are two types: single wrap and double wrap.

❏ Simple and compound servos are used to engage bands. Transmission clutches, either overrunning or multiple-disc, are capable of both holding and driving members.

❏ The valve body is the control center of the automatic transmission. It is made of two or three main parts. Internally, the valve body has many fluid passages called worn tracks.

❏ A shift schedule contains the actual shift points to be used by the computer according to the input data it receives from the sensors. Its logic chooses the proper shift schedule for the current conditions of the transmission.

- Front-wheel drive axles transfer engine torque generally from the transaxle differential to the front wheels.
- Constant velocity (CV) joints provide the necessary transfer of uniform torque and a constant speed while operating through a wide range of angles.
- A differential is a geared mechanism located between the driving axles of a vehicle. Its job is to direct power flow to the driving axles. Differentials are used in all types of powertrains.
- The major components of an air conditioning system are: compressor, condenser, receiver/dryer or accumulator, expansion valve or orifice tube, and evaporator.
- The compressor is the heart of an automotive air conditioning system. It separates the high-pressure and low-pressure sides of the system. The primary purpose of the unit is to draw the low-pressure vapor from the evaporator and compress this vapor into high-temperature, high-pressure vapor. This action results in the refrigerant having a higher temperature than surrounding air, enabling the condenser to condense the vapor back to liquid.
- The secondary purpose of the compressor is to circulate or pump the refrigerant through the condenser under the different pressures required for proper operation. The compressor is located in the engine compartment.
- Among the many controls used to monitor and maintain the compressor during its operational cycle within various systems are the pressure relief valve, the low- and the high-pressure cutout switches, the ambient temperature switch, and the thermostatic switch. Each of these represents the most common protective control devices designed to ensure safe and reliable operation of the compressor.
- Cruise control is used to mechanically or electronically control the position of the throttle during highway operation. These systems help the operator to maintain a constant speed without having to apply foot pressure to the accelerator pedal. A cruise control switch is used to engage or disengage the system.
- There are several common parts used on cruise control systems. The servo is connected to the throttle linkage to control its position. It uses vacuum to move a wire or chain connected to the throttle mechanism. The transducer is used to control the vacuum sent to the servo. A brake-activated switch is used to disengage the system when the brakes are applied. The electrical circuit is used to control valves, which in turn control the vacuum to the transducer.
- Electronic cruise control systems use several additional parts. These include the electronic control module, a vehicle speed sensor (VSS), a clutch switch, and an accumulator used to store vacuum.
- The four factors that determine a vehicle's braking power are pressure, which is provided by the hydraulic system; coefficient of friction, which represents the frictional relationship between pads and rotors or shoes and drums and is engineered to ensure optimum performance; frictional contact surface, meaning that bigger brakes stop a car more quickly than smaller brakes; and head dissipation, which is necessary to prevent brake fade.
- The drum is mounted to the wheel hub. When the brakes are applied, a wheel cylinder uses hydraulic power to press two brake shoes against the inside surface of the drum. The resulting friction between the shoe's lining and drum slows the drum and wheel.
- Caster is the angle of the steering axis of a wheel from the vertical, as viewed from the side of the vehicle. Tilting the wheel forward is negative caster. Tilting backward is positive.
- Camber is the angle represented by the tilt of either the front or rear wheels inward or outward from the vertical as viewed from the front of the car.
- Toe is the distance comparison between the leading edge and trailing edge of the front tires. If the edge distance is less, then there is toe-in. If it is greater, there is toe-out.
- The primary purpose of tires is to provide traction. They also are designed to carry the weight of the vehicle, to withstand side thrust over varying speeds and conditions, to transfer braking and driving torque to the road, and to absorb much of the rock shock from surface irregularities.
- There are two types of wheel balancing: static balance and dynamic balance.

ASE Style Review Questions

Short Answer Essays

1. What determines whether a conventional transmission or a transaxle is used?

2. Explain the relationship between output speed and torque from a set of gears.

3. Define final drive gear.

4. What component in an electronic cruise control system is used to monitor or sense vehicle speed?

5. Define dynamic and static wheel balance.

6. Briefly describe how an automatic transmission knows when to shift gears.

7. Explain the primary purpose of the torsional coil springs in the transmission's clutch disc.

8. Describe the purpose of a brake wheel cylinder.

9. Explain what happens during the coupling stage of conventional torque converter operation.

10. Explain what is necessary for torque converter clutch lockup engagement to take place.

Fill-in-the-Blanks

1. The _____ converts the force on the brake pedal to hydraulic pressure that is sent to the individual wheel brake units.

2. The most commonly used brake system on the front wheels of a vehicle is _____ brakes.

3. The device on a cruise control system used to monitor the amount of vacuum in the system is the _____.

4. The main job of tires is to provide _____.

5. The drive shaft component that provides a means of connecting two or more shafts together is the _____.

6. The clutch, or friction, disc is connected to the _____.

7. When the clutch is disengaged, the power flow stops at the _____.

8. The friction that is generated between a vehicle's tires and the road when the vehicle is in a non-skidding state is called _____ friction.

9. Automatic transmissions use a _____ _____ instead of a clutch to transfer power from the flywheel to the transmission's input shaft.

10. A _____ _____ is electrically connected in series with the compressor electromagnetic clutch to control it.

ASE Style Questions

1. While discussing automatic transmissions:
Technician A says improper shifting can be caused by a defective vacuum modulator.
Technician B says improper shifting can be caused by a malfunction in the governor.
Who is correct?
   A. A only
   B. B only
   C. Both A and B
   D. Neither A nor B

2. *Technician A* says shift solenoids direct fluid flow to and away from the various apply devices in the transmission.
Technician B says shift solenoids are used to mechanically apply a friction band or multiple-disc clutch assembly.
Who is correct?
   A. A only
   B. B only
   C. Both A and B
   D. Neither A nor B

3. *Technician A* says the electronically controlled lockup torque converter improves fuel economy.
Technician B says it prevents lockup from occurring during certain engine modes.
Who is correct?
   A. A only
   B. B only
   C. Both A and B
   D. Neither A nor B

4. *Technician A* says throttle position is an important input in most electronic shift control systems.
Technician B says vehicle speed is an important input for most electronic shift control systems.
Who is correct?
   A. A only
   B. B only
   C. Both A and B
   D. Neither A nor B

5. *Technician A* says that throttle pressure influences the vehicle speed at which automatic shifts take place.
Technician B says that throttle pressure is opposed by governor pressure.
Who is correct?
   A. A only
   B. B only
   C. Both A and B
   D. Neither A nor B

6. While discussing cruise control systems:
Technician A says the servo is used to adjust and control the position of the throttle.
Technician B says a transducer is used to adjust and control the position of the throttle.
Who is correct?
   A. A only
   B. B only
   C. Both A and B
   D. Neither A nor B

7. While discussing tires:
Technician A says tire inflation pressure directly affects traction.
Technician B says the recommended tire pressures are often lower than maximum pressures.
Who is correct?
   A. A only
   B. B only
   C. Both A and B
   D. Neither A nor B

8. While discussing the purpose of a clutch pressure plate:
Technician A says that the pressure plate assembly squeezes the clutch disc onto the flywheel.
Technician B says that the pressure plate moves away from the clutch disc so that the disc can stop rotating.
Who is correct?
   A. A only
   B. B only
   C. Both A and B
   D. Neither A nor B

9. *Technician A* says that a transaxle is used in combination with a drive shaft in front-wheel drive vehicles.
Technician B says the transaxle eliminates the need for a final drive gearset.
Who is correct?
   A. A only
   B. B only
   C. Both A and B
   D. Neither A nor B

10. *Technician A* says that a gearset can decrease speed and increase torque.
Technician B says a gearset can increase speed and increase torque.
Who is correct?
   A. A only
   B. B only
   C. Both A and B
   D. Neither A nor B

APPENDIX A

Abbreviations

The following abbreviations are some of the more common ones used today in the automotive industry.

TABLE 1—CROSS REFERENCE AND LOOK UP

Existing Usage	Acceptable Usage	Acceptable Acronized Usage
A/C (Air Conditioning)	Air Conditioning	A/C
A/C Cycling Switch	Air Conditioning Cycling Switch	A/C Cycling Switch
A/T (Automatic Transaxle)	Automatic Transaxle[1]	A/T[1]
A/T (Automatic Transmission)	Automatic Transmission[1]	A/T[1]
AAT (Ambient Air Temperature)	**Ambient Air Temperature**	**AAT**
AC (Air Conditioning)	Air Conditioning	A/C
ACC (Air Conditioning Clutch)	Air Conditioning Clutch	A/C Clutch
Accelerator	Accelerator Pedal	AP
Accelerator Pedal Position	**Accelerator Pedal Position[1]**	**APP[1]**
ACCS (Air Conditioning Cyclic Switch)	Air Conditioning Cycling Switch	A/C Cycling Switch
ACH (Air Cleaner Housing)	Air Cleaner Housing[1]	ACL Housing1
ACL (Air Cleaner)	Air Cleaner[1]	ACL[1]
ACL (Air Cleaner) Element	Air Cleaner Element[1]	ACL Element[1]
ACL (Air Cleaner) Housing	Air Cleaner Housing[1]	ACL Housing[1]
ACL (Air Cleaner) Housing Cover	Air Cleaner Housing Cover[1]	ACL Housing Cover[1]
ACS (Air Conditioning System)	Air Conditioning System	A/C System
ACT (Air Charge Temperature)	Intake Air Temperature[1]	IAT[1]
Adaptive Fuel Strategy	Fuel Trim[1]	FT[1]
AFC (Air Flow Control)	Mass Air Flow	MAF
AFC (Air Flow Control(	Volume Air Flow	VAF
AFS (Air Flow Sensor)	Mass Air Flow Sensor	MAF Sensor
AFS (Air Flow Sensor)	Volume Air Flow Sensor	VAF Sensor
After Cooler	Charge Air Cooler[1]	CAC[1]
AI (Air Injection)	Secondary Air Injection[1]	AIR[1]
AIP (Air Injection Pump)	Secondary Air Injection Pump[1]	AIR Pump[1]
AIR (Air Injection Reactor)	Pulsed Secondary Air Injection[1]	PAIR[1]
AIR (Air Injection Reactor)	Secondary Air Injection[1]	AIR[1]
AIRB (Secondary Air Injection Bypass)	Secondary Air Injection Bypass[1]	AIR Bypass[1]
AIRD (Secondary Air Injection Diverter)	Secondary Air Injection Diverter[1]	AIR Diverter[1]
Air Cleaner	Air Cleaner[1]	ACL[1]
Air Cleaner Element	Air Cleaner Element[1]	ACL Element[1]
Air Cleaner Housing	Air Cleaner Housing[1]	ACL Housing[1]
Air Cleaner Housing Cover	Air Cleaner Housing Cover[1]	ACL Housing Cover[1]
Air Conditioning	Air Conditioning	A/C
Air Conditioning Sensor	Air Conditioning Sensor	A/C Sensor
Air Control Valve	Secondary Air Injection Control Valve[1]	AIR Control Valve[1]
Air Flow Meter	Mass Air Flow Sensor[1]	MAF Sensor[1]
Air Flow Meter	Volume Air Flow Sensor[1]	VAF Sensor[1]
Air Intake System	Intake Air System[1]	IA System[1]
Air Flow Sensor	Mass Air Flow Sensor[1]	MAF Sensor[1]
Air Management 1	Secondary Air Injection Bypass[1]	AIR Bypass[1]
Air Management 2	Secondary Air Injection Diverter[1]	AIR Diverter[1]
Air Temperature Sensor	Intake Air Temperature Sensor[1]	IAT Sensor[1]
Air Valve	Idle Air Control Valve[1]	IAC Valve[1]
AIV (Air Injection Valve)	Pulsed Secondary Air Injection[1]	PAIR[1]
ALCL (Assembly Line Communication Link)	Data Link Connector[1]	DLC[1]
Alcohol Concentration Sensor	Flexible Fuel Sensor[1]	FF Sensor[1]
ALDL (Assembly Line Diagnostic Link)	Data Link Connector[1]	DLC[1]

TABLE 1—CROSS REFERENCE AND LOOK UP (CONTINUED)

Existing Usage	Acceptable Usage	Acceptable Acronized Usage
ALT (Alternator)	Generator	GEN
Alternator	Generator	GEN
Ambient Air Temperature	**Ambient Air Temperature**	**AAT**
AM1 (Air Management 1)	Secondary Air Injection Bypass[1]	AIR Bypass[1]
AM2 (Air Management 2)	Secondary Air Injection Diverter[1]	AIR Diverter[1]
APP (Accelerator Pedal Position)	**Accelerator Pedal Position[1]**	**APP[1]**
APS (Absolute Pressure Sensor)	Barometric Pressure Sensor[1]	BARO Sensor[1]
ATS (Air Temperature Sensor)	Intake Air Temperature Sensor[1]	IAT Sensor[1]
Automatic Transaxle	Automatic Transaxle[1]	A/T[1]
Automatic Transmission	Automatic Transmission[1]	A/T[1]
B+ (Battery Positive Voltage)	Battery Positive Voltage	B+
Backpressure Transducer	Exhaust Gas Recirculation Backpressure Transducer[1]	EGR Backpressure Transducer[1]
BARO (Barometric Pressure)	Barometric Pressure[1]	BARO[1]
Barometric Pressure Sensor	Barometric Pressure Sensor[1]	BARO Sensor[1]
Battery Positive Voltage	Battery Positive Voltage	B+
BLM (Block Learn Memory)	Long Term Fuel Trim[1]	Long Term FT[1]
BLM (Block Learn Multiplier)	Long Term Fuel Trim[1]	Long Term FT[1]
BLM (Block Learn Matrix)	Long Term Fuel Trim[1]	Long Term FT[1]
Block Learn Integrator	**Long Term Fuel Trim[1]**	**Long Term FT[1]**
Block Learn Matrix	Long Term Fuel Trim[1]	Long Term FT[1]
Block Learn Memory	Long Term Fuel Trim[1]	Long Term FT[1]
Block Learn Multiplier	Long Term Fuel Trim[1]	Long Term FT
BP (Barometric Pressure) Sensor	Barometric Pressure Sensor[1]	BARO Sensor[1]
BPP (Brake Pedal Position)	**Brake Pedal Position[1]**	**BPP[1]**
Brake Pressure	**Brake Pressure**	**Brake Pressure**
Brake Pedal Position	**Brake Pedal Position[1]**	**BPP[1]**
C3I (Computer Controlled Coil Ignition)	Electronic Ignition[1]	EI[1]
CAC (Charge Air Cooler)	Charge Air Cooler[1]	CAC[1]
Calculated Load Value	**Calculated Load Value**	**LOAD**
Camshaft Position	Camshaft Position[1]	CMP[1]
Camshaft Position Actuator	**Camshaft Position Actuator[1]**	**CMP Actuator[1]**
Camshaft Position Controller	**Camshaft Position Actuator[1]**	**CMP Actuator[1]**
Camshaft Position Sensor	Camshaft Position Sensor[1]	CMP Sensor[1]
Camshaft Sensor	Camshaft Position Sensor[1]	CMP Sensor[1]
Camshaft Timing Actuator	**Camshaft Position Actuator[1]**	**CMP Actuator[1]**
Canister	Canister[1]	Canister[1]
Canister	Evaporative Emission Canister[1]	EVAP Canister[1]
Canister Purge	**Evaporative Emission Canister Purge[1]**	**EVAP Canister Purge[1]**
Canister Purge Vacuum Switching Valve	Evaporative Emission Canister Purge Valve[1]	EVAP Canister Purge Valve[1]
Canister Purge Valve	Evaporative Emission Canister Purge Valve[1]	EVAP Canister Purge Valve[1]
Canister Purge VSV (Vacuum Switching Valve)	Evaporative Emission Canister Purge Valve[1]	EVAP Canister Purge Valve[1]
CANP (Canister Purge)	Evaporative Emission Canister Purge[1]	EVAP Canister Purge[1]
CARB (Carburetor)	Carburetor[1]	CARB[1]
Carburetor	Carburetor[1]	CARB[1]
Catalytic Converter Heater	**Catalytic Converter Heater**	**Catalytic Converter Heater**
CCC (Converter Clutch Control)	Torque Converter Clutch[1]	TCC[1]
CCO (Converter Clutch Override)	Torque Converter Clutch[1]	TCC[1]
CCS (Coast Clutch Solenoid)	**Coast Clutch Solenoid**	**CCS**

TABLE 1—CROSS REFERENCE AND LOOK UP (CONTINUED)

Existing Usage	Acceptable Usage	Acceptable Acronized Usage
CCS (Coast Clutch Solenoid) Valve	**Coast Clutch Solenoid Valve**	**CCS Valve**
CCRM (Constant Control Relay Module)	**Constant Control RM**	**Constant Control RM**
CDI (Capacitive Discharge Ignition)	Distributor Ignition[1]	DI[1]
CDROM (Compact Disc Read Only Memory)	Compact Disc Read Only Memory[1]	CDROM[1]
CES (Clutch Engage Switch)	Clutch Pedal Position Switch[1]	CPP Switch[1]
Central Multiport Fuel Injection	Central Multiport Fuel Injection[1]	Central MFI[1]
Central Sequential Multiport Fuel Injection	**Central Sequential Multiport Fuel Injection**	**Central SFI**
CFI (Continuous Fuel Injection)	Continuous Fuel Injection[1]	CFI[1]
CFI (Central Fuel Injection)	Throttle Body Fuel Injection[1]	TBI[1]
CFV	**Critical Flow Venturi**	**CFV**
Charcoal Canister	Evaporative Emission Canister	EVAP Canister[1]
Charge Air Cooler	Charge Air Cooler	CAC[1]
Check Engine	Service Reminder Indicator[1]	SRI[1]
Check Engine	Malfunction Indicator Lamp[1]	MIL[1]
CID (Cylinder Identification) Sensor	Camshaft Position Sensor	CMP Sensor[1]
CIS (Continuous Injection System)	Continuous Fuel Injection[1]	CFI[1]
CIS-E (Continuous Injection System Electronic)	Continuous Fuel Injection[1]	CFI[1]
CKP (Crankshaft Position)	Crankshaft Position[1]	CKP[1]
CKP (Crankshaft Position) Sensor	Crankshaft Position Sensor[1]	CKP Sensor[1]
CL (Closed Loop)	Closed Loop[1]	CL[1]
Closed Bowl Distributor	Distributor Ignition[1]	DI[1]
Closed Throttle Position	Closed Throttle Position[1]	CTP[1]
Closed Throttle Switch	Closed Throttle Position Switch[1]	CTP Switch[1]
CLS (Closed Loop System)	Closed Loop[1]	CL[1]
CLV	**Calculated Load Value**	**LOAD**
Clutch Engage Switch	Clutch Pedal Position Switch[1]	CPP Switch[1]
Clutch Pedal Position Switch	Clutch Pedal Position Switch[1]	CPP Switch[1]
Clutch Start Switch	Clutch Pedal Position Switch[1]	CPP Switch[1]
Clutch Switch	Clutch Pedal Position Switch[1]	CPP Switch[1]
CMFI (Central Multiport Fuel Injection)	Central Multiport Fuel Injection[1]	Central MFI[1]
CMP (Camshaft Position)	Camshaft Position[1]	CMP[1]
CMP (Camshaft Position) Sensor	Camshaft Position Sensor[1]	CMP Sensor[1]
COC (Continuous Oxidation Catalyst)	Oxidation Catalytic Converter[1]	OC[1]
Coast Clutch Solenoid	**Coast Clutch Solenoid**	**CCS**
Coast Clutch Solenoid Valve	**Coast Clutch Solenoid Valve**	**CCS Valve**
Condenser	Distributor Ignition Capacitor[1]	DI Capacitor[1]
Constant Control Relay Module	**Relay Module**	**RM**
Constant Volume Sampler	**Constant Volume Sampler**	**CVS**
Continuous Fuel Injection	Continuous Fuel Injection[1]	CFI[1]
Continuous Injection System	Continuous Fuel Injection System[1]	CFI System[1]
Continuous Injection System-E	Electronic Continuous Fuel Injection System[1]	Electronic CFI System[1]
Continuous Trap Oxidizer	Continuous Trap Oxidizer[1]	CTOX[1]
Coolant Temperature Sensor	Engine Coolant Temperature Sensor[1]	ECT Sensor[1]
CP (Crankshaft Position)	Crankshaft Position[1]	CKP[1]
CPP (Clutch Pedal Position)	Clutch Pedal Position[1]	CPP[1]
CPP (Clutch Pedal Position) Switch	Clutch Pedal Position Switch	CPP Switch[1]
CPS (Camshaft Position Sensor)	Camshaft Position Sensor[1]	CMP Sensor[1]
CPS (Crankshaft Position Sensor)	Crankshaft Position Sensor[1]	CKP Sensor[1]
Crank Angle Sensor	Crankshaft Position Sensor[1]	CKP Sensor[1]

TABLE 1—CROSS REFERENCE AND LOOK UP (CONTINUED)

Existing Usage	Acceptable Usage	Acceptable Acronized Usage
Crankshaft Position	Crankshaft Position[1]	CKP[1]
Crankshaft Position Sensor	Crankshaft Position Sensor[1]	CKP Sensor[1]
Crankshaft Speed	Engine Speed[1]	RPM[1]
Crankshaft Speed Sensor	Engine Speed Sensor[1]	RPM Sensor[1]
Critical Flow Venturi	**Critical Flow Venturi**	**CFV**
CTO (Continuous Trap Oxidizer)	Continuous Trap Oxidizer[1]	CTOX[1]
CTOX (Continuous Trap Oxidizer)	Continuous Trap Oxidizer[1]	CTOX[1]
CTP (Closed Throttle Position)	Closed Throttle Position[1]	CTP[1]
CTS (Coolant Temperature Sensor)	Engine Coolant Temperature Sensor[1]	ECT Sensor[1]
CTS (Coolant Temperature Switch)	Engine Coolant Temperature Switch[1]	ECT Switch[1]
CVS	**Constant Volume Sampler**	**CVS**
Cylinder ID (Identification) Sensor	Camshaft Position Sensor[1]	CMP Sensor[1]
D-Jetronic	Multiport Fuel Injection[1]	MFI[1]
Data Link Connector	Data Link Connector[1]	DLC[1]
Detonation Sensor	Knock Sensor[1]	KS[1]
DFI (Direct Fuel Injection)	Direct Fuel Injection[1]	DFI[1]
DFI (Digital Fuel Injection)	Multiport Fuel Injection[1]	MFI[1]
DI (Direct Injection)	Direct Fuel Injection[1]	DFI[1]
DI (Distributor Ignition)	Distributor Ignition[1]	DI[1]
DI (Distributor Ignition) Capacitor	Distributor Ignition Capacitor[1]	DI Capacitor[1]
Diagnostic Test Mode	Diagnostic Test Mode[1]	DTM[1]
Diagnostic Trouble Code	Diagnostic Trouble Code[1]	DTC[1]
DID (Direct Injection - Diesel)	Direct Fuel Injection[1]	DFI[1]
Differential Pressure Feedback EGR (Exhaust Gas Recirculation) System	Differential Pressure Feedback Exhaust Gas Recirculation System[1]	Differential Pressure Feedback EGR System[1]
Digital EGR (Exhaust Gas Recirculation)	Exhaust Gas Recirculation[1]	EGR[1]
Direct Fuel Injection	Direct Fuel Injection[1]	DFI[1]
Direct Ignition System	Electronic Ignition System[1]	EI System[1]
DIS (Distributorless Ignition System)	Electronic Ignition System[1]	EI System[1]
DIS (Distributorless Ignition System) Module	Ignition Control Module[1]	ICM[1]
Distance Sensor	Vehicle Speed Sensor[1]	VSS[1]
Distributor Ignition	Distributor Ignition[1]	DI[1]
Distributorless Ignition	Electronic Ignition[1]	EI[1]
DLC (Data Link Connector)	Data Link Connector[1]	DLC[1]
DLI (Distributorless Ignition)	Electronic Ignition[1]	EI[1]
Driver	**Driver**	**Driver**
DS (Detonation Sensor)	Knock Sensor[1]	KS[1]
DTC (Diagnostic Trouble Code)	Diagnostic Trouble Code[1]	DTC[1]
DTM (Diagnostic Test Mode)	Diagnostic Test Mode[1]	DTM[1]
Dual Bed	Three Way + Oxidation Catalytic Converter[1]	TWC+OC[1]
Duty Solenoid for Purge Valve	Evaporative Emission Canister Purge Valve	EVAP Canister Purge Valve[1]
Dynamic Pressure Control	**Dynamic Pressure Control**	**Dynamic PC**
Dynamic Pressure Control Solenoid	**Dynamic Pressure Control Solenoid[1]**	**Dynamic PC Solenoid[1]**
Dynamic Pressure Control Solenoid Valve	**Dynamic Pressure Control Solenoid Valve[1]**	**Dynamic PC Solenoid Valve[1]**
E2PROM (Electrically Erasable Programmable Read Only Memory)	Electrically Erasable Programmable Read Only Memory[1]	EEPROM[1]
Early Fuel Evaporation	Early Fuel Evaporation[1]	EFE[1]
EATX (Electronic Automatic Transmission/ Transaxle)	Automatic Transmission[1]	A/T[1]
EC (Engine Control)	Engine Control[1]	EC[1]

TABLE 1—CROSS REFERENCE AND LOOK UP (CONTINUED)

Existing Usage	Acceptable Usage	Acceptable Acronized Usage
ECA (Electronic Control Assembly)	Powertrain Control Module[1]	PCM[1]
ECL (Engine Coolant Level)	Engine Coolant Level	ECL
ECM (Engine Control Module)	Engine Control Module[1]	ECM[1]
ECT (Engine Coolant Temperature)	Engine Coolant Temperature[1]	ECT[1]
ECT (Engine Coolant Temperature) Sender	Engine Coolant Temperature Sensor[1]	ECT Sensor[1]
ECT (Engine Coolant Temperature) Sensor	Engine Coolant Temperature Sensor[1]	ECT Sensor[1]
ECT (Engine Coolant Temperature) Switch	Engine Coolant Temperature Switch[1]	ECT Switch[1]
ECU4 (Electronic Control Unit 4)	Powertrain Control Module[1]	PCM[1]
EDF (Electro-Drive Fan) Control	Fan Control	FC
EDIS (Electronic Distributor Ignition System)	Distributor Ignition System[1]	DI System[1]
EDIS (Electronic Distributorless Ignition System)	Electronic Ignition System[1]	EI System[1]
EDIS (Electronic Distributor Ignition System) Module	Distributor Ignition Control Module[1]	Distributor ICM[1]
EEC (Electronic Engine Control)	Engine Control[1]	EC[1]
EEC (Electronic Engine Control) Processor	Powertrain Control Module[1]	PCM[1]
EECS (Evaporative Emission Control System)	Evaporative Emission System[1]	EVAP System[1]
EEPROM (Electrically Erasable Programmable Read Only Memory)	Electrically Erasable Programmable Read Only Memory[1]	EEPROM[1]
EFE (Early Fuel Evaporation)	Early Fuel Evaporation[1]	EFE[1]
EFI (Electronic Fuel Injection)	Multiport Fuel Injection[1]	MFI[1]
EFI (Electronic Fuel Injection)	Throttle Body Fuel Injection[1]	TBI[1]
EGO (Exhaust Gas Oxygen) Sensor	Oxygen Sensor[1]	O2S[1]
EGOS (Exhaust Gas Oxygen Sensor)	Oxygen Sensor[1]	O2S[1]
EGR (Exhaust Gas Recirculation)	Exhaust Gas Recirculation[1]	EGR[1]
EGR (Exhaust Gas Recirculation) Diagnostic Valve	Exhaust Gas Recirculation Diagnostic Valve[1]	EGR Diagnostic Valve[1]
EGR (Exhaust Gas Recirculation) System	Exhaust Gas Recirculation System[1]	EGR System[1]
EGR (Exhaust Gas Recirculation) Thermal Vacuum Valve	Exhaust Gas Recirculation Thermal Vacuum Valve[1]	EGR TVV[1]
EGR (Exhaust Gas Recirculation) Valve	Exhaust Gas Recirculation Valve[1]	EGR Valve[1]
EGR TVV (Exhaust Gas Recirculation Thermal Vacuum Valve)	Exhaust Gas Recirculation Thermal Vacuum Valve[1]	EGR TVV[1]
EGRT (Exhaust Gas Recirculation Temperature)	Exhaust Gas Recirculation Temperature	EGRT[1]
EGRT (Exhaust Gas Recirculation Temperature) Sensor	Exhaust Gas Recirculation Temperature Sensor[1]	EGRT Sensor[1]
EGRV (Exhaust Gas Recirculation Valve)	Exhaust Gas Recirculation Valve[1]	EGR Valve[1]
EGRVC (Exhaust Gas Recirculation Valve Control)	Exhaust Gas Recirculation Valve Control[1]	EGR Valve Control[1]
EGS (Exhaust Gas Sensor)	Oxygen Sensor[1]	O2S[1]
EI (Electronic Ignition) (With Distributor)	Distributor Ignition[1]	DI[1]
EI (Electronic Ignition) (Without Distributor)	Electronic Ignition[1]	EI[1]
Electrically Erasable Programmable Read Only Memory	Electrically Erasable Programmable Read Only Memory[1]	EEPROM[1]
Electronic Engine Control	Electronic Engine Control[1]	Electronic EC[1]
Electronic Ignition	Electronic Ignition[1]	EI[1]
Electronic Spark Advance	Ignition Control[1]	IC[1]
Electronic Spark Timing	Ignition Control[1]	IC[1]
EM (Engine Modification)	Engine Modification[1]	EM[1]
EMR (Engine Maintenance Reminder)	Service Reminder Indicator[1]	SRI[1]
Engine Control	Engine Control[1]	EC[1]
Engine Coolant Fan Control	Fan Control	FC
Engine Coolant Level	Engine Coolant Level	ECL
Engine Coolant Level Indicator	Engine Coolant Level Indicator	ECL Indicator

TABLE 1—CROSS REFERENCE AND LOOK UP (CONTINUED)

Existing Usage	Acceptable Usage	Acceptable Acronized Usage
Engine Coolant Temperature	Engine Coolant Temperature[1]	ECT[1]
Engine Coolant Temperature Sender	Engine Coolant Temperature Sensor[1]	ECT Sensor[1]
Engine Coolant Temperature Sensor	Engine Coolant Temperature Sensor[1]	ECT Sensor[1]
Engine Coolant Temperature Switch	Engine Coolant Temperature Switch[1]	ECT Switch[1]
Engine Modification	Engine Modification[1]	EM[1]
Engine Oil Pressure Sender	**Engine Oil Pressure** Sensor	EOP Sensor
Engine Oil Pressure Sensor	**Engine Oil Pressure** Sensor	EOP Sensor
Engine Oil Pressure Switch	**Engine Oil Pressure** Switch	EOP Switch
Engine Oil Temperature	**Engine Oil Temperature**	EOT
Engine Speed	Engine Speed[1]	RPM[1]
EOS (Exhaust Oxygen Sensor)	Oxygen Sensor[1]	O2S[1]
EOT (Engine Oil Temperature)	**Engine Oil Temperature**	EOT
EP (Exhaust Pressure)	**Exhaust Pressure**	EP
EPROM (Erasable Programmable Read Only Memory)	Erasable Programmable Read Only Memory[1]	EPROM[1]
Erasable Programmable Read Only Memory	Erasable Programmable Read Only Memory[1]	EPROM[1]
ESA (Electronic Spark Advance)	Ignition Control[1]	IC[1]
ESAC (Electronic Spark Advance Control)	Distributor Ignition[1]	DI[1]
EST (Electronic Spark Timing)	Ignition Control[1]	IC[1]
EVAP (Evaporate Emission) CANP (Canister Purge)	**Evaporative Emission** Canister Purge[1]	**EVAP** Canister Purge[1]
EVAP (Evaporative Emission)	Evaporative Emission[1]	EVAP[1]
EVAP (Evaporative Emission) Canister	Evaporative Emission Canister[1]	EVAP Canister[1]
EVAP (Evaporative Emission) Purge Valve	Evaporative Emission Canister Purge Valve[1]	EVAP Canister Purge Valve[1]
Evaporative Emission	Evaporative Emission[1]	EVAP[1]
Evaporative Emission Canister	Evaporative Emission Canister[1]	EVAP Canister[1]
EVP (Exhaust Gas Recirculation Valve Position) Sensor	Exhaust Gas Recirculation Valve Position Sensor[1]	EGR Valve Position Sensor[1]
EVR (Exhaust Gas Recirculation Vacuum Regulator) Solenoid	Exhaust Gas Recirculation Vacuum Regulator Solenoid[1]	EGR Vacuum Regulator Solenoid[1]
EVRV (Exhaust Gas Recirculation Vacuum Regulator Valve)	Exhaust Gas Recirculation Vacuum Regulator Valve[1]	EGR Vacuum Regulator Valve[1]
Exhaust Gas Recirculation	Exhaust Gas Recirculation[1]	EGR[1]
Exhaust Gas Recirculation Temperature	Exhaust Gas Recirculation Temperature[1]	EGRT[1]
Exhaust Gas Recirculation Temperature Sensor	Exhaust Gas Recirculation Temperature Sensor[1]	EGRT Sensor[1]
Exhaust Gas Recirculation Vacuum Solenoid Valve Regulator	**Exhaust Gas Recirculation** Vacuum Regulator Solenoid Valve[1]	**EGR Vacuum Regulator Solenoid Valve**[1]
Exhaust Gas Recirculation Vacuum Regulator Valve	**Exhaust Gas Recirculation** Vacuum Regulator Valve[1]	**EGR Vacuum Regulator Valve**[1]
Exhaust Gas Recirculation Valve	Exhaust Gas Recirculation Valve[1]	EGR Valve[1]
Exhaust Pressure	**Exhaust Pressure**	EP
4GR (Fourth Gear)	Fourth Gear	4GR
4WD (Four Wheel Drive)	**Full Time Four Wheel Drive**	F4WD
4WD (Four Wheel Drive)	**Selectable Four Wheel Drive**	S4WD
F4WD	**Full Time Four Wheel Drive**	F4WD
Fan Control	Fan Control	FC
Fan Control Module	Fan Control Module	FC Module
Fan Control Relay	Fan Control Relay	FC Relay
Fan Motor Control Relay	Fan Control Relay	FC Relay
Fast Idle Thermo Valve	Idle Air Control Thermal Valve[1]	IAC Thermal Valve[1]
FBC (Feed Back Carburetor)	Carburetor[1]	CARB[1]
FBC (Feed Back Control)	Mixture Control[1]	MC[1]

TABLE 1—CROSS REFERENCE AND LOOK UP (CONTINUED)

Existing Usage	Acceptable Usage	Acceptable Acronized Usage
FC (Fan Control)	Fan Control	FC
FC (Fan Control) Relay	Fan Control Relay	FC Relay
FEEPROM (Flash Electrically Erasable Programmable Read Only Memory)	Flash Electrically Erasable Programmable Read Only Memory[1]	FEEPROM[1]
FEPROM (Flash Erasable Programmable Read Only Memory)	Flash Erasable Programmable Read Only Memory[1]	FEPROM[1]
FF (Flexible Fuel)	Flexible Fuel[1]	FF[1]
FI (Fuel Injection)	Central Multiport Fuel Injection[1]	Central MFI[1]
FI (Fuel Injection)	Continuous Fuel Injection[1]	CFI[1]
FI (Fuel Injection)	Direct Fuel Injection[1]	DFI[1]
FI (Fuel Injection)	Indirect Fuel Injection[1]	IFI[1]
FI (Fuel Injection)	Multiport Fuel Injection[1]	MFI[1]
FI (Fuel Injection)	Sequential Multiport Fuel Injection[1]	SFI[1]
FI (Fuel Injection)	Throttle Body Fuel Injection[1]	TBI[1]
Flame Ionization Detector	**Flame Ionization Detector**	**FID**
Flash EEPROM (Electrically Erasable Programmable Read Only Memory)	Flash Electrically Erasable Programmable Read Only Memory[1]	FEEPROM[1]
Flash EPROM (Erasable Programmable Read Only Memory)	Flash Erasable Programmable Read Only Memory[1]	FEPROM[1]
Flexible Fuel	Flexible Fuel[1]	FF[1]
Flexible Fuel Sensor	Flexible Fuel Sensor[1]	FF Sensor
Fourth Gear	Fourth Gear	4GR
FP (Fuel Pump)	Fuel Pump	FP
FP (Fuel Pump) Module	Fuel Pump Module	FP Module
Freeze Frame	**Freeze Frame**	See Table 4
Front Wheel Drive	**Front Wheel Drive**	**FWD**
FRZF (Freeze Frame)	**Freeze Frame**	See Table 4
FT (Fuel Trim)	Fuel Trim[1]	FT[1]
Fuel Charging Station	Throttle Body[1]	TB[1]
Fuel Concentration Sensor	Flexible Fuel Sensor[1]	FF Sensor[1]
Fuel Injection	Central Multiport Fuel Injection[1]	Central MFI[1]
Fuel Injection	Continuous Fuel Injection[1]	CFI[1]
Fuel Injection	Direct Fuel Injection[1]	DFI[1]
Fuel Injection	Indirect Fuel Injection[1]	IFI[1]
Fuel Injection	Multiport Fuel Injection[1]	MFI[1]
Fuel Injection	Sequential Multiport Fuel Injection[1]	SFI[1]
Fuel Injection	Throttle Body Fuel Injection[1]	TBI[1]
Fuel Level Sensor	Fuel Level Sensor	Fuel Level Sensor
Fuel Module	Fuel Pump Module	FP Module
Fuel Pressure	Fuel Pressure[1]	Fuel Pressure[1]
Fuel Pressure	**Fuel Pressure**	See Table 4
Fuel Pressure Regulator	Fuel Pressure Regulator[1]	Fuel Pressure Regulator[1]
Fuel Pump	Fuel Pump	FP
Fuel Pump Relay	Fuel Pump Relay	FP Relay
Fuel Quality Sensor	Flexible Fuel Sensor[1]	FF Sensor[1]
Fuel Regulator	Fuel Pressure Regulator[1]	Fuel Pressure Regulator[1]
Fuel Sender	Fuel Pump Module	FP Module
Fuel Sensor	Fuel Level Sensor	Fuel Level Sensor
Fuel System Status	**Fuel System Status**	See Table 4
FUEL SYS	**Fuel System Status**	See Table 4
Fuel Tank Unit	Fuel Pump Module	FP Module

TABLE 1—CROSS REFERENCE AND LOOK UP (CONTINUED)

Existing Usage	Acceptable Usage	Acceptable Acronized Usage
Fuel Trim	Fuel Trim[1]	FT[1]
Full Time Four Wheel Drive	**Full Time Four Wheel Drive**	**F4WD**
Full Throttle	Wide Open Throttle[1]	WOT[1]
FWD	**Front Wheel Drive**	**FWD**
GCM (Governor Control Module)	Governor Control Module	GCM
GEM (Governor Electronic Module)	Governor Control Module	GCM
GEN (Generator)	Generator	GEN
Generator	Generator	GEN
Glow Plug	**Glow Plug[1]**	**Glow Plug[1]**
GND (Ground)	Ground	GND
Governor	Governor	Governor
Governor Control Module	Governor Control Module	GCM
Governor Electronic Module	Governor Control Module	GCM
Gram Per Mile	**Gram Per Mile**	**GPM**
GRD (Ground)	Ground	GND
Ground	Ground	GND
Heated Oxygen Sensor	Heated Oxygen Sensor[1]	HO2S[1]
HEDF (High Electro-Drive Fan) Control	Fan Control	FC
HEGO (Heated Exhaust Gas Oxygen) Sensor	Heated Oxygen Sensor[1]	HO2S[1]
HEI (High Energy Ignition)	Distributor Ignition[1]	DI[1]
High Speed FC (Fan Control) Switch	High Speed Fan Control Switch	High Speed FC Switch
HO2S (Heated Oxygen Sensor)	Heated Oxygen Sensor[1]	HO2S[1]
HOS (Heated Oxygen Sensor)	Heated Oxygen Sensor[1]	HO2S[1]
Hot Wire Anemometer	Mass Air Flow Sensor[1]	MAF Sensor[1]
IA (Intake Air)	Intake Air	IA
IA (Intake Air) Duct	Intake Air Duct	IA Duct
IAC (Idle Air Control)	Idle Air Control[1]	IAC[1]
IAC (Idle Air Control) Thermal Valve	Idle Air Control Thermal Valve[1]	IAC Thermal Valve[1]
IAC (Idle Air Control) Valve	Idle Air Control Valve[1]	IAC Valve[1]
IACV (Idle Air Control Valve)	Idle Air Control Valve[1]	IAC Valve[1]
IAT (Intake Air Temperature)	Intake Air Temperature[1]	IAT[1]
IAT (Intake Air Temperature) Sensor	Intake Air Temperature Sensor[1]	IAT Sensor[1]
IATS (Intake Air Temperature Sensor)	Intake Air Temperature Sensor[1]	IAT Sensor[1]
IC (Ignition Control)	Ignition Control[1]	IC[1]
ICM (Ignition Control Module)	Ignition Control Module[1]	ICM[1]
ICP (Injection Control Pressure)	**Injection Control Pressure[1]**	**ICP[1]**
IDFI (Indirect Fuel Injection)	Indirect Fuel Injection[1]	IFI[1]
IDI (Integrated Direct Ignition)	Electronic Ignition[1]	EI[1]
IDI (Indirect Diesel Injection)	Indirect Fuel Injection[1]	IFI[1]
Idle Air Bypass Control	Idle Air Control[1]	IAC[1]
Idle Air Control	Idle Air Control[1]	IAC[1]
Idle Air Control Valve	Idle Air Control Valve[1]	IAC Valve[1]
Idle Speed Control	Idle Air Control[1]	IAC[1]
Idle Speed Control	Idle Speed Control[1]	ISC[1]
Idle Speed Control Actuator	Idle Speed Control Actuator[1]	ISC Actuator[1]
IFI (Indirect Fuel Injection)	Indirect Fuel Injection[1]	IFI[1]
IFS (Inertia Fuel Shutoff)	Inertia Fuel Shutoff	IFS
Ignition Control	Ignition Control[1]	IC[1]
Ignition Control Module	Ignition Control Module[1]	ICM[1]
I/M (Inspection and Maintenance)	**Inspection and Maintenance**	**I/M**

TABLE 1—CROSS REFERENCE AND LOOK UP (CONTINUED)

Existing Usage	Acceptable Usage	Acceptable Acronized Usage
IMRC (Intake Manifold Runner Control)	**Intake Manifold Runner Control**	**IMRC**
In Tank Module	Fuel Pump Module	FP Module
Indirect Fuel Injection	Indirect Fuel Injection[1]	IFI[1]
Inertia Fuel Shutoff	Inertia Fuel Shutoff	IFS
Inertia Fuel - Shutoff Switch	Inertia Fuel Shutoff Switch	IFS Switch
Inertia Switch	Inertia Fuel Shutoff Switch	IFS Switch
Injection Control Pressure	**Injection Control Pressure[1]**	**ICP[1]**
Input Shaft Speed	**Input Shaft Speed**	**ISS**
INT (Integrator)	Short Term Fuel Trim[1]	Short Term FT[1]
Inspection and Maintenance	**Inspection and Maintenance**	**I/M**
Intake Air	Intake Air	IA
Intake Air Duct	Intake Air Duct	IA Duct
Intake Air Temperature	Intake Air Temperature[1]	IAT[1]
Intake Air Temperature Sensor	Intake Air Temperature Sensor[1]	IAT Sensor[1]
Intake Manifold Absolute Pressure Sensor	Manifold Absolute Pressure Sensor[1]	MAP Sensor[1]
Intake Manifold Runner Control	**Intake Manifold Runner Control**	**IMRC**
Integrated Relay Module	Relay Module	RM
Integrator	Short Term Fuel Trim[1]	Short Term FT[1]
Inter Cooler	Charge Air Cooler[1]	CAC[1]
ISC (Idle Speed Control)	Idle Air Control[1]	IAC[1]
ISC (Idle Speed Control)	Idle Speed Control[1]	ISC[1]
ISC (Idle Speed Control) Actuator	Idle Speed Control Actuator[1]	ISC Actuator[1]
ISC BPA (Idle Speed Control By Pass Air)	Idle Air Control[1]	IAC
ISC (Idle Speed Control) Solenoid Vacuum Valve	Idle Speed Control Solenoid Vacuum Valve[1]	ISC Solenoid Vacuum Valve[1]
ISS (Input Shaft Speed)	**Input Shaft Speed**	**ISS**
K-Jetronic	Continuous Fuel Injection[1]	CFI[1]
KAM (Keep Alive Memory)	Non Volatile Random Access Memory[1]	NVRAM[1]
KAM (Keep Alive Memory)	Keep Alive Random Access Memory[1]	Keep Alive RAM[1]
KE-Jetronic	Continuous Fuel Injection[1]	CFI[1]
KE-Motronic	Continuous Fuel Injection[1]	CFI[1]
Knock Sensor	Knock Sensor[1]	KS[1]
KS (Knock Sensor)	Knock Sensor[1]	KS[1]
L-Jetronic	Multiport Fuel Injection[1]	MFI[1]
Lambda	Oxygen Sensor[1]	O2S[1]
LH-Jetronic	Multiport Fuel Injection[1]	MFI[1]
Light Off Catalyst	Warm Up Three Way Catalytic Converter[1]	WU-TWC[1]
Light Off Catalyst	Warm Up Oxidation Catalytic Converter[1]	WU-OC[1]
Line Pressure Control Solenoid Valve	**Line Pressure Control Solenoid Valve**	**Line PC Solenoid Valve**
LOAD (Calculated Load Value)	**Calculated Load Value**	**LOAD**
Lock Up Relay	Torque converter Clutch Relay[1]	TCC Relay[1]
Long Term FT (Fuel Trim)	Long Term Fuel Trim[1]	Long Term FT[1]
Long Term Fuel Trim	**Long Term FT**	**Long Term FT**
LONG FT	**Long Term Fuel Trim**	**See Table 4**
Low Speed FC (Fan Control) Switch	Low Speed Fan Control Switch	Low Speed FC Switch
LUS (Lock Up Solenoid) Valve	Torque Converter Clutch Solenoid Valve[1]	TCC Solenoid Valve[1]
M/C (Mixture Control)	Mixture Control[1]	MC[1]
MAF (Mass Air Flow)	Mass Air Flow[1]	MAF[1]
MAF (Mass Air Flow) Sensor	Mass Air Flow Sensor[1]	MAF Sensor[1]

TABLE 1—CROSS REFERENCE AND LOOK UP (CONTINUED)

Existing Usage	Acceptable Usage	Acceptable Acronized Usage
Malfunction Indicator Lamp	Malfunction Indicator Lamp[1]	MIL[1]
Manifold Absolute Pressure	Manifold Absolute Pressure[1]	MAP[1]
Manifold Absolute Pressure Sensor	Manifold Absolute Pressure Sensor	MAP Sensor[1]
Manifold Differential Pressure	Manifold Differential Pressure[1]	MDP[1]
Manifold Surface Temperature	Manifold Surface Temperature[1]	MST[1]
Manifold Vacuum Zone	Manifold Vacuum Zone[1]	MVZ[1]
Manual Lever Position Sensor	Transmission Range Sensor[1]	TR Sensor[1]
MAP (Manifold Absolute Pressure)	Manifold Absolute Pressure[1]	MAP[1]
MAP (Manifold Absolute Pressure) Sensor	Manifold Absolute Pressure Sensor[1]	MAP Sensor[1]
MAPS (Manifold Absolute Pressure Sensor)	Manifold Absolute Pressure Sensor[1]	MAP Sensor[1]
Mass Air Flow	Mass Air Flow[1]	MAF[1]
Mass Air Flow Sensor	Mass Air Flow Sensor[1]	MAF Sensor[1]
MAT (Manifold Air Temperature)	Intake Air Temperature[1]	IAT[1]
MATS (Manifold Air Temperature Sensor)	Intake Air Temperature Sensor[1]	IAT Sensor[1]
MC (Mixture Control)	Mixture Control[1]	MC[1]
MCS (Mixture Control Solenoid)	Mixture Control Solenoid[1]	MC Solenoid[1]
MCU (Microprocessor Control Unit)	Powertrain Control Module[1]	PCM[1]
MDP (Manifold Differential Pressure)	Manifold Differential Pressure[1]	MDP[1]
MFI (Multiport Fuel Injection)	Multiport Fuel Injection[1]	MFI[1]
MIL (Malfunction Indicator Lamp)	Malfunction Indicator Lamp[1]	MIL[1]
Mixture Control	Mixture Control[1]	MC[1]
MLPS (Manual Lever Position Sensor)	**Transmission Range Sensor[1]**	**TR Sensor[1]**
Modes	Diagnostic Test Mode[1]	DTM[1]
Mono-Jetronic	**Throttle Body Injection[1]**	**TBI[1]**
Mono-Motronic	**Throttle Body Injection[1]**	**TBI[1]**
Monotronic	Throttle Body Fuel Injection[1]	TBI[1]
Motronic-Pressure	**Multiport Fuel Injection[1]**	**MFI[1]**
Motronic	Multiport Fuel Injection[1]	MFI[1]
MPI (Multipoint Injection)	Multiport Fuel Injection[1]	MFI[1]
MPI (Multiport Injection)	Multiport Fuel Injection[1]	MFI[1]
MRPS (Manual Range Position Switch)	Transmission Range Switch	TR Switch
MST (Manifold Surface Temperature)	Manifold Surface Temperature[1]	MST[1]
Multiport Fuel Injection	Multiport Fuel Injection[1]	MFI[1]
MVZ (Manifold Vacuum Zone)	Manifold Vacuum Zone[1]	MVZ[1]
NDS (Neutral Drive Switch)	Park/Neutral Position Switch[1]	PNP Switch[1]
Neutral Safety Switch	Park/Neutral Position Switch[1]	PNP Switch[1]
NGS (Neutral Gear Switch)	Park/Neutral Position Switch[1]	PNP Switch[1]
Non Dispersive Infrared	**Non Dispersive Infrared**	**NDIR**
Non Volatile Random Access Memory	Non Volatile Random Access Memory[1]	NVRAM[1]
NPS (Neutral Position Switch)	Park/Neutral Position Switch[1]	PNP Switch[1]
NVM (Non Volatile Memory)	Non Volatile Random Access Memory[1]	NVRAM[1]
NVRAM (Non Volatile Random Access Memory)	Non Volatile Random Access Memory[1]	NVRAM[1]
O2 (Oxygen) Sensor	Oxygen Sensor[1]	O2S[1]
O2S (Oxygen Sensor)	Oxygen Sensor[1]	O2S[1]
Oxygen Sensor Location	**Oxygen Sensor Location**	**See Table 4**
OBD (On Board Diagnostic)	On Board Diagnostic[1]	OBD[1]
OBD Status	**OBD Status**	**see Table 4**
OBD STAT	**OBD Status**	**see Table 4**
OC (Oxidation Catalyst)	Oxidation Catalytic Converter[1]	OC[1]

TABLE 1—CROSS REFERENCE AND LOOK UP (CONTINUED)

Existing Usage	Acceptable Usage	Acceptable Acronized Usage
Oil Pressure Sender	Engine Oil Pressure Sensor	EOP Sensor
Oil Pressure Sensor	Engine Oil Pressure Sensor	EOP Sensor
Oil Pressure Switch	Engine Oil Pressure Switch	EOP Switch
OL (Open Loop)	Open Loop[1]	OL[1]
On Board Diagnostic	On Board Diagnostic[1]	OBD[1]
Open Loop	Open Loop[1]	OL[1]
OS (Oxygen Sensor)	Oxygen Sensor[1]	O2S[1]
OSS (Output Shaft Speed) Sensor	**Output Shaft Speed Sensor[1]**	**OSS Sensor[1]**
Output Driver	**Driver**	**Driver**
Output Shaft Speed Sensor	**Output Shaft Speed Sensor[1]**	**OSS Sensor[1]**
Oxidation Catalytic Converter	Oxidation Catalytic Converter[1]	OC[1]
OXS (Oxygen Sensor) Indicator	Service Reminder Indicator[1]	SRI[1]
Oxygen Sensor	Oxygen Sensor[1]	O2S[1]
P/N (Park/Neutral)	Park/Neutral Position[1]	PNP[1]
P/S (Power Steering) Pressure Switch	Power Steering Pressure Switch	PSP Switch
P- (Pressure) Sensor	Manifold Absolute Pressure Sensor[1]	MAP Sensor[1]
PAIR (Pulsed Secondary Air Injection)	Pulsed Secondary Air Injection[1]	PAIR[1]
Parameter Identification	**Parameter Identification**	**PID**
Parameter Identification Supported	**Parameter Identification Supported**	**See Table 4**
Park/Neutral Position	Park/Neutral Position[1]	PNP[1]
PC (Pressure Control) Solenoid Valve	**Pressure Control Solenoid Valve[1]**	**PC Solenoid Valve[1]**
PCM (Powertrain Control Module)	Powertrain Control Module[1]	PCM[1]
PCV (Positive Crankcase Ventilation)	Positive Crankcase Ventilation[1]	PCV[1]
PCV (Positive Crankcase Ventilation) Valve	Positive Crankcase Ventilation Valve[1]	PCV Valve[1]
Percent Alcohol Sensor	Flexible Fuel Sensor[1]	FF Sensor[1]
Periodic Trap Oxidizer	Periodic Trap Oxidizer[1]	PTOX[1]
PFE (Pressure Feedback Exhaust Gas Recirculation Sensor	**Feedback Pressure** Exhaust Gas Recirculation Sensor[1]	**Feedback Pressure** EGR Sensor[1]
PFI (Port Fuel Injection)	Multiport Fuel Injection[1]	MFI[1]
PG (Pulse Generator)	Vehicle Speed Sensor[1]	VSS[1]
PGM-FI (Programmed Fuel Injection)	Multiport Fuel Injection[1]	MFI[1]
PID (Parameter Identification)	**Parameter Identification**	**PID**
PID SUP	**Parameter Identification Supported**	**See Table 4**
PIP (Position Indicator Pulse)	Crankshaft Position[1]	CKP[1]
PNP (Park/Neutral Position)	Park/Neutral Position[1]	PNP[1]
Positive Crankcase Ventilation	Positive Crankcase Ventilation[1]	PCV[1]
Positive Crankcase Ventilation Valve	Positive Crankcase Ventilation Valve[1]	PCV Valve[1]
Power Steering Pressure	Power Steering Pressure	PSP
Power Steering Pressure Switch	Power Steering Pressure Switch	PSP Switch
Powertrain Control Module	Powertrain Control Module[1]	PCM[1]
Pressure Control Solenoid Valve	**Pressure Control Solenoid Valve[1]**	**PC Solenoid Valve[1]**
Pressure Feedback EGR (Exhaust Gas Recirculation)	Feedback Pressure Exhaust Gas Recirculation[1]	Feedback Pressure EGR[1]
Pressure Sensor	Manifold Absolute Pressure Sensor[1]	MAP Sensor[1]
Pressure Feedback EGR (Exhaust Gas **Recirculation**) **System**	**Feedback Pressure** Exhaust Gas Recirculation System[1]	**Feedback Pressure** EGR System[1]
Pressure Transducer EGR (Exhaust Gas Recirculation) System	Pressure Transducer Exhaust Gas Recirculation System[1]	Pressure Transducer EGR System[1]
PRNDL (Park- Reverse- Neutral- Drive- Low)	Transmission Range	TR
Programmable Read Only Memory	Programmable Read Only Memory[1]	PROM[1]

TABLE 1—CROSS REFERENCE AND LOOK UP (CONTINUED)

Existing Usage	Acceptable Usage	Acceptable Acronized Usage
PROM (Programmable Read Only Memory)	Programmable Read Only Memory[1]	PROM[1]
PSP (Power Steering Pressure)	Power Steering Pressure	PSP
PSP (Power Steering Pressure) Switch	Power Steering Pressure Switch	PSP Switch
PSPS (Power Steering Pressure Switch)	Power Steering Pressure Switch	PSP Switch
PTOX (Periodic Trap Oxidizer)	Periodic Trap Oxidizer[1]	PTOX[1]
Pulsair	Pulsed Secondary Air Injection[1]	PAIR[1]
Pulsed Secondary Air Injection	Pulsed Secondary Air Injection[1]	PAIR[1]
Pulse Width Modulation	**Pulse Width Modulation**	**PWM**
PWM	**Pulse Width Modulation**	**PWM**
QDM (Quad Driver Module)	**Driver**	**Driver**
Quad Driver Module	**Driver**	**Driver**
Radiator Fan Control	Fan Control	FC
Radiator Fan Relay	Fan Control Relay	FC Relay
RAM (Random Access Memory)	Random Access Memory[1]	RAM[1]
Random Access Memory	Random Access Memory[1]	RAM[1]
Read Only Memory	Read Only Memory[1]	ROM[1]
Rear Wheel Drive	**Rear Wheel Drive**	**RWD**
Recirculated Exhaust Gas Temperature Sensor	Exhaust Gas Recirculation Temperature Sensor	EGRT Sensor[1]
Reed Valve	Pulsed Secondary Air Injection Valve[1]	PAIR Valve[1]
REGTS (Recirculated Exhaust Gas Temperature Sensor)	Exhaust Gas Recirculation Temperature Sensor[1]	EGRT Sensor[1]
Relay Module	Relay Module	RM
Remote Mount TFI (Thick Film Ignition)	Distributor Ignition[1]	DI[1]
Revolutions per Minute	Engine Speed[1]	RPM[1]
RM (Relay Module)	Relay Module	RM
ROM (Read Only Memory)	Read Only Memory[1]	ROM[1]
RPM (Revolutions per Minute)	Engine Speed[1]	RPM[1]
RWD	**Rear Wheel Drive**	**RWD**
S4WD	**Selectable Four Wheel Drive**	**S4WD**
SABV (Secondary Air Bypass Valve)	Secondary Air Injection Bypass Valve[1]	AIR Bypass Valve[1]
SACV (Secondary Air Check Valve)	Secondary Air Injection Control Valve[1]	AIR Control Valve[1]
SASV (Secondary Air Switching Valve)	Secondary Air Injection Switching Valve[1]	AIR Switching Valve[1]
SBEC (Single Board Engine Control)	Powertrain Control Module[1]	PCM[1]
SBS (Supercharger Bypass Solenoid)	Supercharger Bypass Solenoid[1]	SCB Solenoid[1]
SC (Supercharger)	Supercharger[1]	SC[1]
Scan Tool	Scan Tool[1]	ST[1]
SCB (Supercharger Bypass)	Supercharger Bypass[1]	SCB[1]
Secondary Air Bypass Valve	Secondary Air Injection Bypass Valve[1]	AIR Bypass Valve[1]
Secondary Air Check Valve	Secondary Air Injection Check Valve[1]	AIR Check Valve[1]
Secondary Air Injection	Secondary Air Injection[1]	AIR[1]
Secondary Air Injection Bypass	Secondary Air Injection Bypass[1]	AIR Bypass[1]
Secondary Air Injection Diverter	Secondary Air Injection Diverter[1]	AIR Diverter[1]
Secondary Air Switching Valve	Secondary Air Injection Switching Valve[1]	AIR Switching Valve[1]
Selectable Four Wheel Drive	**Selectable Four Wheel Drive**	**S4WD**
SEFI (Sequential Electronic Fuel Injection)	Sequential Multiport Fuel Injection[1]	SFI[1]
Self Test	On Board Diagnostic[1]	OBD[1]
Self Test Codes	Diagnostic Trouble Code[1]	DTC[1]
Self Test Connector	Data Link Connector[1]	DLC[1]
Sequential Multiport Fuel Injection	Sequential Multiport Fuel Injection[1]	SFI[1]

TABLE 1—CROSS REFERENCE AND LOOK UP (CONTINUED)

Existing Usage	Acceptable Usage	Acceptable Acronized Usage
Service Engine Soon	Service Reminder Indicator[1]	SRI[1]
Service Engine Soon	Malfunction Indicator Lamp[1]	MIL[1]
Service Reminder Indicator	Service Reminder Indicator[1]	SRI[1]
SFI (Sequential Fuel Injection)	Sequential Multiport Fuel Injection[1]	SFI[1]
Shift Solenoid	**Shift Solenoid[1]**	**SS[1]**
Shift Solenoid Valve	**Shift Solenoid Valve[1]**	**SS Valve[1]**
Short Term FT (Fuel Trim)	Short Term Fuel Trim[1]	Short Term FT[1]
Short Term Fuel Trim	**Short Term Fuel Trim[1]**	**Short Term FT[1]**
SHRT FT	**Short Term Fuel Trim[1]**	**See Table 4**
SLP (Selection Lever Position)	Transmission Range	TR
SMEC (Single Module Engine Control)	Powertrain Control Module[1]	PCM[1]
Smoke Puff Limiter	Smoke Puff Limiter[1]	SPL[1]
SPARK ADV	**Spark Advance**	**See Table 4**
Spark Advance	**Spark Advance**	**See Table 4**
Spark Plug	**Spark Plug[1]**	**Spark Plug[1]**
SPI (Single Point Injection)	Throttle Body Fuel Injection[1]	TBI[1]
SPL (Smoke Puff Limiter)	Smoke Puff Limiter[1]	SPL[1]
SS (Shift Solenoid)	**Shift Solenoid[1]**	**SS[1]**
SRI (Service Reminder Indicator)	Service Reminder Indicator[1]	SRI[1]
SRT (System Readiness Test)	System Readiness Test[1]	SRT[1]
ST (Scan Tool)	Scan Tool[1]	ST[1]
Supercharger	Supercharger[1]	SC[1]
Supercharger Bypass	Supercharger Bypass[1]	SCB[1]
Sync Pickup	Camshaft Position[1]	CMP[1]
System Readiness Test	System Readiness Test[1]	SRT[1]
3-2TS (3-2 Timing Solenoid)	**3-2 Timing Solenoid**	**3-2TS**
3-2TS Valve (3-2 Timing Solenoid)Valve	**3-2 Timing Solenoid Valve**	**3-2TS Valve**
3-2 Timing Solenoid	**3-2 Timing Solenoid**	**3-2TS**
3-2 Timing Solenoid Valve	**3-2 Timing Solenoid Valve**	**3-2TS Valve**
3GR (Third Gear)	Third Gear	3GR
TAB (Thermactor Air Bypass)	Secondary Air Injection Bypass[1]	AIR Bypass[1]
TAC (Throttle Actuator Control)	**Throttle Actuator Control**	**TAC**
TAC (Throttle Actuator Control) Module	**Throttle Actuator Control Module[1]**	**TAC Module[1]**
TAD (Thermactor Air Diverter)	Secondary Air Injection Diverter[1]	AIR Diverter[1]
TB (Throttle Body)	Throttle Body[1]	TB[1]
TBI (Throttle Body Fuel Injection)	Throttle Body Fuel Injection[1]	TBI[1]
TBT (Throttle Body Temperature)	Intake Air Temperature[1]	IAT[1]
TC (Turbocharger)	Turbocharger[1]	TC[1]
TC (Turbocharger) Wastegate	**Turbocharger Wastegate[1]**	**TC Wastegate[1]**
TC (Turbocharger) Wastegate Regulating Valve	**Turbocharger Wastegate Regulating Valve[1]**	**TC Wastegate Regulating Valve[1]**
TCC (Torque Converter Clutch)	Torque Converter Clutch[1]	TCC[1]
TCC (Torque Converter Clutch) Relay	Torque Converter Clutch Relay[1]	TCC Relay[1]
TCC (Torque Converter Clutch) Solenoid	**Torque Converter Clutch Solenoid[1]**	**TCC Solenoid[1]**
TCC (Torque Converter Clutch) Solenoid Valve	**Torque Converter Clutch Solenoid Valve[1]**	**TCC Solenoid Valve[1]**
TCM (Transmission Control Module)	Transmission Control Module	TCM
TCCP (Torque Converter Clutch Pressure)	**Torque Converter Clutch Pressure**	**TCCP**
TFI (Thick Film Ignition)	Distributor Ignition[1]	DI[1]
TFI (Thick Film Ignition) Module	Ignition Control Module[1]	ICM[1]

TABLE 1—CROSS REFERENCE AND LOOK UP (CONTINUED)

Existing Usage	Acceptable Usage	Acceptable Acronized Usage
TFP (Transmission Fluid Pressure)	Transmission Fluid Pressure	TFP
TFT (Transmission Fluid Temperature) Sensor	Transmission Fluid Temperature Sensor	TFT Sensor
Thermac	Secondary Air Injection[1]	AIR[1]
Thermac Air Cleaner	Air Cleaner[1]	ACL[1]
Thermactor	Secondary Air Injection[1]	AIR[1]
Thermactor Air Bypass	Secondary Air Injection Bypass[1]	AIR Bypass[1]
Thermactor Air Diverter	Secondary Air Injection Diverter[1]	AIR Diverter[1]
Thermactor II	Pulsed Secondary Air Injection[1]	PAIR[1]
Thermal Vacuum Switch	Thermal Vacuum Valve[1]	TVV[1]
Thermal Vacuum Valve	Thermal Vacuum Valve[1]	TVV[1]
Third Gear	Third Gear	3GR
Three Way + Oxidation Catalytic Converter	Three Way + Oxidation Catalytic Converter[1]	TWC+OC[1]
Three Way Catalytic Converter	Three Way Catalytic Converter[1]	TWC[1]
Throttle Actuator Control	**Throttle Actuator Control**	**TAC**
Throttle Actuator Control Module	**Throttle Actuator Control** Module	**TAC** Module
Throttle Body	Throttle Body[1]	TB[1]
Throttle Body Fuel Injection	Throttle Body Fuel Injection[1]	TBI[1]
Throttle Opener	Idle Speed Control[1]	ISC[1]
Throttle Opener Vacuum Switching Valve	Idle Speed Control Solenoid Vacuum Valve[1]	ISC Solenoid Vacuum Valve[1]
Throttle Opener VSV (Vacuum Switching Valve)	Idle Speed Control Solenoid Vacuum Valve[1]	ISC Solenoid Vacuum Valve[1]
Throttle Position	Throttle Position[1]	TP
Throttle Position Sensor	Throttle Position Sensor[1]	TP Sensor[1]
Throttle Position Switch	Throttle Position Switch[1]	TP Switch[1]
Throttle Potentiometer	Throttle Position Sensor[1]	TP Sensor[1]
TOC (Trap Oxidizer - Continuous)	Continuous Trap Oxidizer[1]	CTOX[1]
TOP (Trap Oxidizer - Periodic)	Periodic Trap Oxidizer[1]	PTOX[1]
Torque Converter Clutch	Torque Converter Clutch[1]	TCC[1]
Torque Converter Clutch Pressure	**Torque Converter Clutch Pressure**	**TCCP**
Torque Converter Clutch Relay	Torque Converter Clutch Relay[1]	TCC Relay[1]
Torque Converter Clutch Solenoid	**Torque Converter Clutch** Solenoid[1]	**TCC** Solenoid[1]
Torque Converter Clutch Solenoid Valve	**Torque Converter Clutch** Solenoid Valve[1]	**TCC** Solenoid Valve[1]
TP (Throttle Position)	Throttle Position[1]	TP[1]
TP (Throttle Position) Sensor	Throttle Position Sensor[1]	TP Sensor[1]
TP (Throttle Position) Switch	Throttle Position Switch[1]	TP Switch[1]
TPI (Tuned Port Injection)	Multiport Fuel Injection[1]	MFI[1]
TPNP (Transmission Park Neutral Position)	**Park/Neutral Position**[1]	**PNP**[1]
TPS (Throttle Position Sensor)	Throttle Position Sensor[1]	TP Sensor[1]
TPS (Throttle Position Switch)	Throttle Position Switch[1]	TP Switch[1]
TR (Transmission Range)	Transmission Range	TR
Track Road Load Horsepower	**Track Road Load Horsepower**	**TRLHP**
Transmission Control Module	Transmission Control Module	TCM
Transmission Fluid Pressure	**Transmission Fluid Pressure**	**TFP**
Transmission Fluid Temperature Sensor	**Transmission Fluid Temperature** Sensor	**TFT** Sensor
Transmission Park Neutral Position	**Park/Neutral Position**[1]	**PNP**[1]
Transmission Position Switch	Transmission Range Switch	TR Switch
Transmission Range Selection	Transmission Range	TR
Transmission Range Sensor	**Transmission Range** Sensor	**TR** Sensor
TRS (Transmission Range Selection)	Transmission Range	TR

TABLE 1—CROSS REFERENCE AND LOOK UP (CONTINUED)

Existing Usage	Acceptable Usage	Acceptable Acronized Usage
TRSS (Transmission Range Selection Switch)	Transmission Range Switch	TR Switch
TSS (Turbine Shaft Speed) Sensor	**Turbine Shaft Speed Sensor**[1]	**TSS Sensor**[1]
Tuned Port Injection	Multiport Fuel Injection[1]	MFI[1]
Turbine Shaft Speed Sensor	**Turbine Shaft Speed Sensor**[1]	**TSS Sensor**[1]
Turbo (Turbocharger)	Turbocharger[1]	TC[1]
Turbocharger	Turbocharger[1]	TC[1]
Turbocharger Wastegate	**Turbocharger Wastegate**[1]	**TC Wastegate**[1]
Turbocharger Wastegate Regulating Valve	**Turbocharger Wastegate Regulating Valve**[1]	**TC Wastegate Regulating Valve**[1]
TVS (Thermal Vacuum Switch)	Thermal Vacuum Valve[1]	TVV[1]
TVV (Thermal Vacuum Valve)	Thermal Vacuum Valve[1]	TVV[1]
TWC (Three Way Catalytic Converter)	Three Way Catalytic Converter[1]	TWC[1]
TWC + OC (Three Way + Oxidation Catalytic Converter)	Three Way + Oxidation Catalytic Converter[1]	TWC+OC[1]
VAC (Vacuum) Sensor	Manifold Differential Pressure Sensor[1]	MDP Sensor[1]
Vacuum Switches	Manifold Vacuum Zone Switch	MVZ Switch[1]
VAF (Volume Air Flow)	Volume Air Flow[1]	VAF[1]
Valve Position EGR (Exhaust Gas Recirculation) System	**Valve Position Exhaust Gas Recirculation System**[1]	**Valve Position EGR System**[1]
Vane Air Flow	Volume Air Flow[1]	VAF[1]
Variable Control Relay Module	**Variable Control Relay Module**	**VCRM**
Variable Fuel Sensor	Flexible Fuel Sensor	FF Sensor[1]
VAT (Vane Air Temperature)	Intake Air Temperature[1]	IAT[1]
VCC (Viscous Converter Clutch)	Torque Converter Clutch[1]	TCC[1]
VCM	**Vehicle Control Module**	**VCM**
VCRM	**Variable Control Relay Module**	**VCRM**
Vehicle Control Module	**Vehicle Control Module**	**VCM**
Vehicle Identification Number	**Vehicle Identification Number**	**VIN**
Vehicle Speed Sensor	Vehicle Speed Sensor[1]	VSS[1]
VIN (Vehicle Identification Number)	**Vehicle Identification Number**	**VIN**
VIP (Vehicle In Process) Connector	Data Link Connector[1]	DLC[1]
Viscous Converter Clutch	Torque Converter Clutch[1]	TCC[1]
Voltage Regulator	Voltage Regulator	VR
Volume Air Flow	Volume Air Flow[1]	VAF[1]
VR (Voltage Regulator)	Voltage Regulator	VR
VSS (Vehicle Speed Sensor)	Vehicle Speed Sensor[1]	VSS[1]
VSV (Vacuum Solenoid Valve) (Canister)	Evaporative Emission Canister Purge Valve[1]	EVAP Canister Purge Valve[1]
VSV (Vacuum Solenoid Valve) (EVAP)	Evaporative Emission Canister Purge Valve[1]	EVAP Canister Purge Valve[1]
VSV (Vacuum Solenoid Valve) (Throttle)	Idle Speed Control Solenoid Vacuum Valve[1]	ISC Solenoid Vacuum Valve[1]
Warm Up Oxidation Catalytic Converter	Warm Up Oxidation Catalytic Converter[1]	WU-OC[1]
Warm Up Three Way Catalytic Converter	Warm Up Three Way Catalytic Converter[1]	WU-OC[1]
Wide Open Throttle	Wide Open Throttle[1]	WOT[1]
WOT (Wide Open Throttle)	Wide Open Throttle[1]	WOT[1]
WOTS (Wide Open Throttle Switch)	Wide Open Throttle Switch[1]	WOT Switch[1]
WU-OC (Warm Up Oxidation Catalytic Converter)	Warm Up Oxidation Catalytic Converter[1]	WU-OC[1]
WU-TWC (Warm Up Three Way Catalytic Converter)	Warm Up Three Way Catalytic Converter[1]	WU-TWC[1]

Recommended Terms and Recommended Acronyms See Table 2
[1] Emission-Related Term
Bold indicates new/revised entry

GLOSSARY

Accelerator pump system A carburetor system that prevents a hesitation in engine operation during low-speed acceleration.

Sistema de la bomba del acelerador Sistema del carburador que evita la vacilación del funcionamiento del motor durante una aceleración de baja velocidad.

Actuator A control device that delivers mechanical action in response to an electrical signal.

Actuador Un dispositivo de control que suministra una acción mecánica en respuesta a una señal eléctrica.

Adaptive strategy The ability of a computer to adapt to certain defects in the system.

Estrategia de adaptación Capacidad de una computadora para adaptarse a ciertas fallas en el sistema.

After top dead center (ATDC) Refers to the position of the piston while it is moving away from its upmost position.

Después del punto muerto superior Se refiere a la posición del pistón mientras se está moviendo después de su punto más alto en su carrera hacia la posición inferior.

Air charge temperature (ACT) sensor An input sensor that sends a voltage signal to the computer in relation to intake air temperature.

Sensor de la temperatura de la carga de aire Sensor de entrada que le envía una señal de tensión a la computadora referente a la temperatura del aire aspirado.

Air cleaner ducts Ducts connected from the air intake source to the air cleaner.

Conductos del filtro de aire Conductos conectados desde la fuente del aire aspirado hasta el filtro de aire.

Air-cooled cooling system A cooling system that provides engine cooling by passing air over the outside of the engine.

Sistema de enfriamiento enfriado por aire Sistema de enfriamiento que enfría el motor al pasar aire sobre la parte exterior del mismo.

Air door A butterfly-type valve in the heated air inlet system, usually positioned in the air cleaner snorkel.

Puerta de ventilación Válvula tipo mariposa en el sistema de admisión de aire calentado, que normalmente se coloca en el tubo de respiración del filtro de aire.

Air door vacuum motor diaphragm A vacuum diaphragm that opens and closes the air door.

Diafragma del motor de vacío de la puerta de ventilación Diafragma de vacío que abre y cierra la puerta de ventilación.

Air-fuel ratio The ratio of the amount of air to the amount of fuel entering the cylinders.

Relación de aire y combustible Relación entre la cantidad de aire y la cantidad de combustible que entra en los cilindros.

Alternating Current An electric current that reverses its direction at regularly recurring intervals.

Corriente Alterna Una corriente eléctrica que reserva su dirección a intervalos de repetición regular.

Ammeter A device used to measure electrical current.

Amperímetro Un dispositivo para medir la corriente eléctrica.

Ampere A measurement for the amount of electron movement or current flow.

Amperio Medida de la cantidad del movimiento de electrones o del flujo de corriente.

Amp-hour rating A battery rating indicating the amount of amperes a battery will deliver over a longer time period with a small electrical load.

Clasificación de amperios-horas Clasificación de una batería que indica la cantidad de amperios que la misma podrá generar durante un mayor espacio de tiempo con una pequeña carga eléctrica.

Amplitude The difference between the highest and lowest voltage in a waveform signal.

Amplitud La diferencia entre el voltaje más alto y el voltaje más bajo en una señal en forma de onda.

Analog A nondigital measuring method that uses a needle to indicate readings. A typical dashboard gauge with a moving needle is an analog instrument.

Análogo Un método de medir en forma no digital, que usa una aguja para indicar la lectura. Un reloj en el tablero del vehículo con una aguja que se mueve es un instrumento análogo.

Analog/digital conversion The process of changing analog voltage signals to digital signals.

Conversión analógica/digital Proceso de convertir señales de tensión analógicas en señales digitales.

Analog voltage signal A voltage signal that is usually produced by input sensors and is continuously variable within a certain voltage range.

Señal de tensión analógica Señal de tensión que normalmente es producida por sensores de entrada y que varía de modo continuo en proporción a cierto margen de tensión.

Antidrainback valve A valve that prevents oil drainback from components in the cylinder heads into the lubrication system.

Válvula que evita la filtración del aceite Válvula que evita que el aceite de los componentes en las culatas de los cilindros se filtre a través del sistema de lubrificación.

Antifoaming agents Oil additives that prevent oil foaming.

Agentes antiespumantes Aditivos de aceite que evitan que éste se espume.

Antiknock Refers to a fuel's ability to resist self ignition.

Antidetonante Se refiere a la habilidad de un combustible de resistir la tendencia de que se incinere por si mismo.

API American Petroleum Institute. An organization that sets standards for petroleum based products, such as engine oil.

API Instituto Americano del petróleo. Una organización que establece estándares para productos basados en petróleo.

Atmospheric pressure The pressure exerted on the earth by the atmosphere.

Presión atmosférica Presión que la atmósfera ejerce sobre la tierra.

Atom The smallest particle of an element.

Átomo La partícula más pequeña de un elemento.

Atomization The process of breaking up a liquid into small particles or droplets.

Atomización El proceso de romper un líquido adentro de partículas pequeñas o gotas.

Automatic shutdown relay A computer-controlled relay that opens and closes the circuit to the fuel pump, coil primary winding, and other components on Chrysler products.

Relé de parada automática Relé controlado por computadora que abre y cierra el circuito que conduce hacia la bomba del combustible, el bobinado primario, y otros componentes de productos fabricados por la Chrysler.

Back pressure Pressure created by restriction in an exhaust system.

Contrapresión La presión creada por la restricción en un sistema de escape.

Balanced carburetor A carburetor in which the float bowl is vented to the air horn, and may also be vented to the atmosphere.

Carburador equilibrado Carburador en el que el depósito del flotador es ventilado hacia la bocina neumática; puede ser ventilado también hacia la atmósfera.

Ballast resistor A resistor connected in series between the ignition switch and the coil positive primary terminal on some DI systems.

Resistor de compensación Resistor conectado en serie entre el botón conmutador de encendido y el borne primario positivo de la bobina en algunos sistemas de encendido con distribuidor.

Band A holding member for an automatic transmission.

Banda Un miembro de retención de la transmisión.

Barometric pressure switch A sensor or signal circuit that sends a varying frequency signal to the processor relating actual barometric pressure.

Interruptor para la presión barométrica Un sensor o su circuito de señal que envía una señal de frecuencia variable al procesador relacionándole la presión barométrica actual.

Barrel A common term used to describe the number of throttle bores a carburetor has.

Barril Un termino común usado para describir el número de los barriles que un carburador tiene.

Baud rate The rate at which a PCM is able to transfer and receive data. Baud rate is measured in bits per second.

Velocidad Baud La velocidad a la cual el PCM (módulo de control de la potencia del motor) es capaz de transferir y recibir data. La velocidad Baud es medida en mordidas por segundo.

Bearing crush Bearing inserts are slightly longer than the connecting rod bore in which they are mounted. This design crushes the bearing slightly when the rod bolts are tightened to provide improved contact between the bearing insert and the connecting rod bore.

Quiebra de cojinete Piezas insertas de cojinetes que son ligeramente más largas que el calibre de las bielas en las que van montadas. Este diseño quebranta un poco el cojinete cuando se aprietan los pernos de la biela a fin de proporcionar un mejor contacto entre la pieza inserta del cojinete y el calibre de la biela.

Bearing inserts Circular bearings mounted between the connecting rod bore and the crankshaft journal.

Piezas insertas de cojinetes Cojinetes circulares montados entre el calibre de la biela y el gorrón del cigüeñal.

Bearing spread A bearing design in which the bearing curvature is slightly larger than the bore in which it is mounted.

Extensión del cojinete Cojinete diseñado de forma tal que su curvatura es un poco más grande que el calibre en el que está montado.

Before top dead center (BTDC) Refers to the position of the piston while it is moving toward its upmost position.

Antes del punto muerto superior Se refiere a la posición del pistón mientras se está moviendo hasta su punto más alto.

Belt-driven cooling fan A cooling fan driven by a belt from the crankshaft.

Ventilador de enfriamiento accionado por correa Ventilador de enfriamiento accionado por una correa desde el cigüeñal.

Bimetal sensor A sensor that controls the vacuum supplied to the air door diaphragm in a heated air inlet system.

Sensor bimetal Sensor que controla el vacío suministrado al diafragma de la puerta de ventilación en un sistema de admisión de aire calentado.

Binary code A group of numbers assigned to digital voltage signals.

Código binario Grupo de números asignados a las señales de tensión digitales.

Blowby Compression and exhaust gases that blow by the piston rings and enter into the engine's crankcase.

Fuga en la cámara de la combustión Compresión y gases de escape que se escapan atraves de los anillos del pistón y entran adentro de la cacerola del aceite.

Boost pressure The amount of pressure in the intake manifold created by a turbocharger or supercharger.

Presión de sobrealimentación Cantidad de presión en el colector de aspiración producida por un turbocompresor o un compresor.

Boost venturi A smaller diameter venturi placed in a larger venturi.

Venturi auxiliar Venturi de diámetro más pequeño ubicado dentro de un Venturi más grande.

Bore The diameter of a hole, commonly used to describe the dimensions of an engine.

Diámetro del cilindro El diámetro del orificio, comúnmente usado para describir las dimensiones de un motor.

Bore and stroke The diameter of the cylinder bore and the length of the piston stroke.

Calibre y carrera El diámetro del calibre del cilindro y el largo de la carrera del pistón.

Bottom dead center (BDC) Piston position at the bottom of the cylinder.

Punto muerto inferior La posición del pistón en la parte inferior del cilindro.

Breaker plate The assembly inside a distributor that holds the breaker points in position.

Plato ruptor El ensamblaje adentro del distribuidor que contiene los platinos en su posición.

Breaker point The triggering device used in early ignition systems. These mechanical switches opened and closed the primary circuit of the ignition system.

Platinos El dispositivo de disparar usado en los sistemas de ignición más antiguos. Estos interruptores mecánicos abren y cierran el circuito primario del sistema de ignición.

Burn time The length of the spark line while the spark plug is firing measured in milliseconds.

Duración del encendido Espacio de tiempo que la línea de chispas de la bujía permanece encendida, medido en milisegundos.

By-pass oil filter An engine oil filter in which a portion of the flow from the pump circulates through the filter.

Filtro del aceite de paso Filtro del aceite de un motor en el que parte del flujo de la bomba circula a través del filtro.

Calibrator package (CALPAK) A removable chip in some computers, usually contains a fuel backup program.

Paquete del calibrador Pastilla desmontable en algunas computadoras; normalmente contiene un programa de reserva para el combustible.

California Air Research Board (CARB) The part of the California government that monitors vehicle emissions and sets standards for the state to reduce vehicle emissions.

Junta de recursos del aire de California La parte del gobierno de California que revisa las emisiones de los vehículos y establece los estándares del estado para reducir las emisiones de los vehículos.

Camber The attitude of a wheel and tire assembly when viewed from the front of a car. If it leans outward, away from the car at the top, the wheel is said to have positive camber. If it leans inward, it is said to have negative camber.

Comba La aptitud de un ensamblaje de rueda y el neumático cuando se observa desde la parte delantera del vehículo. Si se inclina hacia la parte de afuera del vehículo en la parte superior, es dicho que la rueda tiene comba positiva. Si se inclina hacia la parte de adentro del vehículo en la parte superior, es dicho que la rueda tiene comba negativa.

Cam-ground piston A piston with the skirt designed in the shape of a cam rather than being perfectly round.

Pistón de leva excéntrica Pistón con una faldilla diseñada en forma de leva en vez de ser perfectamente redonda.

Camshaft The component in the engine that opens and closes the valves.

Árbol de levas Componente en el motor que abre y cierra las válvulas.

Camshaft lobes Rotating high points on the camshaft that open the valves.

Lóbulos del árbol de levas Puntos altos giratorios en el árbol de levas que abren las válvulas.

Camshaft reference sensor A sensor that sends a voltage signal to the ignition module; this signal is often used for injector sequencing.

Sensor de referencia del árbol de levas Sensor que le envía una señal de tensión al módulo del encendido; dicha señal se utiliza con frecuencia para el ordenamiento del inyector.

Canister purge valve The valve that controls the release of fuel fumes from the charcoal canister into the intake.

Válvula para la purga del canasto La válvula que controla la liberación de los vapores del combustible desde el canasto de carbón hacia adentro del múltiple de escape.

Carbon dioxide (CO_2) A gas that is a by-product of the combustion process.

Bióxido de carbono (CO_2) Gas que es un producto derivado del proceso de combustión.

Carbon monoxide (CO) A gas formed as a by-product of the combustion process in the engine cylinders. This gas is very dangerous or deadly to the human body in high concentrations.

Monóxido de carbono (CO) Gas que es un producto derivado del proceso de combustión en los cilindros del motor. Este gas es muy peligroso y en altas concentraciones podría ocasionar la muerte.

Carburetor icing Ice forming around the carburetor throttles or in the venturi.

Congelación del carburador Formación de hielo alrededor de las mariposas del carburador o en el Venturi.

Carburetor inlet needle and seat A valve assembly that controls the flow of fuel from the fuel pump into the float bowl.

Aguja y asiento de admisión del carburador Conjunto de válvulas que controla el flujo del combustible desde la bomba del combustible hasta el interior del depósito del flotador.

Caster Angle formed between the kingpin axis and a vertical axis as viewed from the side of the vehicle. Caster is considered positive when the top of the kingpin axis is behind the vertical axis.

Inclinación del eje delantero El ángulo formado entre el axis del eje pivote de la dirección y un axis vertical según es observado desde el lado del vehículo. La inclinación del eje es considerada positiva cuando la parte superior del axis del eje pivote de la dirección está detrás del axis vertical.

Catalyst efficiency monitor A monitor in OBD II systems that checks the oxygen content of the exhaust before it enters the catalytic converter and after it leaves the converter.

Monitor de la eficiencia del catalítico Un monitor en los sistemas OBD II (diagnostico abordo del vehículo II) que chequea el contenido del oxigeno en el escape antes de que entre en el convertidor catalítico y después que sale del convertidor.

Catalytic converter A device installed in a vehicle's exhaust system which changes undesired gases into harmless gases. An emission control device.

Convertidor catalítico Un dispositivo instalado en el sistema de escape de un vehículo que cambia los gases indeseables a gases no dañinos. Un dispositivo de control para las emisiones.

Cell group The collection of positive and negative plates in one cell of a battery. Each group typically provides 2.1 volts.

Grupo de celdas La colección de los plato negativos y positivos en una celda de la batería. Cada grupo típicamente provee 2.1 volteos.

Cellular radiator core A type of core made from many small interconnected cells.

Núcleo del radiador celular Tipo de núcleo compuesto de muchas células conectadas entre sí.

Central port injection (CPI) A fuel injection system with one central, computer-controlled injector that supplies fuel to pressure-operated poppet nozzles in the intake ports.

Inyección central con lumbrera Sistema de inyección de combustible con un inyector central controlado por computadora que les suministra combustible a las toberas de movimiento vertical accionadas por presión en las lumbreras de aspiración.

Centrifugal advance A mechanism that controls spark advance in relation to engine speed.

Avance centrífugo Dispositivo que controla el avance de la chispa de acuerdo a la velocidad del motor.

Cetane A rating used to classify diesel fuel, refers to the volatility of the fuel.

Cetano Una relación usada para clasificar el combustible diesel, se refiere a la votalidad del combustible.

Charcoal canister A emission control device that collects gasoline vapors and prevents them from entering into the atmosphere.

Canasto de carbón Un dispositivo para el control de las emisiones que colecta los vapores de gasolina y los previene de que entren adentro de la atmósfera.

Choke A plate located at the top of a carburetor to restrict air flow into the carburetor when the engine is cold.

Estrangulador Un plato localizado en la parte superior del carburador para restringir el flujo de aire adentro del carburador cuando el motor está frío.

Choke unloader A mechanism designed to allow the choke plate to open when extra air is needed and the choke plate is closed.

Descargador del estrangulador Un mecanismo diseñado para permitir de que el plato del estrangulador se abra cuando aire adicional es necesitado y el plato del estrangulador está cerrado.

Circuit opening relay A relay that opens and closes the fuel pump circuit on some Toyota products.

Relé para la apertura del circuito Relé que abre y cierra el circuito de la bomba del combustible en algunos productos fabricados por Toyota.

Clear flood mode A computer operating mode that supplies a leaner air-fuel ratio if an engine becomes flooded. This mode is entered by holding the throttle wide open while cranking the engine.

Modo para disminuir una inundación Modo de funcionamiento de una computadora que suministra una relación de aire y combustible más pobre si el motor se inunda. Este modo se activa manteniendo la mariposa abierta de par en par mientras se arranca el motor.

Closed loop A computer system operating mode in which the computer uses the oxygen sensor to help control the air-fuel ratio.

Bucle cerrado Modo de funcionamiento de una computadora en el que se utiliza el sensor de oxígeno para ayudar a controlar la relación de aire y combustible.

Clutch disc The part of a clutch that receives the driving motion from the flywheel and pressure plate assembly and transmits that motion to the transmission input shaft.

Disco del embrague La parte de un embrague que recibe el movimiento desde el volante del motor, el ensamblaje del plato de presión y transmite ese movimiento al eje de entrada de la transmisión.

Clutch fork A forked lever that moves the clutch release bearing and hub back and forth.

Tenedor del embrague Una palanca en configuración a un tenedor que libera el balero de liberación del embrague y el cubo hacia adelante y hacia atrás.

Clutch pack A common term for the multiple-disc clutch pack used to hold a member of the planetary gearset in an automatic transmission.

Páguete de embrague Un termino común para un embrague con múltiple discos, usados para detener un miembro del juego de los engranes planetarios en una transmisión automática.

Clutch release bearing A sealed, prelubricated ball bearing that moves the pressure plate release levers or diaphragm spring through the engagement and disengagement of the clutch.

Balero de liberación del embrague Un balero prelubricado y sellado que mueve la palanca de liberación del plato de presión, resortes del diafragma atraves del enganche y desenganche del embrague.

CNG Compressed Natural Gas. An alternative fuel source for engines.

CNG Gas Natural Comprimido. Una fuente de combustible alternativo para los motores.

Coil pack A common term for the ignition coil assembly in a distributorless ignition system.

Juego de bobinas Un termino común usado para el ensamblaje de la bobina de la ignición en un sistema de ignición sin distribuidor.

Coil sequencing Firing coils on an EI system to match the engine firing order.

Ordenamiento de bobina Bobinas del encendido en un sistema de encendido electrónico para equilibrar la secuencia del encendido del motor.

Cold cranking ampere rating A battery rating indicating the amount of amperes a battery will deliver at a specific temperature.

Clasificación de amperios de arranque en frío Clasificación de una batería que indica la cantidad de amperios que la misma podrá generar a una temperatura específica.

Cold spark plug A spark plug designed so the electrodes operate at lower temperatures.

Bujía en frío Bujía diseñada para que los electrodos funcionen a temperaturas más bajas.

Cold start injector An injector that delivers more fuel into the intake manifold when the engine is cold.

Inyector de arranque en frío Inyector que le envía una mayor cantidad de combustible al colector de aspiración cuando el motor está frío.

Combustion chamber The area formed in the cylinder head when the piston is at TDC.

Cámara de combustión El área formada adentro de la cabeza del cilindro cuando el pistón está en el TDC (punto muerto superior).

Compound A material with two or more types of atoms.

Compuesto Material que tiene dos o más tipos de átomos.

Comprehensive component monitor One of the monitor systems in an OBD II system that checks the activity and efficiency of several systems and major components.

Componente de monitor comprensivo Uno de los monitores en el sistema OBD II (diagnostico abordo del vehículo II) que chequea las actividades y eficiencias de varios sistemas y componentes mayores.

Compressibility The ability of a material to become smaller when pressure is applied.

Compresibilidad Capacidad de un material para empequeñecerse al aplicársele presión.

Compression ignition (CI) An engine in which the air-fuel mixture is ignited by the heat of compression.

Encendido por compresión Motor en el que la mezcla de aire y combustible se enciende por medio del calor de compresión.

Compression ratio The ratio of the volume in the cylinder above the piston when the piston is at bottom dead center to the volume in the cylinder above the piston when the piston is at top dead center.

Relación de compresión La relación del volumen en el cilindro encima del pistón cuando el pistón está en el punto muerto inferior, comparado con el volumen en la parte superior del cilindro, cuando el pistón está en el punto muerto superior.

Compression rings Rings mounted near the top of the piston to seal the compression and combustion gases in the cylinder.

Anillos de compresión Anillos montados cerca de la parte superior del pistón para atrapar los gases de compresión y de combustión dentro del cilindro.

Compressor wheel A vaned wheel mounted in the air intake and connected to one end of the turbocharger shaft.

Rueda compresora Rueda con paletas montada en el aire aspirado y conectada a un extremo del árbol turbocompresor.

Computer command control (3C) system A computer-controlled carburetor system used on some General Motors vehicles.

Sistema de mando controlado por computadora Sistema del carburador controlado por computadora que se utiliza en algunos vehículos fabricados por la General Motors.

Conduction A method of heat transfer in which heat from a warmer object is transferred to a cooler object.

Conducción Método de transferencia de calor en el que el calor de un objeto más tibio se transfiere a un objeto más frío.

Conductor An element with one, two, or three valence electrons that easily conducts electric current.

Conductor Elemento con uno, dos o tres electrones de valencia que conduce una corriente eléctrica fácilmente.

Connecting rod bore The circular opening in the lower end of the connecting rod that retains the connecting rod bearings.

Calibre de biela Abertura circular en el extremo inferior de la biela que retiene los cojinetes de la biela.

Connecting rod eye A circular opening in the top of the connecting rod in which the piston pin is located.

Ojal de biela Abertura circular en la parte superior de la biela en la que está ubicado el pistón.

Continuous injection system (CIS) A system that uses fuel under pressure to modulate or change the fuel injection area.

Sistema de inyección continuo Un sistema que usa combustible debajo de presión para modular o cambiar el área de inyección de combustible.

Controllers A term used to describe electronic devices that control the activity of various actuators.

Controladores Un termino usado para describir dispositivos electrónicos que controlan las actividades de varios actuadores.

Convection A method of heat transfer in which heated atoms or molecules rise upward and cooler atoms or molecules sink downward where they are heated.

Convección Método de transferencia de calor en el que los átomos o moléculas calentados ascienden y los átomos o moléculas más fríos descienden a donde son calentados.

Coolant by-pass passage A coolant passage or hose through which coolant is circulated when the thermostat is closed.

Tubo para el paso del refrigerante Tubo o manguera por el que circula el refrigerante cuando el termóstato está cerrado.

Coolant recovery system A container that catches any coolant that comes out the radiator overflow hose and returns this coolant to the radiator.

Sistema para la recuperación del refrigerante Recipiente que atrapa el refrigerante que se escapa de la manguera de rebose del radiador y lo devuelve al radiador.

Core plugs Metal plugs in a cast component such as the engine block which seal openings required by the casting tools.

Tapones para núcleo Tapones de metal en un componente fundido, como por ejemplo el bloque de un motor, que sellan las aberturas requeridas por las herramientas de fundición.

Corporate average fuel economy (CAFE) Fleet average fuel economy standards imposed on car manufacturers by the US Congress.

Economía promedio de combustible para fabricantes Normas sobre la economía promedio de combustible para flotas de vehículos que el Congreso de los Estados Unidos les impone a los fabricantes de automóviles.

Corrosion and rust inhibitors Oil additives that neutralize acids in the oil that cause corrosion of engine components.

Inhibidores de corrosión y oxidación Aditivos de aceite que neutralizan los ácidos en el aceite que pueden provocar la corrosión de los componentes del motor.

Corrosive A material that dissolves metals or burns the skin.

Corrosivo Material que disuelve metales o que quema la piel.

Counterweights Heavy weights that are part of the crankshaft and help to provide proper balance.

Contrapesos Pesos pesados que forman parte del cigüeñal y que ayudan a proporcionar un equilibrio adecuado.

Coupling point This is the period of time when the speed of a torque converter's turbine approaches the speed of the impeller. At this point torque multiplication is minimized.

Punto de acoplación Este es el periodo de tiempo cuando la velocidad de la turbina del convertidor de torque se acerca a la velocidad del impulsor. En este punto la multiplicación de torque es minimizada.

Crankshaft A rotating component mounted in the lower side of the block that changes vertical piston motion to rotary motion.

Cigüeñal Componente giratorio montado en la parte inferior del bloque que convierte el movimiento vertical del pistón en movimiento giratorio.

Crankshaft timing sensor A pickup assembly that produces a voltage signal used for ignition triggering in an EI system.

Sensor de regulación del cigüeñal Conjunto de captación que produce una señal de tensión utilizada para el arranque del encendido en un sistema de encendido electrónico.

Crank throw The distance from the crankshaft main bearing centerline to the connecting rod journal centerline.. The stroke of any engine is the crank throw.

Tiro del cigüeñal La distancia central desde el cojinete principal del cigüeñal a la línea central del muñón de la biela. La carrera de cualquier motor es el tiro del cigüeñal.

Crossflow radiator A radiator in which the tanks are positioned on each side of the core and the coolant flows horizontally through the core.

Radiador de flujo transversal Radiador en el que los tanques se colocan a cada lado del núcleo y donde el refrigerante fluye horizontalmente a través de éste.

Crude oil Petroleum products that are unrefined. This term normally is used to describe oil that has just been pumped from an oil well.

Aceite crudo Productos de petróleo que no están refinados. Este termino es normalmente usado para describir el aceite que se han acabado de ser bombeado desde un pozo de petróleo.

Cycle One complete set of changes in a recurring signal.

Ciclo Un juego de cambio completo en una señal recurrente.

Cycling clutch A type of air conditioning system in which system pressure is controlled by turning the compressor clutch on and off.

Embrague de ciclo Un sistema de tipo de aire acondicionado en cual la presión del sistema es controlada apagando y encendiendo el embrague del compresor.

Dashpot A device used to slow down the closing of the throttle plates during deceleration.

Válvula amortiguadora Un dispositivo usado para disminuir la velocidad del cierre de las mariposas del carburador durante la deceleración.

Data link connector (DLC) A computer system connector to which the computer supplies data for diagnostic purposes.

Conector de enlace de datos Conector de computadora al que ésta suministra datos para propósitos diagnósticos.

Decel valve A vacuum-operated valve that delivers additional air to the intake manifold during deceleration to improve emission levels.

Válvula de desaceleración Válvula accionada por vacío que le envía una cantidad adicional de aire al colector de aspiración durante la desaceleración a fin de reducir los niveles de emisiones.

Deep skirt block An engine with the block skirt extending well below the bottom of the cylinders.

Faldilla profunda de bloque Motor con una faldilla de bloque que se extiende hasta más abajo de la parte inferior de los cilindros.

Delay valve A valve used to delay or cause a vacuum signal to hesitate. This allows the vacuum operated device to operate after the signal has arrived.

Válvula de retardo Una válvula usada para retardar o causar un titubeo en una señal de vacío. Esto permite que el dispositivo operado por vacío opere después de que la señal haya llegado.

Detergents and dispersants Oil additives that help the oil to clean carbon particles off engine components and disperse large carbon particles in the oil.

Detergentes y dispersores Aditivos de aceite que ayudan al aceite a remover las partículas de carbón de los componentes del motor y a dispersar las partículas grandes de carbón en el aceite.

Detonation Abnormal combustion. Refers to the ignition of the air/fuel mixture inside the combustion chamber prior to the firing of a spark plug.

Detonación Combustión anormal. Se refiere a la ignición de la mezcla de aire/combustible adentro de la cámara de combustión antes de que la chispa de la bujía ocurra.

Diesel A compression ignited engine.

Diesel Un motor que hace su combustión con su compresión.

Diesel particulate emissions Diesel exhaust emissions that consist mainly of small carbon particles.

Emisiones de partículas de diesel Emisiones del escape de diesel que consisten principalmente en pequeñas partículas de carbón.

Digital Normally refers to a signal that is either on or off. Digital signals are required by most automotive computers.

Digital Normalmente se refiere a una señal que no está ni apagada ni encendida. Las señales digitales son requeridas por la mayorías de las computadoras automotrices.

Digital EGR valve An EGR valve that contains a computer-operated solenoid, or solenoids.

Válvula EGR digital Una válvula EGR que contiene un solenoide o solenoides accionados por computadora.

Digital voltage signal A voltage signal that is either on or off, high or low.

Señal de tensión digital Una señal de tensión que está encendida o apagada, o que es alta o baja.

Diode A one-way flow valve for electricity.

Diodo Válvula de flujo de una vía para la electricidad.

Direct Current An electrical current that flows in one direction.

Corriente Directa Una corriente eléctrica que fluye solamente en una dirección.

Dispatch sheets Contain appointment records in the shop.

Hojas de servicios Contienen información sobre las reparaciones de vehículos que se llevan a cabo en el taller mecánico.

Displacement The volume of each cylinder multiplied by the number of cylinders, and usually expressed in liters, cubic centimeters, or cubic inches.

Desplazamiento Volumen de cada cilindro multiplicado por el número de cilindros, y expresado normalmente en litros, centímetros cúbicos, o pulgadas cúbicas.

Display scope pattern A scope pattern in which the voltage traces from the cylinders are displayed one after the other across the screen.

Modelo de radio de visualización Modelo de radio en el que las pequeñas cantidades de tensión provenientes de los cilindros se proyectan una tras otra a través de la pantalla.

Distributor cap and rotor A mechanical device that distributes the ignition sparks from the coil secondary winding to the spark plugs.

Tapa y rotor del distribuidor Dispositivo mecánico que distrubuye las chispas del encendido desde el bobinado secundario hasta las bujías.

Distributor ignition (DI) system SAE J1930 terminology for an ignition system with a distributor.

Sistema de encendido con distribuidor Término utilizado por la SAE J1930 para referirse a un sistema de encendido que tiene un distribuidor.

Distributorless ignition system (DIS) An ignition system that does not use a distributor for spark distribution, rather spark distribution is controlled by the vehicle's computer.

Sistema de ignición sin distribuidor Un sistema de ignición que no usa un distribuidor para la chispa de la ignición, envés la distribución de la chispa es controlada por la computadora del vehículo.

Diverter valve Part of a typical air injection system. This valve allows the air from the air pump to be diverted into the atmosphere rather than the exhaust to prevent backfiring.

Válvula de diversión Una parte típica del sistema de inyección de aire. Esta válvula permite que el aire desde la bomba de aire sea divertido adentro de la atmósfera envés de enviarlo adentro del sistema de escape para prevenir una contra explosión.

Downflow radiator A radiator in which the tanks are positioned on the top and bottom of the core, and the coolant flows vertically through the core.

Radiador con flujo descendente Radiador en el que los tanques se colocan sobre las partes superior e inferior del núcleo y donde el refrigerante fluye verticalmente a través de éste.

Drive cycle A specified set of driving conditions. The drive cycle is important to adaptive strategies of a computer. The proper drive cycle is also required of OBD II systems to complete monitor tests on certain systems.

Ciclo de ejecución Un juego especifico de condiciones de manejar. Estas condiciones son importantes para la estrategia adaptiva de la computadora. El ciclo de manejo apropiado es también requerido para el sistema OBD II (diagnóstico abordo del vehículo II) para competir con las pruebas del monitor en algunos sistemas.

Dry-type cylinder sleeve A cylinder sleeve installed in the block so coolant does not contact the outside area of the sleeve.

Manguito de cilindro tipo seco Manguito de cilindro instalado en el bloque para que el refrigerante no entre en contacto con la parte exterior del mismo.

Dry-type intake manifold An intake manifold without coolant circulation through the manifold.

Colector de aspiración tipo seco Colector de aspiración a través del cual no circula refrigerante.

Dual overhead camshaft (DOHC) An engine with two camshafts positioned above each cylinder head.

Árbol de levas superpuesto doble Motor con dos árboles de levas colocados sobre cada cada una de las culatas de los cilindros.

Duty cycle On-time to off-time ratio, as measured in a percentage of pulse width or degrees of dwell.

Ciclo de duración La relación de apagado y encendido, según es medido en porcentaje de la amplitud del pulso o grados Dwell (tiempo en que los puntos están cerrados medidos en grados).

Dwell The amount of time the current is flowing through a circuit. Most often, this term is applied to ignition systems.

Tiempo en que los puntos están cerrados medidos en grados La cantidad de tiempo que la corriente está fluyendo atraves de un circuito. Más común, este termino es aplicado a sistemas de ignición.

Dynamic balance Refers to the balance of a wheel and tire assembly when it is in motion.

Balanceo dinámico Se refiere al balance del ensamblaje de una llanta y un neumático cuando está en moción.

EGR vacuum regulator (EVR) A vacuum solenoid that is pulsed on and off by the computer to supply the proper amount of vacuum to the EGR valve.

Regulador de vacío EGR Solenoide de vacío que es encendido y apagado por la computadora para suministrarle una cantidad adecuada de vacío a la válvula EGR.

Electric assist choke A choke that is heated by heat from a heat pipe and heat supplied by an electric heating element.

Estrangulador auxiliado eléctricamente Estrangulador que se calienta por medio del calor proveniente de un tubo de calor y el calor suministrado por un elemento de calefacción eléctrica.

Electric choke A choke that is heated by an electric heating element.

Estrangulador eléctrico Estrangulador que se calienta por medio de un elemento de calefacción eléctrica.

Electric-drive cooling fan A cooling fan driven by an electric motor.

Ventilador de enfriamiento accionado eléctricamente Ventilador de enfriamiento accionado por un motor eléctrico.

Electrolyte A sulfuric acid and water solution in automotive batteries.
Electrolito Solución de ácido sulfúrico y agua en baterías de automóviles.

Electromagnet A magnet created by the flow of electricity through a coil of wire.
Electroimán Imán producido por el flujo de electricidad a través de un alambre.

Electromagnetic induction The process of inducing a voltage by moving a conductor through a magnetic field or vice versa.
Inducción electromagnética Proceso de inducir una tensión moviendo un conductor a través de un campo magnético o viceversa.

Electromechanical regulator An alternator voltage regulator containing one or more mechanical relays.
Regulador electromecánico Regulador alternador de tensión que contiene uno o más relés mecánicos.

Electron A negatively charged particle that orbits around the nucleus of an atom.
Electrón Partícula de carga negativa que orbita alrededor del núcleo de un átomo.

Electronic continuous injection system (CIS-E) A computer-controlled fuel injection system that continually delivers fuel from the injectors.
Sistema electrónico de inyección continua Sistema de inyección de combustible controlado por computadora que continuamente envía combustible desde los inyectores.

Electronic control unit (ECU) An electronic device that opens and closes the primary ignition circuit.
Unidad de control electrónico Dispositivo electrónico que abre y cierra el circuito primario de encendido.

Electronic data system Service bulletins and other information contained on CD discs.
Sistema electrónico de datos Publicaciones y otra información referente al servicio mecánico; dicha información se guarda en discos compactos.

Electronic engine control V (EEC V) A term applied to a fourth-generation Ford computer system.
Sistema de control electrónico del motor V Término aplicado a una computadora de cuarta generación de la Ford.

Electronic fuel injection (EFI) A generic term applied to various fuel injection systems.
Inyección electrónica de combustible Término general aplicado a varios sistemas de inyección de combustible.

Electronic ignition (EI) system SAE J1930 terminology for an ignition system without a distributor.
Sistema de encendido electrónico Término utilizado por la SAE J1930 para referirse a un sistema de encendido que no tiene distribuidor.

Electronic muffler A muffler that suppresses noise with electronic sound waves.
Silenciador electrónico Silenciador que suprime el ruido por medio de ondas sónicas electrónicas.

Electronic spark timing (EST) A term applied to computer-controlled spark advance on some ignition systems.
Regulación electrónica de chispas Término aplicado al avance de chispas controlado por computadora en algunos sistemas de encendido.

Electronically erasable programmable read only memory (EEPROM) A computer chip that may be easily erased and reprogrammed with special equipment.
Memoria de solo lectura borrable y programable electrónicamente (EEPROM) Pastilla de memoria que puede borrarse fácilmente y reprogramarse con equipo especial.

Element A liquid, solid, or gas containing only one type of atom.
Elemento Líquido, sólido o gas que contiene sólo un tipo de átomo.

Emulsion The mixture of air and fuel in the carburetor.
Emulsión La mezcla de aire y combustible en el carburador.

Energy The ability to do work.
Energía Capacidad para realizar un trabajo.

Energy-Conserving oil Engine oil that is specially formulated to reduce power losses from friction.
Aceite de conservación de energía Aceite de motor que es especialmente formulado para reducir la perdida de fuerza debido a la fricción.

Engine coolant temperature (ECT) sensor An input sensor that sends a voltage signal to the computer in relation to coolant temperature.
Sensor de la temperatura del refrigerante del motor Sensor de entrada que le envía una señal de tensión a la computadora referente a la temperatura del refrigerante.

Environmental Protection Agency (EPA) The Federal government agency in charge of air and water quality in the United States.
Agencia para la Protección del Medio Ambiente (EPA) Agencia del gobierno federal que tiene a su cargo todos los aspectos relacionados a la calidad del agua y del aire en los Estados Unidos.

Ethylene glycol The basic chemical used in automotive antifreeze.
Glicol de etileno Producto químico básico utilizado en anticongelantes de automóviles.

Evaporative (EVAP) emission control systems Emission systems that control evaporative emissions from the fuel tank.
Sistemas de control de emisiones de evaporación Sistemas de emisión que controlan las emisiones de evaporación del tanque del combustible.

Exhaust gas recirculation valve position (EVP) sensor A sensor that sends a voltage signal to the computer in relation to EGR valve position.
Sensor de la posición de la válvula de recirculación del gas del escape Sensor que le envía una señal de tensión a la computadora referente a la posición de la válvula EGR.

Exhaust gas temperature sensor An input sensor that sends a voltage signal to the computer in relation to exhaust gas temperature.
Sensor de la temperatura del gas del escape Sensor de entrada que le envía una señal de tensión a la computadora referente a la temperatura del gas del escape.

Expansion tank Part of the cooling system that allows high-pressure antifreeze to escape the system. This tank allows the antifreeze to leave and relieve the system of some pressure without spilling the antifreeze onto the ground or into the atmosphere.
Tanque de expansión Parte del sistema de enfriamiento que permite que el anticongelante de alta presión se escape del sistema. Este tanque permite que el anticongelante se salga y libere la presión del sistema sin derramar el anticongelante en el piso o en la atmósfera.

Extreme pressure resistance An oil additive that helps to prevent high pressure from squeezing the oil out of the engine bearings.
Resistencia a presión extrema Aditivo de aceite que ayuda a evitar que la alta presión fuerce el aceite fuera de los cojinetes del motor.

Fast idle thermo valve A valve operated by a thermo-wax element that allows more air into the intake manifold to increase idle speed when the engine is cold.
Termoválvula de la marcha lenta rápida Válvula accionada por un elemento de termocera que permite que una mayor cantidad de aire entre en el colector de aspiración para aumentar la velocidad de la marcha lenta cuando el motor está frío.

Fast-start EI system An EI system with the capability to start firing with less crankshaft rotation compared to other systems.
Sistema de encendido electrónico de arranque rápido Sistema de encendido electrónico que, a diferencia de otros sistemas, tiene la capacidad de encenderse con menor giro del cigüeñal.

Federal Test Procedure A transient-speed mass sampling emissions test conducted on a loaded dynamometer. This is the test used by manufacturers to certify vehicles before they can be sold.

Procedimiento de la Prueba Federal Una prueba de la velocidad transitoria para la prueba de la masa en un dinamómetro cargado. Esta es la prueba usada por los fabricantes para certificar los vehículos antes de que ellos puedan ser vendidos.

Feedback A term used to describe a PCM's ability to control the activity of an actuator in response to input from sensors.

Retroalimentación Un termino usado para describir la habilidad del PCM (módulo de control de la potencia del motor) de controlar la actividad de un actuador en respuesta a la información de los sensores de entra.

F-head design In this type of engine, one of the valves in each cylinder is positioned in the block and the other valve is in the cylinder head.

Diseño de culata en F En este tipo de motor, una de las válvulas en cada cilindro está colocada en el bloque y la otra en la culata del cilindro.

Field of flux The lines of force that are emitted by a magnet. The strength of this field decreases as it moves away from the magnet.

Campo del flujo La líneas de fuerza que son emitidas por un magneto. La fuerza de este campo disminuye según se mueve hacia afuera del magneto.

Filler pipe A pipe through which fuel flows when filling the fuel tank.

Tubo de llenado Tubo a través del cual fluye el combustible al llenarse el tanque del combustible.

Filler pipe vent hose A hose connected from the top of the fuel tank to the outer end of the filler pipe.

Manguera de alivio del tubo de llenado Manguera conectada desde la parte superior del tanque del combustible hasta el extremo exterior del tubo de llenado.

Firing line The vertical line on a secondary scope pattern. The top of this line indicates the voltage required to start the spark plug firing.

Línea del encendido Línea vertical en los modelos de radio secundarios. La parte superior de esta línea indica la tensión que se requiere para encender la bujía.

Final drive gear The final set of gears for the powertrain. These are typically fixed torque multiplication gears and are in the differential assembly.

Engrane de conducción final El juego final del engrane para el tren de fuerza transmitida. Estos son los engranes para la multiplicación de torque/par e torsión aplicados fijamente y están en el ensamblaje del diferencial.

Firing order The order in which the cylinders of an engine are on their power stroke.

Orden del encendido El orden en el cual los cilindros de un motor están en su carrera de fuerza.

Flame front The outer edge of the immense heat buildup in a cylinder during combustion.

Frente de la flema La parte exterior del inmenso calor acumulado en un cilindro durante la combustión.

Flathead or side valve An engine with the valves mounted in the block.

Válvula lateral o de cabeza plana Motor con las válvulas montadas en el bloque.

Flat rate manual A manual that lists suggested times for each repair operation on various vehicles.

Manual de tarifa única Manual que especifica la duración aproximada de una reparación según los diferentes modelos de vehículos.

Flex-blade fan Fan blades that straighten out at high speed so less engine power is required to turn the fan.

Ventilador con aletas flexibles Aletas de un ventilador que se enderezan a alta velocidad, por lo que se requiere menos potencia del motor para hacer girar el ventilador.

Flexplate or flywheel The component that is bolted to the crankshaft and connects the crankshaft to the torque converter and supports the ring gear.

Volante de motor Componente que se emperna al cigüeñal, lo conecta al convertidor de torsión y apoya la corona.

Float system A system that controls fuel level in the float bowl.

Sistema del flotador Sistema que controla el nivel del combustible en el depósito del flotador.

Flywheel A heavy wheel attached to an engine's crankshaft used to absorb crankshaft vibrations, mount the clutch assembly, and add inertia to the rotation of the crankshaft.

Volante del motor Una rueda pesada acoplada al cigüeñal del motor usado para absorber las vibraciones del cigüeñal, montada en el ensamblaje del embrague y agregarle inercia a la rotación del cigüeñal.

Four-barrel carburetor A carburetor with four throttle bores.

Carburador de cuatro cilindros Carburador que tiene cuatro calibres de mariposa.

Four-stroke cycle An engine in which a complete cycle requires four piston strokes.

Ciclo de cuatro tiempos Motor en el que un ciclo completo requiere cuatro carreras del pistón.

Four-valve (4-V) cylinder head A cylinder head with four valves per cylinder.

Culata del cilindro con cuatro válvulas Culata del cilindro que tiene cuatro válvulas por cilindro.

Free-wheeling or spin-free engine An engine in which the valves do not contact the pistons if the timing chain or belt slips or breaks.

Motor de rueda libre o giro libre Motor en el que las válvulas no entran en contacto con los pistones si se rompen o se deslizan la cadena o la correa de regulación del encendido.

Freeze Frame A feature of some scan tools and a requirement of OBD II. This feature takes a snapshot of the operating conditions present when the PCM sets a diagnostic trouble code.

Marco congelado Una característica de algunas herramientas exploratorias y un equipo de OBD II (diagnóstico abordo del vehículo II). Esta característica toma una foto de las condiciones de operaciones presente cuando el PCM (módulo de control de la potencia del motor) establece un código de diagnostico de problema.

Frequency The number of complete cycles that occur in a specific period of time.

Frecuencia El número de ciclos completos que ocurren en un periodo especifico de tiempo.

Friction The resistance to motion when one object is moved over another object.

Fricción Resistencia al movimiento cuando un objeto se mueve sobre otro.

Friction modifiers Oil additives that reduce friction in the oil and between moving components that contact each other to improve fuel economy.

Modificadores de fricción Aditivos de aceite que disminuyen la fricción en el aceite y entre componentes en movimiento en contacto el uno con el otro, a fin de mejorar la capacidad para economizar combustible.

Fuel pump impeller A vaned rotating impeller attached to the fuel pump motor shaft that creates fuel pressure.

Impulsor de la bomba del combustible Impulsor giratorio con paletas fijado al árbol motor de la bomba del combustible que produce la presión del combustible.

Fuel pressure regulator A device designed to limit the amount of pressure built-up in a fuel delivery system.

Regulador de presión del combustible Un dispositivo diseñado para limitar la cantidad de acumulación de presión en un sistema de suministro de combustible.

Fuel pump relay A computer-operated relay that opens and closes the fuel pump relay.

Relé de la bomba del combustible Relé accionado por computadora que abre y cierra el relé de la bomba del combustible.

Fuel pump rocker arm An arm that rests against a camshaft lobe to operate the fuel pump.

Brazo de balancín de la bomba del combustible Brazo que se apoya contra un lóbulo del árbol de levas para accionar la bomba del combustible.

Fuel rail A metal or plastic pipe in which the upper end of the injectors is installed in port injection systems.

Carril del combustible Tubo metálico o plástico en el que se monta el extremo superior de los inyectores en sistemas de inyección con lumbrera.

Fuel vaporization Changing fuel from a liquid to a vapor.

Vaporización de combustible Cuando el combustible líquido se convierte en vapor.

Full-flow oil filter An engine oil filter in which all the flow from the pump flows through the filter.

Filtro del aceite de flujo completo Filtro del aceite de un motor a través del cual fluye todo el aceite de la bomba.

G-Lader supercharger A special design of supercharger in which the rotors rotate on an eccentric. The primary benefit of this design is the quietness of operation.

Supercargador con el sobre nombre de escalera G Un diseño especial de supercargador, en el cual los rotores giran en un excéntrico. El beneficio primario de este tipo de diseño, es que la operación es callada.

Gear reduction starting motor A starting motor with a set of reduction gears between the armature and the drive.

Motor de arranque con desmultiplicación de engranajes Motor de arranque con un juego de engranajes de reducción entre la armadura y el mecanismo de mando.

Gear-type oil pump An oil pump with a meshed pair of gears that create oil pressure and move the oil through the lubrication system.

Bomba del aceite con engranajes Bomba del aceite con un par de engranajes endentados que producen la presión del aceite y lo conducen a través del sistema de lubrificación.

General repair manual A repair manual published by a company that is independent from the car manufacturers. This manual usually includes several years or models of vehicles.

Manual para reparaciones generales Manual de reparación publicado por una compañía no relacionada a los fabricantes de automóviles. En dicho manual se incluyen normalmente varios años o modelos de vehículos.

Glitch A momentary disruption in an electrical circuit.

Falla Una interrupción momentánea en un circuito eléctrico.

Glow plug A plug used in diesel engines to warm the intake air to allow for quicker starting and better operation when the engine is cold.

Precalentador para la cámara de combustión Una resistencia eléctrica usada en un motor diesel para calentar el aire de admisión, para permitirlo de que arranque más rápido y que tenga mejor operación cuando el motor está frío.

Grounded circuit An unwanted copper-to-metal connection in an electric circuit.

Circuito puesto a tierra Conexión no deseada entre cobre y metal en un circuito eléctrico.

Hall effect switch A type of pickup assembly containing a Hall element and a permanent magnet with a rotating blade between these components.

Conmutador de efecto Hall Tipo de conjunto de captación que contiene un elemento Hall y un imán permanente entre los cuales está colocada una aleta giratoria.

Halogen and halon fire extinguishers Fire extinguishers that are effective on class B fires, but that give off toxic gases.

Extinctores de incendio de halógeno y de halón Extinctores de incendio que eficazmente apagan incendios de clase B, pero que emiten gases tóxicos.

Hazard Communication Standard The beginning of the right-to-know laws published by the Occupational Safety and Health Administration (OSHA).

Norma de Comunicación de Riesgo Comienzo de las leyes del derecho a saber; documento publicado por la Dirección para la Seguridad y Salud Industrial (OSHA).

Heated air inlet system A system that heats the intake air entering the engine.

Sistema de admisión de aire calentado Sistema que calienta el aire aspirado que entra en el motor.

Heat range A rating of a spark plug. This defines the plug's ability to dissipate heat. The hotter the heat range, the more heat it retains.

Rango de calor Una clasificación de una bujía. Esto define la habilidad de las bujías de disipar en calor. Lo más caliente que sea el rango, la mayor cantidad de calor que retiene.

Heated air inlet A device designed to warm intake air in order to improve cold engine operation.

Aire de entrada calentado Un dispositivo diseñado para calentar el aire de admisión en orden de mejorar la operación del motor en frío.

Heat riser valve A butterfly-type valve in the exhaust system that directs exhaust gas through a passage in the intake manifold, particularly when the engine is cold.

Válvula de camisa de calor de tubería Válvula tipo mariposa en el sistema de escape que conduce el gas del escape a través de un tubo en el colector de aspiración, particularmente cuando el motor está frío.

Heat stove A metal component in or on the exhaust manifold, or in the exhaust crossover passage that supplies heat to the choke.

Aparato de calor Componente de metal en o sobre el colector de escape, o en el tubo transversal de escape que le suministra calor al estrangulador.

Hertz A unit of measurement for counting the number of times an electrical cycle repeats every second. One hertz is one pulse each second.

Hertz Una unidad de medida para contar los números de veces que un ciclo eléctrico se repite cada segundo. Un hertz es un pulso cada segundo.

High energy ignition (HEI) The name given to early GM electronic ignition systems.

Sistema de ignición de alta energía El nombre otorgado a los sistemas de inyección de la GM.

Hold-in winding A winding in the starter solenoid that is energized while the solenoid is activated.

Devanado de retención Devanado en el solenoide de arranque que se energiza cuando se activa el solenoide.

Horsepower A rating that indicates engine power.

Potencia en caballos Clasificación que indica la potencia del motor.

Hot air-type choke A choke that is heated only by hot air.

Estrangulador tipo aire caliente Estrangulador que se calienta solamente por medio de aire caliente.

Hot spark plug A spark plug designed with a longer heat path so the electrodes operate at higher temperatures.

Bujía caliente Bujía diseñada con una trayectoria de calor más larga para que los electrodos funcionen a temperaturas más altas.

Hydrocarbons (HC) Left-over fuel after the combustion process.

Hidrocarburos El combustible restante después del proceso de combustión.

Idle air control by-pass air (IAC BPA) motor An IAC motor that controls idle speed by regulating the amount of air by-passing the throttle.

Motor para el control de la marcha lenta con el paso de aire Un motor IAC que controla la velocidad de la marcha lenta regulando la cantidad de aire que se desvía de la mariposa.

Idle air control by-pass air (IAC BPA) valve A valve operated by the IAC BPA motor that regulates the air by-passing the throttle to control idle speed.

Válvula para el control de la marcha lenta con el paso de aire Válvula accionada por el motor IAC BPA que regula el aire que se desvía de la mariposa para controlar la velocidad de la marcha lenta.

Idle air control (IAC) motor A computer-controlled motor that controls idle speed under all engine operating conditions.

Motor para el control de la marcha lenta con aire Motor controlado por computadora que controla la velocidad de la marcha lenta bajo cualquier condición del funcionamiento del motor.

Idle contact switch, or nose switch A switch in the IAC motor stem that informs the computer when the throttle linkage is contacting the motor stem.

Conmutador de contacto de la marcha lenta Conmutador en el vástago del motor IAC que le advierte a la computadora cuándo la conexión de la mariposa está en contacto con el vástago del motor.

Idle stop solenoid A solenoid that allows the throttle to move toward the closed position to prevent engine dieseling when the ignition is turned off.

Solenoide de detención de la marcha lenta Solenoide que le permite a la mariposa cerrarse para evitar el autoencendido del motor al desconectarse el encendido.

Idle system A carburetor system that supplies fuel when the engine is idling.

Sistema de marcha lenta Sistema del carburador que le suministra combustible al motor cuando éste se encuentra en marcha lenta.

Ignitable If a material burns when contacted by a spark, flame, or a certain degree of heat, it is ignitable.

Inflamable Si un material se quema al ser expuesto a una chispa, a una llama o a cierto grado de calor, se dice que dicho material es inflamable.

Ignition dwell time The number of degrees that the distributor shaft rotates while the primary circuit is closed prior to firing each cylinder.

Duración de retraso del encendido Número de grados que gira el árbol del distribuidor mientras el circuito primario se cierra antes de encender cada cilindro.

Ignition module An electronic device used to open and close the primary ignition circuit.

Módulo del encendido Dispositivo electrónico utilizado para abrir y cerrar el circuito primario del encendido.

Ignition timing Refers to the time, in relationship to piston movement, when spark ignition begins.

Tiempo de regulación de la ignición Se refiere al tiempo, en relación al movimiento del pistón, cuando la chispa de la ignición comienza.

I-head design An engine with the valves mounted in the cylinder head. This design may be classified as an overhead valve.

Diseño de culata en I Motor con las válvulas montadas en la culata del cilindro. Este diseño se puede clasificar como una válvula superpuesta.

Impeller The name given to one of the rotor or fin assemblies in a torque converter, turbocharger, and supercharger.

Propulsor El nombre otorgado a uno de los rotores o ensamblajes de las aletas en un convertidor de torque, turbo cargador o supercargador.

Impeller-type water pump A water pump in which a vaned impeller is driven to create pressure and force coolant through the cooling system.

Bomba del agua tipo impulsor Bomba del agua en la que se acciona un impulsor con paletas para producir presión y hacer circular el refrigerante a través del sistema de enfriamiento.

Induction The spontaneous creation of an electrical current in a conductor as the conductor passes through a magnetic field or a magnetic field passes across the conductor.

Inducción La creación espontanea de una corriente eléctrica en un conductor según el conductor atraviesa un campo magnético o un campo magnético atraviesa el conductor.

Inductive reluctance Refers to the opposition to current flow that takes places when a conductor is generating electricity as it is passed by a magnetic field.

Reluctancia inductiva Se refiere a la posición del flujo de la corriente que toma lugar cuando un conductor está generando electricidad según pasa atraves de un campo magnético.

Inertia The tendency of an object at rest to remain at rest, or the tendency of an object to remain in motion.

Inercia La tendencia de un objeto inmóvil a permanecer inmóvil, o la tendencia de un objeto a permanecer en movimiento.

Inertia switch A switch used in some fuel pump circuits that opens when it is impacted by moderate collision force.

Conmutador por inercia Conmutador utilizado en algunos circuitos de bombas de combustible que se abre al ser impactado por una colisión moderada.

Information retrieval Refers to the microprocessor retrieving information from some of the computer memories.

Recuperación de información Se refiere a la recuperación de información que lleva a cabo el microprocesador de algunas de las memorias de la computadora.

Information storage Refers to the microprocessor sending information to some of the memories where it is stored for future reference.

Almacenamiento de información Se refiere al envío de información que lleva a cabo el microprocesador hacia algunas de las memorias; es aquí donde la información se guarda para uso futuro.

Infrared analyzer An analyzer used to measure oxygen, carbon monoxide, carbon dioxide, and hydrocarbons in the exhaust.

Analizador infrarrojo Analizador utilizado para medir el oxígeno, el monóxido de carbono, el bióxido de carbono, y los hidrocarburos en el escape.

Injector sequencing Opening the injectors in the proper order to match the cylinder firing order.

Ordenamiento del inyector La apertura de los inyectores en una secuencia adecuada para equilibrar la secuencia del encendido del motor.

In-line block An engine block in which the cylinders are arranged in line with each other from the front to the back.

Bloque en línea Bloque de un motor en el que los cilindros están arreglados en relación el uno con el otro de la parte delantera a la trasera.

Input signal amplification The process of increasing input voltage signal power.

Amplificación de la señal de entrada Proceso de aumentar la potencia de tensión de la señal de entrada.

Inspection maintenance (I/M) programs Programs that are administered by various states to monitor emission levels on vehicles.

Programas de inspección y mantenimiento Programas administrados por varios estados para controlar los niveles de emisiones provenientes de vehículos.

Insulator An element with five or more valence electrons that does not conduct electric current.

Aislador Elemento con cinco o más electrones de valencia que no conduce corriente eléctrica.

Intake manifold heating grid An electrically heated grid mounted under the carburetor to heat the intake air on a cold engine.

Parrillas de calefacción del colector de aspiración Parrillas calentadas eléctricamente que se montan debajo del carburador para calentar el aire aspirado en un motor frío.

Intake manifold tuning valve (IMTV) A computer-operated valve that changes the length of the intake manifold air passages.

Válvula de ajuste del colector de aspiración Válvula accionada por computadora que varía el largo de los tubos de aire del colector de aspiración.

Integrated circuit A silicon chip with many components such as diodes and transistors etched on it.

Circuito integrado Pastilla de silicio en la que se han grabado una variedad de componentes, como por ejemplo diodos y transistores.

Integrated circuit (IC) regulator A voltage regulator designed on a silicon chip, which is usually integral with the alternator.

Regulador del circuito integrado Regulador de tensión diseñado en una pastilla de silicio, que normalmente es un complemento del alternador.

Intercooler A device that cools the intake air on a supercharged or turbocharged engine.

Interenfriador Dispositivo que enfría el aire aspirado en un motor de compresor o de turbocompresor.

Intermediate pipe Part of the exhaust system, primarily used to connect different majors parts of the system together.

Pipa intermedia Parte del sistema de escape, primariamente usado para acoplar diferentes partes mayores del sistema de escape.

Job estimate A cost estimate for repair work given to the customer.

Estimado del servicio Precio estimado que se le ofrece al cliente sobre la reparación de un vehículo.

Keep alive memory (KAM) The microprocessor can read information from the KAM, but it cannot write information into this chip. KAM memory is retained when the ignition switch is turned off, but it is erased when battery power is disconnected from the computer.

Memoria de entretenimiento (KAM) El microprocesador puede leer la información que contiene la KAM, pero no puede escribir información en esta pastilla. La memoria de entretenimiento se retiene cuando se desconecta el botón conmutador de encendido, pero se borra cuando la potencia de la batería se desconecta de la computadora.

Knock sensor A sensor that sends a voltage signal to the computer in relation to engine detonation.

Sensor de golpeteo Sensor que le envía una señal de tensión a la computadora referente a la detonación del motor.

Lead (Pb) An element used in paste form on negative battery plates.

Plomo (Pb) Elemento utilizado en forma de pasta en placas negativas de baterías.

Lead peroxide (PbO$_2$) A chemical paste on positive battery plates in a fully charged battery.

Peróxido de plomo (PbO$_2$) Producto químico en forma de pasta utilizado en placas positivas de baterías en una batería completamente cargada.

Lead sulfate (PbSO$_4$) A compound on the plates of a discharged battery.

Sulfato de plomo (PbSO$_4$) Compuesto utilizado en las placas de una batería descargada.

Lean A condition where there is a larger amount of air in the air-fuel mixture than would be if the mixture was stoichiometric.

Pobre Una condición donde hay una cantidad mayor de aire en la mezcla de aire combustible que si hubiera una mezcla estequiometrica.

L-head design An engine with the valves mounted in the block, with all the valves on one side of the cylinders.

Motor de válvulas al costado Motor con las válvulas montadas en el bloque; todas las válvulas se encuentran en un solo lado de los cilindros.

Light emitting diode (LED) A diode that gives off light when it conducts electric current.

Diodo emisor de luz Diodo que emite luz cuando conduce una corriente eléctrica.

Linear EGR valve An EGR valve containing a solenoid that is pulsed on and off by the computer to provide a precise EGR flow.

Válvula EGR lineal Válvula EGR con un solenoide que la computadora enciende y apaga para proporcionar un flujo exacto de EGR.

Linear TPS A TPS with a movable contact on a horizontal or vertical resistance coil.

TPS lineal TPS con un contacto móvil en una bobina de resistencia horizontal o vertical.

Load A term used to describe the weight or force an engine must overcome or an electric device that consumes electricity.

Carga Un termino usado para describir el peso o la fuerza que un motor debe de superar o la electricidad que consume un dispositivo con un motor eléctrico.

Lobe An eccentric rise from a base circle. Typically refers to the shape of a camshaft.

Lóbulo Una elevación desde la base de un circulo. Típicamente se refiere a la figura de un árbol de levas.

Logic module Older term used by Chrysler Corporation for an engine computer.

Módulo lógico Término que antiguamente utilizaba la Chrysler Corporation para referirse a la computadora de un motor.

Longitudinal engine mounting An engine mounted lengthways in the chassis.

Montaje longitudinal del motor Motor montado longitudinalmente en el chasis.

Long Term Adaptive Fuel Trim Long term fuel injector pulse width compensation determined by the PCM according to operating conditions. This fuel trim is set to maintain minimum emissions output and is the base point for short term fuel trim.

Restricción del combustible por tiempo largo Compensación amplia de los pulsos del inyector de combustible determinado por el PCM (módulo de control de la potencia del motor) de acuerdo con las condiciones de operaciones. Esta restricción de combustible es calibrada para mantener las emisiones a un mínimo y es el punto de base para la restricción del combustible por tiempo corto.

Look-up tables Contained in control computers, these tables contain the specifications and calibrations for the powertrain and its systems.

Tabla de observación Contenido en los controles de computadoras, estas tablas contienen las especificaciones y calibraciones para el tren de fuerza transmitida y sus sistemas.

Low-emission vehicle (LEV) Emission standards required by 1997.

Vehículo de baja emisión Normas sobre emisiones a establecerse antes del año 1997.

Low-maintenance battery A battery designed to minimize electrolyte loss, but water can be added to the battery.

Batería de bajo mantenimiento Batería diseñada para disminuir la pérdida de electrolitos, pero a la que se le puede agregar agua.

Low-tension rings Piston rings designed with reduced tension on the cylinder walls to decrease friction and improve fuel economy.

Anillos de baja tensión Anillos de pistón diseñados con menos tensión en las paredes de los cilindros a fin de disminuir la fricción y mejorar la capacidad para economizar combustible.

LP gas Liquefied petroleum gas, often referred to as propane, which burns clean in the engine and can be precisely controlled.

Gas LP Gas de petróleo líquido, comúnmente referido como propano, que quema limpio en el motor y puede ser precisamente controlado.

Magnetic field collapse Refers to magnetic field movement across conductors such as the ignition coil windings.

Derrumbe del campo magnético Se refiere al movimiento del campo magnético a través los conductores, como por ejemplo los bobinados del encendido.

Magnetic sensor A triggering device containing a winding and a permanent magnet.

Sensor magnético Dispositivo accionador que contiene un devanado y un imán permanente.

Main bearing bores Circular openings in the lower side of the engine block that contain the crankshaft main bearings and support the crankshaft.

Calibres de cojinetes principales Aberturas circulares en la parte inferior del bloque del motor que contienen los cojinetes principales del cigüeñal y que apoyan a éste.

Main system The carburetor system that supplies fuel from part throttle to full throttle.

Sistema principal Sistema del carburador que suministra combustible desde la apertura parcial de la mariposa hasta la apertura completa de la misma.

Maintenance-free battery A battery to which water cannot be added because there are no filler caps in the top of the case. Special design reduces electrolyte loss.

Batería libre de mantenimiento Batería a la que no se le puede agregar agua ya que no hay tapones de llenado en la parte superior de la caja. Este diseño especial disminuye la pérdida de electrolitos.

Malfunction indicator light (MIL) A light in the instrument panel that is illuminated if a defect occurs in the computer system.

Luz indicadora de funcionamiento defectuoso Luz en el panel de instrumentos que se ilumina si ocurre una falla en la computadora.

Manifold absolute pressure (MAP) sensor An input sensor that sends a voltage signal to the computer in relation to intake manifold vacuum.

Sensor de la presión absoluta del colector Sensor de entrada que le envía una señal de tensión a la computadora referente al vacío del colector de aspiración.

Manufacturer's service manuals Manuals published by the car manufacturer for a specific vehicle.

Manuales de servicio del fabricante Manuales publicados por el fabricante de automóviles para un vehículo particular.

Mass air flow (MAF) sensor An input sensor that sends a voltage signal to the computer in relation to the total volume of air entering the engine.

Sensor del flujo de aire en masa Sensor de entrada que le envía una señal de tensión a la computadora referente al volumen total de aire que entra en el motor.

Mass, weight, and volume Mass is the measurement of an object's inertia. Weight is the measurement of the earth's gravitational pull on an object. Volume is the length, width, and height of a space occupied by an object.

Masa, peso, y volumen La masa es la medida de la inercia de un objeto. El peso es la medida de la fuerza gravitacional de la tierra sobre un objeto. El volumen es el largo, el ancho, y la altura del espacio ocupado por un objeto.

Material Safety Data Sheets (MSDS) Sheets that provide information regarding hazardous materials.

Hojas de datos sobre la seguridad de un material Hojas de información sobre materiales peligrosos.

Maximum secondary coil voltage The maximum voltage that the coil is capable of producing.

Tensión máxima de la bobina secundaria Tensión máxima que la bobina es capaz de generar.

Mechanical efficiency The relationship between the engine power delivered and the power that would be delivered if the engine operated without any power loss.

Rendimiento mecánico Relación entre la potencia generada por el motor y la potencia que podría generarse si el motor funcionase sin pérdida de potencia alguna.

Memory calibrator (MEM-CAL) A removable chip in some computers that replaces the PROM and CAL-PAK chips.

Calibrador de memoria Pastilla desmontable en algunas computadoras que reemplaza las pastillas PROM y CAL-PAK.

Metering The process of controlling the flow of something. Metering of gasoline is controlled by the time it is allowed to flow, by the size of the opening it is flowing through, or by the pressure causing it to flow.

Regulación El proceso de controlar el flujo de algo. La regulación de la gasolina es controlada por el tiempo que es permitida que fluya, por el tamaño de la apertura que está fluyendo, o por la presión que la está causando que fluya.

Metering rods Small rods that are used in some carburetors to control the air/fuel mixture. Often these rods are tapered and move in and out of an orifice to control the flow of fuel through the orifice.

Varillas de regulación Pequeñas varillas que son usadas en algunos carburadores para controlar la mezcla de aire combustible. Comúnmente estas varillas son escalonadas y se mueven hacia adentro y hacia afuera de un orificio para controlar el flujo de combustible atraves de un orificio.

Methanol The lightest and simplest of the alcohols; also known as wood alcohol.

Metanol El más liviano y el más simple de los alcoholes; también conocido como alcohol de madera.

Microprocessor The decision-making chip in a computer.

Microprocesador Pastilla sobre la cual están implementadas las funciones aritméticas y lógicas de una computadora.

Microprocessor control unit (MCU) A term applied to some computer-controlled carburetor systems.

Unidad de control del microprocesador Término aplicado a algunos sistemas de carburador controlados por computadora.

Mid-engine transverse engine mounting An engine mounted crossways near the center of the vehicle.

Montaje transversal central del motor Motor montado transversalmente cerca del centro del vehículo.

Mixture control (MC) solenoid A computer-operated solenoid that controls air-fuel ratio in a carburetor.

Solenoide de control de mezcla Solenoide accionado por computadora que controla la relación de aire y combustible en un carburador.

Mixture heater An electrically operated heater mounted between the carburetor and the intake manifold that heats the air-fuel mixture on a cold engine.

Calentador de la mezcla Calentador eléctrico que se monta entre el carburador y el colector de aspiración para calentar la mezcla de aire y combustible en un motor frío.

Molecule The smallest particle of a compound.

Molécula La partícula más pequeña de un compuesto.

Momentum An object gains momentum when a force overcomes static energy and moves the object.

Impulso Un objeto cobra impulso cuando una fuerza supera la energía estática y mueve el objeto.

Monolithic-type catalytic converter A catalytic converter in which a honeycomb-type element is coated with the catalyst materials.

Convertidor catalítico tipo monolítico Convertidor catalítico en el que un elemento parecido a un panal está cubierto de materiales catalizadores.

Muffler A part of the exhaust system in which the pulses from the exhaust are dampened and the noise reduced.

Silenciador Una parte del sistema del escape en el cual los pulsos desde el escape son amortiguados y el sonido reducido.

Multiple-disc clutch *See* Clutch pack

Embrague de discos múltiples Vea clutch pack (páguete de embrague)

Multiport fuel injection (MFI) An electronic fuel injection system in which the injectors are grounded in the computer in pairs or groups of three or four.

Inyección de combustible de paso múltiple Sistema de inyección electrónico de combustible en el que los inyectores se ponen a tierra en la computadora en pares o en grupos de tres o cuatro.

Multipurpose dry chemical fire extinguisher A common type of extinguisher that may be used on various types of fires.

Extinctor de incendio de producto químico seco para aplicaciones múltiples Tipo común de extinctor que puede utilizarse en varios tipos de incendios.

Multiviscosity oil A chemically modified oil that has been tested for viscosity at cold and hot temperatures.

Aceite de viscosidad múltiple Un aceite modificado químicamente que ha sido probado por viscosidad a temperaturas frías y calientes.

Negative backpressure EGR valve An EGR valve with a vacuum bleed valve operated by negative pressure pulses in the exhaust system.

Válvula EGR de contrapresión negativa Válvula EGR con una válvula de descarga de vacío accionada por impulsos de presión negativa en el sistema de escape.

Negative pressure A pressure less than atmospheric pressure.

Presión negativa Presión más baja que la presión atmosférica.

Neutral drive switch (NDS) A switch that informs the computer regarding gear selector position.

Conmutador de mando neutral Conmutador que le advierte a la computadora sobre la posición del selector de velocidades.

Neutral safety switch A switch connected in the starter circuit that prevents the starting motor from operating except in park or neutral.

Conmutador de seguridad neutral Conmutador conectado en el circuito de arranque que evita que el motor de arranque funcione si el selector de velocidades no se encuentra en las posiciones PARK o NEUTRAL.

Neutron A particle with no electric charge positioned in the nucleus of an atom.

Neutrón Partícula desprovista de carga eléctrica ubicada en el núcleo de un átomo.

Noncompressibility A material with the capability to remain the same size when pressure is applied.

No compresibilidad Un material que tiene la capacidad de retener su tamaño al aplicársele presión.

Normal required secondary coil voltage The secondary coil voltage required to fire the spark plugs with the engine operating at idle or low speed.

Tensión de la bobina secundaria normal requerida Tensión de la bobina secundaria que se requiere para encender las bujías durante el funcionamiento del motor a una velocidad de marcha lenta o baja.

Occupational Safety and Health Administration (OSHA) The Federal government agency in the United States in charge of safety and healthful working conditions.

Dirección para la Seguridad y Salud Industrial Agencia del gobierno federal en los Estados Unidos que tiene a su cargo el establecimiento de condiciones seguras y saludables en el trabajo.

Octane A rating used to classify gasoline, refers to the volatility of the fuel.

Octano Una clasificación usada para clasificar la gasolina, se refiere a que tan volátil es el combustible.

Octane number A unit of measurement on a scale intended to indicate the tendency of a fuel to detonate or knock.

Número de octano Una unidad de medida en una escala con la intención para indicar la tendencia de un combustible de detonar o producir golpe de chispa.

Ohm A measurement for electrical resistance.

Ohmio Medida de resistencia eléctrica.

Oil classifications Various oil categories in relation to the type of use.

Clasificaciones de aceite Clases de aceite divididos en diferentes categorías de acuerdo a su utilización.

Oil rings Usually a single ring positioned below the compression rings on the piston to control the amount of oil on the cylinder wall.

Anillos de aceite Normalmente un solo anillo colocado debajo de los anillos de compresión en el pisión para controlar la cantidad de aceite en la pared del cilindro.

Oil viscosity ratings Various oil categories in relation to the oil's ability to flow.

Clasificaciones de la viscosidad del aceite Clases de aceite divididos en diferentes categorías de acuerdo a su capacidad de flujo.

Oil-wetted, resin-impregnated pleated paper element A type of air cleaner element.

Elemento de papel plegado, impregnado de resina y saturado de aceite Tipo de elemento del filtro de aire.

On-board diagnostics II (OBD II) A diagnostic system mandated by the California Air Resources Board (CARB) and other US agencies that illuminates the MIL light if emission levels exceed 1.5 times the specified limit. OBD II systems must be installed on 1996 cars.

Diagnóstico instalado en el vehículo Sistema diagnóstico bajo mandato de la Comisión para Recursos del Aire de California (CARB) y otras agencias estadounidenses; dicho sistema enciende la luz MIL si los niveles de emisión superan 1,5 veces el límite especificado. El sistema diagnóstico deberá instalarse en vehículos fabricados a partir del año 1996.

Open circuit An unwanted break in an electric circuit.

Circuito abierto Interrupción no deseada en un circuito eléctrico.

Open loop A computer system operating mode in which the computer ignores the oxygen sensor signal as it controls the air-fuel ratio.

Bucle abierto Modo de funcionamiento de una computadora en el que se pasa por alto la señal del sensor de oxígeno mientras se controla la relación de aire y combustible.

Opposed-type block An engine block designed with the cylinders opposite each other.

Bloque tipo opuesto Diseño de bloque de motor en el que los cilindros se encuentran en lados opuestos.

Optical-type pickup A pickup assembly containing a photo diode and a light emitting diode with a slotted plate rotating between these components.

Capacitación tipo óptico Conjunto de capacitación que contiene un fotodiodo y un diodo emisor de luz entre los cuales gira una placa ranurada.

Output Driver A transistor in the output area of a control device that is used to turn various output devices off and on.

Ejecutador de salida Un transistor en el área de salida de un dispositivo de control, que es usado para apagar y prender varios dispositivos de ejecución.

Output drivers and actuators Computer system outputs including relays and solenoids.

Excitadores y accionadores de salida Datos producidos por una computadora que incluyen relés y solenoides.

Overhead cam engine An engine with the camshaft located above or in the cylinder head.

Motor con árbol de levas superpuesto Motor con el árbol de levas ubicado sobre o en la culata del cilindro.

Overhead valve engine An engine with the valves positioned in the cylinder head.

Motor con válvulas superpuestas Motor con las válvulas colocadas en la culata del cilindro.

Overrunning clutch drive A starter drive that connects and disconnects the armature and the ring gear.

Mando del embrague de rueda libre Mecanismo de mando de arranque que conecta y desconecta la armadura y la corona.

Oxidation inhibitors Oil additives that reduce sticky tar-like substances in the oil caused by oxidation of the oil when oxygen in the air combines with hot oil.

Inhibidores de oxidación Aditivos de aceite que disminuyen sustancias pegajosas, parecidas al alquitrán. Dichas sustancias se encuentran en el aceite y son el producto de la oxidación del aceite cuando el oxígeno en el aire se mezcla con el aceite caliente.

Oxides of nitrogen (NO$_x$) A gas formed by the combining of oxygen and nitrogen at high combustion temperatures in the engine cylinders.

Oxidos de nitrógeno (NO$_x$) Gas formado por la combinación de oxígeno y nitrógeno a altas temperaturas de combustión en los cilindros del motor.

Oxygen (O$_2$) An gaseous element that is present in air.

Oxígeno (O$_2$) Elemento gaseoso presente en el aire.

Oxygen (O$_2$) feedback solenoid A computer-controlled solenoid in a carburetor.

Solenoide de realimentación de oxígeno (O$_2$) Solenoide controlado por computadora en un carburador.

Oxygen (O$_2$) feedback system A computer-controlled carburetor system on some Chrysler vehicles.

Sistema de realimentación de oxígeno (O$_2$) Sistema del carburador controlado por computadora incluído en algunos vehículos fabricados por la Chrysler.

Oxygen (O$_2$) sensor An input sensor that sends a voltage signal to the computer in relation to the amount of oxygen in the exhaust stream.

Sensor de oxígeno (O$_2$) Sensor de entrada que le envía una señal de tensión a la computadora referente a la cantidad de oxígeno en el caudal del escape.

Parallel circuit In this type of circuit, there is more than one path for the current to follow.

Circuito paralelo En este tipo de circuito, hay más de un camino para que fluya la corriente.

Particulates Part of the exhaust from a diesel engine. Typically particulates are caused by the melting of the paraffin present in diesel fuel.

Partículas Parte del escape de un motor diesel. Típicamente estas partículas son causadas por el derretimiento de la parafina presente en el combustible diesel.

Parts requisition An order requesting the necessary parts for a vehicle.

Pedido de piezas Encargo de las piezas necesarias para un vehículo.

Pellet-type catalytic converter A converter that contains small pellets coated with catalyst materials.

Convertidor catalítico tipo grano gordo Convertidor que contiene pequeños granos gordos cubiertos de materiales catalizadores.

Photo diode A diode that produces a voltage signal when light shines on it.

Fotodiodo Diodo que produce una señal de tensión cuando es iluminado por la luz.

Photoelectric sensor A sensor that uses a LED, photo cell, and a slotted disc to measure rotational speeds.

Sensor fotoeléctrico Un sensor que usa un LED (diodo emisor de luz o electroluminiscente,) foto celda y un disco ranurado para medir la velocidad de las rotaciones.

Pickup coil A winding and permanent magnet assembly mounted in the distributor and used for ignition triggering.

Bobina de captación Conjunto de devanado e imán permanente montado en el distribuidor y utilizado para el arranque del encendido.

Pilot bearing A bearing inserted in the flywheel to support and guide the movement of the transmission's input shaft.

Balero piloto Un balero insertado en el volante para soportar y guiar el movimiento del eje de entrada de la transmisión.

Pintle The center pin used to control a fluid passing through a hole; a small pin or pointed shaft used to open or close a passageway.

Pivote central La clavija central usada para controlar un flúido pasando atraves de un orificio; una clavija pequeña o un eje con una punta usado para abrir o cerrar un pasaje.

Piston boss The area around the piston pin bore.

Cubo de pistón El área alrededor del calibre del pasador de pistón.

Piston pin bore The circular opening in the piston in which the piston pin is located.

Calibre del pasador de pistón Abertura circular en el pistón en la que se encuentra el pasador del pistón.

Piston skirt The area of the piston below the rings.

Faldilla del pistón El área del pistón debajo de los anillos.

Piston stroke Piston movement from top dead center to bottom dead center.

Carrera del pistón Movimiento del pistón desde el punto muerto superior hasta el punto muerto inferior.

Pleated paper fuel filter element The cleaning element in some fuel filters.

Elemento de papel plegado del filtro del combustible Elemento depurador en algunos filtros de combustible.

Point-type ignition systems An ignition system in which a set of contacts is used to open and close the primary circuit.

Sistemas de encendido tipo contacto Sistema de encendido en el que se utiliza un juego de contactos para abrir y cerrar el circuito primario.

Polyurethane air cleaner element cover A circular polyurethane ring surrounding the air cleaner element for improved cleaning.

Cubierta de poliuretano del elemento del filtro de aire Anillo circular de poliuretano que rodea el elemento del filtro de aire para facilitar la limpieza.

Poppet nozzles Mechanically operated nozzles in the intake ports of a central port injection system.

Toberas de movimiento vertical Toberas accionadas mecánicamente en las lumbreras de aspiración de un sistema de inyección de lumbrera central.

Ported vacuum switch (PVS) A vacuum switch controlled by ported (above the throttle plates) vacuum.

Interruptor de vacío de puerto Un interruptor de vacío controlado por vacío de puerto (encima de las mariposas del carburador.)

Port EGR valve An EGR valve that is operated by a ported vacuum source above the throttles.

Válvula EGR de lumbrera Válvula EGR accionada por una fuente de vacío con lumbreras sobre las mariposas.

Port fuel injection (PFI) A fuel injection system with an injector mounted in each intake port.

Inyección de combustible de lumbrera Sistema de inyección de combustible que tiene un inyector montado en cada una de las lumbreras de aspiración.

Positive backpressure EGR valve An EGR valve containing a vacuum bleed valve that is operated by positive pressure in the exhaust system.

Válvula EGR de contrapresión positiva Válvula EGR que contiene una válvula de descarga de vacío accionada por presión positiva en el sistema de escape.

Positive crankcase ventilation (PCV) A system that moves crankcase vapors into the intake manifold rather than having them escape to the atmosphere.

Ventilación positiva del cárter Sistema que conduce los vapores del cárter hacia el colector de aspiración en vez de permitir que los mismos se escapen hacia la atmósfera.

Positive pressure A pressure greater than atmospheric pressure.

Presión positiva Presión más alta que la presión atmosférica.

Power The measurement of the rate at which work is done.

Potencia Medida de la capacidad a la que se realiza un trabajo.

Power enrichment system A carburetor system that supplies more fuel to the main system when the engine is operating under load or at high speed.

Sistema de enriquecimiento de potencia Sistema del carburador que le suministra mayor cantidad de combustible al sistema principal cuando el motor funciona bajo presión o a alta velocidad.

Power module An older term used by Chrysler Corporation for an engine computer that worked with a logic module.

Módulo de potencia Término que antiguamente utilizaba la Chrysler Corporation para referirse a una computadora de motor que funcionaba con un módulo lógico.

Power train control module (PCM) Common term in SAE J1930 terminology for an automotive engine computer.

Módulo del control del tren transmisor de potencia Término utilizado comúnmente por la SAE J1930 para referirse a la computadora del motor de un automóvil.

Power (watt) rating A battery rating indicating the total amount of electrical power a battery will deliver.

Clasificación de potencia (watio) Clasificación de una batería que indica la cantidad total de energía eléctrica que la misma puede generar.

Power valve A valve in a carburetor that allows for extra fuel during heavy load operation.

Válvula de poder Una válvula en el carburador que suministra combustible adicional durante operaciones de cargas pesadas.

Pressure feedback electronic (PFE) sensor An input sensor that sends a voltage signal to the computer in relation to the exhaust pressure in a chamber under the EGR valve.

Sensor electróncio de realimentación de presión Sensor de entrada que le envía una señal de tensión a la computadora referente a la presión del escape en una cámara debajo de la válvula EGR.

Pressure regulator A mechanical device that controls fuel pressure in an electronic fuel injection system.

Regulador de presión Dispositivo mecánico que controla la presión del combustible en un sistema de inyección electrónica de combustible.

Pressure transducer A vacuum switching device operated by exhaust pressure that opens and closes the vacuum supply to the EGR valve.

Transconductor de presión Dispositivo de conmutación de vacío accionado por la presión del escape que le abre y cierra el suministro de vacío a la válvula EGR.

Primary circuit The low-voltage circuit of an ignition system.

Circuito primario El circuito de bajo voltaje en un sistema de ignición.

Primary coil winding An ignition coil winding with a few hundred turns of heavy wire.

Bobinado primario Bobinado del encendido enrollado con varios cientos de vueltas de alambre pesado.

Programmable read only memory (PROM) A computer chip containing some of the computer program. This chip is removable in some computers.

Memoria de solo lectura programable (PROM) Pastilla de memoria que contiene parte del programa de la computadora. Esta pastilla es desmontable en algunas computadoras.

Protocol The language or method used by a PCM to communicate to other computers.

Protocolo El lenguaje o método usado por una PCM (módulo de control de la potencia del motor) para comunicarse con las otras computadoras.

Proton A positively charged particle located in the nucleus of an atom.

Protón Partícula con carga positiva ubicada en el núcleo de un átomo.

Pull-in winding A winding in the starter solenoid that is energized while the starter is engaging.

Devanado energizado Devanado en el solenoide de arranque que se energiza cuando se acopla el motor de arranque.

Pulse A voltage signal that increases from a constant value and then decreases back to its original value.

Pulso Una señal de voltaje que incrementa desde un valor constante y después disminuye de regreso a su valor original.

Pulse modulated A circuit that maintains average voltage levels by pulsing the voltage on and off.

Pulso modulado Un circuito que mantiene niveles promedios de voltaje pulsando en una frecuencia de voltaje de apagado a encendido.

Pulse rate The number of pulses that take place over a specific period of time.

Velocidad del pulso Los números de pulso que toman lugar sobre un periodo de tiempo especifico.

Pulse transformer An electrical device that transforms or changes low voltage into high voltage on a pulse basis.

Transformador del pulso Un dispositivo eléctrico que transforma o cambia el voltaje bajo a voltaje alto en una base de pulso.

Pulse width The duration from the beginning to the end of a signal's on-time or off-time.

Amplitud del pulso La duración desde el principio al final de una señal prendida y apagada.

Pulsed secondary air injection A system in which negative exhaust pressure pulses are used to move air into the exhaust ports.

Inyección secundaria de aire por impulsos Sistema en el que se utilizan los impulsos de la presión negativa del escape para conducir el aire hacia las lumbreras del escape.

Pulse width The time an injector is open in milliseconds.

Duración de impulsos Espacio de tiempo que un inyector permanece abierto, medido en milisegundos.

Pyrometer A tool used to measure the temperature of something.

Pirómetro Una herramienta usada para medir la temperatura de algo.

Quick-disconnect fuel line fittings Fuel line fittings that may be disconnected without using a wrench.

Conexiones de la línea del combustible de desmontaje rápido Conexiones de la línea del combustible que se pueden desmontar sin la utilización de una llave de tuerca.

Radial Ply A type of tire that has body cords that extend from bead to bead at an angle of about 90 degrees.

Placas radiales Un tipo de neumático que tiene un cuerpo de cordón que se extiende desde un borde al otro borde en un ángulo de 90 grados.

Radiation A method of heat transfer that occurs when heat waves travel through the atmosphere and strike another object.

Radiación Método de transferencia de calor que occure cuando ondas de calor viajan a través de la atmósfera y chocan con otro objeto.

Radiator cap sealing gasket A gasket that seals the radiator cap to the radiator filler neck.

Guarnición de estanqueidad de la tapa del radiador Guarnición que sella herméticamente la tapa del radiador al cuello de llenado del radiador.

Radiator cap vacuum valve A valve that opens and allows coolant to flow from the coolant recovery container to the radiator when the engine coolant temperature decreases. This valve prevents radiator hose collapse.

Válvula de vacío de la tapa del radiador Válvula que se abre para permitir que el refrigerante fluya desde el recipiente de recuperación de refrigerante hasta el radiador cuando la temperatura del refrigerante del motor disminuye. Esta válvua evita que la manguera del radiador se plegue.

Random access memory (RAM) A computer chip that the microprocessor can write information into and retrieve information from.

Memoria de acceso aleatorio (RAM) Pastilla de memoria en la que el microprocesador puede escribir y de la que puede leer información.

Raster scope pattern A scope pattern in which all the voltage traces from the cylinders are displayed one above the other on the screen.

Modelo de radio de rastreo Modelo de radio en el que las pequeñas cantidades de tensión provenientes de los cilindros se proyectan una sobre otra a través de la pantalla.

Reactive The capability of a material to react violently when it comes in contact with another material.

Reactivo Capacidad de un material para producir una reacción violenta cuando entra en contacto con otro material.

Read only memory (ROM) The microprocessor can read information from the ROM chip, but it cannot write information into this chip.

Memoria de sólo lectura (ROM) El microprocesador puede leer la información que contiene la pastilla ROM, pero no puede escribir información en la misma.

Recovery tank Part of the cooling system in which coolant spillage or excess is collected to be reused when the system is low.

Tanque de recuperación Parte del sistema de enfriamiento en el cual el derramamiento del anticongelante o exceso es coleccionado para ser usado nuevamente cuando el sistema está bajo.

Recycled oil Waste oil that has been reconditioned to original standards.

Aceite reciclado Residuos de aceite que han sido reacondicionados a su estado original.

Reference pickup A type of pickup assembly that is often used for ignition triggering.

Captación de referencia Tipo de conjunto de captación que se utiliza con frecuencia para el arranque del encendido.

Reference voltage (Vref) A voltage provided by a voltage regulator to operate potentiometers and other sensors at a constant level.

Voltaje de referencia Un voltaje suministrado por un regulador de voltaje para operar un potenciómetro y otros sensores a un nivel constante.

Relay An electrical device used to switch a high current circuit with a low current circuit.

Relé Un dispositivo eléctrico usado para cambiar una corriente alta con un circuito de corriente baja.

Reluctor ring A metal ring that is integral with the crankshaft; slots on this ring are used for ignition triggering.

Anillo de reluctancia Anillo de metal que es un complemento del cigüeñal; las ranuras de este anillo se utilizan para el arranque del encendido.

Remote air cleaner An air cleaner mounted separately from the engine.

Filtro remoto de aire Filtro de aire montado lejos del motor.

Repair order An order written by the service writer or shop foreman, detailing the repairs to be completed on a customer's vehicle.

Solicitud de reparación Solicitud que llenan el mecánico o el encargado del taller donde se especifican las reparaciones que se llevarán a cabo en el vehículo de un cliente.

Reserve capacity rating A battery rating that indicates the length of time a battery will deliver current with a 25-ampere load.

Clasificación de capacidad en reserva Clasificación de una batería que indica el espacio de tiempo que una batería podrá generar corriente con una carga de 25 amperios.

Reserve secondary coil voltage The difference between normal required secondary coil voltage and maximum secondary coil voltage.

Tensión de la bobina secundaria en reserva La diferencia entre la tensión de la bobina secundaria normal requerida y la tensión máxima de la bobina secundaria.

Resonator Part of some exhaust systems. Best described as a back-up muffler. The resonator reduces the noise level of the exhaust after it has passed through the muffler.

Resonador Una parte de algunos sistemas de escape. Mejor descrito como un silenciador auxiliar. El resonador reduce el nivel del ruido del sistema de escape después de que haya pasado atraves del silenciador.

Resource Conservation and Recovery Act (RCRA) An act that controls hazardous waste disposal in the United States.

Ley de Conservación y Recuperación de Recursos Ley que controla la eliminación de residuos peligrosos en los Estados Unidos.

Returnless fuel system A fuel injection system with the pressure regulator mounted on top of the fuel tank so fuel is returned directly from this regulator to the tank.

Sistema de combustible sin retorno Sistema de inyección de combustible en el que el regulador de presión está montado sobre el tanque del combustible para que se pueda devolver el combustible directamente del regulador al tanque.

Reverse-flow cooling system A cooling system in which the engine coolant flows through the cylinder heads before it flows through the block.

Sistema de enfriamiento de flujo inverso Sistema de enfriamiento en el que el refrigerante del motor fluye a través de las culatas de los cilindros antes de fluir a través del bloque.

Reverse-flow muffler A muffler that reverses the flow of exhaust through the muffler to provide quieter operation.

Silenciador de flujo inverso Silenciador que invierte la dirección del flujo del escape a través del silenciador para proporcionar un funcionamiento con menos ruido.

Rheostat A variable resistor with two leads, an input and an output.

Reóstato Un resistor variable con dos cables, uno de entrada y uno de salida.

Rich A condition in which there is more fuel in the air/fuel mixture than would be if the mixture was stoichiometric.

Rico Una condición en la cual hay más combustible en la mezcla de aire/combustible que abría si la mezcla fuera estequiometrica.

Right-to-know laws Laws that state the workers have a right to know when the materials they use at work are hazardous.

Leyes del derecho a saber Estas leyes establecen que los empleados tienen el derecho a saber cuando los materiales utilizados en su trabajo son peligrosos.

Rotary engine An engine design that relies on the movement of a rotor instead of pistons to cycle through the four strokes.

Motor rotatorio Un diseño de motor que depende en el movimiento de un rotor envés de pistones para ciclar atraves de los cuatro tiempos.

Rotary throttle position sensor (TPS) A TPS with an electrical contact moving on a circular resistance coil.

Sensor de la posición giratoria de la mariposa Sensor de posición del acelerador con un contacto eléctrico que se mueve en una bobina circular de resistencia.

Rotor-type oil pump An oil pump in which two meshed rotors create pressure to force oil through the lubrication system.

Bomba del aceite tipo rotor Bomba del aceite en la que dos rotores engranados producen una presión para hacer circular el aceite a través del sistema de lubrificación.

Saturation The point reached when current flowing through a coil or wire has built up the maximum magnetic field.

Saturación El punto alcanzado cuando la corriente fluye atraves de una bobina o alambre que haya llenado el campo magnético a su máximo.

SAE J1930 terminology An attempt to standardize electronic terminology in the automotive industry.

Terminología de la SAE J1930 Intento de uniformar los términos electrónicos utilizados en la industria automotriz.

Sealed-terminal battery A battery with the terminals on the side of the case.

Batería con terminales sellados Batería en la que los terminales se encuentran en el lado de la caja.

Secondary air bypass (AIRB) Part of the air injection cycle when air is send to the atmosphere.

Desvío secundario del aire Parte del ciclo de la inyección de aire cuando el aire es enviado a la atmósfera.

Secondary air diverter (AIRD) Part of the air injection cycle when air to sent to the converter or the exhaust manifold.

Diversión secundaria del aire Parte del ciclo de la inyección de aire cuando el aire es enviado al convertidor o al múltiple de escape.

Secondary air injection A system that injects air into the exhaust system with a belt-driven air pump.

Inyección secundaria de aire Sistema que inyecta aire dentro del sistema de escape mediante una bomba de aire accionada por una correa.

Secondary circuit The part of the ignition system that produces and delivers high voltage to the spark plugs.

Circuito secundario La parte del sistema de la ignición que produce y suministra voltaje alto a las bujías.

Secondary coil winding A coil winding with many turns of fine wire in which the high voltage is induced to fire the spark plugs.

Bobinado secundario Bobinado enrollado con varias vueltas de alambre fino en el que se induce alta tensión para encender las bujías.

Semiconductor An element with four valence electrons that has unusual characteristics when combined with some other elements.

Semiconductor Elemento con cuatro electrones de valencia que muestra características inusuales al combinarse con algunos otros elementos.

Sequential fuel injection (SFI) An electronic fuel injection system in which the injectors are grounded individually in the computer.

Inyección de combustible en ordenamiento Sistema de inyección electrónica de combustible en el que los inyectores se ponen individualmente a tierra en la computadora.

Series circuit An electrical circuit that has two or more resistors connected in line so that the same amount of current must pass through them.

Circuito en serie Un circuito eléctrico que tiene dos o más resistores conectados en línea para que la misma cantidad de corriente deba de pasar através de ellos.

Series-parallel circuit An electrical circuit that contains at least one parallel circuit connected in series to a resistor.

Circuito en serie-paralelo Un circuito eléctrico que contiene por lo menos un circuito paralelo conectado en serie al resistor.

Service bulletin Bulletins published by the car manufacturers or independent sources which provide information on service changes and procedures.

Folleto de servicio Folletos publicados por los fabricantes de automóviles o compañías no relacionados a la empresa automotriz que proveen información sobre procedimientos o revisiones acerca del servicio.

Servo unit The part of a cruise control system that maintains the desired car speed by receiving a controlled amount of vacuum from the transducer.

Unidad de servo La parte de un sistema de control de crucero que mantiene la velocidad deseada del vehículo recibiendo una cantidad controlada de vacío desde el transductor.

Shorted circuit An unwanted copper-to-copper connection in an electric circuit.

Cortocircuito Conexión no deseada entre cobre y cobre en un circuito eléctrico.

Short Term Adaptive Fuel Trim Short term fuel injector pulse width compensation determined by the PCM according to operating conditions. This fuel trim is set to minimize emissions output and represents minor adjustments to the long term fuel trim strategy.

Tiempo corto para la adaptación de la restricción del combustible Compensación determinada del pulso corto para la amplitud del inyector de combustible de acuerdo con el PCM (módulo de control de la potencia del motor) para las condiciones de operaciones. Esta restricción del combustible es establecida para minimizar las emisiones y representan pequeños ajustes a la restricción del combustible por tiempo largo.

Shutter blades Rotating blades in the distributor that rotate past the pickup assembly.

Aletas enrejadas Aletas giratorias en el distribuidor que sobrepasan el conjunto de captación.

Side-terminal battery A battery with the terminals located on the side of the case.

Batería con terminales laterales Batería en la que los terminales se encuentran en el lado de la caja.

Single-barrel carburetor A carburetor with a single throttle bore.

Carburador de un solo cilindro Carburador que tiene un solo calibre de mariposa.

Single overhead camshaft (SOHC) An engine with one camshaft positioned above or in the cylinder head.

Árbol de levas superpuesto sencillo Motor con un árbol de levas colocado sobre o en la culata del cilindro.

Sintered brass fuel filter A type of fuel filter with a sintered brass filtering element.

Filtro del combustible de latón sinterizado Tipo de filtro del combustible que tiene un elemento filtrante de latón sinterizado.

Slant-type block An in-line engine mounted on a slant rather than at the true vertical position.

Bloque tipo inclinado Motor en línea montado de forma oblicua en vez de en una posición completamente vertical.

Slow-start EI system An EI system that may require two crankshaft revolutions before it begins firing the spark plugs.

Sistema de encendido electrónico de arranque lento Sistema de encendido electrónico que podría requerir dos revoluciones del cigüeñal antes de encender las bujías.

Snapshot A feature of OBD II systems that captures all data present when the computer detects a problem and sets a trouble code.

Foto instantánea Una característica del sistema OBD II (diagnóstico abordo del vehículo) que captura toda la data presente cuando la computadora detecta un problema y establece un código de problema.

Solenoid An electromagnetic device with a moveable core. Primarily used to cause some mechanical action or to act like a switch.

Solenoide Un dispositivo electromagnético con un núcleo removible. Primariamente usado para causar alguna acción mecánica o para actuar como un interruptor.

Spark ignition (SI) An engine in which the air fuel mixture in the cylinders is ignited by a spark at the spark plug electrodes.

Encendido por chispa Motor en el que la mezcla de aire y combustible en los cilindros se enciende por una chispa en los electrodos de la bujía.

Spark line The horizontal line on a secondary scope pattern that indicates the voltage required to keep the spark plug firing.

Línea de chispa Línea horizontal en un modelo de radio secundario que indica la tensión que se requiere para mantener la bujía encendida.

Spark plug heat range The rate at which a spark plug conducts heat away from the center electrode.

Intervalo de calor de la bujía Velocidad a la que una bujía conduce el calor fuera del centro de un electrodo.

Speed density system A computer-controlled fuel injection system in which the computer calculates the amount of air entering the engine from the engine rpm and manifold absolute pressure (MAP) sensor signals.

Sistema de densidad de velocidad Sistema de inyección de combustible controlado por computadora en el que la misma calcula la cantidad de aire que entra en el motor mediante las señales de las rpm del motor y de los sensores de la presión absoluta del colector.

Starting air valve A vacuum-operated valve that supplies more air into the intake manifold when starting the engine.

Válvula de aire para el arranque Válvula accionada por vacío que le suministra mayor cantidad de aire al colector de aspiración durante el arranque del motor.

Static balance Balance at rest, or still balance. it is the equal distribution of weight of the wheel and tire around the axis of rotation such that the wheel assembly has no tendency to rotate by itself regardless of its position.

Balanceo estático Balanceo en descanso o balanceo sin movimiento. Es la distribución igual de peso de la rueda y el neumático alrededor del axis de rotación, tal como el ensamblaje de la rueda no tiene tendencia de girar por si misma sin considerar su posición.

Stator The name given to many automotive parts, such as the stationary windings of an alternator, the third member of a torque converter, and the trigger wheel of an electronic distributor-type ignition system.

Estator El nombre otorgado a una parte automotriz, tal como el embobinado estacionario de un alternador, el tercer miembro de un convertidor de torque y la rueda disparadora de un distribuidor electrónico sistema de inyección electrónico.

Steering gear The device that converts the movement of the steering wheel to the movement of the linkages that turn the wheels.

Caja de la dirección El dispositivo que convierte el movimiento del volante de la dirección al movimiento de la varilla que gira las ruedas.

Steering linkage The arms that connect the steering gear to the wheels.

Varilla de la dirección Los brazos que conectan la caja de la dirección a las ruedas.

Stepper motor An electric motor that moves a pintle or valve horizontally in steps.

Motor por etapas Motor eléctrico que mueve una clavija o válvula horizontalmente por etapas.

Stoichiometric air-fuel ratio The ideal air-fuel ratio at which combustion is most complete.

Relación estequiométrica de aire y combustible Relación ideal de aire y combustible donde la combustión es más completa.

Strategy A plan. Typically refers to the programs of a PCM that insure low emissions levels.

Estrategia Un plan. Típicamente se refiere a los programas del PCM (módulo de control de la potencia del motor) que aseguran niveles bajos de emisiones.

Stroke Normally refers to the movement of an engine's piston from TDC to BDC.

Carrera Normalmente se refiere a los movimientos de los pistones de un motor desde el punto muerto superior al punto muerto inferior.

Sulfuric acid (H_2SO_4) A corrosive acid mixed with water and used in automotive batteries.

Ácido sulfúrico (H_2SO_4) Ácido sumamente corrosivo mezclado con agua y utilizado en las baterías de automóviles.

Supercharger A device driven by the engine's crankshaft used to force more air into the cylinder during its intake stroke.

Supercargador Un dispositivo conducido por el cigüeñal de los motores usado para forzar más aire adentro de los cilindros durante su carrera de admisión.

Superimposed scope pattern A scope pattern in which the voltage traces from all the cylinders are superimposed on top of each other.

Modelo de radio sobrepuesto Modelo de radio en el que las pequeñas cantidades de tensión provenientes de todos los cilindros se sobreponen una sobre otra.

Synchronizer (SYNC) pickup A pickup assembly that produces a voltage signal for ignition triggering or injector sequencing.

Captación del sincronizador Conjunto de captación que produce una señal de tensión para el arranque del encendido o para el ordenamiento del inyector.

Synthetic oil Oil that is formulated in laboratories.

Aceite sintético Aceite creado en laboratorios.

Tail pipe The part of the exhaust system that allows the exhaust gases to leave the rear of the vehicle and into the atmosphere.

Tubo de escape La parte del sistema de escape que permite que los gases de escape salgan de la parte de atrás del vehículo y adentro de la atmósfera.

Temperature vacuum switch (TVS) A switch that is controlled by temperature to control vacuum signals from one point to another.

Interruptor de vacío por temperatura Un interruptor que es controlado por temperatura para controlar las señales de vacío desde un punto a otro.

T-head design An engine with the valves in the block, and one valve on each side of the cylinder.

Diseño de culata en T Motor con válvulas en el bloque, y con una válvula a cada lado del cilindro.

Theft deterrent computer A computer that controls the theft deterrent system and activates specific warning devices if the vehicle is entered improperly.

Computadora para el sistema anti-robo Computadora que controla el sistema anti-robo y activa dispositivos de advertencia específicos en caso de que alguien intente adueñarse del vehículo indebidamente.

Thermal efficiency The relationship between the engine power output and the heat energy available in the fuel.

Rendimiento térmico Relación entre la salida de la potencia del motor y la energía térmica disponible en el combustible.

Thermal vacuum valve (TVV) A vacuum valve that is opened and closed by a thermo-wax element in the cooling system.

Válvula térmica de vacío Válvula de vacío que un elemento de termocera en el sistema de enfriamiento abre y cierra.

Thermostatic air cleaner An assembly that either allows air from the outside or from the exhaust manifold to enter the air cleaner assembly.

Filtro de aire termostatico Un ensamblaje que puede fluir aire desde la parte exterior o desde el múltiple de escape para el ensamblaje del purificador de aire.

Thermostatic element A heat-sensitive wax pellet that may be used to open and close the air door in a heated air inlet system.

Elemento termostático Grano gordo de cera sensible al calor que puede utilizarse para abrir y cerrar la puerta de ventilación en un sistema de admisión de aire calentado.

Thermostatic or bimetallic coil A coil made from two different metals with different expansion rates. The coil winds and unwinds as it is heated and cooled.

Bobina termostática o bimetálica Bobina compuesta de dos metales diferentes con diferentes capacidades de expansión. La bobina se enrolla y desenrolla al calentarse y enfriarse.

Thermo time switch A switch that controls a component such as the cold start injector in relation to temperature and time.

Conmutador de termotiempo Conmutador que controla un componente, como por ejemplo el inyector de arranque en frío, de acuerdo a la temperatura y al tiempo.

Three-way catalyst A catalytic converter that reduces three pollutants: carbon monoxide, unburned hydrocarbons, and oxides of nitrogen.

Catalizador triple Convertidor catalítico que disminuye tres sustancias contaminantes: el monóxido de carbono, los hidrocarburos no quemados, y los óxidos de nitrógeno.

Throttle body injection (TBI) A fuel injection system in which the injector, or injectors, is positioned above the throttles.

Inyección del cuerpo de la mariposa Sistema de inyección de combustible en el que el inyector, o los inyectores, están colocados sobre las mariposas.

Throttle kicker An electric- or vacuum-operated device that holds the throttle open a certain amount under specific engine operating conditions.

Nivelador de la mariposa Dispositivo accionado eléctricamente o por vacío que mantiene la mariposa algo abierta bajo ciertas condiciones del funcionamiento del motor.

Throttle stop solenoid A solenoid used to keep the throttle plates from fully closing during deceleration and under certain conditions.

Solenoide para detener el acelerador Un solenoide usado para mantener las mariposas del acelerador que se cierren completamente durante deceleración y debajo de ciertas condiciones.

Thrust angle A line that divides the total toe angle of the rear wheels.

Ángulo del eje en relación al chasis Una línea que divide el ángulo total de la convergencia o la divergencia de las ruedas traseras.

Timing chain tensioner A mechanical or hydraulic device used to maintain tension on the engine timing chain.

Mecanismo tensor de la cadena de regulación del encendido Dispositivo mecánico o hidráulico utilizado para mantener la tensión en la cadena de regulación del encendido del motor.

Timer core The trigger wheel or armature in some electronic distributors.

Cronometro del núcleo La rueda disparadora o armadura en algunos distribuidores electrónicos.

Toe A wheel geometry concern that deals with the direction the front of the wheels is facing compared to the direction of the rear of the wheels.

Convergencia o divergencia Una geometría que concierne con la dirección de como las ruedas delanteras apuntan comparado con la dirección de las ruedas traseras.

Top dead center (TDC) Piston position at the top of the cylinder.

Punto muerto superior La posición del pistón en la parte superior del cilindro.

Torque A force that does work with a turning force.

Par de torsión Fuerza que realiza trabajo mediante una fuerza de torsión.

Torque converter clutch (TCC) lockup A system to lock the turbine to the front of the converter.

Fijador del embrague del convertidor del par motor Sistema para fijar la turbina en la parte delantera del convertidor.

Toxic Poisonous to the human body.

Tóxico Dañino para el ser humano.

Transducer An electronic device that monitor conditions, primarily pressure conditions, and sends out a voltage signal that represents the changes.

Transductor Un dispositivo electrónico que monitorea las condiciones, primariamente las condiciones de la presión y envía una señal de voltaje de salida que representa los cambios.

Transfer slot A slot in the lower throttle bore of a carburetor. Its purpose is to help the carburetor respond to the change of engine speeds from idle to fast idle.

Ranura de transferencia Una ranura en la parte inferior del barril del carburador. Su propósito es de ayudar al carburador a responder a los cambios de las velocidades del motor desde la marcha mínima a la marcha mínima rápida.

Transistor An automatic electronic switch with no moving parts.

Transistor Conmutador electrónico automático desprovisto de piezas móviles.

Transitional low emission vehicle (TLEV Emission standards required by 1994.

Vehículo transicional de baja emisión Normas sobre emisiones a establecerse antes del año 1994.

Transverse engine mounting An engine mounted crossways.

Montaje transversal del motor Motor montado transversalmente.

Tube and fin radiator core A core with tubes surrounded by fins connected between the radiator tanks.

Núcleo del radiador de tubo y aleta Núcleo en el que los tubos están rodeados por aletas conectas entre los tanques del radiador.

Tuned exhaust headers Headers are connected between the cylinder head exhaust outlets and the exhaust pipe. Each exhaust passage, or pipe, in the header is the same length to improve exhaust flow.

Pipas afinadas de escape Las pipas están conectadas entre las salidas del escape de la culata del cilindro y el tubo de escape. Cada paso o tubo de escape en la pipa es de igual largo para un mejor flujo de escape.

Tune-up procedure The procedure followed by a technician while performing a tune-up.

Procedimiento de afinación del motor Procedimiento que un mecánico lleva a cabo durante la afinación del motor.

Tune-up purpose To restore or maintain the original performance and economy of the engine.

Propósito de la afinación del motor Restaurar o mantener el rendimiento original y la economía del motor.

Tune-up requirements Satisfactory compression, ignition, and fuel system.

Requisitos para una afinación Compresión, encendido y sistema de combustible satisfactorios.

Tune-up revolution Massive changes in the tune-up business created by electronic controls such as the replacement of carburetors with electronic fuel injection.

Revolución de la afinación Cambios notables en la afinación producidos por controles electrónicos, como por ejemplo el reemplazo de carburadores por la inyección electrónica de combustible.

Turbine wheel A vaned wheel mounted in the exhaust system and connected to the opposite end of the turbocharger shaft from the compressor wheel.

Rueda de la turbina Rueda con paletas montada en el sistema de escape y conectada al extremo opuesto del árbol turbocompresor desde la rueda compresora.

Turbo lag The term used to describe the condition that exists when full engine power is needed and the turbocharger is not developing air pressure boost.

Demora del turbocargador Un termino usado para describir la condición que existe cuando la fuerza completa del motor es necesitada y el turbocargador no está desempeñando amplificación para la presión de aire adentro del múltiple de admisión.

Turbocharger A device, driven by exhaust gas flow, that force more air into the engine's cylinders.

Turbocargador Un dispositivo, conducido por el flujo de los gases de escape, que fuerza más aire adentro de los cilindros del motor.

Two-barrel carburetor A carburetor with two throttles on a common shaft.

Carburador de doble cilindro Carburador que tiene dos mariposas en un árbol común.

Two-stage, two-barrel carburetor A carburetor with two throttles that open in stages.

Carburador de doble cilindro de dos etapas Carburador con dos mariposas que se abren en etapas.

Two-stroke cycle An engine in which a complete cycle requires two piston strokes.

Ciclo de dos tiempos Motor en el que un ciclo completo requiere dos carreras del pistón.

Two-valve (2-V) cylinder head A cylinder head with two valves per cylinder.

Culata del cilindro de dos válvulas Culata del cilindro que tiene dos válvulas por cilindro.

Two-way catalyst A catalytic converter that oxidizes carbon monoxide and unburned hydrocarbons into carbon dioxide and water vapor.

Catalizador doble Convertidor catalítico que oxida monóxido de carbono e hidrocarburos no quemados en bióxido de carbono y vapor de agua.

Ultra low emission vehicle (ULEV) Emission standards required by the year 2000.

Vehículo de emisión ultra baja Normas sobre emisiones a establecerse antes del año 2000.

Unleaded fuel nozzle A small nozzle on unleaded fuel dispensers that fits in the filler pipe on vehicles requiring unleaded fuel.

Tobera para combustible sin plomo Tobera pequeña en la bomba medidora de combustible sin plomo que se ajusta al tubo de llenado en vehículos que utilizan combustible sin plomo.

Vacuum A pressure less than atmospheric pressure.

Vacío Presión más baja que la presión atmosférica.

Vacuum advance An ignition timing control unit that allows timing to change in response to engine load.

Avance de vacío Un control de la unidad de avance de la ignición, que permite cambio del tiempo en respuesta a la carga del motor.

Vacuum amplifier A device used to apply a strong vacuum signal to a component in response to a weak vacuum signal.

Amplificador de vacío Un dispositivo usado para aplicar una señal de vacío fuerte a un componente en respuesta a una señal de vacío débil.

Vacuum delay valve A vacuum valve with a restrictive port that delays a vacuum increase through the valve.

Válvula de retardo de vacío Válvula de vacío con una lumbrera restrictiva que retarda el aumento de vacío a través de la válvula.

Valence ring The outer ring on an atom.

Anillo de valencia Anillo exterior de un átomo.

Valve overlap The few degrees of crankshaft rotation when both valves are open with the piston near top dead center on the exhaust stroke.

Solape de la válvula Los pocos grados que gira el cigüeñal cuando ambas válvulas están abiertas y el pistón se encuentra cerca del punto muerto superior durante la carrera de escape.

Vaporization The process in which a liquid changes to a gas or vapor.

Vaporización El proceso en el cual un líquido cambia de un gas a un vapor.

Variable dwell A feature of some electronic distributor-type ignition systems in which dwell time can vary according to engine conditions.

Dwell variable (tiempo en que la bobina se está saturando medido en grados) Una característica de algunos sistemas de ignición de tipos de distribuidores electrónicos en el cual el dwell puede variar de acuerdo con las condiciones del motor.

Variable nozzle turbine (VNT) The variable nozzle turbine unit designed to allow a turbocharger to accelerate quickly,, thus reducing lag time.

Turbina con tobera variable La unidad de la turbina con la unidad de la tobera variable diseñada para permitir que un turbocargador acelere rápidamente, reduciendo su tiempo de suministrar presión al múltiple de admisión.

Variable venturi carburetor A carburetor with a venturi that varies its opening in relation to engine speed.

Carburador de venturi variable Carburador con un venturi que varía su apertura de acuerdo a la velocidad del motor.

Vehicle speed sensor (VSS) A sensor that is usually mounted in the transmission and sends a voltage signal to the computer in relation to vehicle speed.

Sensor de la velocidad del vehículo Sensor montado normalmente en la transmisión y que le envía una señal de tensión a la computadora referente a la velocidad del vehículo.

Venturi A narrowing of a passage through which a liquid or gas is flowing.

Venturi Estrechamiento de un paso a través del que fluye un líquido o un gas.

Viscosity A term used to describe a liquid's ability of a liquid to flow.

Viscocidad Un termino usado para describir la habilidad de un líquido en fluir.

Viscous fan clutch A cooling fan clutch that drives the fan faster at high temperatures and slower at low temperatures.

Embrague viscoso del ventilador Embrague del ventilador de enfriamiento que acciona el ventilador para que gire más rápido a temperaturas altas y más despacio a temperaturas bajas.

Volt A measurement of electrical pressure difference.

Voltio Medida de una diferencia en presión eléctrica.

Voltage generating devices Electronic devices that generate voltage in response to movement or a change in conditions.

Dispositivo generador de voltaje Dispositivos eléctricos que generan voltaje en respuesta a los movimientos o un cambio en las condiciones.

Volumetric efficiency The relationship between the amount of air actually taken into the engine on the intake stroke compared to the amount of air required to fill the cylinder at atmospheric pressure.

Rendimiento volumétrico Relación entre la cantidad de aire realmente aspirado en el motor durante la carrera de aspiración y la cantidad de aire requerido para llenar el cilindro a una presión atmosférica.

V-type block An engine with two banks of cylinders arranged in a V configuration.

Bloque tipo V Motor con dos bancos de cilindros arreglados en forma de V.

Wankel engine A rotary cycle engine.

Motor Wankel Un motor de ciclo rotatorio.

Wastegate diaphragm and valve A valve operated by boost pressure that controls the amount of turbocharger boost pressure.

Diafragma y válvula de la compuerta de desagüe Válvula accionada por presión de sobrealimentación que controla la cantidad de presión de sobrealimentación del turbocompresor.

Waste spark system A term that may be applied to any EI system because each pair of spark plugs fires with one cylinder on the compression stroke and the other cylinder on the exhaust stroke.

Sistema de chispa residual Término que puede aplicarse a cualquier sistema de encendido electrónico ya que cada par de bujías se enciende con un cilindro durante la carrera de compresión y con el otro cilindro durante la carrera de escape.

Water (H_2O) A compound containing hydrogen and oxygen.

Agua (H_2O) Compuesto que contiene hidrógeno y oxígeno.

Wax pellet A small sealed container filled with wax that expands and contracts when it is heated and cooled to open and close components such as the engine thermostat.

Grano gordo de cera Pequeño recipiente sellado, lleno de cera, que se expande y contrae al calentarse y enfriarse para abrir y cerrar componentes, como por ejemplo el termostato del motor.

Wet-type cylinder sleeve A cylinder sleeve with engine coolant contacting the outside area of the sleeve.

Manguito de cilindro tipo húmedo Manguito de cilindro donde el refrigerante del motor entra en contacto con la parte exterior del mismo.

Wet-type intake manifold An intake manifold with coolant circulated through the manifold.

Colector de aspiración tipo húmedo Colector de aspiración a través del cual fluye el refrigerante.

Wheel tramp The bouncing action of rotating wheels that is caused by static imbalance.

Rebote de la rueda La acción de saltar de una rueda girando, que es causado cuando está fuera de balance estático.

Work place hazardous materials information systems (WHMIS) Sheets that provide information regarding hazardous materials.

Sistemas de información relacionados a materiales peligrosos en el lugar de trabajo Hojas de información sobre materiales peligrosos.

Y-pipe Three pipes connected in a Y configuration.

Tubo en Y Tres tubos conectados en forma de Y.

Zener diode A special type of diode that allows current to flow in the desired direction only when certain conditions are met.

Diodo Zener Un tipo especial de diodo que permite que la corriente fluya en la dirección deseada solamente cuando ciertas condiciones son cumplidas.

Zero emission vehicle (ZEV) Emission standards required on a certain percentage of vehicles beginning in 1998.

Vehículo sin emisiones Normas sobre emisiones a establecerse en cierto número de vehículos a partir del año 1998.

INDEX

oil pump, 89
oil pump pickup, 89
oil seals and gaskets, 93
pressure relief valve, 90
radiator, 97-98
radiator pressure caps, 98
temperature indicator, 105-106
thermostat, 99-101
water jackets, 104
water outlet, 99
water pump, 96

M

Magnetic pulse generator, 126-128, 286, 385
Magnetism, 32, 39
Magnets, 39-40
Main metering circuit, 154-155
Malfunction indicator lamp (MIL), 434, 451-453
Manifold absolute pressure (MAP) sensor, 303-304, 337
Manifold heat control valves, 267-268
Manual transmission, 470-471
Mass, 20
Mass air flow sensors, 336-337
    heated resistor-type, 305-306
    hot wire-type, 306
    vane-type, 305
Material safety data sheets (MSDS), 13
Mechanical fuel pumps, 187-189
Metal deactivators, 145
Metal detection sensors, 128, 385-386
Metering, 150
Metering rods, 156
Methanol, 146
Microprocessor, 283
    design of, 292-293
    information retrieval, 293-294
    information storage, 293
    program, 293
Miller-cycle engine, 84
Misfire monitor, 438-440
Misfires, 455
Mixture control (M/C) solenoids, 316
Mixture heater system
    computer-controlled, 269
    mechanically controlled, 268-269
MMT, 145
Molecular energy, 21
Molecule, 17
Momentum, 20
MON, 143
MPMT switch, 38
MTBE, 146
Muffler, 218-219
Multiple-disc clutch, 477
Multiport injection (MPI), 177
Multipurpose dry chemical fire extinguishers, 8
Multiviscosity oils, 87

N

Natural gas, 147
Negative backpressure EGR valve, 260-261
Negative pressure, 24
Negative temperature coefficient (NTC) thermistor, 285
Neutral drive switch (NDS), 308-309
Neutral safety switch operation, 53-54
Neutrons, 17
Newton, Sir Isaac
    laws of motion, 18
Newton meter, 18
Nitrile rubber, 152
Nonadjustable oxygen feedback solenoid systems, 316-320
NPN transistor, 43, 384

O

OBD II trip, 446
Occupational Safety and Health Act (OSHA), 1-2
Octane, 143
O_2 feedback solenoid, 316-320
Off-idle circuit, 153-154
Ohm, 33
Ohmmeter, 33
Ohm's law, 34
Oil cooler, 93, 106
Oil filter, 90
Oil pan/sump, 89
Oil pressure indicator, 93
Oil pressure switch, 193
Oil pump, 89
Oil pump pickup, 89
Oil seals/gaskets, 93
On-Board Diagnostic II systems, 433-434
    data links, 453-454
    drive cycle, 446-447
    Ford's electronic engine control V (EEC V) system, 448-450
    introduction and implementation, 434-436
    malfunction indicator lamp (MIL), 451-453
    monitoring capabilities, 436-446
    OBD II trip, 446
    snapshots, 456
    system readiness mode, 446
    terminology list, 456-461
    test connector, 450-451
    test modes, 453, 455-456
One-way check valve, 274
Open loop, 168, 315
Opposed cylinder engine, 73
Output driver, 47, 297
Overhead cam (OHC), 68, 74
Overhead valve (OHV), 68, 74
Oxidation inhibitors, 87, 145
Oxides of nitrogen, 167, 215-216, 238, 242, 314
Oxygen, 242-243
Oxygen (O_2) sensor, 299-301

P

Palladium, 272
Paraffin, 147
Parallel circuit, 36
Parallel flow system, 104
PCV systems, 251-254
Periodic motor vehicle inspection (PMVI), 237-238
Personal protection, 2
Photoelectric sensor, 131, 387
Pickup coil, 286
Piezoresistive, 285
Pilot bushing, 467
Pintle, 340
PIV, 42
Planetary gear controls, 477
Platinum, 272
PNP transistor, 43
Pollutants, 215-216, 240-244
Poppet nozzles, 370
Ported vacuum switch (PVS), 211, 258, 267
Port fuel injection (PFI), 169, 174-177, 271, 352
    cold start injector, 356
    design of, 353-355
    import multiport system, 358-360
    multiport system, 357-358
    pressure regulators, 357
Positive backpressure EGR valve, 260
Positive displacement pump, 90
Positive pressure, 24
Positive temperature coefficient (PTC) thermistor, 285
Post-combustion control systems, 240
    air injection systems, 272-278
    catalytic converters, 271-272

Potentiometers, 37
Pour point, 148
Power, 21
Power enrichment circuit, 156
Power stroke, 70
Powertrain control module (PCM), 211, 325, 339-340. *See also* Idle speed control systems
Power valves, 156
Precombustion control systems, 240, 248
    EGR systems, 256-265
    engine design changes, 249-251
    intake heat control systems, 266-271
    PCV systems, 251-254
    spark control systems, 254-255
Pressure, 22
Pressure feedback electronic (PFE) sensor, 264
Pressure regulator, 176-177
Pressure relief valve, 90
Pressure transducer, 264-265
Primary circuit, 114-115, 382-383
Primary stage, 155
Profile ignition pickup (PIP) sensor, 409-410
Programmable read only memory (PROM), 296
Protocol, 453
Protons, 17
psi, 24
psig, 24
Pulse transformer, 118
Pulse width, 298, 347

R

Rack-and-pinion system, 490
Radiator, 97-98
Radiator pressure caps, 98
Radio frequency interference (RFI), 120
Random access memory (RAM), 294
Read only memory (ROM), 295-296
Rear alignment, 494
Recirculating ball system, 490
Recovery tank, 96
Recycled oils, 89
Reference voltage (Vref) sensors, 45, 284-285
Relays, 39
Resistance, 33
Resistors, 36-37, 381
Resonator, 219
Resource Conservation and Recovery Act (RCRA), 12
Reverse flow system, 104
Rheostats, 37
Rhodium, 272
Rich mix, 150
Right-to-know laws, 12
RON, 143
Roots supercharger, 231
Rotary engine, 82-83
Rotor, 59, 116
Rust inhibitors, 86, 145

S

Safety glasses, 9
Safety practices
    electrical safety, 3
    fire safety, 6-7
    gasoline safety, 3-4
    general shop safety, 5-6
    hazardous waste disposal, 10-14
    housekeeping safety, 4-5
    Occupational Safety and Health Act (OSHA), 1-2
    personal protection, 2
    shop hazards, 2
    shop safety equipment, 7-10
    smoking, alcohol, and drugs, 3
Schmitt trigger, 288
Schrader valve, 353